John W. G. Cairney · Susan M. Chambers (Eds)
Ectomycorrhizal Fungi: Key Genera in Profile

Springer

*Berlin
Heidelberg
New York
Barcelona
Hong Kong
London
Milan
Paris
Singapore
Tokyo*

John W. G. Cairney · Susan M. Chambers (Eds)

Ectomycorrhizal Fungi
Key Genera in Profile

With 30 Figures

 Springer

Dr. John W.G. Cairney
Dr. Susan M. Chambers
Mycorrhiza Research Group
School of Science
University of Western Sydney (Nepean)
PO Box 10 Kingswood NSW 2074
Australia

ISBN 3-540-65609-X Springer-Verlag Berlin Heidelberg New York

Library of Congress CIP-data applied for

Die Deutsche Bibliothek – CIP-Einheitsaufnahme
  Ectomycorrhizal fungi : key genera in profile / J. W. G. Cairney ; S.
  M. Chambers (eds.). – Berlin ; Heidelberg ; New York ; Barcelona ;
  Hong Kong ; London ; Milan ; Paris ; Singapore ; Tokyo ; Springer,
  1999
  ISBN 3-540-65609-X

Typesetting: BestSet Typesetter Ltd. Hong Kong
Cover design: design & production GmbH, Heidelberg

SPIN: 10643460   31/3136 – 5 4 3 2 1 0 – printed on acid-free paper

Preface

It is becoming increasingly clear that communities of ectomycorrhizal fungi in natural and man-managed habitats are far more diverse than has previously been recognised. The development of morphotype analysis in the 1980s led to an appreciation that above-ground sporocarp diversity is a poor reflection on below-ground diversity at the root-tip level. The availability of the much higher resolving power of molecular methods in more recent years has revealed still further diversity in below-ground ectomycorrhizal fungal communities. Indeed, from data presented in a number of papers at the Second International Conference on Mycorrhiza in Uppsala, Sweden, in July 1998, it is apparent that we currently have a very poor appreciation of ectomycorrhizal fungal diversity in most habitats. As mycorrhiza researchers struggle to come to grips with the functional significance of such diversity, we thought it timely to bring together a series of chapters, each dedicated to a particular ectomycorrhizal fungal genus. While previous reviews of mycorrhizal biology have sought to generalise aspects of the symbioses, the emphasis in this book is very much upon highlighting diversity in taxonomy, ecology, physiology and, where applicable, practical aspects of ectomycorrhizal fungi. We hope that this approach will facilitate a wider appreciation of ectomycorrhizal functioning, and stimulate further research into the functional significance of fungal diversity in forest ecosystems.

It will be clear to the reader that only a few genera have, for reasons of economics or ease of laboratory manipulation, been investigated extensively. We thus have a reasonable knowledge of the biology of *Cenococcum*, *Hebeloma*, *Laccaria*, *Paxillus*, *Pisolithus*, *Rhizopogon*, *Suillus*, *Thelephora* and *Tuber*. Even within these genera, however, work has often concentrated on only single or a few taxa, and in many cases on a limited number of isolates of each. Several other genera, which are clearly widespread and important in ecological and/or economic terms, such as *Amanita*, *Cantharellus*, *Hysterangium*, *Lactarius* and *Scleroderma*, have received rather less attention so far. In each case, we considered that present knowledge was sufficient for a useful genus-specific chapter. Many other taxa such as *Cortinarius* and *Russula*, although probably key genera in many habitats, have not been included in this volume. While these taxa are clearly geographically widespread, we currently have an insufficient understanding of their biology to merit a broad review. In this context, some may question the inclusion of a chapter on resupinate genera.

Recent work, however, indicates that taxa forming resupinate sporocarps are important components of below-ground communities and we hope that their inclusion here will encourage further work on their biology. Finally, in the interests of saving space, and in order to allow as much data as possible to be presented in each chapter, we have used the abbreviation ECM throughout the book to replace ectomycorrhiza, ectomycorrhizas, ectomycorrhizae and ectomycorrhizal.

October 1998   *John W. G. Cairney*
                                                       Susan M. Chambers

Contents

List of Contributors

J. W. G. Cairney, Mycorrhiza Research Group, School of Science, University of Western Sydney (Nepean), P.O. Box 10, Kingswood, NSW 2747, Australia

M. A. Castellano, USDA Forest Service, Pacific Northwest Research Station, Forestry Sciences Laboratory, 3200 Jefferson Way, Corvallis, Oregon 97331, USA

S. M. Chambers, Mycorrhiza Research Group, School of Science, University of Western Sydney (Nepean), P.O. Box 10, Kingswood, NSW 2747, Australia

J. V. Colpaert, Laboratory of Plant Ecology, Katholieke Universiteit Leuven, Kard. Mercierlaan 92, B-3001 Leuven, Belgium

O. Comandini, Dipartimento di Scienze Ambientali, Università, 67100 L'Aquila, Italy

A. Dahlberg, Department of Forest Mycology and Pathology, Swedish University of Agricultural Sciences, Box 7026, S-750 07 Uppsala, Sweden

E. Danell, Department of Forest Mycology and Pathology, Swedish University of Agricultural Sciences, Box 7026, S-750 07 Uppsala, Sweden

J.-C. Debaud, Laboratoire d'Ecologie Microbienne du Sol (UMR CNRS 5557), Bât. 405, 43 Bd du 11 Novembre 1918, 69622 Villeurbanne Cedex, France

S. Erland, Department of Microbial Ecology, Lund University, Ecology Building, S-223 62 Lund, Sweden

R. D. Finlay, Department of Forest Mycology and Pathology, Swedish University of Agricultural Sciences, Box 7026, S-750 07 Uppsala, Sweden

L. Fraissinet-Tachet, Laboratoire d'Ecologie Microbienne du Sol (UMR CNRS 5557), Bât. 405, 43 Bd du 11 Novembre 1918, 69622 Villeurbanne Cedex, France

G. Gay, Laboratoire d'Ecologie Microbienne du Sol (UMR CNRS 5557), Bât. 405, 43 Bd du 11 Novembre 1918, 69622 Villeurbanne Cedex, France

L. C. Grubisha, Department of Botany and Plant Pathology, Oregon State University, Corvallis, Oregon 97331, USA

H. Gryta, Laboratoire d'Ecologie Microbienne du Sol (UMR CNRS 5557), Bât. 405, 43 Bd du 11 Novembre 1918, 69622 Villeurbanne Cedex, France

M. Guttenberger, Universität Tübingen, Physiologische Ökologie der Pflanzen, Auf der Morgenstelle 1, D-72076 Tübingen, Germany

R. Hampp, Universität Tübingen, Physiologische Ökologie der Pflanzen, Auf der Morgenstelle 1, D-72076 Tübingen, Germany

L. J. Hutchison, Agriculture Canada Research Station, PO Box 3000, Lethbridge, Alberta, Canada T1J 4B1

P. Jargeat, Laboratoire d'Ecologie Microbienne du Sol (UMR CNRS 5557), Bât. 405, 43 Bd du 11 Novembre 1918, 69622 Villeurbanne Cedex, France
P. Jeffries, Research School of Biosciences, University of Kent, Canterbury, Kent, CT2 6NJ, UK
I. Kottke, Universität Tübingen, Spezielle Botanik und Mykologie, Auf der Morgenstelle 1, D-72076 Tübingen, Germany
B. R. Kropp, Department of Biology, Utah State University, Logan, Utah 84322-5303, USA
K. F. LoBuglio, Department of Plant and Microbial Biology, University of California, Berkeley, Berkeley, California 94720, USA
R. Marmiesse, Laboratoire d'Ecologie Microbienne du Sol (UMR CNRS 5557), Bât. 405, 43 Bd du 11 Novembre 1918, 69622 Villeurbanne Cedex, France
R. Molina, USDA Forest Service, Pacific Northwest Research Station, Forestry Sciences Laboratory, 3200 Jefferson Way, Corvallis, Oregon 97331, USA
G. M. Mueller, Department of Botany, The Field Museum, Chicago, Illinois 60605-2496, USA
U. Nehls, Universität Tübingen, Physiologische Ökologie der Pflanzen, Auf der Morgenstelle 1, D-72076 Tübingen, FRG
F. Oberwinkler, Universität Tübingen, Spezielle Botanik und Mykologie, Auf der Morgenstelle 1, D-72076 Tübingen, Germany
G. Pacioni, Dipartimento di Scienze Ambientali, Università, 67100 L'Aquila, Italy
B. Söderström, Department of Microbial Ecology, University of Lund, 223 62 Lund, Sweden
J. W. Spatafora, Department of Botany and Plant Pathology, Oregon State University, Corvallis, Oregon 97331, USA
A. F. S. Taylor, Department of Forest Mycology and Pathology, Swedish University of Agricultural Sciences, Box 7026, S-750 07 Uppsala, Sweden
J. M. Trappe, Department of Forest Science, Oregon State University, Corvallis, Oregon 97331, USA
H. Wallander, Department of Microbial Ecology, University of Lund, 223 62 Lund, Sweden
M. Weiß, Universität Tübingen, Spezielle Botanik und Mykologie, Auf der Morgenstelle 1, D-72076 Tübingen, Germany
Z.-L. Yang, Kunming Institute of Botany, Academia Sinica, Heilongtan, Kunming 650204, P.R. China

Chapter 1

Pisolithus

S. M. Chambers and J. W. G. Cairney

1.1
Introduction

Pisolithus spp. are ectomycorrhizal (ECM) gasteromycetes with a widespread global distribution (Marx 1977, Table 1.1). Having been championed during the 1970s for use in forestry inoculation programmes and a number of inoculation protocols having been developed (reviewed by Marx and Kenney 1982), there exists a considerable body of literature relating to host plant responses to *Pisolithus* infection under a range of conditions. The ease with which the fungus can be grown in vitro has facilitated extensive study of its physiology, and the simplicity of ECM synthesis under controlled conditions with a range of host plants has ensured that the ontogeny and ultrastructure of *Pisolithus* ECM have been well studied. The availability of detailed information on the development, compatibility and physiology of the symbiosis has also made the *Pisolithus–Eucalyptus* interaction the preferred system for current investigation of the molecular basis of the fungus–root interaction (Tagu and Martin 1996).

1.2
Taxonomy and Phylogenetic Relationships

Although considerable heterogeneity exists in terms of sporocarp (Fig. 1.1), spore and isolated culture morphology, taxa within the genus *Pisolithus* have been widely regarded as conspecific, and grouped as *Pisolithus tinctorius* (Pers.) Coker and Couch [syn. = *P. arhizus* (Scop.: Pers.) Rauschert] (Coker and Couch 1928; and see Watling et al. 1995). *Pisolithus* isolates display considerable variation in sporocarp and basidiospore morphology and, while this has to some extent hampered systematic clarification of the genus, it has been suggested that several morphologically distinct species can further be recognised within the group currently described as *P. tinctorius* (Bronchart et al. 1975; Calonge and Demoulin 1975). For example, within Australia, *P.*

Mycorrhiza Research Group, School of Science, University of Western Sydney (Nepean), P.O. Box 10, Kingswood, New South Wales 2747, Australia

Table 1.1. Host genera with which *Pisolithus* spp. isolates have been confirmed as forming ECM in mycorrhiza synthesis experiments

Genus	Source
Abies	Marx (1977)
Acacia	Ba et al. (1994)
Aflezia[a]	Ba and Thoen (1990)
Allocasuarina	Theodorou and Reddell (1991)
Alnus	Godbout and Fortin (1983)
Arbutus[b]	Zak (1976)
Arctostaphylos[b]	Molina and Trappe (1982a)
Betula	Marx (1977)
Carya	Marx (1977)
Castanea	Martins et al. (1996)
Castanopsis	Tam and Griffiths (1994)
Casuarina	Theodorou and Reddell (1991)
Eucalyptus	Marx (1977)
Hopea	Yazid et al. (1994)
Larix	Molina and Trappe (1982b)
Pinus	Marx (1977)
Populus	Godbout and Fortin (1985)
Pseudotsuga	Marx (1977)
Quercus	Marx (1977)
Tsuga	Marx (1977)

[a] No Hartig net formed, only a sheath.
[b] Arbutoid mycorrhizas.

microcarpus (Cke. and Mass.) Cunn has been proposed as a separate species (Cunningham 1942). Furthermore, several *Pisolithus* species, including *P. kisslingi* E. Fisch, *P. pusillum* Pat. and *P. aurantioscabrosus* Watling et al., have been described as occurring in tropical South-East Asia, based on distinctive sporocarp and basidiospore morphology (Watling et al. 1995). There is thus growing consensus that the genus *Pisolithus* is more genetically diverse than was previously assumed and that taxonomic revision on a global scale is required. Molecular approaches are facilitating rapid progress in this respect.

Burgess et al. (1995b) analysed electrophoretic patterns of expressed mycelial proteins from *Pisolithus* isolates collected from different geographical regions of Australia. Results from this study indicated that, although much variability in polypeptide patterns exists, groupings based on similarities in polypeptide patterns display correlation with geographical origin and basidiospore morphotypes. More compelling evidence of the existence of multiple species within the current *P. tinctorius* grouping is provided by recent molecular studies. Random amplified polymorphic DNA (RAPD) analyses indicate that considerable genetic variation exists within *Pisolithus* isolates from Australia (Anderson et al. 1998a), Brazil, Europe, North America,

Fig. 1.1. A typical sporocarp of *Pisolithus* sp.

Scandinavia and the Philippines (Junghans et al. 1998; Sims et al. 1999). More detailed examination of restriction fragment length polymorphism (RFLP) patterns of internal transcribed spacer (ITS) and intergenic spacer (IGS) regions of the rDNA gene complex indicate that distinct polymorphism patterns can be seen in isolates collected from Australia, Kenya and Indonesia (Anderson et al. 1998a; Martin et al. 1998; I. C. Anderson, S. M. Chambers and J. W. G. Cairney, unpubl. data), strongly suggesting that at least five *Pisolithus* spp. exist within the sampled carpophore population in these countries. Where ITS sequences have been obtained for these isolates, they indicate considerable divergence and strongly support the existence of multiple species (Anderson et al. 1998a; Martin et al. 1998). A comparison of ITS sequences for *Pisolithus* isolates in the GenBank Nucleotide Database (as of June 1998) supports this idea (Fig. 1.2).

It is of interest also that Martin et al. (1998) reported that each of the putative species they obtained ITS sequence data for was consistently associated with a different host plant species (either *Afzelia*, *Eucalyptus* or *Pinus*) in

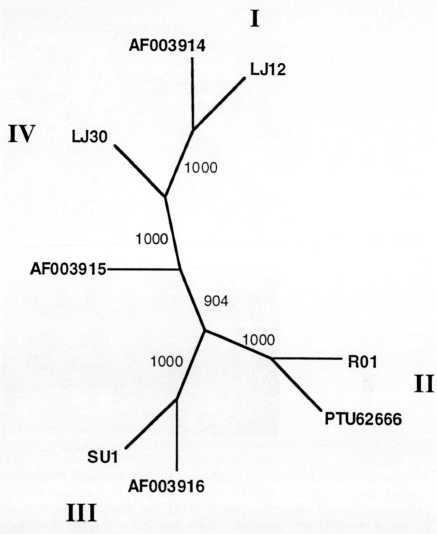

Fig. 1.2. Radial tree showing relationships between three New South Wales *Pisolithus* isolates from mixed sclerophyll forest (*LJ12, LJ30, R01*), a Brazilian isolate from a eucalypt plantation (*PTU62666*), an isolate collected under *Pinus* in Indonesia (*SU1*) and isolates collected from pine (*AF003916*), eucalypt (*AF003914*) and *Afzelia* (*AF003915*) stands in Kenya. ITS DNA sequences were analysed and the tree produced as outlined by Anderson et al. (1998a). Isolates borne on same terminal branches had >98% sequence similarity and are considered to be conspecific. Numbers on internal branches are bootstrap values, indicating number of replicate bootstrap samples (out of 1000) that placed isolates on particular branches. Very high bootstrap values (987–1000) strongly support the existence of five *Pisolithus* species within isolates from Australia, Indonesia, Kenya and Brasil (*AF003915* plus groups I–IV)

Kenya. While the species associated with *Afzelia* is thought to be a native Kenyan species, those associated with the introduced plantation tree hosts have probably been introduced during forestry activities. This is supported by strong ITS sequence similarity (>98%) between the Kenyan *Pisolithus* isolates from under *Eucalypus* and an isolate from Australia (and probably associated with *Eucalyptus*) (Fig. 1.2). We have also found a similar level of sequence similarity between a *Pisolithus* isolate collected under *Pinus* in Indonesia and an isolate (Martin et al. 1998) associated with *Pinus* in Kenya (Fig. 1.2; I. C. Anderson, S. M. Chambers and J. W. G. Cairney, unpubl. data).

1.3
Ecology

Pisolithus has been recorded in a range of habitats including forest, urban and orchard sites, along with eroded and mine-site soils (Marx 1977; Malloch and Kuja 1979), but sporocarps are more usually found in relatively dry sites with little humus or along roadside areas (e.g. Castellano and Trappe 1991). Information from Western Australian reforestation sites indicates that the fungus is an early coloniser (Gardner and Malajczuk 1988) and it is generally regarded as being poorly competitive with other ECM fungi (Marx et al. 1984; McAfee and Fortin 1986). It is perhaps for these reasons that *Pisolithus* persists best in forestry inoculation programmes on sites subject to edaphic stresses (see below). Most *Pisolithus* isolates, although not all (see for example Reid and Woods 1969), produce an extensive extramatrical mycelial phase which in many cases differentiates into linear organs (Kammerbauer et al. 1989; Agerer 1991; Lamhamedi and Fortin 1991). Based on the descriptions of these authors, and adopting the nomenclature of Cairney et al. (1991), linear mycelial organs of *Pisolithus* can be described as "apically diffuse, simple rhizomorphs". From what is known about the structure and function of rhizomorphs in other fungal systems, their formation may be important in channelling nutrients and water to and from the host and in conferring protection against adverse environmental conditions upon extramatrical mycelium (Thompson 1984; Cairney 1992). A recent population study indicates that mycelial individuals (= genets) of a *Pisolithus* species in the field can exceed 30 m in diameter (Anderson et al. 1998b).

Pisolithus also produces sclerotia as part of its extramatrical mycelial phase (Dennis 1980; Piché and Fortin 1982; Fortin et al. 1983). While environmental conditions may influence the shape and structural detail of sclerotia (Grenville et al. 1985), in all cases they comprise an external melanised rind surrounding a cortex and medulla (Dennis 1980; Piché and Fortin 1982; Grenville et al. 1985). In common with other fungal sclerotia, histochemical staining indicates that the cortex and medulla are rich in protein, lipids and carbohydrate, implying a storage function and the ability to allow *Pisolithus* to withstand edaphic stresses in a vegetative state (Piché and Fortin 1982; Grenville et al. 1985). The production of sclerotia may thus, along with aggregation of extramatrical mycelium into rhizomorphs, be important in the reported success of *Pisolithus* in stressful soil conditions.

1.4
ECM Formation

1.4.1
Host–Fungus Specificity

Laboratory synthesis experiments have confirmed that *Pisolithus* forms ECM with a range of host plant genera (see Table 1.1). There is, however, considerable evidence of specificity in the interaction between host root and the fungus. It has been frequently noted, for example, that *Pisolithus* isolated from sporocarps collected in association with *Pinus* spp. are poor colonisers of *Eucalyptus* spp. (e.g. Chilvers 1973; Lei et al. 1990a; Malajczuk et al. 1990; Burgess et al. 1994). Even isolates from different conifer species are not necessarily intercompatible (Marx and Bryan 1970; Marx 1981). Further, some *Pisolithus* isolates show ECM compatibility with clones derived from mature eucalypt trees, but are poorly compatible with clonal host plants generated from young seedlings, suggesting that the developmental maturity of host material can also influence compatibility (Tonkin et al. 1989). A detailed study of interactions between 20 *Pisolithus* isolates from different geographical regions and *Eucalyptus grandis* (Burgess et al. 1994) indicates wide variation in both rate and extent of ECM formation by different isolates. The extent of ECM formation varied from a fully developed sheath and Hartig net in compatible isolates, through isolates that formed only a superficial sheath, to isolates that formed no identifiable ECM structures. Incompatibility between *Pisolithus* and host roots can also be expressed in polyphenol accumulation in host tissue and thickening of host cell walls abutting the incompatible isolate (Tonkin et al. 1989; Lei et al. 1990b). The data of Burgess et al. (1994) are also significant in that they show intraspecific variation within the genus *Eucalyptus* in terms of intercompatibility with *Pisolithus* isolates. Furthermore, the isolates used in their study were derived not only from a variety of geographical origins, but also from sporocarps having different morphological characteristics. The extent to which such characteristics reflect taxonomic variation within *Pisolithus* remains unclear (see above), but it further highlights the need for detailed systematic studies.

1.4.2
Genetics of the Interaction

In vitro germination of spores collected from mature *Pisolithus* sporocarps can be achieved within a few weeks under axenic conditions, and has permitted investigation of aspects of the importance of *Pisolithus* genetics in the fungus–plant interaction. Successful germination requires either the presence of a germination activator (in the form of a yeast colony) (Bulmer 1964; Lamb and Richards 1974) or a host plant seedling (Kope and Fortin 1990). Germination rates are at best 0.38% (Lamb and Richards 1974), but can be much lower in the absence of activator colonies (Kope and Fortin 1990). Based on spores isolated from sporocarps originating in North America, South Africa, Australia and Europe, it appears that tetrapoplar incompatibility (four mating types) exists within *Pisolithus* monokaryons (Kope and Fortin 1990; Rosado et al. 1994b). Crosses between some monokaryons and dikaryons have also been achieved in the laboratory (Kope 1992). Both monokaryons derived as single spore isolates and reconstituted dikaryons are capable of ECM formation with *Pinus* spp. (Lamhamedi et al. 1990). Individual mono- and dikaryons derived from a single sporocarp show differential affinities for mycorrhiza formation on *Pinus* spp., but monokaryons are generally less efficient in ECM formation than dikaryotic mycelia (Lamhamedi et al. 1990). Dikaryosis is thus thought to be required for the full expression of ECM-forming abilities.

Reconstituted *Pisolithus* dikaryons show variability in the growth form of the extramatrical mycelia, particularly in extension rates and the degree to which rhizomorphs are formed (Lamhamedi and Fortin 1991). They also show variable abilities to improve host plant growth, drought tolerance and mineral nutrient content (Lamhamedi et al. 1992a), although the extent to which these reflect the relative abilities of dikaryons to infect roots or differences in extramatrical mycelial characteristics remains unclear. In some cases, monokaryon crosses show additive abilities for ECM formation; in others the interactions are non-additive (Lamhamedi et al. 1990; Rosado et al. 1994b). Such variability may facilitate directed fungal breeding for enhanced ECM-forming abilities, particularly if coupled with the known potential for enhancing host tree receptiveness to *Pisolithus* via a plant breeding programme (Rosado et al. 1994a).

1.4.3
Developmental Aspects of the Fungus–Host Interaction

The interaction between *Pisolithus* and a host root begins prior to fungus–root contact. Diffusible substances released from the host appear to act as a stimulus for a chemotropic growth response of compatible *Pisolithus* hyphae towards the host (Horan and Chilvers 1990). Within 1 day of introduction of the fungus to a eucalypt root system, and prior to fungus–root contact, there is evidence of a chemical interaction between fungus and host in the form of a browning reaction in outer root cap cells and accumulation of the indole compound hypaphorine in colonising hyphae (Horan et al. 1988; Béguiristain and Lapeyrie 1997). Shortly after contact, fibrils (believed to be glycoproteins of fungal origin) can be observed at the fungus–root interface in compatible, but not incompatible, *Pisolithus*–host interactions (Piché et al. 1983; Lei et al. 1990b). The glycoproteins may be important in either recognition and/or attachment processes. *Pisolithus* hyphae may invade moribund root cap cells at this stage, perhaps providing an essential nutrient source for the process of ECM morphogenesis (Horan et al. 1988; Stephanie et al. 1996).

Molecular investigations of the *Pisolithus–Eucalyptus* interaction indicate that there is altered polypeptide expression in both partners, including the appearance of mycorrhiza-specific polypeptides ("ectomycorrhizins") during ECM formation (Hilbert and Martin 1988; Hilbert et al. 1991; Burgess et al. 1995a). Altered patterns of polypeptide expression are more apparent with *Pisolithus* isolates that form ECM rapidly than in less-compatible isolates; indeed, *Pisolithus* isolates that do not form ECM with *Eucalyptus* spp. fail to alter expression patterns of host or self-polypeptides (Burgess et al. 1995a). In particular, several acidic polypeptides appear to be enhanced in *Pisolithus* during ECM formation. From the high level of synthesis that they display, these are regarded largely as structural (cell wall) fungal proteins. At the same time, a major cell wall mannoprotein is downregulated, being almost undetectable in hyphae following root contact (Tagu and Martin 1996). During ECM formation there is a general downregulation of host polypeptides, which is thought to reflect reduced root-system metabolism during symbiosis (Burgess et al. 1995a; Burgess and Dell 1996). While demonstrated changes in wall proteins are taken as indicating that the *Pisolithus* cell wall plays a major role in the fungus–root interaction (Tagu and Martin 1996), identification of the roles of up- and downregulated proteins in the symbiotic interaction remains a major challenge for mycorrhizal researchers.

Changes in polypeptide patterns can be clearly identified within hours of symbiont contact, indicating that the molecular interaction begins prior to visible signs of ECM development (Hilbert et al. 1991). There is even evidence that some *Pisolithus* polypeptides are upregulated prior to fungus–root contact and that these may represent components of fungal surface glycoproteins observed during early *Pisolithus*–root interactions (Burgess et al. 1995a). More direct evidence for altered gene expression during *Pisolithus–Eucalyptus* ECM formation also exists. By screening cDNA libraries derived from free-living and symbiotic material, Tagu et al. (1993), Tagu and Martin (1995) and Carnero Diaz et al. (1997) have shown enhanced transcription in *Pisolithus* during ECM development, while several *Pisolithus* cDNA transcripts encoding polypeptides in the hydrophobin family have now been isolated (Tagu et al. 1996). This is significant since hydrophobins are thought to be involved in differentiation and/or adhesion processes in other fungal systems (Wessels 1994) and may be a critical component in establishment of the *Pisolithus–Eucalyptus* ECM. A compatible isolate of *Pisolithus* has also been shown to increase expression of α-tubulin in eucalypt roots, suggesting a role of the structural protein in ECM formation (Carnero Diaz et al. 1996).

Although host polypeptides appear in general to be down regulated during *Pisolithus* ECM establishment, there are a number of reports that host root chitinase and peroxidase activities are stimulated during the early stages of *Pisolithus* infection and that expression remains at a high level during development of the ECM organ (Albrecht et al. 1994a,b,c). Activities of both enzymes were shown to be higher during infection by highly compatible *Pisolithus* isolates compared to uninfected roots challenged by poorly compatible *Pisolithus* isolates. Although Albrecht et al. (1994a) report fungal extracts to elicit the chitinase response, Albrecht et al. (1994b) found that hypha–root contact was required for increased activity. While it is tempting to speculate a role for chitinase and peroxidase activities in the infection process, possibly in differentiation of the Hartig net (Albrecht et al. 1994b), it must be noted that Hodge et al. (1995) observed no increase in chitinase activity in roots of either *Pinus* or *Eucalyptus* spp. during infection with a single isolate of *Pisolithus*. The latter authors, however, provided no information regarding the efficacy of the isolate for ECM formation with either host. *Pisolithus* infection has also been shown to induce a systemic chitinase increase in *Eucalyptus* sp., although this response was apparently independent of isolate aggressiveness (Albrecht et al. 1994c).

   Pisolithus typically forms bright yellow ECM with a thick fungal sheath and well-developed Hartig net, yet there are inter-isolate differences in the degree of sheath development (Marx et al. 1970). The bright yellow colour develops during the initial process of hyphal aggregation to form the sheath (Massicotte et al. 1990). The ontogeny of *Pisolithus* ECM formation has been studied in most detail for *Eucalyptus* spp. Using the growth pouch or paper sandwich synthesis systems, ECM formation on first order lateral roots occurs within 10 days of inoculation (Massicotte et al. 1987b; Horan et al. 1988). ECM formation can be initiated by either diffuse or rhizomorphic hyphae and will occur on lateral roots as they develop or after initial development has occurred, the latter becoming infected only in the apical portion and roots taking on a match-like appearance (Massicotte et al. 1987a,b). Studies of the early stages of *Pisolithus* ECM formation on *Eucalyptus* indicate that sheath formation can commence within approximately 2 days of inoculation, being more rapid at the apex, and during the subsequent 2 days Hartig net formation may be initiated (Horan et al. 1988; Lei et al. 1990a,b). Colonisation of *Pinus* roots by *Pisolithus* occurs within a similar time-frame (Piché and Peterson 1988). *Pisolithus* hyphae may penetrate moribund root cap cells during the early days of eucalypt root colonisation; however, intercellular penetration in Hartig net establishment is confined to epidermal cells that are formed subsequent to fungal colonisation (Horan et al. 1988). During initial contact with the host surface, *Pisolithus* hyphae undergo a morphogenetic shift and produce repeated apical branchings that result in a labyrinthine growth form on the host surface (Jacobs et al. 1989). Root hairs that have developed prior to colonisation appear to collapse (Massicotte et al. 1987b; Thomson et al. 1989), but where root hairs develop during infection (within the zone of Hartig net development) they may become ensheathed by *Pisolithus* (Thomson et al. 1989; Regvar and Gogola 1996). Root hairs colonised in this way clearly cease to function in a normal fashion and appear to degenerate as the ECM matures (Thomson et al. 1989). There is evidence that *Pisolithus* can stimulate host plant ethylene production during early ECM formation and it has been suggested that this may be important in ECM morphogenesis (Rupp et al. 1989). It is assumed to occur via production of IAA, a known effector of ethylene, by *Pisolithus* (Rupp et al. 1989), there being evidence for production of IAA and other indolic compounds by *Pisolithus* in culture (Frankenberger and Poth 1987; Ho 1987; Gruhn et al. 1992; Béguiristain et al. 1995). Hypaphorine production by *Pisolithus* may also reduce root hair development (Béguiristain and Lapeyrie 1997). Host plant jasmonic acid has also been suggested to play a role in *Pisolithus* infection (Regvar and Gogala 1996), although there is little indication of a potential role for jasmonate at present.

In longitudinal section, the *E. pilularis–Pisolithus* ECM comprises a pre-Hartig net zone, Hartig net zone and older Hartig net zone which are sequentially formed proximal to the root tip (Massicotte et al. 1987b). Although all three regions of the ECM possess a thick, compact mantle there is no Hartig net development in the pre-Hartig net zone while in the older Hartig net zone there is evidence of host tissue degeneration and penetration of epidermal cells by the fungus (Massicotte et al. 1987b). Penetration of host cortical cell walls also occurs in the senescent stages of *Pisolithus–Pinus* ECM (Nylund et al. 1982). The Hartig net is narrow and penetrates only to the root exodermis in angiosperms. During Hartig net formation, *Pisolithus* continues to branch extensively to produce the labyrinthine structure that typifies the Hartig net in many ECM fungi (Massicotte et al. 1987a,b,c, 1990). Such hyphae frequently contain large lipid bodies (Massicotte et al. 1987a). In conifer hosts, the Hartig net produced by *Pisolithus* generally penetrates two to three layers of cortical cells (Molina and Trappe 1982b). The extent to which fungal extracellular enzymes (as opposed to mechanical forces) are involved in penetration of *Pisolithus* hyphae between cortical cells during Hartig net formation remains unclear; however, the shape and small size of some penetration points as seen in scanning electron micrographs of *Pinus taeda* roots have been taken to imply that enzymic digestion may be involved in some cases (Warrington et al. 1981). It is noteworthy also that penetration of host cells by the fungus during the early and advanced stages implies an ability to produce wall-degrading enzymes (see below). No evidence exists that the host wall proliferates in response to *Pisolithus* infection (Massicotte et al. 1987a,b,c, 1990), the surface area at the interface between the symbionts being maximised simply by labyrinthine growth of *Pisolithus* and the marked radial enlargement of host epidermal cells in response to infection (Massicotte et al. 1987b). The recent observation that a single isolate of *Pisolithus* forms a typical sheath and Hartig net (along with radially enlarged host epidermal cells) in *Quercus acutissima*, but in *Q. serrata* forms only a sheath (Oh et al. 1995) strongly suggests that there is a degree of host control over the nature of the exchange interface that develops with *Pisolithus*. It may thus represent an excellent model system for more detailed investigation of aspects of the *Pisolithus*–host interaction.

Pisolithus can increase root branching to form second and third order laterals in both conifer and angiosperm hosts (Sohn 1981; Wullschleger and Reid 1990; Oh et al. 1995). There is also evidence that ECM formation by the fungus results in increased lateral root length in a conifer host (Regvar and Gogala 1996). In *E. pilularis* inoculated with *Pisolithus*, first order ECM laterals give rise to second order ECM laterals which develop acropetally along the first order roots. Second order ECM laterals in turn yield third order ECM laterals, leading to the formation of *Pisolithus* ECM clusters (Massicotte et al. 1987b). Second order laterals are produced in the mature region of the Hartig net zone; however, the Hartig net does not spread internally from the parent root; rather developing laterals are infected by inward growth of surface hyphae (Massicotte et al. 1987b).

1.4.4
Symbiotic Functioning

In compatible *Pisolithus*–*Eucalyptus* ECM at least, the fungal sheath surrounding short lateral roots forms a selectively permeable barrier between the root surface and the soil solution (Ashford et al. 1989). Specifically, carbohydrate deposited in the interhyphal spaces of the sheath appears to be of low permeability to solutes, preventing movement through the sheath apoplast. In this way, the fungus–root interface in the *Pisolithus*–*Eucalyptus* ECM has been described as a sealed apoplastic compartment bounded by the impermeable sheath on one side and the root exodermis on the other, within which the physico-chemical environment (and so presumably nutrient exchange) can be controlled by the two partners (Ashford et al. 1989).

The discovery of a motile, tubular vacuolar system within *Pisolithus* hyphae that extends through dolipore septa and can act as a vehicle for inter-cell transport has provided fresh insight into mechanisms of translocation in fungi (Shepherd et al. 1993a,b; Orlovich and Ashford 1994). The vacuolar system within an individual *Pisolithus* hypha extends into mature hyphal regions (Hyde and Ashford 1997) and can display spatial variation in internal pH, suggesting heterogeneity and functional diversity within the interconnected system (Rost et al. 1995). An anion transport mechanism (presumed to be mediated by either proton symport or OH^- antiport, and linked to H^+-ATPase activity) has recently been shown to exist on the tonoplast of such tubular vacuoles and may be important in detoxification of the fungal cytoplasm (Cole et al. 1997). Equally, whereas polyphosphate accumulation in vacuoles of *Pisolithus* has previously been described as being insoluble and/or granular in nature (e.g. Ashford et al. 1986; Orlovich et al. 1989, 1990), recent data suggest that polyphosphate granules may reflect an artefact of specimen preparation and that polyphosphates in vacuoles of *Pisolithus* are in fact soluble (Orlovich and Ashford 1993; Ashford et al. 1994). While the existence of polyphosphate granules in living *Pisolithus* hyphae remains a possibility (Bucking et al. 1998), taken together, the above data implicate translocation of soluble polyphosphate along *Pisolithus* hyphae as the major means for P transport from the soil solution to the fungus–root interface.

Both absorption and efflux of P from *Pisolithus* mycelia appear to be governed strongly by the intracellular (presumed to be vacuolar) inorganic phosphate (P_i) concentration, absorption being maximal at low intracellular P_i concentrations, with net efflux occurring under conditions where high intracellular P_i concentrations are predicted (Cairney and Smith 1992, 1993). As recently suggested by Cairney and Burke (1996), differential expression of polyphosphate kinase and polyphosphatase activities in the extramatrical mycelium and in hyphae at the fungus–root interface respectively might maximise the efficiency of absorption from soil and transfer to the host root at the exchange interface (by regulation of P_i versus polyphosphate concentrations). Although intracellular polyphosphatase activities are known to be produced by *Pisolithus* (Tillard et al. 1989), differential expression in spatially separated regions of an individual mycelium has yet to be demonstrated.

1.5
Host Plant Growth Responses and Fungus-Derived Benefits

1.5.1
Growth Responses

Growth responses following seedling inoculation under controlled conditions with *Pisolithus* (both gymnosperm and angiosperm hosts) have been repeatedly observed (e.g. Marx and Bryan 1970; Beckjord et al. 1985; Heinrich et al. 1988; Bougher and Malajczuk 1990; Burgess et al. 1994). Growth responses of *Eucalyptus* and *Pinus* spp. to inoculation with *Pisolithus* are, however, strongly influenced by fungal genotype (Dixon et al. 1987; Lamhamedi et al. 1990; Burgess et al. 1994; Thomson et al. 1994). Growth stimulation in *E. grandis*, for example, varied from 2 to 45 times that of controls in a single study using 20 *Pisolithus* genotypes, the extent of the growth response being correlated with the degree of mycorrhization (Burgess et al. 1994). It is noteworthy, though, that in some instances, *Pisolithus* infection under controlled conditions has been associated with significantly reduced host growth, particularly under semi-hydroponic conditions where nutrient depletion zones do not occur (Tonkin et al. 1989; Eltrop and Marschner 1996a).

Edaphic factors can also influence host growth responses under controlled conditions. Although *Pisolithus* displays intraspecific variation with respect to temperature optima for growth in axenic culture and mycorrhization, root colonisation under controlled conditions is generally better at relatively high temperatures (above about 19°C) (Marx and Davey 1969a; Marx et al. 1970; Cline et al. 1987), and growth responses of *Pinus* spp. can be enhanced at such temperatures (Marx and Bryan 1971). Under relatively cool conditions, slow growth and consequent poor infection of newly forming short lateral roots of *Eucalyptus diversicolor* by *Pisolithus* have also been shown to preclude a growth response (Bougher et al. 1990). Increasing soil moisture content or nutrient status (particularly P) can similarly reduce infection levels and *Pisolithus*-induced growth responses (Marx et al. 1982; Beckjord et al. 1985; Bougher and Malajczuk 1990). In the case of N, NO_3^- has been shown to decrease *Pisolithus* infection in *Picea abies*, although the extent to which this reflects the poor ability of *Pisolithus* to utilise NO_3^- as an N source or some indirect effect on the symbiotic interaction remains to be determined (Eltrop and Marschner 1996a). Atmospheric gas composition can also influence rates of ECM formation by *Pisolithus*. While elevated atmospheric CO_2 concentrations can increase colonisation and extramatrical mycelium production, along with host P status and growth (O'Neill et al. 1987; Ineichen et al. 1995; Walker et al. 1995; Delucia et al. 1997), NO_x has no apparent effect (Näsholm et al. 1991). The enhanced colonisation under conditions of elevated atmospheric CO_2 may result in enhanced host tolerance to toxic metals such as Al (Schier and McQuattie 1998; and see below). Conversely, increasing ozone concentration or soil acidification can result in decreased infection (Adams and O'Neill 1991; McQuattie and Schier 1992; Maehara et al. 1993), although the latter has not been recorded in all cases (Mahoney et al. 1985; Keane and Manning 1988).

Aggregate data from a number of field studies indicate that, although responses are variable, specific isolates of *Pisolithus* can enhance tree growth, particularly under relatively warm, dry conditions (e.g. Roland and Albaladejo 1994). The degree of root colonisation by *Pisolithus* at outplanting appears to be a good indicator of subsequent field performance, especially in drier years (Marx and Hatchell 1986). In temperate regions, growth stimulation in *Pisolithus*-infected plants over controls (infected by indigenous fungi) have been reported to extend for up to 7 years in *Pinus* spp. outplanted in sandy soils and mine spoils in south eastern USA (Hatchell and Marx 1987; Walker et al. 1989). This effect was species-specific, with *P. taeda*, for example, showing a *Pisolithus*-induced growth stimulation for only a single growing season. Clearly such growth responses are also site-specific, with *Pisolithus* having stimulated growth of *P. taeda* for over 8 years in another study (Marx and Cordell 1988). Outplanting trials in the tropics indicate that growth responses in *Eucalyptus* spp. and *Pinus* spp. can be obtained in the short term following *Pisolithus* inoculation. The degree of growth enhancement is again variable, but in general appears more marked at drier sites with high soil temperatures and in times of lower precipitation (Momoh and Gbadegesign 1980; Marx et al. 1985; Le Tacon et al. 1988). The rapid and prolific sporulation of *Pisolithus* may also be of benefit at some tropical sites in promoting infection and thus enhanced growth of neighbouring, previously uninfected plantation pines (Le Tacon et al. 1988). In contrast, outplanting trials with conifers in cooler parts of the USA indicate that *Pisolithus* provides no overall growth response when compared with seedlings infected by indigenous fungi

(Danielson and Visser 1989; Castellano and Trappe 1991). It is thus regarded as being of little value in boreal forestry (Navratil et al. 1981).

The lack of sustained growth enhancement in outplanted seedlings under cool conditions is widely held to reflect the relatively poor ability of *Pisolithus* to compete with indigenous ECM fungi in plantation soils (Marx et al. 1984). Thus, complete replacement of inoculated *Pisolithus* on conifer roots by indigenous mycobionts has been reported within 3–5 years following out-planting (Riffle and Tinus 1982; Grossnickle and Reid 1983; Danielson and Visser 1989). Replacement of inoculated *Pisolithus* can be much more rapid than these data might suggest. Where ECM fungal communities on outplanted seedlings have been monitored over a shorter time frame, complete loss of *Pisolithus* has been observed to occur within 2 months of outplanting at some sites (McAfee and Fortin 1986). Persistence of *Pisolithus* appears to be best in soils that are more similar to the natural habitats for *Pisolithus* and in situa-tions, such as on extreme acid mine sites, where edaphic stresses result in less intense competition from other ECM mycobionts (Schramm 1966; Berry 1982; McAfee and Fortin 1986, 1988). The general soil microflora may also influence ECM formation and persistence of *Pisolithus*, with enhancement or depres-sion of ECM formation occurring depending on microflora composition (Bowen and Theodorou 1979; Aggangan et al. 1996).

1.5.2
Nutritional Benefits to the Host

Growth enhancement of conifers and eucalypts associated with *Pisolithus* infection has frequently been correlated with increased P accumulation in the host (e.g. Heinrich et al. 1988; Rousseau et al. 1992; Burgess et al. 1993; Thomson et al. 1994), although this is not always the case (Walker et al. 1989). Enhanced P accumulation appears to relate to the level of ECM infection and the surface area of the extramatrical mycelial phase (Rousseau and Reid 1990; Rousseau et al. 1992, 1994; Thomson et al. 1994). There is direct evidence that *Pisolithus* can absorb orthophosphate from solution in the external environment and transfer absorbed P to the host following translo-cation through extramatrical mycelium (Kammerbauer et al. 1989). The extent to which absorbed P is translocated to the host shoot appears to be isolate-specific, with ECM formed by some *Pisolithus* isolates retaining relatively large proportions of absorbed P in the root system (Cumming 1996). *Pisolithus* has been shown to solubilise relatively insoluble forms of inorganic phosphate (Al and Ca phosphates) in vitro, although this ability appears to be isolate-specific (Lapeyrie et al. 1991). It has further been shown that infection with *Pisolithus* can increase host access to insoluble inorganic phosphate sources, such as Fe/Al PO_4, in *Eucalyptus pilularis* (Heinrich et al. 1988). As is the case for most ECM basidiomycetes, isolates of *Pisolithus* can utilise soluble salts of inositol hexaphosphate as a sole P source (Mousain and Salsac 1986). Although there exists interspecific variation in their relative importance, the basis for utilisation appears to be production of extracellular acid and alkaline phosphomonoesterase and phosphodiesterase enzymes (Ho 1987). Acid phosphomonoesterase production is stimulated by deficiency of inorganic phosphate (Berjaud and d'Auzac 1986; Mousain and Salsac 1986). While it is generally assumed to occur, transfer of P to the host plant, and so host plant benefit, derived from organic phosphate sources remains to be demonstrated for *Pisolithus*.

Enhanced seedling growth in response to *Pisolithus* inoculation can in some instances be correlated with increased foliar N content (Wullschleger and Reid 1990). Since foliar N concentration has in turn been positively correlated with endogenous plant cytokinin levels it has been suggested that N may act via cytokinins as a metabolic regulator (Wullschleger and Reid 1990). *Pisolithus* mycelia growing in axenic culture and conifer roots infected by the fungus can absorb inorganic N as either NH_4^+ or NO_3^-, although rates of NH_4^+ absorption are greater than for NO_3^- (France and Reid 1983, 1984; Eltrop and Marschner 1996a). *Pisolithus* has been shown to transfer N absorbed in the form of NH_4^+ to *Pinus sylvestris* host plants (Finlay et al. 1988). In common with several other ECM fungi, absorbed NH_4^+ appears to be rapidly incorporated into amino acid precursors within the extramatrical mycelium and is translocated to the host largely in this form (Finlay et al. 1988). While an enzymological study of NH_4^+ metabolising enzymes has suggested that *Pisolithus* produces only low levels of glutamine synthetase (GS) [and no glutamate synthase (GOGAT)] activity (Vézina et al. 1989), studies of $^{15}NH_4^+$ metabolism provide strong evidence that NH_4^+ is metabolised via the GS/GOGAT pathways (Kershaw and Stewart 1992; Turnbull et al. 1996). This apparent discrepancy may reflect that single (and different) isolates of *Pisolithus* were used in each study, and that intraspecific differences in inorganic N assimilation exist. More likely, however, they reflect the relative lack in sensitivity of enzyme assays when compared with the direct assessment of ^{15}N assimilation possible at the mass spectrometer level.

When grown in axenic culture, *Pisolithus* can release and utilise NH_4^+ (and also Ca^{2+}) ionically bound to vermiculite, suggesting a potential to partially weather phyllosilicates in soil (Paris et al. 1995b). While soluble fungal exudates appear to be responsible for such weathering, it remains to be determined to what extent the host benefits from the potential N source. Data relating to the ability of *Pisolithus* to enhance host plant acquisition of N from organic sources are rather more equivocal. *Pisolithus* has been classified as a "non-protein" fungus based on the relatively poor ability of certain isolates to produce extracellular protease activity in axenic culture (Abuzinadah and Read 1986; Cao and Crawford 1993a). This notwithstanding, it is clear that some *Pisolithus* isolates are capable of protease secretion under some conditions (Dahm and Strzelczyk 1995). Similarly, while Abuzinadah et al. (1986) concluded that *Pisolithus* had little ability to enhance acquisition of N from protein by *Pinus contorta*, Turnbull et al. (1995) have recently shown enhanced utilisation of N in protein and histidine sources by

Eucalyptus spp. in symbiosis with *Pisolithus*. Such disparate results may indicate an influence of different host plants, perhaps, as suggested by Turnbull et al. (1995), mediated by a differential availability of C compounds from each host. Equally, since in each case only single isolates of *Pisolithus* were utilised, it may simply indicate that considerable intraspecific variation exists within *Pisolithus* with regard to the potential for facilitating organic N utilisation. Recent results from our laboratory indicate that wide intraspecific variation does indeed exist in *Pisolithus* with regard to the ability of individual isolates to utilise organic N sources (J. M. Sharples and J. W. G. Cairney, unpubl. data).

1.5.3
Mineral Transformations

Pisolithus can bring about several mineral transformations that may be important in increasing the availability of certain elements in soil. In axenic culture, isolates of *Pisolithus* are known to bring about oxidation of elemental sulphur (Grayston and Wainwright 1988) and to produce an extracellular substance capable of reducing higher oxides of manganese (Cairney and Ashford 1991). The latter activity is apparently also produced during symbiosis with the host (Cairney and Ashford 1989) and there is direct evidence that *Pisolithus* can enhance manganese accumulation from soil by *Pinus virginiana* seedlings (Miller and Rudolph 1986). *Pisolithus* has also been shown to displace and render available K^+ from non-exchangeable sites in phlogopite mica, probably via secretion of oxalate (Paris et al. 1995a, 1996). Isolates of *Pisolithus* can produce hydroxamate-like siderophores which may be important in chelating scarcely available soil iron compounds (Szaniszlo et al. 1981; Leyval and Reid 1991). Such chelating compounds can be absorbed by both ECM and non-ECM host roots (Leyval and Reid 1991).

Although the fungus is generally assumed to have a poor ability to decompose carbohydrate components of the plant cell wall, recent evidence indicates that components of the cellulase complex are produced differentially by different *Pisolithus* isolates. Thus, β-glucosidase production may be a common feature of many *Pisolithus* isolates while only some isolates, but not others, produce β-galactosidase and/or both endo- and exo-acting glucanases (Cao and Crawford 1993a,b). *Pisolithus* can also grow well on pectin as a C source; however, it appears to produce only small amounts of extracellular polygalacturonase activity. This suggests production of other pectin degrading activity, such as pectin lyase (Perotto et al. 1997). Production of cellulases during symbiosis with the host has not yet been shown, but it is possible that the enzymes might be involved in establishment of the symbiosis (see above), along with hyphal penetration of host walls as the symbiosis ages and possible interactions between extramatrical mycelium and moribund plant material in soil (see Cairney and Burke 1994).

1.5.4
Influence on Host Plant Carbon Economy

Infection of conifer seedlings with *Pisolithus* generally results in increased rates of net photosynthesis (Ekwebelam and Reid 1983; Reid et al. 1983; Rousseau and Reid 1990); under some conditions, increases over uninfected seedlings can be in the order of 75% (Rousseau and Reid 1989). Where soil has a relatively high P status, *Pisolithus* may have a neutral effect on host photosynthesis and result in poorer host growth compared with uninfected controls (Rousseau and Reid 1989), the increased respiratory cost of the fungus having a negative effect on host C balance. Where increased photosynthetic rate has been observed, it has generally been associated with an increase in host plant biomass (Reid et al. 1983; Rousseau and Reid 1990). The extent of increases in net photosynthesis and host biomass shows a strong correlation with the degree of root system infection with *Pisolithus* (Rousseau and Reid 1990). By comparing *Pisolithus*-infected seedlings with seedlings fertilised with different levels of phosphate, Rousseau and Reid (1990) concluded that, at low infection rates, increased photosynthesis in *P. taeda* attributable to infection are probably a result of enhanced phosphate accumulation. Where infection rates are high, however, it appears that the increased C sink created by the fungus results in increased photosynthesis in a more direct manner (Rousseau and Reid 1990). *Pisolithus* has also been shown to maintain high rates of gas exchange and photosynthesis in *Eucalyptus* sp. during drought stress (Dixon and Hiol-Hiol 1992). In circumstances where *Pisolithus* infection has reportedly reduced host plant growth, infection may still increase rates of CO_2 assimilation in the host (Eltrop and Marschner 1996b). In such circumstances, however, it is likely that increased below-ground respiration resulting from the presence of ECM mycelium imparts a significant drain on host C resources in the absence of a significant nutritional benefit.

Clear evidence exists that *Pisolithus* acts as a significant sink for host-derived C, at least during the early phase of the symbiosis. Cairney et al. (1989) used $^{14}CO_2$ pulse labelling to show that some 18 times more C can accumulate in *Pisolithus*-infected *Eucalyptus pilularis* short lateral roots compared with uninfected roots in the same root system. These data are likely to underestimate the increased sink created by *Pisolithus* since they take into consideration neither C translocated from ECM tips into extramatrical mycelium nor fungal respiration. These may be significant given that the autoradiographs produced by Cairney et al. (1989) showed considerable labelling of extramatrical mycelia and the reported three times greater root-derived respiration in *Pisolithus*-infected *P. contorta* recorded by Reid et al. (1983). As the *Pisolithus*–eucalypt association ages, there appears to be a progressive decrease in the degree to which it acts as a sink for host photosynthetic products (Cairney et al. 1989). There is also evidence that small quantities of C compounds can be transferred between host plants interconnected by a common *Pisolithus* mycelium (Finlay and Read 1986). The main soluble carbohydrate in *Pisolithus* extramatrical mycelia and mycelia grown in axenic culture is arabitol, but trehalose and mannitol have also been shown to be present in significant quantities (Söderström et al. 1988; Ineichen and Wiemken 1992). While the main fungal carbohydrate in *Picea abies* ECM is trehalose (Ineichen and Wiemken 1992), the preponderance of arabitol in *Pisolithus* mycelia strongly supports the latter as the major translocatory carbohydrate in *Pisolithus*.

1.5.5
Non-nutritional Benefits to the Host

Infection of both angiosperm and gymnosperm seedlings with *Pisolithus* can reduce host water deficit under conditions of mild drought (Dixon et al. 1983; Parke et al. 1983; Walker et al. 1989). The ability of *Pisolithus* to ameliorate drought stress appears to be strongly isolate-specific, there being a demonstrated correlation between the ability of isolates to produce extensive rhizomorph systems and their ability to enhance host water status (Lamhamedi et al. 1992a,b). While temperature and drought stresses may be important in determining the survival of *Pisolithus* on mine sites, persistence of the fungus will also require a low sensitivity to toxic metals. *Pisolithus* can reduce Zn accumulation in conifer shoots when grown on contaminated coalspoils (Walker et al. 1989) or in Zn-supplemented soil in the glasshouse (Miller and Rudolph 1986). Reduced Zn accumulation in the shoot is accompanied by an increase in Zn accumulation in *Pisolithus*-infected versus non-ECM roots (Miller and Rudolph 1986). The ability of *Pisolithus* to ameliorate Zn sensitivity seems likely to be isolate-specific, since in some instances infection can result in an increase in foliar Zn (Berry and Marx 1976). Similarly, while a single isolate of *Pisolithus* has recently been shown to be ineffective in ameliorating Pb toxicity in *Picea abies* and in preventing entry of the element to the host root cortex (Marschner et al. 1996; Jentschke et al. 1997), screening with multiple isolates may reveal less sensitive isolates.

Because of the potential toxicity to forest trees of Al in acid soils, particularly under the influence of acid precipitation, there has been interest in the potential ability of *Pisolithus* to ameliorate the problem. Infection with *Pisolithus* can reduce Al accumulation in the host shoot and can partially alleviate Al sensitivity in *Pinus* (Berry and Marx 1976; Cumming and Weinstein 1990a; Schier and McQuattie 1995). This may apply only up to certain Al concentrations, since ECM formation by *Pisolithus* can be inhibited at high concentrations (McQuattie and Schier 1992). The precise mechanism of the reduced host sensitivity is not clear, although reduced Al uptake by the host (perhaps due to the diffusion barrier presented by the sheath) is thought to be involved to some extent (Cumming and Weinstein 1990b; Schier and McQuattie 1995; Godbold et al. 1996). Increased host P status (and so enhanced host vigour), arising from an ability of the fungus to prevent precipitation of $AlPO_4$ in the rhizosphere and root apoplast, may also play a role (Cumming and Weinstein 1990b; Schier and McQuattie, 1995). The preference of *Pisolithus* for NH_4^+ rather than NO_3^- as an N source might be further involved. Since nitrate reductase is Al-sensitive, a switch to the less sensitive NH_4^+ assimilation pathways in *Pisolithus* ECM could be important in ameliorating toxicity (Cumming 1990; Cumming and Weinstein 1990a,b). Godbold et al. (1996) have further suggested that the lower pH of the apoplast arising during NH_4^+ utilisation may also be important in preventing Al accumulation in walls of the host cortex. From a recent study of 21 *Pisolithus* isolates originating from soils of differing Al and pH status, it is clear that considerable intraspecific variation in Al sensitivity in axenic culture exists (Egerton-Warburton and Griffin 1995). The degree of sensitivity was inversely correlated with the relative availability of Al in the soil of origin, although the authors were careful to point out that the number of isolates screened was relatively small, that most of the isolates came from a heavily contaminated site and that a larger number of samples from sites contaminated to a lesser degree would need to be included for a proper correlation to be derived. The differential abilities of these isolates to ameliorate plant sensitivity to Al remains to be investigated.

The mechanisms involved in metal detoxification by *Pisolithus* have not been investigated in detail. However, intracellular metallothionein-like proteins can be induced by toxic metal exposure in a single *Pisolithus* isolate (Morselt et al. 1986), while an increase in intracellular tyrosinase activity in another isolate (in response to Cu exposure) has been implicated in chelation of intracellular Cu (Gruhn and Miller 1991). Polysaccharides and cysteine-rich proteins have also been shown to accumulate on the outer cell wall in a further *Pisolithus* isolate in response to extracellular Cd exposure, and electron energy loss spectroscopy (at the TEM level) has been used to demonstrate apparent accumulation of Cd on the outer region of hyphal walls (Turnau et al. 1994). While the latter observation appears to suggest a role for the modified cell wall in Cd detoxification, it must be viewed with a degree of caution since material was prepared for electron microscopy using conventional (hydrated) preparative techniques that do not preclude redistribution of ions during specimen preparation. Where ECM have been prepared for electron microscopy/X-ray microanalysis using cryo-methods, Al appears to accumulate specifically in the *Pisolithus* sheath (Egerton-Warburton et al. 1993). In the case of Al detoxification, it may also be that enhanced Ca and Mg accumulation in tolerant *Pisolithus* isolates alters the ratio of the divalent cations to Al, reducing binding and absorption of Al and so decreasing toxicity (Egerton-Warburton and Griffin 1995).

Pisolithus may also be of value in remediating sites contaminated by xenobiotic organic chemicals. Although only two investigations of the fungus in axenic culture have been conducted to date, Donnelly and Fletcher (1995) have identified an isolate of *Pisolithus* that possesses some ability to degrade polychlorinated biphenyls, while Meharg et al. (1997) indicate an ability to biotransform 2,4,6-trinitrotoluene. Further screening of *Pisolithus* growing in symbiosis with the host is required to determine the real potential for using the fungus in remediation of sites contaminated by organic pollutants.

It has been known for some considerable time that *Pisolithus* has the potential to protect seedlings against a variety of soil-borne pathogens (see Marx 1972). For example, *Pisolithus* has been reported to confer a degree of protection to *Pinus* spp. against (among others) *Phytophthora*, *Fusarium*, *Rhizoctonia* and *Cylindrocarpon* spp. in glasshouse trials (Ross and Marx 1972; Chakravarty and Unestam 1987a,b). In some instances, the protective effect of *Pisolithus* isolates has been attributed solely to the provision of a physical barrier by the fungal sheath, there having been no evidence of production of antimicrobial activities by the fungus (Marx and Davey 1969b; Marx 1970). There have, however, been many reports of in vitro inhibition of the growth of a range of pathogens in the presence of *Pisolithus* mycelium, strongly implying production of antimicrobial metabolites (e.g. Marx 1969; Kope and Fortin 1989a,c), although the degree of this effect is isolate-specific (Kope and Fortin 1989b; Suh et al. 1991). Unidentified phenolic compounds (Suh et al. 1991) were implicated in one study, while two specific antifungal compounds [*p*-hydroxybenzoylformic acid (pisolithin A) and (*R*)-(−)-*p*-hydroxymandelic acid (pisolithin B)] have been isolated by other workers and their effectiveness demonstrated against a range of pathogens in vitro (Kope and Fortin 1989a; Kope et al. 1991). Production of these secondary metabolic products has not yet been confirmed during symbiosis with the host. However, given the physiological heterogeneity that exists within individual ECM mycelia, and the potential for idiophase (secondary metabolism) onset in different spatio-temporal regions therein (see Cairney and Burke 1994, 1996), it is certainly conceivable that such products can be expressed in symbiotic *Pisolithus* mycelia in soil. It is further possible that the extracellular antimicrobial effect of *Pisolithus* may be enhanced by non-specific acidification of the rhizosphere in some instances (Rasanayagam and Jeffries 1992).

1.6
Conclusions

A considerable body of literature exists relating to aspects of the ecology, physiology and molecular biology of interactions between *Pisolithus* and its plant hosts. From work conducted to date, it is clear that considerable intraspecific variation exists within *Pisolithus* in terms of host specificity, growth form of extramatrical mycelia or organic N utilisation. Significant progress is currently being made towards an understanding of the infection process and compatibility between *Pisolithus* and various hosts using multiple *Pisolithus* isolates, although the extent to which differential host compatibility reflects taxonomic variation within the *Pisolithus* group is not yet clear. It may be, however, that careful examination of phylogeny within *Pisolithus* at the molecular level will reveal a genetic basis for differential host specificity.

In many instances, physiological aspects of the fungus–host interaction have been studied simply in the form of observations of individual *Pisolithus* mycelia or comparisons between single *Pisolithus* isolates and other ECM fungi. Given the level of variation displayed within *Pisolithus* in other aspects of the symbiosis, such data should be used with caution to extrapolate the physiological capabilities of the fungus. Physiological screening of a range of *Pisolithus* isolates, preferably those isolates upon which current molecular and host compatability investigation is focused, will be required in order to develop a true picture of *Pisolithus* ECM symbioses and their effectiveness in enhancing host plant growth and survival.

Acknowledgement. This chapter represents a revised and updated version of an article originally published in Mycorrhiza 7:117–131.

References

Abuzinadah RA, Read DJ (1986) The role of proteins in the nitrogen nutrition of ectomycorrhizal plants. I. Utilization of peptides and proteins by ectomycorrhizal fungi. New Phytol 103:481–493

Abuzinadah RA, Finlay RD, Read DJ (1986) The role of proteins in the nitrogen nutrition of ectomycorrhizal plants. II. Utilization of protein by mycorrhizal plants of *Pinus contorta*. New Phytol 103:495–506

Adams MB, O'Neill EG (1991) Effects of ozone and acidic deposition on carbon allocation and mycorrhizal colonization of *Pinus taeda* L. seedlings. For Sci 37:5–16

Agerer R (1991) Comparison of the ontogeny of hyphal and rhizoid strands of *Pisolithus tinctorius* and *Polytrichum juniperinum*. Crypt Bot 2/3:85–92

Aggangan NS, Dell B, Malajczuk N, De la Cruz RE (1996) Soil fumigation and phosphorus supply affect the formation of *Pisolithus–Eucalyptus urophylla* ectomycorrhizas in two acid Philippine soils. Plant Soil 180:259–266

Albrecht C, Asselin A, Piché Y, Lapeyrie F (1994a) Chitinase activities are induced in *Eucalyptus globulus* roots by ectomycorrhizal or pathogenic fungi, during early colonization. Physiol Plant 91:104–110

Albrecht C, Burgess T, Dell B, Lapeyrie F (1994b) Chitinase and peroxidase activities are induced in eucalyptus roots according to aggressiveness of Australian ectomycorrhizal strains of *Pisolithus* sp. New Phytol 127:217–222

Albrecht C, Laurent P, Lapeyrie F (1994c) *Eucalyptus* root and shoot chitinases, induced following root colonisation by pathogenic versus ectomycorrhizal fungi, compared on one- and two-dimensional activity gels. Plant Sci 100:157–164

Anderson IC, Chambers SM, Cairney JWG (1998a) Molecular determination of genetic variation in *Pisolithus* isolates from a defined region in New South Wales, Australia. New Phytol 138:151–162

Anderson IC, Chambers SM, Cairney JWG (1998b) Use of molecular methods to estimate the size and distribution of mycelial individuals of the ectomycorrhizal basidiomycete *Pisolithus tinctorius*. Mycol Res 102:295–300

Ashford AE, Peterson RL, Dwarte D, Chilvers GA (1986) Polyphosphate granules in eucalypt mycorrhizas: determination by energy dispersive X-ray microanalysis. Can J Bot 64:677–687

Ashford AE, Allaway WG, Peterson CA, Cairney JWG (1989) Nutrient transfer and the fungus-root interface. Aust J Plant Physiol 16:85–97

Ashford AE, Ryde S, Barrow KD (1994) Demonstration of a short chain polyphosphate in *Pisolithus tinctorius* and implications for phosphorus transport. New Phytol 126:239–247

Ba AM, Thoen D (1990) First syntheses of ectomycorrhizas between *Afzelia africana* Sm. (Caesalpinioideae) and native fungi from West Africa. New Phytol 114:99–103

Ba AM, Balaji B, Piché Y (1994) Effect of time of inoculation on in vitro ectomycorrhizal colonization and nodule initiation in *Acacia holosericea* seedlings. Mycorrhiza 4:109–119

Beckjord PR, Melhuish JH, McIntosh MS (1985) Effects of nitrogen and phosphorus fertilization on growth and formation of ectomycorrhizae of *Quercus alba* and *Q. rubra* seedlings by *Pisolithus tinctorius* and *Scleroderma aurantium*. Can J Bot 63:1677–1680

Béguiristain T, Lapeyrie F (1997) Host plant stimulates hypaphorine accumulation in *Pisolithus tinctorius* hyphae during ectomycorrhizal infection while excreted hypaphorine controls root hair development. New Phytol 136:525–532

Béguiristain T, Cote R, Rubini P, Jay-Allemand C, Lapeyrie F (1995) Hypaphorine accumulation in hyphae of the ectomycorrhizal fungus, *Pisolithus tinctorius*. Phytochemistry 40:1089–1091

Berjaud C, d'Auzac J (1986) Isolement et caractérisation des phosphatases d'un champignon ectomycorhizogène typique, *Pisolithus tinctorius*. Effets de la carence en phosphate. Physiol Vég 24:163–172

Berry CR (1982) Survival and growth of pine hybrid seedlings with *Pisolithus* ectomycorrhiza on coal spoils in Alabama and Tennessee. J Environ Qual 11:709–715

Berry CR, Marx DH (1976) Sewage sludge and *Pisolithus tinctorius* ectomycorrhiza: their effect on growth of pine seedlings. For Sci 22:351–358

Bougher NL, Malajczuk N (1990) Effects of high soil moisture on formation of ectomycorrhizas and growth of karri (*Eucalyptus diversicolor*) seedlings inoculated with *Descolea maculata*, *Pisolithus tinctorius* and *Laccaria laccata*. New Phytol 114:87–91

Bougher NL, Grove TS, Malajczuk N (1990) Growth and phosphorus acquisition of karri (*Eucalyptus diversicolor* F. Muell.) seedlings inoculated with ectomycorrhizal fungi in relation to phosphorus supply. New Phytol 114:77–85

Bowen GD, Theodorou C (1979) Interactions between bacteria and ectomycorrhizal fungi. Soil Biol Biochem 11:119–126

Bronchart R, Calonge FD, Demoulin V (1975) Nouvelle contribution à l'étude de l'ultrastructure de la paroi sporale des Gastéromycètes. Bull Soc Mycol Fr 91:232–246

Bucking H, Beckmann S, Heyser W, Kottke I (1998) Elemental contents in vacuolar granules of ectomycorrhizal fungi measured by EELS and EDXS – a comparison of different methods and preparation techniques. Micron 29:53–61

Bulmer GS (1964) Spore germination of forty-two species of puffballs. Mycologia 56:630–632

Burgess T, Dell B (1996) Changes in protein biosynthesis during the differentiation of *Pisolithus–Eucalyptus grandis* ectomycorrhiza. Can J Bot 74:553–560

Burgess T, Malajczuk N, Grove TS (1993) The ability of 16 ectomycorrhizal fungi to increase growth and phosphorus uptake by *Eucalyptus globulus* Labill and *E. diversicolor* F. Muell. Plant Soil 153:155–164

Burgess T, Dell B, Malajczuk N (1994) Variation in mycorrhizal development and growth stimulation by 20 *Pisolithus* isolates inoculated on to *Eucalyptus grandis* W. Hill ex Maiden. New Phytol 127:731–739

Burgess T, Pascal L, Dell B, Malajczuk N, Martin F (1995a) Effect of fungal-isolate aggressivity on the biosynthesis of symbiotic-related polypeptides in differentiating eucalypt ectomycorrhizas. Planta 195:408–417

Burgess T, Malajczuk N, Dell B (1995b) Variation in *Pisolithus* based on basidiome and basidiospore morphology, culture characteristics and analysis of polypeptides using 1D SDS-PAGE. Mycol Res 99:1–13

Cairney JWG (1992) Translocation of solutes in ectomycorrhizal and saprotrophic rhizomorphs. Mycol Res 96:135–141

Cairney JWG, Ashford AE (1989) Reducing activity at the root surface in *Eucalyptus pilularis–Pisolithus tinctorius* ectomycorrhizas. Aust J Plant Physiol 16:99–105

Cairney JWG, Ashford AE (1991) Release of a reducing substance by the ectomycorrhizal fungi *Pisolithus tinctorius* and *Paxillus involutus*. Plant Soil 135:147–150

Cairney JWG, Burke RM (1994) Fungal enzymes degrading plant cell walls: their possible significance in the ectomycorrhizal symbiosis. Mycol Res 98:1345–1356

Cairney JWG, Burke RM (1996) Physiological heterogeneity within fungal mycelia: an important concept for a functional understanding of the ectomycorrhizal symbiosis. New Phytol 134:685–695

Cairney JWG, Smith SE (1992) Influence of intracellular phosphorus concentration on phosphate absorption by the ectomycorrhizal basidiomycete *Pisolithus tinctorius*. Mycol Res 96:673–676

Cairney JWG, Smith SE (1993) Efflux of phosphate from the ectomycorrhizal basidiomycete *Pisolithus tinctorius*: general characteristics and the influence of intracellular phosphorus concentration. Mycol Res 97:1261–1266

Cairney JWG, Ashford AE, Allaway WG (1989) Distribution of photosynthetically fixed carbon within root systems of *Eucalyptus pilularis* plants ectomycorrhizal with *Pisolithus tinctorius*. New Phytol 112:495–500

Cairney JWG, Jennings DH, Agerer R (1991) The nomenclature of fungal multi-hyphal linear aggregates. Crypt Bot 2/3:246–251

Calonge FD, Demoulin V (1975) Les Gastéromycètes d'Espagne. Bull Soc Mycol Fr 91:247–292

Cao W, Crawford DL (1993a) Carbon nutrition and hydrolytic and cellulolytic activities in the ectomycorrhizal fungus *Pisolithus tinctorius*. Can J Microbiol 39:529–535

Cao W, Crawford DL (1993b) Purification and some properties of β-glucosidase from the ectomycorrhizal fungus *Pisolithus tinctorius* strain SMF. Can J Microbiol 39:125–129

Carnero Diaz E, Martin F, Tagu D (1996) Eucalypt α-tubulin: cDNA cloning and increased level of transcripts in ectomycorrhizal root system. Plant Mol Biol 31:905–910

Carnero Diaz E, Tagu D, Martin F (1997) Ribosomal DNA internal transcribed spacers to estimate the proportion of *Pisolithus tinctorius* and *Eucalyptus globulus* RNAs in ectomycorrhiza. Appl Environ Microbiol 63:840–843

Castellano MA, Trappe JM (1991) *Pisolithus tinctorius* fails to improve plantation performance of inoculated conifers in southwestern Oregon. New For 5:349–358

Chakravarty P, Unestam T (1987a) Differential influence of ectomycorrhizae on plant growth and disease resistance in *Pinus sylvestris* seedlings. J Phytopathol 120:104–120

Chakravarty P, Unestam T (1987b) Mycorrhizal fungi prevent disease in stressed pine seedlings. J Phytopathol 118:335–340

Chilvers GA (1973) Host range of some eucalypt mycorrhizal fungi. Aust J Bot 21:103–111

Cline ML, France RC, Reid CPP (1987) Intraspecific and interspecific growth variation of ectomycorrhizal fungi at different temperatures. Can J Bot 65:869–875

Coker WC, Couch JN (1928) The Gasteromycetes of the eastern United States and Canada. University of North Carolina Press, Chapel Hill

Cole L, Hyde GJ, Ashford AE (1997) Uptake and compartmentalisation of fluorescent probes by *Pisolithus tinctorius* hyphae: evidence for an anion transport mechanism at the tonoplast but not for fluid-phase endocytosis. Protoplasma 199:18–29

Cumming JR (1990) Nitrogen source effects on Al toxicity in nonmycorrhizal and mycorrhizal pitch pine (*Pinus rigida*) seedlings. II. Nitrate reduction and NO_3^- uptake. Can J Bot 68:2653–2659

Cumming JR (1996) Phosphate-limitation physiology in ectomycorrhizal pitch pine (*Pinus rigida*) seedlings. Tree Physiol 16:977–983

Cumming JR, Weinstein LH (1990a) Aluminum–mycorrhizal interactions in the physiology of pitch pine seedlings. Plant Soil 125:7–18

Cumming JR, Weinstein LH (1990b) Nitrogen source effects on Al toxicity in nonmycorrhizal and mycorrhizal pitch pine (*Pinus rigida*) seedlings. I. Growth and nutrition. Can J Bot 68:2644–2652

Cunningham GH (1942) The Gasteromycetes of Australia and New Zealand. John McIndoe, Dunedin

Dahm H, Strzelczyk E (1995) Impact of vitamins on cellulolytic, pectolytic and proteolytic activity of mycorrhizal fungi. Symbiosis 18:233–250

Danielson RM, Visser S (1989) Host response to inoculation and behaviour of introduced and indigenous ectomycorrhizal fungi of jack pine grown on oil-sand tailings. Can J For Res 19:1412–1421

Delucia EH, Callaway RM, Thomas EM, Schlesinger WH (1997) Mechanisms of phosphorus acquisition for ponderosa pine seedlings under high CO_2 and temperature. Ann Bot 79:111–120

Dennis JJ (1980) Sclerotia of the Gasteromycete *Pisolithus*. Can J Microbiol 26:1505–1507

Dixon RK, Hiol-Hiol F (1992) Gas exchange and photosynthesis of *Eucalyptus camaldulensis* seedlings inoculated with different ectomycorrhizal symbionts. Plant Soil 147:143–149

Dixon RK, Pallardy SG, Garrett HE, Cox GS, Sander IL (1983) Comparative water relations of container-grown and bare-root ectomycorrhizal and non-mycorrhizal *Quercus velutina* seedlings. Can J For Res 61:1559–1565

Dixon RK, Garrett HE, Stelzer HE (1987) Growth and development of Loblolly pine progenies inoculated with three isolates of *Pisolithus tinctorius*. Silvae Genet 36:240–245

Donnelly PK, Fletcher JS (1995) PCB metabolism by ectomycorrhizal fungi. Bull Environ Contam Toxicol 54:507–513

Egerton-Warburton LM, Griffin BJ (1995) Differential responses of *Pisolithus tinctorius* isolates to aluminium in vitro. Can J Bot 73:1229–1233

Egerton-Warburton LM, Kuo J, Griffin BJ, Lamont BB (1993) The effect of aluminium on the distribution of calcium, magnesium and phosphorus in mycorrhizal and non-mycorrhizal seedlings of *Eucalyptus rudis*: a cryo-microanalytical study. Plant Soil 156:481–484

Ekwebelam SA, Reid CPP (1983) Effect of light, nitrogen fertilization and mycorrhizal fungi on growth and photosynthesis of lodgepole pine seedlings. Can J For Res 13:1099–1106

Eltrop L, Marschner H (1996a) Growth and mineral nutrition of non-mycorrhizal and mycorrhizal Norway spruce (*Picea abies*) seedlings grown in semi-hydroponic sand culture. I. Growth and mineral nutrient uptake in plants supplied with different forms of nitrogen. New Phytol 133:469–478

Eltrop L, Marschner H (1996b) Growth and mineral nutrition of non-mycorrhizal and mycorrhizal Norway spruce (*Picea abies*) seedlings grown in semi-hydroponic sand culture. II. Carbon partitioning in plants supplied with ammonium or nitrate. New Phytol 133:479–486

Finlay RD, Read DJ (1986) The structure and function of the vegetative mycelium of ectomycorrhizal plants. I. Translocation of ^{14}C-labelled carbon between plants interconnected by a common mycelium. New Phytol 103:143–156

Finlay RD, Ek H, Odham G, Söderström B (1988) Mycelial uptake, translocation and assimilation of nitrogen from ^{15}N-labelled ammonium by *Pinus sylvestris* plants infected with four different ectomycorrhizal fungi. New Phytol 110:59–66

Fortin JA, Piché Y, Godbout C (1983) Methods for synthesising ectomycorrhizas and their effect on mycorrhizal development. New Phytol 71:275–284

France RC, Reid CPP (1983) Interactions of nitrogen and carbon in the physiology of ectomycorrhizae. Can J Bot 61:964–984

France RC, Reid CPP (1984) Pure culture growth of ectomycorrhizal fungi on inorganic nitrogen sources. Microbial Ecol 10:187–195

Frankenberger WT, Poth M (1987) Biosynthesis of indole-3-acetic acid by the pine ectomycorrhizal fungus *Pisolithus tinctorius*. Appl Environ Microbiol 53:2908–2913

Gardner JH, Malajczuk N (1988) Recolonisation of rehabilitated bauxite mine sites in Western Australia by mycorrhizal fungi. For Ecol Manage 24:27–42

Godbold DL, Jentschke G, Marschner P (1996) Solution pH modifies the response of Norway spruce seedlings to aluminium. Plant Soil 171:175–178

Godbout C, Fortin JA (1983) Morphological features of synthesized ectomycorrhizae of *Alnus crispa* and *A. rugosa*. New Phytol 94:249–262

Godbout C, Fortin JA (1985) Synthesized ectomycorrhizae of aspen: fungal genus level of structural characterisation. Can J Bot 63:252–262

Grayston SJ, Wainwright M (1988) Sulphur oxidation by soil fungi including some species of mycorrhizae and wood-rotting basidiomycetes. FEMS Microbiol Ecol 53:1–8

Grenville DJ, Peterson RL, Piché Y (1985) The development, structure, and histochemistry of sclerotia of ectomycorrhizal fungi. I. *Pisolithus tinctorius*. Can J Bot 63:1402–1411

Grossnickle SC, Reid CPP (1983) Ectomycorrhiza formation and root development patterns of conifer seedlings on a high-elevation mine site. Can J For Res 13:1145–1158

Gruhn CM, Miller OK (1991) Effect of copper on tyrosinase activity and polyamine content of some ectomycorrhizal fungi. Mycol Res 95:268–272

Gruhn CM, Gruhn AV, Miller OK (1992) *Boletinellus meruloides* alters root morphology of *Pinus densiflora* without mycorrhizal formation. Mycologia 84:528–533

Hatchell GE, Marx DH (1987) Response of longleaf, sand and loblolly pines to *Pisolithus* ectomycorrhizae and fertilizer on a sandhills site in South Carolina. For Sci 33:301–315

Heinrich PA, Mulligan DR, Patrick JW (1988) The effect of ectomycorrhizas on the phosphorus and dry weight acquisition of *Eucalyptus* seedlings. Plant Soil 109:147–149

Hilbert J-L, Martin F (1988) Regulation of gene expression in ectomycorrhizas. I. Protein changes and the presence of ectomycorrhiza-specific polypeptides in the *Pisolithus– Eucalyptus* symbiosis. New Phytol 110:339–346

Hilbert J-L, Costa G, Martin F (1991) Ectomycorrhizin synthesis and polypeptide changes during the early stages of eucalypt mycorrhiza development. Plant Physiol 97:977–984

Ho I (1987) Comparison of eight *Pisolithus tinctorius* isolates for growth rate, enzyme activity, and phytohormone production. Can J For Res 17:31–35

Hodge A, Alexander IJ, Gooday GW (1995) Chitinolytic activities of *Eucalyptus pilularis* and *Pinus sylvestris* root systems challenged with mycorrhizal and pathogenic fungi. New Phytol 131:255–261

Horan DP, Chilvers GA (1990) Chemotropism – the key to ectomycorrhizal formation? New Phytol 116:297–301

Horan DP, Chilvers GA, Lapeyrie FF (1988) Time sequence of the infection process in eucalypt ectomycorrhizas. New Phytol 109:451–458

Hyde GJ, Ashford AE (1997) Vacuole motility and tubule-forming activity in *Pisolithus tinc torius* hyphae are modified by environmental conditions. Protoplasma 198:85–92

Ineichen K, Wiemken V (1992) Changes in fungus-specific, soluble-carbohydrate pool during rapid and synchronous ectomycorrhiza formation of *Picea abies* with *Pisolithus tinctorius*. Mycorrhiza 2:1–17

Ineichen K, Wiemken V, Wiemken A (1995) Shoots, roots and ectomycorrhizal formation of pine seedlings at elevated atmospheric carbon dioxide. Plant Cell Environ 18:703–707

Jacobs PF, Peterson RL, Massicotte HB (1989) Altered fungal morphogenesis during early stages of ectomycorrhiza formation in *Eucalyptus pilularis*. Scanning Electron Microsc 3:249–255

Jentschke G, Fritz E, Marschner P, Rapp C, Wolters V, Godbold DL (1997) Mycorrhizal colonization and lead distribution in root tissues of Norway spruce seedlings. Z Pflanzenernähr Bodenkd 160:317–321

Junghans DT, Gomes EA, Guimarães WV, Barros EG, Araúgo EF (1998) Genetic diversity of the ectomycorrhizal fungus *Pisolithus tinctorius* based on RAPD-PCR analysis. Mycorrhiza 7:243–248

Kammerbauer H, Agerer R, Sandermann H (1989) Studies on ectomycorrhiza. XXII. Mycorrhizal rhizomorphs of *Thelephora terrestris* and *Pisolithus tinctorius* in association with Norway spruce (*Picea abies*): formation in vitro and translocation of phosphate. Trees 3:78–84

Keane KD, Manning WJ (1988) Effects of ozone and simulated acid rain on birch seedling growth and formation of ectomycorrhizae. Environ Pollut 52:55–65

Kershaw JL, Stewart GR (1992) Metabolism of ^{15}N-labelled ammonium by the ectomycorrhizal fungus *Pisolithus tinctorius* (Pers.) Coker and Couch. Mycorrhiza 1:71–77

Kope HH (1992) Interactions of heterokaryotic and homokaryotic mycelium of sibling isolates of the ectomycorrhizal fungus *Pisolithus arhizus*. Mycologia 84:659–667

Kope HH, Fortin JA (1989a) Antifungal activity in culture filtrates of the ectomycorrhizal fungus *Pisolithus tinctorius*. Can J Bot 68:1254–1259

Kope HH, Fortin JA (1989b) Genetic variation in antifungal activity by sibling isolates of the ectomycorrhizal fungus *Pisolithus arhizus*. Soil Biol Biochem 23:1047–1051

Kope HH, Fortin JA (1989c) Inhibition of phytopathogenic fungi in vitro by cell free culture media of ectomycorrhizal fungi. New Phytol 113:57–63

Kope HH, Fortin JA (1990) Germination and comparative morphology of basidiospores of *Pisolithus arhizus*. Mycologia 82:350–357

Kope HH, Tsantrizos YS, Fortin JA, Ogilvie KK (1991) *p*-Hydroxybenzoylformic acid and (*R*)-(−)-*p*-hydroxymandelic acid, two antifungal compounds isolated from liquid culture of the ectomycorrhizal fungus *Pisolithus arhizus*. Can J Microbiol 37:258–264

Lamb RJ, Richards BN (1974) Survival potential of sexual and asexual spores of ectomycorrhizal fungi. Trans Br Mycol Soc 62:181–191

Lamhamedi MS, Fortin JA (1991) Genetic variations of ectomycorrhizal fungi: extramatrical phase of *Pisolithus* sp. Can J Bot 69:1927–1934

Lamhamedi MS, Fortin, JA, Kope HH, Kropp BR (1990) Genetic variation in ectomycorrhiza formation by *Pisolithus arhizus* on *Pinus pinaster* and *Pinus banksiana*. New Phytol 115:689–697

Lamhamedi MS, Bernier PY, Fortin JA (1992a) Growth, nutrition and response to water stress of *Pinus pinaster* inoculated with ten dikaryotic strains of *Pisolithus* sp. Tree Physiol 10:153–167

Lamhamedi MS, Bernier PY, Fortin JA (1992b) Hydraulic conductance and soil water potential at the soil–root interface of *Pinus pinaster* seedlings inoculated with different dikaryons of *Pisolithus* sp. Tree Physiol 10:231–244

Lapeyrie F, Ranger J, Vairelles D (1991) Phosphate-solubilising activity of ectomycorrhizal fungi in vitro. Can J Bot 69:342–346

Lei J, Lapeyrie F, Malajczuk N, Dexheimer J (1990a) Infectivity of pine and eucalypt isolates of *Pisolithus tinctorius* (Pers.) Coker and Couch on roots of *Eucalyptus urophylla* S. T. Blake in vitro. New Phytol 114:627–631

Lei J, Lapeyrie F, Malajczuk N, Dexheimer J (1990b) Infectivity of pine and eucalypt isolates of *Pisolithus tinctorius* (Pers.) Coker and Couch on roots of *Eucalyptus urophylla* S. T. Blake in vitro. II. Ultrastructural and biochemical changes at the early stage of mycorrhiza formation. New Phytol 116:115–122

Le Tacon F, Garbaye J, Carr G (1988) The use of mycorrhizas in tropical forests. In: Ng FSP (ed) Trees and mycorrhiza. Forest Research Institute of Malaysia, Kuala Lumpur, pp 15–32

Leyval C, Reid CPP (1991) Utilization of microbial siderophores by mycorrhizal and non-mycorrhizal pine roots. New Phytol 119:93–98

Maehara N, Kikuchi J, Futai K (1993) Mycorrhizae of Japanese black pine (*Pinus thunbergii*). Protection of seedlings from acid mist and effect of acid mist on mycorrhiza formation. Can J Bot 71:1562–1567

Mahoney MJ, Chevone BI, Skelly JM, Moore LD (1985) Influence of mycorrhizae on the growth of loblolly pine seedlings exposed to ozone and sulfur dioxide. Phytopathology 75:679–682

Malajczuk N, Lapeyrie F, Garbaye J (1990) Infectivity of pine and eucalypt isolates of *Pisolithus tinctorius* on roots of *Eucalyptus urophylla* in vitro. New Phytol 114:627–631

Malloch D, Kuja AL (1979) Occurrence of the ectomycorrhizal fungus *Pisolithus tinctorius* in Ontario. Can J Bot 57:1848–1849

Marschner P, Godbold DL, Jentschke G (1996) Dynamics of lead accumulation in mycorrhizal and non-mycorrhizal Norway spruce (*Picea abies* (L.) Karst.). Plant Soil 178:239–245

Martin F, Delaruelle C, Ivory M (1998) Genetic variability in intergenic spacers of ribosomal DNA in *Pisolithus* isolates associated with pine, *Eucalyptus* and *Afzelia* in lowland Kenyan forests. New Phytol 139:341–352

Martins A, Barroso J, Pais MS (1996) Effect of ectomycorrhizal fungi on survival and growth of micropropagated plants and seedlings of *Castanea sativa* mill. Mycorrhiza 6:265–270

Marx DH (1969) The influence of ectotrophic mycorrhizal fungi on the resistance of pine roots to pathogenic infections. I. Antagonism of mycorrhizal fungi to root pathogenic fungi and soil bacteria. Phytopathology 59:153–163

Marx DH (1970) The influence of ectotropic mycorrhizal fungi on the resistance to pathogenic infections. V. Resistance of mycorrhizae to infection by vegetative mycelium of *Phytophthora cinnamomi*. Phytopathology 60:1472–1473

Marx DH (1972) Ectomycorrhizae as biological deterrents to pathogenic root infections. Annu Rev Phytopathol 10:429–454

Marx DH (1977) Tree host range and world distribution of the ectomycorrhizal fungus *Pisolithus tinctorius*. Can J Microbiol 23:217–223

Marx DH (1981) Variability in ectomycorrhizal development and growth among isolates of *Pisolithus tinctorius* as affected by source, age and reisolation. Can J For Res 11:168–174

Marx DH, Bryan WC (1970) Pure culture synthesis of ectomycorrhizae by *Thelephora terrestris* and *Pisolithus tinctorius* on different conifer hosts. Can J Bot 48:639–643

Marx DH, Bryan WC (1971) Influence of ectomycorrhizae on survival and growth of aseptic seedlings of loblolly pine at high temperature. For Sci 17:37–41

Marx DH, Cordell CE (1988) Specific ectomycorrhizae improve reforestation and reclamation in the eastern United States. In: Lalonde M, Piché Y (eds) Canadian workshop on mycorrhizae in forestry. Centre de recherche en biologie forestière, Université Laval, Sainte-Foy, pp 75–86

Marx DH, Davey CB (1969a) The influence of ectotrophic mycorrhizal fungi on the resistance of pine roots to pathogenic infections. III. Resistance of aseptically formed mycorrhizae to infection by *Phytophthora cinnamomi*. Phytopathology 59:549–558

Marx DH, Davey CB (1969b) The influence of ectotrophic mycorrhizal fungi on the resistance of pine roots to pathogenic infections. IV. Resistance of normally occurring mycorrhizae to infections by *Phytophthora cinnamomi*. Phytopathology 59:559–565

Marx DH, Hatchell GE (1986) Root stripping of ectomycorrhizae decreases field performance of loblolly and longleaf pine seedlings. South J Appl For 10:173–179

Marx DH, Kenney DS (1982) Production of ectomycorrhizal fungus inoculum. In: Schenck NC (ed) Methods and principles of mycorrhizal research. The American Phytopathological Society, St Paul, pp 131–146

Marx DH, Bryan WC, Davey CB (1970) Influence of temperature on aseptic synthesis of ectomycorrhizae by *Thelephora terrestris* and *Pisolithus tinctorius* on loblolly pine. For Sci 16:424–431

Marx DH, Ruehle JL, Kenney DS, Cordell CE, Riffle JW, Molina RJ, Pawuk WH, Navratil S, Tinus RW, Goodwin OC (1982) Commercial vegetative inoculum of *Pisolithus tinctorius* and inoculation techniques for development of ectomycorrhizae on container-grown tree seedlings. For Sci 28:373–400

Marx DH, Cordell CE, Kenney DS, Mexal JG, Artman JD, Riffle JW, Molina RJ (1984) Commercial vegetative inoculum of *Pisolithus tinctorius* and inoculation techniques for development of ectomycorrhizae on bare-rooted tree seedlings. For Sci Monogr 25:1–101

Marx DH, Hedin A Toe SFP (1985) Field performance of *Pinus caribaea* var. *hondurensis* seedlings with specific ectomycorrhizae and fertilizer after three years on a savanna site in Liberia. For Ecol Manage 13:1–25

Massicotte HB, Ackerley CA, Peterson R (1987a) The root–fungus interface as an indicator of symbiont interaction in ectomycorrhizae. Can J For Res 17:846–854

Massicotte HB, Peterson RL, Ashford AE (1987b) Ontogeny of *Eucalyptus pilularis-Pisolithus tinctorius* ectomycorrhizae. I. Light microscopy and scanning electron microscopy. Can J Bot 65:1927–1939

Massicotte HB, Peterson RL, Ackerley CA, Ashford AE (1987c) Ontogeny of *Eucalyptus pilularis-Pisolithus tinctorius* ectomycorrhizae. II. Transmission electron microscopy. Can J Bot 65:1940–1947

Massicotte HB, Peterson RL, Ackerley CA, Melville LH (1990) Structure and ontogeny of *Betula alleghaniensis-Pisolithus tinctorius* ectomycorrhizae. Can J Bot 68:579–593

McAfee BJ, Fortin JA (1986) Competitive interactions of ectomycorrhizal mycobionts under field conditions. Can J Bot 64:848–852

McAfee BJ, Fortin JA (1988) Comparative effects of the soil microflora on ectomycorrhizal inoculation of conifer seedlings. New Phytol 108:443–449

McQuattie CJ, Schier GA (1992) Effect of ozone and aluminium on pitch pine (*Pinus rigida*) seedlings: anatomy of mycorrhizae. Can J For Res 22:1901–191

Meharg AA, Dennis GR, Cairney JWG (1997) Biotransformation of 2,4,6-trinitrotoluene (TNT) by ectomycorrhizal basidiomycetes. Chemosphere 35:513–521

Miller FA, Rudolph ED (1986) Uptake and distribution of manganese and zinc in *Pinus virginiana* seedlings infected with *Pisolithus tinctorius*. Ohio J Sci 86:22–25

Molina R, Trappe JM (1982a) Lack of mycorrhizal specificity by the ericaceous hosts *Arbutus menziesii* and *Arctostaphylos uva-ursi*. New Phytol 90:495–509

Molina R, Trappe JM (1982b) Patterns of ectomycorrhizal host specificity and potential among Pacific Northwest conifers and fungi. For Sci 28:423–458

Momoh ZO, Gbadegesign RA (1980) Field performance of *Pisolithus tinctorius* as a mycorrhizal fungus of pines in Nigeria. In: Mikola P (ed) Tropical mycorrhizal research. Clarendon Press, Oxford, pp 72–79

Morselt AFW, Smits WTM, Limonard T (1986) Histochemical demonstration of heavy metal tolerance in ectomycorrhizal fungi. Plant Soil 96:417–420

Mousain D, Salsac L (1986) Utilisation du phytate et activités phosphatases acides chez *Pisolithus tinctorius*, basidiomycète mycorhizien. Physiol Vég 24:193–200

Näsholm T, Högberg P, Edfast A-B (1991) Uptake of NO_x by mycorrhizal and non-mycorrhizal Scots pine seedlings: quantities and effects on amino acid and protein concentrations. New Phytol 119:83–92

Navratil S, Phillips NJ, Wynia A (1981) Jack pine seedling performance improved by *Pisolithus tinctorius*. For Chron 57:212–217

Nylund J-E, Kasimir A, Arveby AS (1982) Cell wall penetration and papilla formation in senescent cortical cells during ectomycorrhiza synthesis in vitro. Physiol Plant Pathol 21:71–73

Oh KI, Melville LH, Peterson RL (1995) Comparative structural study of *Quercus serrata* and *Q. acutissima* formed by *Pisolithus tinctorius* and *Hebeloma cylindrosporum*. Trees 9:171–179

O'Neill EG, Luxmoore RJ, Norby RJ (1987) Increases in mycorrhizal colonisation and seedling growth in *Pinus echinata* and *Quercus alba* in an enriched CO_2 atmosphere. Can J For Res 17:878–883

Orlovich DA, Ashford AE, Cox GC (1989) A reassessment of polyphosphate granule composition in the ectomycorrhizal fungus *Pisolithus tinctorius*. Aust J Plant Physiol 16:107–115

Orlovich DA, Ashford A (1993) Polyphosphate granules are an artefact of specimen preparation in the ectomycorrhizal fungus *Pisolithus tinctorius*. Protoplasma 173:91–102

Orlovich DA, Ashford AE (1994) Structure and development of the dolipore septum in *Pisolithus tinctorius*. Protoplasma 178:66–80

Orlovich DA, Ashford AE, Cox GC, Moore AEP (1990) Freeze-substitution and X-ray microanalysis of polyphosphate granules in the mycorrhizal fungus *Pisolithus tinctorius* (Pers.) Coker and Couch. Endocytobiology 4:139–143

Paris F, Bonnaud P, Ranger J, Lapeyrie F (1995a) In vitro weathering of phlogopite by ectomycorrhizal fungi. I. Effect of K and Mg deficiency on phyllosilicate evolution. Plant Soil 177:191–201

Paris F, Bonnaud P, Ranger J, Robert M, Lapeyrie F (1995b) Weathering of ammonium- or calcium-saturated 2:1 phyllosilicates by ectomycorrhizal fungi in vitro. Soil Biol Biochem 27:1237–1244

Paris F, Botton B, Lapeyrie F (1996) In vitro weathering of phlogopite by ectomycorrhizal fungi. II. Effect of K^+ and Mg^{2+} deficiency and N sources on accumulation of oxalate and H^+. Plant Soil 179:141–150

Parke JL, Linderman RG, Black CH (1983) The role of ectomycorrhizas in drought tolerance of Douglas fir seedlings. New Phytol 95:83–95

Perotto S, Coisson JD, Perugini I, Cometti V, Bonfante P (1997) Production of pectin-degrading enzymes by ericoid mycorrhizal fungi. New Phytol 135:151–162

Piché Y, Fortin JA (1982) Development of mycorrhizal extramatrical mycelium and sclerotia on *Pinus strobus* seedlings. New Phytol 91:211–220

Piché Y, Peterson RL (1988) Mycorrhiza initiation: an example of plant–microbial interactions. In: Valentine FA (ed) Forest and crop biotechnology progress and prospects. Springer, Berlin Heidalberg New York, pp 298–313

Piché Y, Peterson RL, Howarth MJ, Fortin JA (1983) A structural study of the interaction between the ectomycorrhizal fungus *Pisolithus tinctorius* and *Pinus strobus* roots. Can J Bot 61:1185–1193

Rasanayagam S, Jeffries P (1992) Production of acid is responsible for antibiosis by some ectomycorrhizal fungi. Mycol Res 96:971–976

Regvar M, Gogala N (1996) Changes in root growth patterns of (*Picea abies*) spruce roots by inoculation with an ectomycorrhizal fungus *Pisolithus tinctorius* and jasmonic acid treatment. Trees 10:410–414

Reid CPP, Woods FW (1969) Translocation of ^{14}C-labeled compounds in mycorrhizae and its implications in interplant nutrient cycling. Ecology 50:179–187

Reid CPP, Kidd FA, Ekwebelam SA (1983) Nitrogen nutrition, photosynthesis and carbon allocation in ectomycorrhizal pine. Plant Soil 71:415–432

Riffle JW, Tinus RW (1982) Ectomycorrhizal characteristics, growth, and survival of artificially inoculated ponderosa and Scots pine in a greenhouse and plantation. For Sci 28:646–660

Roland A, Albaladejo J (1994) Effect of mycorrhizal inoculation and soil restoration on the growth of *Pinus halapensis* seedlings in a semiarid soil. Biol Fert Soils 18:143–149

Rosado SCS, Kropp BR, Piché Y (1994a) Genetics of ectomycorrhizal symbiosis. I. Host plant variability and heritability of ectomycorrhizal and root traits. New Phytol 126:105–110

Rosado SCS, Kropp BR, Piché Y (1994b) Genetics of ectomycorrhizal symbiosis. II. Fungal variability and heritability of ectomycorrhizal traits. New Phytol 126:111–117

Ross EW, Marx DH (1972) Susceptibility of sand pine to *Phytophthora cinnamomi*. Phytopathology 62:1197–1200

Rost FWD, Shepherd VA, Ashford AE (1995) Estimation of vacuolar pH in actively growing hyphae of the fungus *Pisolithus tinctorius*. Mycol Res 99:549–553

Rousseau JVD, Reid CPP (1989) Carbon and phosphorus relations in mycorrhizal and non-mycorrhizal pine seedlings. Ann Sci For 46 (Suppl):715s–171s

Rousseau JVD, Reid CPP (1990) Effects of phosphorus and ectomycorrhizas on the carbon balance of loblolly pine seedlings. For Sci 36:101–112

Rousseau JVD, Reid CPP, English RJ (1992) Relationship between biomass of the mycorrhizal fungus *Pisolithus tinctorius* and phosphorus uptake in loblolly pine seedlings. Soil Biol Biochem 24:183–184

Rousseau JVD, Sylvia DM, Fox AJ (1994) Contribution of ectomycorrhiza to the potential nutrient-absorbing surface of pine. New Phytol 128:639–644

Rupp LA, Mudge KW, Negm FB (1989) Involvement of ethylene in ectomycorrhiza formation and dichotomous branching of roots of mugo pine seedlings. Can J Bot 67:477–482

Schier GA, McQuattie CJ (1995) Effect of aluminum on the growth, anatomy, and nutrient content of ectomycorrhizal and nonmycorrhizal eastern white pine seedlings. Can J For Res 25:1252–1262

Schier GA, McQuattie CJ (1998) Effects of carbon dioxide enrichment on response of mycorrhizal pitch pine (*Pinus rigida*) to aluminium – growth and mineral nutrition. Trees 12:340–346

Schramm JE (1966) Plant colonization studies on black wastes from anthracite mining in Pennsylvania. Trans Am Philos Soc 56:1–194

Shepherd VA, Orlovich DA, Ashford AE (1993a) A dynamic continuum of pleiomorphic tubules and vacuoles in growing hyphae of a fungus. J Cell Sci 104:495–507

Shepherd VA, Orlovich DA, Ashford AE (1993b) Cell-to-cell transport via motile tubules in growing hyphae of a fungus. J Cell Sci 105:1173–1178

Sims KP, Sen R, Watling R, Jeffries P (1999) Species and population structures of *Pisolithus* and *Scleroderma* identified by combined phenotypic and genomic marker analysis. Mycol Res 103:449–458

Söderström B, Finlay RD, Read DJ (1988) The structure and function of the vegetative mycelium of ectomycorrhizal plants. IV. Qualitative analysis of carbohydrate contents of mycelium interconnecting host plants. New Phytol 109:163–166

Sohn RF (1981) *Pisolithus tinctorius* forms long ectomycorrhizae and alters root development in seedlings of *Pinus resinosa*. Can J Bot 59:2129–2133

Stephanie A-L, Chalot M, Botton B, Dexheimer J (1996) Morphological and physiological evidences for the involvement of the root-cap in ectomycorrhiza formation between *Eucalyptus globulus* and *Pisolithus tinctorius*. In: Szaro TM, Bruns TD (eds) Abstracts of the 1st Int Conf on Mycorrhizae, August 1996, University of California, Berkeley, p 22

Suh H-W, Crawford DL, Korus RA, Shetty K (1991) Production of antifungal metabolites by the ectomycorrhizal fungus *Pisolithus tinctorius* strain SMF. J Ind Microbiol 8:29–36

Szaniszlo PJ, Powell PE, Reid CPP, Cline GR (1981) Production of hydroxomate siderophore iron chelators by ectomycorrhizal fungi. Mycologia 73:1158–1174

Tagu D, Martin F (1995) Expressed sequence tags of randomly selected cDNA clones from *Eucalyptus globulus–Pisolithus tinctorius* ectomycorrhiza. Mol Plant Microbe Interact 8:781–783

Tagu D, Martin F (1996) Molecular analysis of cell wall proteins expressed during the early steps of ectomycorrhiza development. New Phytol 133:73–85

Tagu D, Python M, Crétin C, Martin F (1993) Cloning symbiosis-related cDNAs from eucalypt ectomycorrhiza by PCR-assisted differential screening. New Phytol 125:339–343

Tagu D, Nasse B, Martin F (1996) Cloning and characterisation of hydrophobins-encoding cDNAs from the ectomycorrhizal basidiomycete *Pisolithus tinctorius*. Gene 168:93–97

Tam PCF, Griffiths DA (1994) Mycorrhizal associations in Hong-Kong Fagaceae. VI. growth and nutrient uptake by *Castanopsis fissa* seedlings inoculated with ectomycorrhizal fungi. Mycorrhiza 4:169–172

Theodorou C, Redell P (1991) In vitro synthesis of ectomycorrhizas on Casuarinaceae with a range of mycorrhizal fungi. New Phytol 118:279–288

Thompson W (1984) Distribution, development and functioning of mycelial cord systems of decomposer basidiomycetes of the deciduous woodland floor. In: Jennings DH, Rayner ADM (eds) The ecology and physiology of the fungal mycelium. Cambridge University Press, Cambridge, pp 185–241

Thomson BD, Grove TS, Malajczuk N, Hardy GEStJ (1994) The effectiveness of ectomycorrhizal fungi in increasing the growth of *Eucalyptus globulus* Labill. in relation to root colonization and hyphal development in soil. New Phytol 126:517–524

Thomson J, Melville LH, Peterson RL (1989) Interactions between the ectomycorrhizal fungus *Pisolithus tinctorius* and root hairs of *Picea mariana* (Pinaceae). Am J Bot 76:632–636

Tillard P, Bousquet N, Mousain D, Martin F, Salsac L (1989) Polyphosphatase activities in the soluble fraction of mycelial homogenates of *Pisolithus tinctorius*. Agric Ecosyst Environ 28:525–528

Tonkin CM, Malajczuk N, McComb JA (1989) Ectomycorrhizal formation by micropropagated clones of *Eucalyptus marginata* inoculated with isolates of *Pisolithus tinctorius*. New Phytol 111:209–214

Turnau K, Kottke I, Dexheimer J, Botton B (1994) Elemental distribution in mycelium of *Pisolithus arhizus* treated with cadmium dust. Ann Bot 74:137–142

Turnbull MH, Goodall R, Stewart GR (1995) The impact of mycorrhizal colonization upon nitrogen source utilization and metabolism in seedlings of *Eucalyptus grandis* Hill ex Maiden and *Eucalyptus maculata* Hook. Plant Cell Environ 18:1386–1394

Turnbull MH, Goodall R, Stewart GR (1996) Evaluating the contribution of glutamate dehydrogenase and the glutamate synthase cycle to ammonia assimilation by four ectomycorrhizal fungal isolates. Aust J Plant Physiol 23:151–159

Vézina L-P, Margolis HA, McAfee BJ, Delaney S (1989) Changes in the activity of enzymes involved with primary nitrogen metabolism due to ectomycorrhizal symbiosis on jack pine seedlings. Physiol Plant 75:55–62

Walker RF, West, DC, McLaughlin SB, Amundsen CC (1989) Growth, xylem pressure potential, and nutrient absorption of loblolly pine on a reclaimed surface mine as affected by an induced *Pisolithus tinctorius* infection. For Sci 35:569–581

Walker RF, Geisinger DR, Johnson DW, Ball JT (1995) Enriched atmospheric CO_2 and soil P effects on growth and ectomycorrhizal colonisation of juvenile ponderosa pine. For Ecol Manage 78:207–215

Warrington SJ, Black HD, Coons LB (1981) Entry of *Pisolithus tinctorius* hyphae into *Pinus taeda* roots. Can J Bot 59:2135–2139

Watling R, Taylor A, Lee SS, Sims K, Alexander I (1995) A rainforest *Pisolithus*; its taxonomy and ecology. Nova Hedwigia 61:417–429

Wessels JGH (1994) Developmental regulation of fungal cell wall formation. Annu Rev Phytopathol 32:413–437

Wullschleger SD, Reid CPP (1990) Implication of ectomycorrhizal fungi in the cytokinin relations of loblolly pine (*Pinus taeda* L.). New Phytol 116:681–688

Yazid SM, Lee SS, Lapeyrie F (1994) Growth stimulation of *Hopea* spp. (Dipterocarpaceae) seedlings following ectomycorrhizal inoculation with an exotic strain of *Pisolithus tinctorius*. For Ecol Manage 67:339–343

Zak B (1976) Pure culture synthesis of Pacific madrone ectendomycorrhizae. Mycologia 68:362–369

Chapter 2

Suillus

A. Dahlberg and R. D. Finlay

2.1
Introduction

Study of the representatives of the ectomycorrhizal (ECM) fungal genus *Suillus* has long since attracted mycologists; the genus consists of species with conspicuous and generally epigeous mushrooms with tubular hymenophores, that commonly contribute a major portion of ECM sporocarp production in conifer forest ecosystems. *Suillus* species exhibit a high degree of host specificity to conifers and their distribution coincides with the natural distribution of pinaceous conifers in the northern hemisphere. In contrast to most ECM species, mycelia of *Suillus* are generally easy to culture and have been frequently used in ECM studies, including physiology, ECM synthesis and population studies. In an analysis of MYCOLIT, a comprehensive bibliographic database of scientific papers of ECM research, *Suillus* was the fourth most frequently encountered genus, following *Pisolithus*, *Tuber* and *Laccaria* (Klironomos and Kendrick 1993). In total, 8.2% of ECM papers in the past 40 years have concerned *Suillus*. In this review, we have tried to compile as complete and broad an analysis of the available literature on *Suillus* as possible.

2.2
Taxonomy and Phylogenetic Relationships

Taxonomically, the genus *Suillus* S. F. Grey is grouped within the order Boletales (Hawksworth et al. 1995), where the majority of species form ECM. The genus *Suillus* gained acceptance among mycologists after the war (Singer 1951; Moser 1953), before which it was distinguished as a subgenus within *Boletus*. Boletes are easily distinguished from other agaric orders, as they mostly possess vertically arranged tubes instead of bearing gills on the underside of the pileus. Within Boletales, Hawksworth et al. (1995) group *Suillus* within the family Boletaceae, whereas Knudsen (1995) uses an alternative taxonomic arrangement in which *Suillus* is grouped together with

Department of Forest Mycology and Pathology, P.O. Box 7026, Swedish University of Agricultural Sciences, 750 07 Uppsala, Sweden

Gomphideus and *Chroogomphus* in the Gomphidaceae. Commonly, boletes are considered to consist of three major groups; the *Suillus* group, the *Strobilomycete* group and the larger group of *Boletus*. The genus *Suillus* is characterised by medium to large fleshy terrestrial boletes associated with conifers. The cap is sometimes dry and scaly, but more commonly viscid to glutinous. The stipe is with or without ring and the hymenium has small or larger, sometimes radial pores. Spores are elongate, smooth and pale brownish to brown. Clamp connections are absent or extremely scarce in the sporocarp, but commonly present in the mycelia (Pantidou and Groves 1966). The presence of distinct cystidia within the hymenium of the sporocarps separates the genus *Suillus* within *Boletales* (Pegler and Young 1981). The *Boletus* group includes the genera *Boletus*, *Leccinum* and *Tylopilus*. One of the characters of this group is a specific morphology of the hymenophoral trama (Singer 1986) and this group also forms ECM with both conifers and dicotyledonous trees. The monogeneric *Strobilomycetes* are characterised by having ornamented, short spores, pileus fleshy squamose to squarose with soft scales or woolly-warty, woolly-fibrous and specific pigments that cause a reddening-blackish colour change in the flesh.

The relationships between *Suillus* and related genera have repeatedly been discussed during the last decade (Singer 1986; Høiland 1987; Kretzer et al. 1996; Bruns et al. 1998). Bruns et al. (1989) gave evidence that *Rhizopogon* is closely related to *Suillus* and suggested that a change in a few developmental genes with strong selection pressure for a hypogeous habitat would lead to rapid morphological divergence. Bruns and Szaro (1992) identified the Gomphidaceae as closely related to *Suillus* and collectively referred to the three groups *Rhizopogon*, *Suillus* and Gomphidaceae as the suilloid radiation of Boletales. Recent chemotaxonomic analyses also suggest that *Suillus sensu lato* is more closely related to the Gomphidaceae and *Rhizopogon* than to the remaining boletes (Besl and Bresinsky 1997; see also Knudsen 1995).

Molecular data (Baura et al. 1992) indicate a very close relationship between the secotioid *Gastrosuillus laricinus* (Sing. & Both) Thiers and *S. grevillei* (Klotzsch) Sing. Secotioid fungi are morphological intermediates between epigeous sporocarps and hypogeous false truffles. Their study argued that *G. laricinus* arose only within the last 60 years from a local population of *S. grevillei*, and considered it merely a local mutant of *Suillus*. Because all secotioid forms are very rare and limited in distribution, and since the interfertility of some secotioid forms with normal agarics is known, Baura et al. (1992) suggested that all secotioid forms may be of recent origin and of short evolutionary duration. It was later shown that three other *Gastrosuillus* species were genetically very closely related to each other, but, in contrast to the *G. laricinus* example, exhibited significant sequence divergence from their closest non-secotioid, *S. variegatus* (Swartz: Fr.) Richon & Roze. This result suggests that at least this *Gastrosuillus* lineage is not a recent mutant (Kretzer and Bruns 1997). In any case, *Gastrosuillus* are clearly derived at least twice from *Suillus*. Due to its polyphyletic and derived nature, they proposed collapsing *Gastrosuillus* into the genus *Suillus* (Kretzer and Bruns 1997).

The genus *Boletinus*, consisting of three species, has been considered as closely related but distinguishable from *Suillus* due to abundant clamp connections, having a decurrant hymenophore and a hollow stipe (in two of the species) and forming ECM exclusively with larch (Pegler and Young 1981; Singer 1986). However, there are species of *Suillus sensu stricto* that form boletinoid hymenophores and also form ECM exclusively with larch, e.g. *S. grevillei*, which may form clamps in sporocarps. The genus *Fuscoboletinus* is also considered as closely related to *Suillus* and has been distinguished by the colour of its spore print which is red to chocolate brown (Pomerleau and Smith 1962). Singer (1986) did not recognise *Fuscoboletinus* as an independent genus. Analyses of characteristics of mycelial cultures (Hutchison 1991) and DNA analyses (Bruns et al. 1988; Bruns and Palmer 1989) revealed no gap between *Suillus*, *Fuscoboletinus* and *Boletinus*. Using chemosystematics, Besl and Bresinsky (1997) suggested *Suillus* and *Fuscoboletinus* should be treated as one genus, but stated that *Suillus* and *Boletinus* can be considered as closely related genera. Based on internal transcribed spacer (ITS) sequences from 38 species of *Suillus sensu lato*, Kretzer et al. (1996) argued that *Boletinus* and *Fuscoboletinus* should be collapsed into *Suillus* as they are paraphyletic and form a group within *Suillus sensu lato*. In this review, we thus include *Boletinus*, *Fuscoboletinus* and *Gastrosuillus* in *Suillus sensu lato*.

Analyses of the molecular clock imply that the appearance of ECM fungi coincided with the earliest known *Pinus* macrofossils, 130 million years ago (Alvin 1960; Berbee and Taylor 1993). In an analysis of evolutionary interferences of a sequence database of ECM basidiomycetes, it was suggested that the suilloid fungi, *Suillus* and *Rhizopogon*, radiated 35–60 Ma, in the mid-Tertiary (Bruns et al. 1998). During this time, the earth's climate became cooler and more temperate, and trees in Pinaceae and Fagales, both obligate ECM taxa, came to dominate temperate forests (Berggren and Prothero 1992). It has been suggested that there has been a convergent radiation of several groups of ECM fungi, including *Suillus*, in response to the expanding geographical range of their host trees during this period (Bruns et al. 1998). Recently, the oldest fossil ECM so far found (50 million years B.P.) were observed among permineralised plant remains in the middle Eocene Princeton chert of British Colombia, Canada. The ECM were associated with roots of *Pinus* and, based on the morphological characters, and the identity of the host, *Suillus* or *Rhizopogon* were suggested as the likely ECM fungi (Lepage et al. 1997).

2.3
Geographic Distribution and Host Specificity

2.3.1
Geographic Distribution

A recent overview reports the number of *Suillus sensu lato* species as approaching 100, of which about 90 are considered as *Suillus sensu stricto*

Table 2.1. Distribution of *Suillus* species and their coniferous host taxa in the world (Critchenfield and Little 1966; Engel 1996; Mabberley 1997)

	No of species						
	North and Central America	South America	Asia	Europe	Africa	Australia New Zealand	Total number
Suillus	68	3	36	31	9	7	99
Pinus	59	0	27	10	4	0	94
Larix	3	0	7	2	0	0	10
Pseudotsuga	2	0	4	0	0	0	6

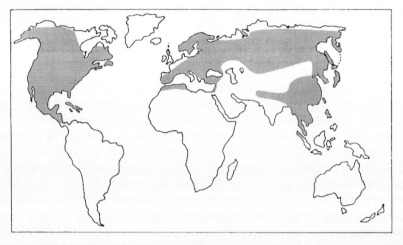

Fig. 2.1. The natural distribution of *Suillus* species coincides with the natural range of their conifer host trees, which are restricted to the northern hemisphere. Map indicates natural distribution of the Pinaceae. (Redrawn from Farjon 1990)

(Engel 1996; Table 2.1). The members of *Suillus* are exclusively ECM associates of *Pinaceae* (Singer 1986; Molina et al. 1992; Engel 1996), and are naturally found only in the northern hemisphere, coinciding with the distribution and diversity of pinaceous conifers (Fig. 2.1; Table 2.1). Some *Suillus* species enter the tropical, usually montane, zone in the palaeo- and neotropics, but are predominately distributed in the temperate and boreal zones. The genus *Suillus* may be classified as having an intermediate host range as they are limited to forming ECM with a single host family (Molina et al. 1992). However, most *Suillus* species have a narrow host range and associate only with a single host genus. Host specificity within *Suilllus sensu lato* does not seem to have evolved through cospeciation with the ECM hosts (Kretzer et al. 1996). While *Larix* is considered to be more recently derived than *Pinus* (Hart 1987), larch associations seem to be more primitive in *Suillus* and associations with pines and Douglas fir seem to be derived. Rather than cospeciation, multiple jumps to new host species and genera seem to have occurred during the evolution of *Suillus sensu lato*. The close relationship between *Suillus* and *Rhizopogon* is supported by the fact that both seem to form ECM exclusively with members of the Pinaceae and, in both genera, the highest species richness corresponds with that of the Pinaceae (Molina and Trappe 1994).

The family Pinaceae encompasses twelve genera with about 220 species distributed in the northern hemisphere (Mabberley 1997; Fig. 2.1). The most important host genera for *Suillus* are *Pinus*, *Larix* and *Pseudotsuga*. The genus *Pinus* comprises 94 species and is one of the most widely distributed genera of trees in the northern hemisphere (Critchenfield and Little 1966; Table 2.1). The pines extend from the polar region to the tropics, and with this enormous range they thus dominate the natural vegetation in many regions. The genus *Larix* comprises ten species, of which two are widely distributed in Eurasia and one in North America, while the remaining species have limited distributions (Farjon 1990; cf. Table 2.1). The genus *Pseudotsuga*, comprising six species (Mabberley 1997), is closely related to larch (Price et al. 1987).

The highest species richness of *Suillus*, almost 70 species, is found in North America, coinciding with the highest species richness of conifers (Engel 1996; Table 2.1). Many *Suillus* species have a limited geographical range and nearly half of the North American species are found in California, where several endemic pines and a high species richness of pines are found. Europe and Asia, with fewer species of *Pinus*, have 31 and 36 species of *Suillus* respectively. A few *Suillus* species are distributed within the natural range of pines in the Canary Islands and in the northern parts of Africa. In addition, a few *Suillus* species occur commonly and fruit abundantly in introduced pine plantations, predominantely *P. radiata* (Dunstan et al. 1998). In the absence of suitable host taxa, *Suillus* species are naturally absent in Australia, New Zealand and South America.

2.3.2
Host Specificity

As pointed out by Harley and Smith (1983) compatibility in ECM associations (potential host range) deduced from laboratory experiments may not reflect that under field conditions. Thus, the concept of ecological specificity (*sensu* Molina et al. 1992), embracing all environmental abiotic and biotic factors that affect the ability of plants to form functional ECM with particular fungi, may explain why *Suillus* species may exhibit a broader host range under experimental conditions than observed in nature. Finlay (1989) reported that larch associates *S. cavipes* (Opat) Smith & Thiers and *S. grevillei*, although capable of forming ECM with *P. sylvestris* in microcosms, were inferior to the pine fungus *S. bovinus* (L.: Fr.) Rouss. in nutrient uptake and translocation. Doudrick et al. (1990) reported sporocarps of *S. cavipes* in pure old stands of

Picea mariana with no *Larix* in the vicinity, and further synthesised ECM between *S. cavipes* and *P. mariana* under aseptic conditions. As early as 1922, Melin reported colonisation by *S. grevillei* on roots of *Pinus sylvestris*, producing a thin mantle, but no regular Hartig net. In aseptic culture, *S. grevillei* has also been found to form ECM with *Pinus contorta* and *Pinus ponderosa* (Molina and Trappe 1982), *Pseudotsuga menziesii* (Grand 1968; Molina and Trappe 1982; Duddridge 1986a,b) and arbutoid mycorrhiza with both *Arctostaphylos uva-ursi* and *Arbutus menziesii* (Molina and Trappe 1982). *S. bovinus* and *S. luteus* (L.: Fr.) Rouss. formed ECM in pure culture synthesis on *P. menziesii* seedlings and two exotic pine species (Parladé et al. 1996). *S. flavidus* (Fr.: Fr.) J. S. Presl., naturally associating with *P. sylvestris* in Europe, readily forms functional ECM and fruits abundantly in Swedish stands of the introduced *P. contorta* from NW America (Blomgren 1994). With few exceptions, field observations and pure-culture syntheses confirm the host specificity of *Suillus*. Reported exceptions include *Quercus* (e.g. Mitchell et al. 1984) and *Picea* (e.g. Göransson and Eldhuset 1991; Repac 1996). Other exceptions are from field observations, which can be erroneous if made in mixed forest (cf. Trappe 1963; Engel 1996). Recently, it has been shown that in some monotrophs, the achlorophyllous mycoheterotrophic ericaceous plants are highly specific in their fungal association, and that so far only suilloid fungi, the monophyletic group including *Suillus* and *Rhizopogon*, have been identified associated with *Monotropa hypopithys* (Cullings et al. 1996).

Roots of many trees produce exudates that induce spore germination of ECM fungi, although this is still often less than a few percent (Fries 1987a). Coating spores of *Suillus* spp. individually on to roots of young pine seedlings enhances germination success to about 30% in both *Suillus* and *Rhizopogon* (Theodorou and Bowen 1987; Miller et al. 1993). One active compound in roots of pine that induces germination of *Suillus* was identified as abietic acid, a diterpene resin acid (Fries et al. 1987). Later, an additional five diterpene resins, together with abietic acid, were found to possess the capacity to induce spore germination in ten European and American *Suillus* species, while ECM species from other genera did not respond (Fries 1988). Spore germination of *Suillus* appears to have evolved to be induced by specific chemical stimuli, with germination occurring in response to specific host roots (Theodorou and Bowen 1987). Abietic acid has subsequently been used to induce germination in experiments with *Suillus* [*S. luteus* and *S. granulatus* (L.: Fr.) Rouss. (Fries and Neuman 1990), *S. bovinus* (Fries and Sun 1992), *S. granulatus* (Jacobson and Miller 1994), *S. variegatus* (Fries 1994), *S. pungens* Thiers & Smith (Bonello et al. 1998)]. Observed germination *in vitro* is low (<1–7%). The production of spores ranges from 1 to 10×10^8 sporocarp^{-1} in *S. bovinus* (Dahlberg and Stenlid 1990). In a study of the persistence of spores, less than 1% was recovered in the litter and upper soil layer in the spring, in comparison to the amount recorded directly under *Suillus* sporocarps the preceding autumn (Miller et al. 1994).

2.4
Community Structure, Population Biology and Genetics

2.4.1
Community Structure

Suillus sporocarps are among the most prominent fruiting ECM species in pine forests. In *P. sylvestris* forests, *S. variegatus* often constitutes 15–25% (sometimes up to 60%) of the total epigeous ECM production, ranging from 0.5 to 1.5 kg, rarely 10 kg, dry wt ha^{-1} year^{-1} (Hintikka 1988a; Ohenoja 1993; Väre et al. 1996; Dahlberg 1997). In young *P. sylvestris* forests, *S. bovinus* may dominate and produce 1–20 kg dry wt sporocarps ha^{-1} year^{-1} (Dahlberg and Stenlid 1994). Romell (1939, cited in Harley and Smith 1983) reported that *S. bovinus* may produce 180 kg sporocarps dry wt ha^{-1} year^{-1}; however, that figure was extrapolated from an observation of 350 sporocarps within 100 m^{2} (Larsen 1934). In a mature *P. contorta* stand in northern Utah, *S. brevipes* (Peck) Kuntze and *S. tomentosum* constituted 32% of the total number of sporocarps recorded for 48 observed species of ECM basidiomycetes (Kropp and Albee 1996).

Several studies suggest that species of *Suillus* alter in abundance and importance in ECM communitites at different forest successional stages. In *P. sylvestris* forests, *S. bovinus* was generally replaced by *S. variegatus* as the forests aged (Hintikka 1988a; Dahlberg 1997). Both species may, however, be present locally as sporocarps at all successional ages of forests. Similarly, *S. brevipes* is more prominent in young than in old jack pine stands, where *S. tomentosum* (Kauffm.) Sing., Snell & Dick may have a prolific production of sporocarps (Danielson and Visser 1989; Danielson 1991; Visser 1995). However, no difference in fruiting of *S. brevipes*, and two other *Suillus* species, was found when comparing stands of *P. contorta* 6–19 years after clearcutting with adjacent old stands (Bradbury et al. 1998).

Fire is an important factor of disturbance in boreal forest ecosystems and *Suillus* species may have adaptations to persist following such events. In low-intensity fires, typical for boreal forests, a high proportion of trees commonly survive and the lower part of the humus layer remains unaffected, enabling mycelia of many ECM fungal species, including *Suillus*, to survive (Jonsson et al. 1999). In contrast, in ecosystems where fire intensity is high, the trees are killed and the organic layer combusted, and the reduction of the ECM fungal community is dramatic. All trees in the Californian stand of *P. muricata* studied by Gardes and Bruns (1996) were later killed by an intensive fire, reducing the richness of ECM fungi (Horton et al. 1998). However, *S. pungens* was found on resprouting young seedlings and it was proposed that this fungus survived as resident inoculum in the soils rather than appearing subsequently from dispersal into the sites after the fire (Horton et al. 1998).

The gastronomic value of several species of *Suillus* is highly appreciated and some areas of planted pine forests in South America are devoted to the production of *S. luteus* sporocarps. Hedger (1986) reported an annual production of *S. luteus* in *P. radiata* plantations of 78 000–156 000 sporocarps (ca. 1 tonne dry wt) ha^{-1}. An almost continuously temperate, humid climate with daily temperatures around 8–16°C results in a perpetual autumn crop at this site. Other recent estimates range between 100 kg and 1 tonne dry wt ha^{-1} year^{-1} (I. Chapela, pers. comm.). The *P. radiata* trees are introduced into a vesicular arbuscular mycorrhiza-associated bunch grass community native to this region of the Andes. It has been hypothesised that the high sporocarp production occurs because of a simplified ECM community, resulting in low competition for tree-derived carbon (I. Chapela, pers. comm.). Since the early 1980s, the huge production of *S. luteus* mushrooms in these Ecuadorian *P. radiata* forests has been collected, dried and exported to Europe. In Chilean *P. radiata* plantations, where conditions are similar, sporocarps of *S. luteus* have been exported to the USA since the 1970s.

Despite the often abundant above-ground fruiting of *Suillus* species, they are rarely detected on roots. Danielson (1984) observed that the most common sporocarps in natural mature jack pine stands were those of *S. tomentosum*. However, he estimated that ECM roots formed by *Suillus* rarely constituted more than 5% of the total. Similarly, two different species of *Suillus* were never encountered in various soil samples collected from different *P. patula* plantations, where sprorcarps of these two species were found in abundance (Natarjan et al. 1992). Gardes and Bruns (1996) found that whereas sporocarps of *S. pungens* during a 3-year period constituted about 40% of all ECM sporocarps in a Californian bishop pine forest, it constituted less than 5% of the mycorrhizal roots. In another study of an old Swedish Scots pine stand, *Suillus* sporocarps constituted 12% of the total sporocarp biomass, the corresponding abundance on roots being about 5% (Jonsson et al. 1999). However, in a study of ECM communities in a chronosequence of jack pine, the correspondence between fruiting and ECM abundance of *S. brevipes* was high (Visser 1995). In the same study, however, it was low for *S. tomentosum*. In a similar study of regenerating stands of lodgepole pine, a high discrepancy between the relative contribution to the ECM community of sporocarps and roots was found for *S. brevipes*, *S. tomentosum* and *S. umbonatus* Dick & Snell (Bradbury et al. 1998). In two additional studies (Kåren 1997; Jonsson 1998), each comprising several Swedish Scots pine stands, the correspondence between the abundance of mycorrhizal roots and sporocarps of *S. variegatus* varied inconsistently between stands. Possible reasons for the apparent discrepancy between the frequency and abundance of sporocarps and roots, which vary among ECM fungal species, may be the pattern of resource allocation or specific ecological requirements (Gardes and Bruns 1996).

2.4.2
Population Biology

It is only recently that awareness of fungal individualism has emerged. It is now clear that genetically different mycelia within a fungal species co-exist as physiologically and spatially discrete individuals or genets (Todd and Rayner 1980). To examine the distribution of fungal individuals, the expression of somatic incompatibility (SI) reactions in interacting hyphal margins of "self" and "non-self" pairings has been used to identify genets in mycelial isolations from sporocarps. With the ability to identify genets, questions about spatial and temporal structures of populations can be addressed. A population can be defined as an arbitrarily delimited assembly of genets within a single species. The first spatial distribution of individual ECM fungal genets was implied in a study of *S. tomentosum* where isolates from sporocarps were identical in eight different enzyme systems (Zhu et al. 1988). The authors suggested that genets, as a result of mycelial growth, might extend for 10 m in soil. However, the first study specifically dealing with spatial population characteristics of an ECM fungus employed SI and revealed that *S. luteus* in a forest stand consisted of several genets, and that single pine trees could be colonised by several genets and that some genets extended at least 2 m (Fries 1987b). It was later shown, in a comparison between populations in first generation forests (old sandpits) and forest of long continuity, that in populations of *S. bovinus* in Scots pine forests, the number of fungal genets decreased over time while the genet size increased. Under suitable conditions there were on average 60 genets ha^{-1} in old forests, with a mean genet size of 6.8 m (max. 20 m), whereas the number of genets approached 1000 ha^{-1}, with a mean size of 2.1 m (max. 3.2 m) in young forests (Dahlberg and Stenlid 1990, 1994). The possible influences of forest age on population characteristics are less pronounced when considering secondary forest successions. It has been suggested that a significant part of ECM fungal community species, probably on the level of genets, survive periods with absence of the tree-layer, due to fire or forest management clearcut, in boreal forests (cf. Jonsson et al. 1999). In a self-regenerated Scots pine stand, clearcut 15 years previously, the maximum genet sizes were found to be 8.8 and 11 m, suggesting that genets had persisted between forest generations (A. Dahlberg, unpubl. data). Similar spatial characteristics as in *S. bovinus* were found in *S. variegatus*, e.g. 60–70 genets ha^{-1}, with a mean size ca. 10 m (max. 27 m) in two studies conducted in old Scots pine stands (Dahlberg 1997; Dahlberg et al. 1997). Significantly, *S. bovinus* populations are more scattered throughout middle-aged to old Scots pine stands compared with *S. variegatus*. In such Scots pine forests, *S. variegatus* may constitute an average of 25% of the total ECM sporocarp biomass and *S. bovinus*, if present, less than 0.5% (Dahlberg 1997). These results suggest that *Suillus* species are competitors that are able to persist in time and expand in space while maintaining reproductive potential. In studies like these, it is important to consider that the observed genets only represent genets that produce sporocarps (the reproductive part of the fungal population visible at the time of study). Furthermore, the production of sporocarps is not proportional to the size of a genet, suggesting that genets may consist of several independent ramets due to fragmentation (Dahlberg and Stenlid 1995). The number of sporocarps for larger genets is typically 10–30, but in *S. bovinus* 300 sporocarps were encountered in a genet covering 300 m^2

(Dahlberg and Stenlid 1994). Due to a low correspondence between spatial distribution of genets and their production of sporocarps, mycelial fragmentation of genets has been suggested to be common (Dahlberg and Stenlid 1995).

A major limitation to SI studies is that this technique can be used only with species that can be cultured in vitro, and most ECM fungi cannot. Thus, to date, only two ECM genera, *Suillus* (Fries 1987b; Dahlberg and Stenlid 1990, 1994; Dahlberg 1997) and *Laccaria* (Baar et al. 1994), have successfully been studied in some detail using this technique. In addition, the outcome of somatic incompatibility can be ambiguous in *Suillus*, sometimes even in self pairings (Jacobson et al. 1993; cf. Bonello et al. 1998). Furthermore, whereas incompatibility is a sufficient condition to define two isolates as different, compatibility does not necessarily imply that two isolates are identical.

Using sequence-based markers, the genetic structure and reproductive biology of the ECM fungus *S. pungens* was studied in a 40-year-old stand of bishop pine in California (Bonello et al. 1998). Six genotypes were distinguished within $1200 \, \text{m}^2$, of which one comprised 13 sporocarps, the outermost 40 m apart and covering an area of $300 \, \text{m}^2$; the remaining five were represented by single sporocarps. It was concluded that the extension of the largest genet was due to mycelial growth, as the possibility of genetic identity from either apomixis or fortuitously indistinguishable recombinant genotypes was excluded by a genetic analysis from single spore isolates (Bonello et al. 1998).

Measured sizes of the sporocarp-producing genets and assumptions of constant mycelial extension between years have been used to obtain conservative age estimates. A mean life expectancy and maximum age of *S. bovinus* was estimated to be 35 and 75 years respectively, the maximum age of *S. variegatus* 30–70 years, and the maximum age of *S. pungens* to be 40 years (Dahlberg and Stenlid 1990, 1994; Dahlberg 1997; Bonello et al. 1998). However, mycelial extension of *Suillus* is probably multidirectional rather than unidirectional, and will vary between years. Thus, it has been speculated that genets of *S. variegatus* in old virgin forests may be several hundred years, rather than decades, old (Dahlberg 1997).

In an experiment with three pine seedlings per pot, each inoculated with either one or two *S. bovinus* isolates or one *S. variegatus* isolate and all isolates present in each pot, the outcome of genet competition was assessed after 6 months using isoenzyme fingerprinting (Timonen et al. 1997a). None of the genets was superior, as reflected by number of ECM, and all were generally present on all seedlings at the end of the experiment.

2.4.3
Genetics

The mating system in *Suillus* is heterothallic and it possesses a bipolar sexual incompatibility (Fries and Neumann 1990; Fries and Sun 1992; Fries 1994).

Other ECM basidiomycetes so far investigated possess a tetrapolar mating system. However, recent studies indicate that several tetrasporic *Suillus* species have both binucleate and uninucleate spores (Jacobson and Miller 1992a, 1994; Treu and Miller 1993; Bonello et al. 1998). In *S. americanius* (Peck) Snell: Slipp & Snell and *S. cothurnatus* Sing. the percentage of binucleate spores was 12% (Treu and Miller 1993), while in *S. granulatus* and *S. pungens* it was 5–10% and 1.3–1.4% respectively (Jacobson and Miller 1994; Bonello et al. 1998). If the binucleate spores are homokaryotic, resulting from post-meiotic mitotic divisions in the spores, then the breeding system is primarily determined by the heterothallic mating system of the fungus, be that bipolar or tetrapolar. Alternatively, if the postmeiotic mitoses occur in the basidia prior to nuclear migration, then some proportion of the spores could also be secondarily homothallic (heterokaryotic), this proportion being determined by the pattern of nuclear migration into the spores. A genetic analysis suggested that the binucleate spores of American *S. granulatus* may result from postmeiotic mitotic divisions in the basidia, prior to nuclear migration into the spores, and are thus potentially secondarily homothallic (Jacobson and Miller 1994). In a study of germinated spores of *S. pungens*, all spores were secondary homothallic; however, their potential ecological significance is not clear (Bonello et al. 1998). These studies suggest that American *S. granulatus* and *S. pungens* potentially have a mixed homothallic–heterothallic mating system, resulting from heterothallic uninucleate spores and secondarily homothallic binucleate spores (Treu and Miller 1993; Jacobson and Miller 1994). By contrast, only mononucleate spores were found in a study of European *S. granulatus* (Fries and Neuman 1990).

In an isoenzymatic study of 43 isolates of *S. tomentosum*, Zhu et al. (1988) demonstrated variability among and within forest regions, and suggested habitat isolation and host selection to be the major sources of genetic variation. Sen (1990a) analysed the pattern of isoenzymes in 11 isolates of *S. variegatus* and 14 isolates of *S. bovinus* in a comparison between two forest stands in Finland. He found a high intraspecific variation between the stands in *S. variegatus*, whereas no such population-related isoenzyme variation could be detected in *S. bovinus*. Another isoenzymatic study revealed variation between populations of *S. plorans* (Roll.) O. Kuntze from two forest regions, whereas no such pattern was obvious in *S. placidus* (Bon) Sing. (Keller 1992). In a study using isozymes of *S. collinitus* (Fr.) O. Kuntze two main groupings were found that largely corresponded to the geographic origin of the 43 fungal isolates (Mediterranean versus the alpine regions) and to the pine species present at these sites (El Karkouri et al. 1996). Analyses of ITS sequences and mating tests suggested that isolates of *S. granulatus* from North America, Europe or Asia represent two different species (Kretzer et al. 1996). This view is supported by a study where American and Asian populations of *S. granulatus* were found to differ in host specificity (Jacobson and Miller 1992b). On the contrary, isolates of *S. luteus* from North America and Europe had almost identical sequences (Kretzer et al. 1996), supporting the view of a recent introduction of *S. luteus* to North America (Fries and Neuman 1990). So far, no detailed analysis of spatiotemporal genetic variation in populations of *Suillus* has been conducted.

2.5
Ectomycorrhizal Morphology and Colonisation

2.5.1
Ectomycorrhizal Morphology

Suillus mycorrhizas are characterized by the same set of general characters, as those of *Rhizopogon*. These include a dichotomous to tuberculate form, conspicuous rhizomorphs, and abundant mycelia. As early as 1923 Melin described this form of tuberculate ECM, *knollenmycorrhiza*. He classified it in 1927 as ECM morphotype C and suggested the mycobiont to be a *Boletus* species (*Suillus* was at this time included in *Boletus*). Since then, the ability to distinguish between ECM morphotypes based on their morphological characters has developed substantially (Agerer 1991). Several species of *Suillus* are now described in detail (Agerer 1987–1993).

Recently, a molecular classification based on polymerase chain reaction-restriction fragment length polymorphism (PCR–RFLP) analysis of the ITS region of the rDNA has been employed for detection of mycobionts in single ECM (Gardes et al. 1991). Kårén et al. (1997) found that *Suillus* species were easily distinguished by this technique. Bonfante et al. (1997) demonstrated interspecific differences for seven additional species of *Suillus* and also revealed that random amplified polymorphic DNA and microsatellite primers are convenient tools for revealing intraspecific differences. Another molecular method for identification of the mycobiont in single ECM is the use of isoenzymes, which was successfully developed by Sen (1990b). He found a high intraspecific similarity within *S. variegatus* and *S. bovinus* and high interspecific variation. Keller (1992) was able, using isoenzymes, to distinguish three *Pinus cembra* associated *Suillus* species [*S. plorans*, *S. placidus* and *S. sibiricus* (Sing.) Sing.] and thereby to identify isolated strains from tuberculate ECM. This technique has been used to confirm the mycobiont identity of ECM (Keller 1992; Dahlberg and Stenlid 1994; Timonen et al. 1997b). Rhizomorphs in *Suillus* appear to be as highly evolved as in their close ally *Rhizopogon*, with highly differentiated, centrally arranged thicker hyphae and partially or completely dissolved septa (Agerer 1995).

2.5.2
Ectomycorrhizal Colonisation

Studies of *Suillus* species have played a central part in improving our understanding of the establishment and development of ECM roots and the role played in developmental processes by the cytoskeleton (e.g. Timonen et al.

1993). Several experiments with *S. bovinus* have contributed to our current knowledge concerning the role of microtubules and microfilaments in the hyphal morphogenesis which takes place during ECM formation (Niini and Raudaskoski 1993, 1997; Niini et al. 1996). Several *Suillus* species are reported to colonise conifer seedlings in experiments and under nursery and natural conditions. Danielson (1991) indicated that *Suillus* may colonise containerised seedlings of *P. banksiana* in seedling nurseries. In a recent survey of spontaneously occurring ECM fungi in 30 Swedish nurseries producing containerised seedlings of Scots pine, both *S. variegatus* and *S. bovinus* were observed to occur spontaneously (A. Dahlberg, unpubl. observ.).

Naturally established seedlings are readily colonised by *Suillus* species. *S. tomentosum* was found on 3-month-old seedlings of *P. contorta*, and *S. umbonatus* on 17-month-old seedlings (Bradbury 1998). In a study of old Scots pine forest stands in northern Sweden, *S. variegatus* was present as ECM both on old pines and on 1-year-old seedlings (Jonsson et al. 1999). Both studies suggest that naturally regenerating seedlings are largely colonised by the same set of ECM fungi as are present on old trees. Danielson and Visser (1989) reported 12-week-old seedlings of *P. contorta* to be colonised by *Suillus*. In natural humus microcosms, *S. bovinus* was identified as colonising young Scots pine seedlings (Timonen et al. 1997b), and *S. pungens* was found on 5-month-old pine seedlings after an intensive fire (Horton et al. 1998).

2.6
Physiology

2.6.1
Nutrient and Water Translocation

The physiology of *Suillus* species has been well studied and includes a number of pioneering investigations. The relative ease of isolation, culture and ECM synthesis and the fact that many species build relatively robust, well-differentiated mycelial systems have probably contributed to the popularity of the genus in studies of nutrient uptake and translocation. The pioneering studies of Melin and Nilsson (1950, 1952) provided the first experimental evidence of P and N translocation through ECM mycelia. Melin and Nilsson (1953) also presented the first experimental evidence of uptake of organic N (glutamate) by the mycelium. The above studies were later extended by Duddridge et al. (1980) and Brownlee et al. (1983) to include uptake and translocation of water and nutrients in larger mycelial systems grown under non-sterile conditions. Translocation of ^{32}P-labelled orthophosphate through differentiated mycelia of *S. bovinus* has since been demonstrated both autoradiographically (Finlay and Read 1986b) and by using a modified β-scanner (Timonen et al. 1996). The above studies suggested fluxes of P through rhizomorph systems consistent with transport via symplastic flow rather than turgor-driven bulk flow (Finlay 1992); however, an apoplastic pathway through the sheath has been shown (Behrmann and Heyser 1992) in ECM roots of *P. sylvestris* colonised by *S. bovinus*. Estimation of the amounts of extramatrical hyphae produced is not easy, but Read and Boyd (1986) estimated that mycelial systems of *S. bovinus* growing in peat could produce 200 mg^{-1} peat. The potential of the mycelium to provide conduits for the transport of water was suggested by experiments in which rhizomorphs connecting seedlings to moist soil were cut, resulting in a marked decline in transpiration rates within minutes (Boyd et al. 1986).

Many experiments have made use of semi-hydroponic systems to study nutrient uptake and *Suillus* species have been used in these systems with varying success. Kähr and Arveby (1986) developed a method for establishing ectomycorhiza on conifer seedlings under conditions of steady-state nutrition, using nutrient solution flowing over sloping plastic plates. The method was used with *S. bovinus* (Ingestad et al. 1986) and typical ECM were formed within a short space of time, although portions of the root systems extending beyond the plates were not well infected. Wallander and Nylund (1992) used a semi-hydroponic system with leca pellets and obtained better mycelial growth by the hydrophilic *Laccaria bicolor* (Maire) Orton and *Hebeloma crustulineforme* (Bull.: St-Amans) Quél., than by *S. bovinus* which only grew superficially in in the upper, drier parts of the system. The above observation is consistent with the finding of Stenström (1991) that *S. flavidus* and *S. bovinus* are highly sensitive to flooding of even 2 min day^{-1} (four times per week) compared with *Thelephora terrestris* Ehrh.: Fr., *Laccaria laccata* (Scop.: Fr.) Cooke and *H. crustuliniforme*. Colpaert et al. (1992, 1996) used drier perlite systems saturated with nutrient solution to only 80% of its water holding capacity and obtained satisfactory growth of the extramatrical mycelium of *S. bovinus* and *S. luteus*. Reduced shoot growth in these experiments was attributed to a drain of C and possibly N to the well-developed *Suillus* mycelium but would presumably be less of a problem under more natural conditions of nutrient supply and mycelial growth.

2.6.2
Nitrogen Metabolism

Following the initial studies of Melin and Nilsson (1952, 1953), *Suillus* species have been used in many studies of N metabolism. Growth on both NO_3^- and NH_4^+ sources has been shown in a number of species such as *S. granulatus*, *S. variegatus* and *S. bovinus* (France and Reid 1984; Littke et al. 1984; Finlay et al. 1992). Sarjala (1990) showed that nitrate reductase activity of *S. variegatus* in pure culture was increased by additions of NO_3^- but not by additions of NH_4^+. These differences were not found for ECM *P. sylvestris* roots synthesised with the same species, but ECM roots had higher nitrate reductase activity than non-ECM root tips in the same root system (Sarjala 1991). Glutamine synthetase (GS) activity has been demonstrated in ECM roots colonised by *S. variegatus* and to increase slightly with increasing concentrations of NH_4^+ (Sarjala 1993). Rudawska et al. (1994) found GS activity in pure cultures of *S. bovinus* which was stimulated by increased ammonium but only negligible levels of glutamate dehydrogenase (GDH). Experiments by Finlay et al. (1988) demonstrated mycelial uptake and assimilation of ^{15}N-labelled ammonium chloride and uniformly high levels of ^{15}N enrichment in mycelial amino acid pools of widely differring size, suggesting high levels of amino-transferase activity.

Utilisation of organic N compounds has been demonstrated in a number of *Suillus* species following the initial experiments of Melin and Nilsson (1953). Lundeberg (1970) showed better growth on media containing asparagine and glycine than on inorganic N sources. Subsequent experiments showed uptake of a range of amino acids (Abuzinadah and Read 1988) and alanine peptides (Abuzinadah et al. 1986) in pure cultures of *S. bovinus* and by ECM plants colonised by the same fungus (Abuzinadah and Read 1986). Utilisation of the protein bovine serum albumin (BSA) has also been shown in pure culture for *S. variegatus* (Finlay et al. 1992), but growth of the high sub-alpine species *S. placidus*, *S. plorans* and *S. sibiricus* on BSA and the plant protein gliadin showed considerable intra-specific variability in another study by Keller (1996).

2.6.3
Enzymatic Activity of Mycelia

Activity of a wide range of enzymes has been demonstrated in mycelial systems of *Suillus* species, both in pure cultures and intact mycelial systems. Acid phosphatase, alkaline phosphatase and nitrate reductase were demonstrated in *S. caerulescens* Smith & Thiers, *S. lakeii* (Murr.) Smith & Thiers, *S. brevipes* and *S. tomentosus* by Ho (1989) who also showed considerable variation in activity between isolates. Wall- and membrane-bound acid phosphatase activities were measured in *S. grevillei* by McElhinney and Mitchell (1993) and found to be optimum over a broad pH range. Phosphatase activities have also been investigated by Dighton (1983) and Mousain et al. (1988). Timonen and Sen (1998) studied the spatial heterogeneity of enzyme expression in intact ECM systems of *P. sylvestris–S. bovinus* and found highest acid phosphatase activities in ECM roots and in fine hyphae at the margins of the extramatrical mycelium. Expression of polyphenol oxidase activity was reduced in ECM roots compared with non-ECM roots and, although the level of peroxidase activity was not altered by the ECM status of roots, differential isozyme expression was detected, indicating a change in root peroxidase activities following ECM formation. Colpaert et al. (1997) studied the use of inositol hexaphosphate by *S. luteus* and found substantial extracellular acid phosphatase activity but only limited utilisation of phytate. Uptake and accumulation of inorganic P as mobile, medium chain length polyphosphate has been demonstrated in *S. bovinus* by Gerlitz and Werk (1994) using comparative in vivo ^{31}P-NMR. Mobilisation of stored polyphosphate was initiated by deficiency of external P, but a high fungal biomass rendered ECM P metabolism less sensitive to external variation in nutrient concentrations. Transformation of mobile polyphosphate into immobile long chain length or granular polyphosphate was also observed. Maximal P uptake occurred at an external pH close to 5.5 (Gerlitz and Gerlitz 1997). Cairney and Burke (1994) reviewed the possible significance of plant cell wall degrading enzymes in ECM symbioses. Generally, the enzymatic competence of ECM fungi to degrade more recalcitrant plant compounds appears to be lower than that of saprotrophic fungi (e.g. Colpaert and Van Laere 1996; Colpaert and Van Tichelen 1996), but presumptive lignolytic activity has been detected in *S. bovinus* and *S. luteus* using decoloration of polyaromatic dyes (Bending and Read 1997). Carboxymethyl cellulose hydrolysing activity has also been detected in *S. bovinus* (Maijala et al. 1991) and seems to be constitutive and expressed in the presence of glucose (Giltrap and Lewis 1982). Experiments on decomposition of organic substrates in the presence of a host plant are less frequent than those in asymbiotic axenic culture, but Dighton et al. (1987) obtained some evidence that ECM formation by *S. luteus* increased the decomposition of organic substrates added to soil microcosms. Experiments by Durall et al. (1994) on decomposition of ^{14}C-labelled cellulose and hemicellulose showed that *S. lakeii* grown in association with *P. menziesii* was able to release significant amounts of ^{14}C from both substrates.

2.6.4
Siderophore Production and Chemical Weathering

Suillus species have played a central role in investigations of the capacity of ECM fungi to contribute to weathering of minerals. Wallander et al. (1997) found that seedlings colonised by *S. variegatus* had higher foliar P concentrations than non-ECM seedlings when grown with apatite as the sole P source. No clear relationship was found between organic acid production and P uptake, but the ability to utilise P from apatite was attributed to a reduction in soil pH which was larger than for other ECM fungi examined. In studies of *S. collinitus* growing in ECM association with *P. pinaster*, Rigou et al. (1995) also concluded that inoculation with the ECM fungus could contribute to improved availability of nutrients through release of protons into growth media with neutral pH. The role of *S. granulatus* in dissolution and immobilisation of P and Cd from rock phosphate was examined by Surtiningsih et al. (1992) who concluded that the fungus tended to solubilise P and immobilise Cd. Production of siderophores has been demonstrated in *S. granulatus* (Watteau and Bethelin 1990, 1994; Leyval et al. 1992) which seems to release more iron chelates than other tested fungi such as *Paxillus involutus* (Batsch: Fr.) Fr. and *Pisolithus tinctorius* (Pers.) Coker and Couch. Unestam and Sun (1995) discussed the ecological consequences of the fact that walls at the tips of *S. bovinus* hyphae are hydrophilic whereas the remaining extramatrical mycelia are hydrophobic.

2.6.5
Carbon Metabolism

The fact that many *Suillus* species produce well-differentiated, extensive mycelial systems has made them a natural choice in studies of ECM of C metabolism and assimilate allocation within mycelial systems. Brownlee et al. (1983) demonstrated allocation of C compounds within mycelial systems of *S. bovinus* growing in microcosms containing *P. sylvestris* plants. The capacity of *S. bovinus* and *S. granulatus* mycelia to interconnect plants and provide potential pathways for the flow of C along concentration gradients was shown by Finlay and Read (1986a). The acropetal translocation velocity of ^{14}C-labelled C compounds within the mycelia in these experiments was in excess of $20\,cm\,h^{-1}$ and suggests that movement of carbohydrates in *Suillus* mycelia may take place along a hydrostatic pressure gradient generated by high internal solute concentrations as has been suggested previously for *Serpula lacrymans* (Wulfen: Fr.) J. Schröt. Distribution of ^{14}C-labelling within *S. bovinus* mycelia interconnecting plants was studied by Duddridge et al. (1988) using microautoradiography and provided evidence of the capacity for movement of C compounds from fungus to plant, probably as C skeletons of amino compounds, as suggested by Finlay et al. (1996). The identity of the C compounds transported in intact mycelial systems of *S. bovinus* was examined by Söderström et al. (1988) who found that the predominant compounds were trehalose, mannitol and arabitol. Experiments with *Suillus* species have also provided novel insights into mycelial function in relation to respiratory turnover of C compounds. Using microcosm experiments it was shown that approximately 30% of below-ground respiration was attributable to the ECM mycelium and that this respiration was highly dependent on the supply of current assimilate from the host plant (Söderström and Read 1987). In systems where the mycelial connections to the roots were severed, there were reductions in the respiration rate of more than 50% within 24h.

2.7
Anthropogenic Influences

2.7.1
Accumulation of Radionuclides

An extensive literature exists on anthropogenic interactions involving the distribution, growth and function of *Suillus* species. The emphasis has often been on the effects of various pollutants on the ECM fungi, but there are also many reports of different ways in which ECM associations involving *Suillus* themselves affect the responses of host trees to different types of anthropogenic stress. Following the Chernobyl accident in 1986, a number of studies have reported the intrinsic capability of fungi to accumulate radiocaesium and that this ability varies significantly between fungal species. Generally, ECM fungi have higher levels of ^{137}Cs than saprophytic fungi, and species of *Suillus* are typically in the upper range of measured values in ECM fungi (Heinrich 1993). In Sweden, 40–150 kBq kg^{-1} dry wt sporocarp are found, and these values have not decreased since 1986 (Dahlberg et al. 1997). Elevated levels of ^{137}Cs have also been found in sporocarps of *S. grevillei* in Scottish grassland sites (Anderson et al. 1997) and in Austrian sites (Haselwandter et al. 1988) where contents of ^{137}Cs are 3.0–4.8 times higher than before the Chernobyl accident.

2.7.2
Nitrogen Deposition

Much of Europe has been subjected to an increased deposition of atmospheric N arising from industrial processes and traffic exhausts, as well as local effects of intensive stock rearing. A decline in the production of sporophores and of species diversity has been documented in the Netherlands (Arnolds 1991) and seems to be correlated to increases in N concentration in forest litter horizons. *Suillus* species have been studied in a number of experiments. Experiments by Termorshuizen and Ket (1991) showed that NH$_4^+$ fertilisation negatively influenced ECM root tips. Arnebrant (1994) examined the extension rates of mycelia of different ECM fungi in response to three different nitrogen sources added at concentrations of 1–4 mg N g^{-1} dry wt of peat. She found that growth of *S. bovinus* mycelium was reduced to 30% of the control and that *S. bovinus* was the second most sensitive of the fungi tested. Wallander and Nylund (1992) also found a similar reduction in growth in a semi-hydroponic system and that *S. bovinus* was the most sensitive of the three species that they tested.

2.7.3
Effects of Toxic Metals

Acidification of forest soils and the accompanying increase in solubility of heavy metals has led to an increased interest in the role of ECM fungi in interactions involving heavy metals. The subject has been reviewed by Wilkins (1991) and Hartley et al. (1997a). A number of studies are available on the long-term impact of smelters on ECM function. Barcan et al. (1998) reported on the metal concentrations in fungi growing near the Severo Ni plant, Monchegorsk, the Russian Federation, on the Kola Peninsula. *Suillus luteus* was found amongst the species in this region and the nickel concentrations of the sporophores were 15–40 times the background level. Studies of other areas affected by smelter pollution have also shown accumulation by *Suillus* species. Rühling and Söderström (1990) found accumulation of Zn, Pb, Cu and Cd in *S. variegatus* along the metal gradient around the primary smelter at Skelleftehamn, Sweden. Available evidence suggests that there are large differences between different ECM symbionts in their effectiveness in providing resistance to metal toxicity. Colpaert and Van Assche (1987) isolated strains of *S. luteus* from Zn-contaminated soil that were able to grow at concentrations of 1000 μg g^{-1}. Strains from uncontaminated soil were less resistant and conferred lower levels of tolerance on *P. sylvestris* host plants. *S. bovinus* was included in two later studies by Colpaert and Van-Assche (1992a,b). When Cd-sensitive and Cd-tolerant strains were grown on combinations of two non-toxic Zn concentrations and three Zn concentrations, adding a higher Zn concentration to the medium resulted in a reduction in the toxic effect of Cd which was associated with a lower concentration in the mycelium. A reduction in Cd uptake by ECM *P. sylvestris* seedlings was also shown (Colpaert and Van Assche 1993), but it was not possible to relate this to growth. Hartley et al. (1997b) also found interactive effects of metals which appeared related to the initial sensitivity of an isolate to a particular metal. One isolate of *S. variegatus* was sensitive to Cd^{2+} in the absence of Zn^{2+} but increased its Cd tolerance in the presence of Zn: by contrast another iosolate of *S. granulatus* was significantly less sensitive to Cd^{2+} in the absence of Zn^{2+}. Dixon (1988) and Dixon and Buschena (1988) reported amelioration of toxicity by *S. luteus* at low levels of Cd, Cu, Ni and Pb as well as reduced tissue concentrations of the metals in *Pinus banksiana* and *Picea glauca*. Bücking and Heyser (1994) reported the ability of *S. bovinus* to maintain shoot concentrations of Zn at a low level when plants were exposed to high levels of the metal and that this effect was additionally improved by pre-incubation of the fungus on a Zn-enriched growth medium. Gruhn and Miller (1991) found increased synthesis of tyrosinase by *S. granulatus* in response to Cu stress as well as decreases in polyamine content and speculated that tyrosine may play a role in Cu detoxification.

Jones and Muehlchen (1994) examined *in vitro* responses to a range of metals by measuring fungal growth on agar plates and found wide variation in the ability of different species to tolerate different metals. *S. variegatus* was the species most tolerant of Al, growing at concentrations of up to 100 μg g^{-1}, supporting earlier observations of Väre (1990) that *S. variegatus* is able to grow on high concentrations of Al, detoxifying the metal by building abundant Al-polyphosphate granules. Al–P interactions have been discussed in a number of studies with other fungal genera such as *Laccaria*, but one interesting study conducted with *S. bovinus* was conducted by Gerlitz (1996) who used NMR to examine polyphosphate mobilisation in relation to Al exposure. In this study an Al-adapted fungus had an average chain length of mobile polyphosphate that was considerably shorter than in a non-Al-adapted fungus. Detoxification of freely mobile Al ions is thus attributed to a higher mobile polyphosphate concentration and capture of intracellular Al by mobile polyphosphate of shorter chain length. High Al tolerance has been reported in a range of fungi by Hintikka (1988b) who found that different species of *Suillus* were less sensitive than species of *Amanita* and *Tricholoma*. These results are supported by pure culture experiments of Thompson and Medve (1984) who found that *S. luteus* was the most Al-tolerant of six ECM isolates tested. Göransson and Eldhuset (1991) grew *P. sylvestris* seedlings in a semihydroponic system and found that seedlings infected with *S. bovinus* tolerated higher Al concentrations than non-ECM plants before showing a reduction in relative growth rate and Ca and Mg uptake. Kasuya et al. (1990) found that formation of ECM by *Suillus* species on *Pinus caribea* was reduced at lower Al concentrations but stimulated at higher concentrations and concluded that inoculation of trees with Al-tolerant ECM fungi could improve reforestation efforts in highly weathered tropical soils.

2.7.4
Herbicides

Ways in which ECM fungi influence the response of plants to different herbicides, or are themselves influenced by herbicides, have been investigated in relatively few studies. Chakravarty and Sidhu (1987) conducted *in vitro* growth tests with glyphosate (Roundup), hexazinone (liquid Velpar L and granule Pronone TM5G) and trichlopyr (Garlon) using five species of ECM fungi. The species, which included *S. tomentosus*, showed varied species reactions to different herbicide concentrations and fungal growth was reduced at concentrations above 10 ppm. Garlon was the most toxic herbicide. In other laboratory tests (Chakravarty and Chatarpaul 1990) *Cenococcum graniforme* (Sow.) Ferd. and Winge, *H. crustuliniforme* and *L. laccata* were more sensitive to glyphosate and hexazinone than *S. tomentosus* which was more susceptible than *P. involutus*. In greenhouse and field tests overall effects of the herbicides were less intense, but seedlings inoculated with *S. tomentosus* were more sensitive to herbicide than non-mycorrhizal ones.

2.7.5
Bioremediation

The potential of ECM basidiomycetes as agents of bioremediation has attracted increasing interest over the past few years, but the number of actual studies performed is still low. Meharg et al. (1997b) investigated the ability of four ECM fungi to degrade 2,4,6-trinitrotoluene in axenic culture and found that *S. variegatus* produced the greatest biomass and thus transformed most TNT. Biotransformation was lower under conditions of N limitation. In other experiments (Meharg et al. 1997a) degradaton of 2,4-dichlorophenol by *P. involutus* and *S. variegatus* was examined in both axenic and liquid culture and both fungi readily degraded the compound, mineralising up to 17% in a 17-day period. Growth of the fungi in symbiosis with *P. sylvestris* stimulated greater mineralisation than when plants were grown without the fungi and *S. variegatus* was more efficient than *P. involutus*.

2.8
Inoculation Experiments and Applications

Attempts to utilise *Suillus* species in large-scale inoculation experiments have been reported on several different continents. Castellano and Molina (1989) list successful fungus-host inoculations for 12 different combinations of *Suillus* species and their effects on growth of container seedlings. Most of the reported experiments show no significant growth response. In one study, growth responses are reported for several oak species, but the identity of the fungal isolate was not confirmed (Dixon et al. 1984). Field trials of *P. radiata* inoculated with *S. luteus* and *S. granulatus* were carried out by Theodorou and Bowen (1970) in Australia who found significant stimulation of growth after 36 months that was related to the degree of ECM development. *S. granulatus* was the most effective species in this experiment. In Swedish experiments (Stenström and Ek 1990) using *S. variegatus* and *S. luteus*, ECM stimulation of growth was not significant and colonisation of the root system was impossible to assess after 12 months. Roldán and Albaladejo (1994) evaluated the growth of ECM *Pinus halipensis* seedlings planted in a semi-arid soil amended with urban refuse in Spain. The results of the inoculation were evaluated after 1 year *S. collitinus* was the most effective fungus in improving plant growth in control plots but had respectively no or negative effects on plant growth at lower and higher doses of refuse addition. Chu-Chou and Grace (1990) studied the relationships between ECM symbionts of *P. radiata* in nurseries and forests of New Zealand. *Suillus* species were the dominant fungi in recently established nurseries but became less common following outplanting in associated forests. In other nurseries *Suillus* spp. were never isolated from the roots of seedlings, but they were commonly isolated from roots of trees in associated forests. In a later study of conifer establishment in New Zealand, Davis et al. (1996) found that inoculation of pine seed was more effective with spores of *Rhizopogon rubescens* Tul. than with *Rhizopogon luteolus* Fr. or *S. luteus*.

2.9
Interactions with Other Organisms

There are several reports of fungi growing only in association with *Suillus*, for example *Gomphidius roseus* (Fr.) P. Karst in association with *S. bovinus* (Singer 1949; Agerer 1990). Agerer (1990) found that mycelia of *Chroogomphus* spp., potentially forming ECM individually, can be detected within the mantle and rhizomorphs of *Suillus* ECM and also as haustoria within root cortex cells. The study provides detailed descriptions of the three-way relationships between *Chroogomphus helveticus* (Sing.) Moser with *S. plorans* and *S. sibiricus* on *Pinus cembra* as well as *Chroogomphus rutilans* (Schaeff.: Fr.) O. K. Miller with *S. bovinus*, *S. collinatis* and *S.* cf. *variegatus* on *P. sylvestris*. Agerer (1990) suggested that the growth of *Chroogomphus* may be dependent upon the presence of hyphae of *Suillus* or *Rhizopogon* (which also forms this kind of three-way association). However, the functional significance of these associations remains to be investigated. Shaw et al. (1995) studied interactions between ECM and saprotrophic fungi on agar and in association with seedlings of *P. contorta*. *Lactarius rufus* (Scop.: Fr.) Fr. stimulated growth of and root colonisation by *S. bovinus*, but *L. laccata* suppressed ECM formation by *S. bovinus*.

About 50% of all *Suillus* sporocarps, as with other macrofungi, are infested by fungivorous insects, particularly dipterous insects (Hackman and Meinander 1979; Yakovlev 1987) but also beetles (Benick 1952). Mycophagous nematodes commonly inhabit tree rhizospheres and *Aphelenchus avenae* can suppress development of *S. granulatus* ECM roots with *P. resinosa* seedlings (Sutherland and Fortin 1968). Suppression of *S. granulatus* ECM formation on *P. ponderosa* by *Aphelenchoides cibolensis* has also been reported by Riffle (1975). Mycophagous collembola have further been shown to reduce ECM colonisation, and in experiments by Hiol et al. (1994) it was shown that *S. luteus* was consumed less than *Rhizoctonia solani* Kühn by *Proisotoma minuta* but more than *P. tinctorius*.

Interaction with pathogenic organisms has been a central area of research within ECM research, but it has often been difficult to separate pure antagonistic or inhibitory effects of ECM fungi from the positive effects arising from improved nutrient status and vigour. Bianco-Coletto and Giardino (1996) tested different basidiomycetes for antibiotic activity and found that *S. bovinus* had some activity against *Bacillus cereus* and *Bacillus subtilis*. Some inhibition of phytopathogenic fungi by cell-free culture media of *S. brevipes* was found by Kope and Fortin (1989). Hwang et al. (1995) tested the effects of *P. involutus* and *S. tomentosus*, as well as *B. subtilis*, against *Fusarium moniliforme* Sheldon and found that *S. tomentosus* had no effect on the *Fusarium*. Olsson et al. (1996) demonstrated reduction of bacterial activity, as measured by ^{3}H-thymidine and ^{14}C-leucine incorporation, by different ECM fungi, including *S. variegatus*.

Sarand et al. (1998) examined the cellular interactions and catabolic activities of ECM-associated bacteria forming a biofilm at the interface between petroleum hydrocarbon (PHC) contaminated soil and dense, long-lived patches of *S. bovinus* hyphae colonising the contaminated soil. Fluorescent pseudomonads were isolated which harboured similarly sized mega-plasmids and exhibited degradative ability, including growth on m-toluate and m-xylene as sole carbon sources, as well as catechol cleavage. The authors suggested that the fungal patch differentiation leads to enrichment and stability of bacterial populations at the fungal–soil interface. Schelkle and Peterson (1996) examined effects of four strains of so-called ECM helper bacteria, together with the ECM species *L. bicolor*, *L. proxima* (Boud.) Pat. and *S. granulatus*, on the pathogens *Fusarium oxysporum* Schlecht. and *Cylindrocarpon* sp. None of the ECM species alone inhibited *Fusarium* but all showed slight inhibition of *Cylindrocarpon* growth. Helper bacterium strain MB3 (*B. subtilis*) inhibited both pathogens and was still more effective at inhibiting *Fusarium* when inoculated together with either *L. proxima* or *S. granulatus*. With *Cylindrocarpon* only *S. granulatus* inoculated together with MB3 showed enhanced inhibition over MB3 alone.

Interactions between bacteria and "*Suillus*-like" ECM fungi were examined by Shishido et al. (1996a,b) and were shown to depend upon the bacterial species involved. Growth promoting strains of *Bacillus* appeared to influence conifer seedlings through a mechanism unrelated to ECM fungi while growth promotion of *P. contorta* by fluorescent *Pseudomonas* strains appeared to be facilitated by an interaction with ECM. Vares et al. (1996) found 27 bacterial species associated with *S. grevillei* sporocarps and ECM roots of *Larix decidua*. The genera *Pseudomonas, Bacillus* and *Streptomyces* were dominant. Gram-positive bacteria seldom stimulated fungal growth, but among the gram-negative bacteria two *Pseudomonas* strains showed the greatest enhancement of growth and *Streptomyces* always caused significant inhibition of the fungus. Nurmiaho-Lassila et al. (1997) studied bacterial colonisation patterns in intact mycorrhizospheres of *P. sylvestris* colonised by *S. bovinus* and *Paxillus involutus*. The mycorrhizospheres formed by these two fungi appeared dissimilar and the spatially and physiologically defined habitats were shown to host distinct populations of bacteria.

Acknowledgements. We greatly appreciate the constructive comments of Tom Bruns, Svengunnar Ryman and Andya Taylor.

References

Abuzinadah RA, Read DJ (1986) The role of proteins in the nitrogen nutrition of ecto-mycorrhizal plants. II. Utilization of protein by mycorrhizal plants of *Pinus contorta*. New Phytol 103:495–506

Abuzinadah RA, Read DJ (1988) Amino acids as nitrogen sources for ectomycorrhizal fungi utilization of individual amino acids. Trans Br Mycol Soc 91:473–480

Abuzinadah RA, Finlay RD, Read DJ (1986) The role of proteins in the nitrogen nutrition of ectomycorrhizal plants. I. Utilization of peptides and proteins by ectomycorrhizal fungi. New Phytol 103:481–593

Agerer R (1990) Studies of ectomycorrhizae. XXIV. Ectomycorrhizae of *Chroogomphus helveticus* and *C. rutilus* (Gomphidaceae, Basidiomycetes) and their relationships to those of *Suillus* and *Rhizopogon*. Nova Hedwigia 50:1–63

Agerer R (1991) Characterisation of ectomycorrhiza. Methods Microbiol 23:25–73

Agerer R (1995) Anatomical characteristics of identified ectomycorrhizas: an attempt towards a natural classification. In: Varma A, Hock B (eds) Mycorrhiza: function, molecular biology and biotechnology. Springer, Berlin Heidelberg New York, pp 685–734

Agerer R (1987–1993) Colour atlas of ectomycorrhizae. Einhorn, Schwäbisch Gmund

Alvin KL (1960) Further conifers of the Pinaceae from Wealden Formation of Belgium. Inst R Sci Nat Belg Mem 146:1–39

Anderson P, Davidson CM, Littlejohn D, Ure AM, Shand CA, Cheshire MV (1997) The translocation of caesium and silver by fungi in some Scottish soils. Comm Soil Sci Plant Anal 28:635–650

Arnebrant K (1994) Nitrogen amendments reduce the growth of extramatrical ectomycorrhizal mycelium. Mycorrhiza 5:7–15

Arnolds E (1991) Decline of ectomycorrhizal fungi in Europe. Agric Ecosyst Environ 35:209–244

Baar J, Ozinga WA, Kuyper TW (1994) Spatial distribution of *Laccaria bicolor* genets as reflected by sporocarps after removal of litter and humus layer in a *Pinus sylvestris* forest. Mycol Res 98:726–728

Barcan VSH, Kovnatsky EF, Smetannikova MS (1998) Absorption of heavy metals in wild berries and edible mushrooms in an area affected by smelter emissions. Water Air Soil Pollut 103:173–195

Baura G, Szaro TM, Bruns TD (1992) *Gastrosuillus laricinus* is a recent derivate of *Suillus grevillei*; molecular data. Mycologia 84:592–597

Behrmann P, Heyser W (1992) Apoplastic transport through the fungal sheath of *Pinus sylvestris-Suillus bovinus* ectomycorrhizae. Bot Acta 105:427–434

Bending GD, Read DJ (1997) Lignin and soluble phenolic degradation by ectomycorrhizal and ericoid mycorrhizal fungi. Mycol Res 101:1348–1354

Benick L (1952) Pilzkäfer und Käferpilze. Ökologische und statistische Untersuchungen. Acta Zool Fenn 70:1–250

Berbee ML, Taylor JW (1993) Dating of evolutionary radiations of true fungi. Can J Bot 71:1114–1127

Berggren WA, Prothero DR (1992) Eocene–Oligocene climatic and biotic evolution. Princeton University Press, Princeton

Besl H, Bresinsky A (1997) Chemosystematics of Suillaceae and Gomphidaceae (suborder Suillineae). Plant Syst Evol 206:223–242

Bianco-Coletto MA, Giardino L (1996) Antibiotic activity in Basidiomycetes. X. Antibiotic activity of mycelia and cultural filtrates of 25 new strains. Allionia 34:39–43

Blomgren M (1994) Studies of macromycetes in stands of Scots pine and lodgepole pine. Doct Thesis, Swedish University of Agricultural Sciences, Uppsala

Bonello P, Bruns TD, Gardes M (1998) Genetic structure of a natural population of the ectomycorrhizal fungus *Suillus pungens*. New Phytol 138:533–542

Bonfante P, Lanfranco L, Cometti V, Genre A (1997) Inter- and intraspecific variability in strains of the ectomycorrhizal fungus *Suillus* as revealed by molecular techniques. Microbiol Res 152:287–292

Boyd R, Furbank RT, Read DJ (1986) Ectomycorrhiza and the water relations of trees. In: Gianinazzi-Pearson V, Gianinazzi S (eds) Physiological and genetical aspects of mycorrhizae. INRA, Paris pp 689–693

Bradbury SM (1998) Ectomycorrhizas of lodgepole pine (*Pinus contorta*) seedlings originating from seeds in southwestern Alberta cut blocks. Can J Bot 76:213–217

Bradbury SM, Danielson RM, Visser S (1998) Ectomycorrhizas of regenerating stands of lodgepole pine (*Pinus contorta*). Can J Bot 76:218–227

Brownlee C, Duddridge JA, MalibariA, Read DJ (1983) The structure and function of mycelial systems of ectomycorrhizal roots with special reference to their role in forming inter plant connections and providing pathways for assimilate and water transport. Plant Soil 71:433–444

Bruns TD, Palmer JD (1989) Evolution of mushroom mitochondrial DNA: *Suillus* and related genera. J Mol Evol 28:349–362

Bruns TD, Szaro T (1992) Rates and mode differences between nuclear and mitochondrial small-subunit rRNA genes in mushrooms. Mol Biol Evol 9:836–855

Bruns TD, Palmer JD, Shumard DS, Grossman LI, Hudspeth MES (1988) Mitochondrial DNAs of *Suillus*: three fold size change in molecules that share a common gene order. Curr Genet 13:49–56

Bruns TD, Fogel R, White TJ, Palmer J (1989) Accelerated evolution of a false truffle from a mushroom ancestor. Nature 339:140–142

Bruns TD, Szaro TM, Gardes M, Cullings KW, Pan JJ, Taylor DL, Horton TR, Kretzer A, Garbelotto M, Li I (1998) A sequence database for the identification of ectomycorrhizal basidiomycetes by phylogenetic analysis. Mol Ecol 7:257–272

Bücking H, Heyser W (1994) The effect of ectomycorrhizal fungi on Zn uptake and distribution in seedlings of *Pinus sylvestris* L. Plant Soil 167:203–212

Cairney JWG, Burke RM (1994) Fungal enzymes degrading plant cell walls: their possible significance in the ectomycorrhizal symbiosis. Mycol Res 98:1345–1356

Castellano MA, Molina R (1989) Mycorrhizae. In: Landis TD, Tinus RW, McDonald SE, Barnett JP (eds) The container tree nursery manual, vol 5. USDA Forest Service, Washington, DC, pp 101–167

Chakravarty P, Chatarpaul L (1990) Non-target effect of herbicides. I. Effect of glyphosate and hexazinone on soil microbial activity microbial population and in-vitro growth of ectomycorrhizal fungi. Pestic Sci 28:233–242

Chakravarty P, Sidhu SS (1987) Effect of glyphosate hexazinone and triclopyr on in-vitro growth of five species of ectomycorrhizal fungi. Eur J For Pathol 17:204–210

Chu-Chou M, Grace LJ (1990) Mycorrhizal fungi of radiata pine seedlings in nurseries and trees in forests. Soil Biol Biochem 22:959–966

Colpaert JV, Van Assche JA (1987) Heavy metal tolerance in some ectomycorrhizal fungi. Funct Ecol 1:415–421

Colpaert JV, Van Assche JA (1992a) Zinc toxicity in ectomycorrhizal *Pinus sylvestris*. Plant Soil 143:201–211

Colpaert JV, Van Assche JA (1992b) The effects of cadmium and the cadmium–zinc interaction on the axenic growth of ectomycorrhizal fungi. Plant Soil 145:237–243

Colpaert JV, Van Assche JA (1993) The effects of cadmium on ectomycorrhizal *Pinus sylvestris* L. New Phytol 123:325–333

Colpaert JV, Van Laere A (1996) A comparison of the extracellular enzyme activities of two ectomycorrhizal and leaf-saprotrophic basidiomycete colonizing beech leaf litter. New Phytol 134: 133–141

Colpaert JV, VanTichelen KK (1996) Decomposition, nitrogen and phosphorus mineralization from beech leaf litter colonized by ectomycorrhizal or litter-decomposing basidiomycetes. New Phytol 134:123–132

Colpaert JV, Van Assche JA, Luijtens K (1992) The growth of the extramatrical mycelium of ectomycorrhizal fungi and the growth response of *Pinus sylvestris* L. New Phytol 120:127–135

Colpaert JV, Van Laere A, Van Assche JA (1996) Carbon and nitrogen allocation in ecto-mycorrhizal and non-mycorrhizal *Pinus sylvestris* L. seedlings. Tree Physiol 16:787–793

Colpaert J, Van Laere A, Van Tichelen KK, Van Assche JA (1997) The use of inositol hexa-phosphate as a phosphorus source by mycorrhizal and non-mycorrhizal Scots pine (*Pinus sylvestris*). Funct Ecol 11:407–415

Critchenfield WB, Little JEL (1966) Geographical distribution of the pines of the world. US Dep Agric For Serv Misc Publ 991

Cullings KW, Szaro TM, Bruns TD (1996) Evolution of extreme specialization within a lineage of ectomycorrhizal epiparasites. Nature 379:63–66

Dahlberg A (1997) Population ecology of *Suillus variegatus* in old Swedish Scots pine forests. Mycol Res 101:47–54

Dahlberg A, Stenlid J (1990) Population structure and dynamics in *Suillus bovinus* as indicated by spatial distribution of fungal clones. New Phytol 115:487–493

Dahlberg A, Stenlid J (1994) Size, distribution and biomass of genets in populations of *Suillus bovinus* (L.: Fr.) Roussel revealed by somatic incompatibility. New Phytol 128:225–234

Dahlberg A, Stenlid J (1995) Spatiotemporal patterns in ectomycorrhizal populations. Can J Bot 73:1222–1230

Dahlberg A, Nikolova I, Johanson KJ (1997) Intraspecific variation in ^{137}Cs activity concentration in sporocarps of *Suillus variegatus* in seven Swedish populations. Mycol Res 101:545–551

Danielson RM (1984) Ectomycorrhizal associations in jack pine stands in northeastern Alberta. Can J Bot 62:932–939

Danielson RM (1991) Temporal changes and effects of amendments on the occurrence of sheath-ing ectomycorrhizas of conifers growing in oil sands tailings and coal spoil. Agric Ecosyst Environ 35:261–281

Danielson RM, Visser S (1989) Host response to inoculation and behavior of introduced and indigenous ectomycorrhizal fungi of jack pine grown on oil-sands tailings. Can J For Res 19:1412–1421

Davis MR, Grace LJ, Horrell RF (1996) Conifer establishment in South Island high country: influence of mycorrhizal inoculation, competition removal, fertiliser application, and animal exclusion during seedling establishment. NZ J For Sci 26:380–394

Dighton J (1983) Phosphatase production by mycorrhizal fungi. Plant Soil 71:455–462

Dighton J, Thomas ED, Latter PM (1987) Interactions between tree roots, mycorrhizas, a saprotrophic fungus and the decomposition of organic substrates in a microcosm. Biol Fertil Soils 4:145–150

Dixon RK (1988) Response of ectomycorrhizal *Quercus rubra* to soil cadmium, nickel and lead. Soil Biol Biochem 20:555–560

Dixon RK, Buschena CA (1988) Response of ectomycorrhizal *Pinus banksiana* and *Picea glauca* to heavy metals in soil. Plant Soil 105:265–272

Dixon RK, Garrett HE, Cox GS, Marx DH, Sander IL (1984) Inoculation of three *Quercus* species with eleven isolates of ectomycorrhizal fungi. I. Inoculation success and seedling growth relationships. For Sci 30:364–372

Doudrick RL, Stewart EL, Alm AA (1990) Survey and ecological aspects of presumed ectomycorrhizal fungi associated with black spruce in northern Minnesota. Can J Bot 68:825–831

Duddridge JA (1986a) The development and ultrastructure of ectomycorrhizas. IV. Compatible and incompatible interactions between *Suillus grevillei* (Klotzsch) Sing. and a number of ectomycorrhizal hosts in vitro in the presence of exogenous carbohydrate. New Phytol 103:465–471

Duddridge JA (1986b) The development and ultrastructure of ectomycorrhizas. III. Compatible and incompatible interactions between *Suillus grevillei* (Klotzsch) Sing. and 11 species of ectomycorrhizal hosts in vitro in the absence of exogenous carbohydrate. New Phytol 103:457–464

Duddridge JA, Malibari A, Read DJ (1980) Structure and function of mycorrhizal rhizomorphs with special reference to their role in water transport. Nature 287:834–836

Duddridge JA, Finlay RD, Read DJ, Söderström B (1988) The structure and function of the vegetative mycelium of ectomycorrhizal plants. III. Ultrastructural and autoradiographic analysis of inter-plant carbon distribution through intact mycelial systems. New Phytol 108:183–188

Dunstan WA, Dell B, Malajczuk N (1998) The diversity of ectomycorrhizal fungi associated with introduced *Pinus* spp. in the southern hemisphere, with particular reference to Western Australia. Mycorrhiza 8:71–79

Durall DM, Todd AW, Trappe JM (1994) Decomposition of ^{14}C-labelled substrates by ectomycorrhizal fungi in association with Douglas fir. New Phytol 127:725–729

El Karkouri K, CleyetMarel JC, Mousain D (1996) Isoenzyme variation and somatic incompatibility in populations of ectomycorrhizal fungus *Suillus collinitus*. New Phytol 134:143–153

Engel H (1996) Schmier- und Filzröhrlinge s.l. in Europe. Verlag H Engel, Weidhausen b. Coburg, Germany

Farjon A (1990) Pinaceae drawings and descriptions of the genera *Abies, Cedrus, Pseudolarix, Keteleeria, Nothotsuga, Tsuga, Cathaya, Pseudotsuga, Larix* and *Picea*. Koeltz Scientific Books, Königstein

Finlay RD (1989) Functional aspects of phosphorus uptake and carbon translocation in incompatible ectomycorrhizal associations between *Pinus sylvestris* and *Suillus grevillei* and *Boletinus cavipes*. New Phytol 112:185–192

Finlay RD (1992) Uptake and mycelial translocation of nutrients by ectomycorrhizal fungi. In: Read DJ, Lewis DH, Fitter AH, Alexander IJ (eds) Mycorrhiza in ecosystems. CAB International, Wallingford, pp 91–97

Finlay RD, Read DJ (1986a) The structure and function of the vegetative mycelium of ectomycorrhizal plants. I. Translocation of ^{14}C-labelled carbon between plants interconnected by a common mycelium. New Phytol 103:143–156

Finlay RD, Read D (1986b) The structure and function of the vegetative mycelium of ectomycorrhizal plants. II. The uptake and distribution of phosphorus by mycelial strands interconnecting host plants. New Phytol 103:157–166

Finlay RD, Ek H, Odham G, Söderström B (1988) Mycelial uptake, translocation and assimilation of N from nitrogen-15 labelled ammonium by *Pinus sylvestris* plants infected with four different ectomycorrhizal fungi. New Phytol 110:59–66

Finlay RD, Frostegård A, Sonnerfeldt A-M (1992) Utilization of organic and inorganic nitrogen sources by ectomycorrhizal fungi in pure culture and in symbiosis with *Pinus contorta* Dougl. Ex Loud. New Phytol 120:105–115

Finlay RD, Chalot M, Brun A, Söderström B (1996) Interactions in the carbon and nitrogen metabolism of ectomycorrhizal associations. In: Azcon-Aguilar C, Barea JM (eds) Mycorrhizas in integrated systems – from genes to plant development. European Commission, Brussels, pp 279–284

France RC, Reid CPP (1984) Pure culture growth of ectomycorrhizal fungi on inorganic nitrogen sources. Microb Ecol 10:187–196

Fries N (1987a) Ecological and evolutionary aspects of spore germination in higher basidiomycetes. Trans Br Mycol Soc 88:1–7

Fries N (1987b) Somatic incompatibility and field distribution of the ectomycorrhizal fungus *Suillus luteus* (Boletaceae). New Phytol 107:735–739

Fries N (1988) Specific effects of diterpene resin acids on spore germination of ectomycorrhizal basidiomycetes. Experientia 44:1027–1030

Fries N (1994) Sexual incompatibility in *Suillus variegatus*. Mycol Res 98:545–546

Fries N, Neumann W (1990) Sexual incompatibility in *Suillus luteus* and *S. granulatus*. Mycol Res 94:64–70

Fries N, Sun YP (1992) The mating system of *Suillus bovinus*. Mycol Res 96:237–238

Fries N, Serck-Hanssen K, Dimberg LH, Theander O (1987) Abietic acid, an activator of basidiospore germination in ectomycorrhizal species of the genus *Suillus* Boletaceae. Exp Mycol 11:360–363

Gardes M, Bruns TD (1996) Community structure of ectomycorrhizal fungi in a *Pinus muricata* forest: above- and below-ground views. Can J Bot 74:11572–1583

Gardes M, White TJ, Fortin JA, Bruns TD, Taylor JW (1991) Identification of indigenous and introduced symbiotic fungi in ectomycorrhizae by amplification of nuclear and mitochondrial ribosomal DNA. Can J Bot 69:180–190

Gerlitz TGM (1996) Effects of aluminium on polyphosphate mobilization of the ectomycorrhizal fungus *Suillus bovinus*. Plant Soil 178:133–140

Gerlitz TGM, Gerlitz A (1997) Phosphate uptake and polyphosphate metabolism of mycorrhizal and nonmycorrhizal roots of pine and of *Suillus bovinus* at varying external pH measured by in vivo ^{31}P-NMR. Mycorrhiza 7:101–106

Gerlitz TGM, Werk WB (1994) Investigations on phosphate uptake and polyphosphate metabolism by mycorrhized and nonmycorrhized roots of beech and pine as investigated by in vivo ^{31}P-NMR. Mycorrhiza 4:207–214

Giltrap NJ, Lewis DH (1982) Catabolic repression of the synthesis of pectin degrading enzymes of *Suillus luteus* (L. ex Fr.) S.F. Gray and *Hebeloma oculatum* Bruchet. New Phytol 90:485–493

Göransson A, Eldhuset TD (1991) Effects of aluminium on growth and nutrient uptake of small *Picea abies* and *Pinus sylvestris* plants. Trees 5:136–142

Grand LF (1968) Conifer associates and mycorrhizal syntheses of some Pacific Northwest *Suillus* spp. For Sci 14:304–312

Gruhn CM, Miller OK (1991) Effect of copper on tyrosinase activity and polyamine content of some ectomycorrhial fungi. Mycol Res 95:268–272

Hackman W, Meiander M (1979) Diptera feeding as larvae on macrofungi in Finland. Ann Zool Fenn 16:50–83

Harley JL, Smith SE (1983) Mycorrhizal symbiosis. Academic Press, London

Hart JA (1987) A cladistic analysis of conifers: preliminary results. J Arnold Arbor 68:269–307

Hartley J, Cairney JWG, Meharg AA (1997a) Do ectomycorrhizal fungi exhibit adaptive tolerance to potentially toxic metals in the environment? Plant Soil 189:303–319

Hartley J, Cairney JWG, Sanders FE, Meharg AA (1997b) Toxic interactions of metal ions (Cd^{2+} Pb^{2+}, Zn^{2+}, Sb^{3-}) on in vitro biomass production of ectomycorrhizal fungi. New Phytol 137:551–562

Haselwandter K, Berreck M, Brunner P (1988) Fungi as bioindicators of radiocesium contamination pre- and post-Chernobyl activities. Trans Br Mycol Soc 90:171–174

Hawksworth DL, Kirk PM, Sutton BC, Pegler DN (1995) Dictionary of the fungi. CAB International, Wallingford

Hedger J (1986) *Suillus luteus* on the equator. Bull Br Mycol Soc 20:53-54

Heinrich G (1993) Distribution of radiocesium in the different parts of mushrooms. J Environ Radioactiv 18:229-245

Hintikka V (1988a) On the macromycete flora in oligotrophic pine forest of different ages in south Finland. Acta Bot Fenn 136:89-94

Hintikka V (1988b) High aluminum tolerance among ectomycorrhizal fungi. Karstenia 28:41-44

Hiol Hiol F, Dixon RK, Curl EA (1994) The feeding preference of mycophagous Collembola varies with the ectomycorrhizal symbiont. Mycorrhiza 5:99-103

Ho I (1989) Acid phosphatase alkaline phosphatase and nitrate reductase activity of selected ectomycorrhizal fungi. Can J Bot 67:750-753

Høiland K (1987) A new approach to the phylogeny of the order Boletales (Basidiomycotina). Nord J Bot 7:705-718

Horton TR, Bruns TD, Cázares E (1998) Ectomycorrhizal, vesicular-arbuscular and dark septate fungal colonization of bishop pine (*Pinus muricata*) seedlings in the first five months of growth after wildfire. Mycorrhiza 8:11-18

Hutchinson LJ (1991) Description and identification of cultures of ectomycorrhizal fungi found in North America. Mycotaxon 7:705-718

Hwang SH, Chakravarty P, Chang KF (1995) The effect of two ectomycorrhizal fungi, *Paxillus involutus* and *Suillus tomentosus*, and of *Bacillus subtilis* on *Fusarium* damping off in jack pine seedlings. Phytoprotection 76:57-66

Ingestad T, Arveby AS, Kähr M (1986) The influence of ectomycorrhiza on nitrogen nutrition and growth of *Pinus sylvestris* seedlings. Physiol Plant 68:575-582

Jacobson KM, Miller OK (1992a) The nuclear status of *Suillus granulatus* spores and implications for dispersal and colonisation. Inoculum 1:37

Jacobson KM, Miller OK (1992b) Physiological variation between tree-associated populations of *Suillus granulatus* as determined by in vitro mycorrhizal synthesis experiments. Can J Bot 70: 26-31

Jacobson KM, Miller OK (1994) Postmeiotic mitosis in the basidia of *Suillus granulatus* – implications, for populations, structure and dispersal biology. Mycologia 86:511-516

Jacobson KM, Miller OK, Turner BJ (1993) Randomly amplified polymorphic DNA markers are superior to somatic incompatibility tests for discriminating genotypes in natural populations of ectomycorrhizal *Suillus granulatus*. Proc Natl Acad Sci 90:9159-9163

Jones D, Muehlchen A (1994) Effects of the potentially toxic metals, aluminium, zinc and copper on ectomycorrhizal fungi. J Environ Sci Health Part A Environ Sci Eng 29:949-966

Jonsson L (1998) Community structure of ectomycorrhizal fungi in Swedish boreal forests. Doct Thesis, Swedish University of Agricultural Sciences, Uppsala

Jonsson L, Dahlberg A, Nilsson M-C, Zackrisson O, Kårén O (1999) Ectomycorrhizal fungal communities in late-successional Swedish boreal forests and composition following wildfire. Mol Ecol 8:205-216

Kähr M, Arveby AS (1986) A method for establishing ectomycorrhiza on conifer seedlings in steady-state conditions of nutrition. Physiol Plant 67:333-339

Kåren O (1997) Effects of air pollution and forest regeneration methods on the community structure of ectomycorrhizal fungi. Doct Thesis, Swedish University of Agricultural Sciences, Uppsala

Kårén O, Högberg N, Dahlberg A, Jonsson L, Nylund J-E (1997) Inter- and intraspecific variation in the ITS region of rDNA of ectomycorrhizal fungi in Fennoscandia as detected by endonuclease analysis. New Phytol 136:313-325

Kasuya MCM, Muchovej RMC, Muchovej JJ (1990) Influence of aluminum on in-vitro formation of *Pinus caribaea* mycorrhizae. Plant Soil 124:73-78

Keller G (1992) Isozymes in isolates of *Suillus* spp. from *Pinus cembra* L. New Phytol 120:351-358

Keller G (1996) Utilization of inorganic and organic nitrogen sources by high-subalpine ectomycorrhizal fungi of *Pinus cembra* in pure culture. Mycol Res 100:989-998

Klironomos JN, Kendrick WB (1993) Research on mycorrhizas – trends in the past 40 years as expressed in the mycolite database. New Phytol 125:595-600

Knudsen H (1995) Taxonomy of the basidiomycetes in Nordic Macromycetes. Acta Univ Ups Symb Bot Ups XXX3:169–208

Kope HH, Fortin JA (1989) Inhibition of phytopathogenic fungi in vitro by cell free culture media of ectomycorrhizal fungi. New Phytol 113:57–64

Kretzer A, Bruns TD (1997) Molecular revisitation of the genus *Gastrosuillus*. Mycologia 89: 586–589

Kretzer A, Li Y, Szaro T, Bruns TD (1996) Internal transcribed spacer sequences from 38 recognised species of *Suillus sensu lato*: phylogenetic and taxonomic implications. Mycologia 88:776–785

Kropp BR, Albee S (1996) The effects of silvicultural treatments on occurrence of mycorrhizal sporocarps in a *Pinus contorta* forest: a preliminary study. Biol Conserv 78:313–318

Larsen P (1934) Undersogelser over storsvamp-vegetationen paa et vestjydsk hedeomraade. Friesia 1:157–193

Lepage BA, Currah RS, Stockey RA, Rothwel GW (1997) Fossil ectomycorrhiza from the middle Eocene. Am J Bot 84:410–412

Leyval C, Watteau F, Berthelin J, Reid CPP (1992) Production of siderophores by ectomycorrhizal fungi. In: Read DJ, Lewis DH, Fitter AH, Alexander IJ (eds) Mycorrhiza in ecosystems. CAB International, Wallingford, pp 389–390

Littke WR, Bledsoe CS, Edmonds RL (1984) Nitrogen uptake and growth in vitro by *Hebeloma crustuliniforme* and other Pacific northwest mycorrhizal fungi. Can J Bot 62:647–652

Lundeberg G (1970) Utilisation of various nitrogen sources, in particular bound soil nitrogen, by mycorrhizal fungi. Stud For Suec 79:1–95

Mabberley DJ (1997) The plant book. Cambridge University Press, Cambridge

Maijala P, Fagerstedt KV, Raudaskoski M (1991) Detection of extracellular cellulolytic and proteolytic activity in ectomycorrhizal fungi and *Heterobasidion annosum* Fr. Bref. New Phytol 117:643–648

McElhinney C, Mitchell DT (1993) Phosphatase activity of four ectomycorrhizal fungi found in a Sitka spruce Japanese larch plantation in Ireland. Mycol Res 97:725–732

Meharg AA, Cairney JWG, Maguire N (1997a) Mineralization of 2,4-dichlorophenol by ectomycorrhizal fungi in axenic culture and in symbiosis with pine. Chemosphere 34: 2495–2504

Meharg AA, Dennis GR, Cairney JWG (1997b) Biotransformation of 2,4,6-trinitrotoluene (TNT) by ectomycorrhizal basidiomycetes. Chemosphere 35:513–521

Melin E (1922). Untersuchungen über die *Larix* mycorrhiza. I. Synthese der mykorrhiza in Reinkulture. Sven Bot Tidskr 16:165–196

Melin E (1923) Experimentelle Untersuchungen über die Konstitutionen und Ökologie der Mykorrhizen von *Pinus sylvestris* L. und *Picea abies* (L.) Karst. Mykol Unters Ber 2:73–331

Melin E (1927) Studier över barrträdsplantans utveckling i råhumus. II. Mykorrhizans utbildning hos tallplantan i olika råhumusformer. Medd Sveriges Skogsförsöksanstalt Stockholm 23:433–487

Melin E, Nilsson H (1950) Transfer of radioactive phosphorus to pine seedlings by means of mycorrhizal hyphae. Physiol Plant 3:88–92

Melin E, Nilsson H (1952) Transport of labelled nitrogen from an ammonium source to pine seedlings through mycorrhizal mycelium. Sven Bot Tidskr 46:281–285

Melin E, Nilsson H (1953) Transfer of labelled nitrogen from glutamic acid to pine seedlings through the mycelium of *Boletus variegatus* (Sw.) Fr. Nature 171:134

Miller SL, Torres P, McClean TM (1993) Basidiospore viability and germination in ectomycorrhizal and saprotrophic basidiomycetes. Mycol Res 97:141–149

Miller SL, Torres P, McClean TM (1994) Persistence of basidiospores and sclerotia of ecto-mycorrhizal fungi and *Morchella* in soil. Mycologia 86:89–95

Mitchell RJ, Cox GS, Dixon RK, Garrett HE, Sander IL (1984) Inoculation of 3 *Quercus* species with 11 isolates of ectomycorrhizal fungi 2. Foliar nutrient content and isolate effectiveness. For Sci 30:563–572

Molina R, Trappe JM (1982) Lack of mycorrhizal specificity in the ericaceous hosts *Arbutus menziezii* and *Arctostaphylos uva-ursi*. New Phytol 90:495–509

Molina R, Trappe JM (1994) Biology of the ectomycorrhizal genus *Rhizopogon*. I. Host associations, host-specificity and pure culture synthesis. New Phytol 126: 653–675

Molina R, Massicotte H, Trappe JM (1992) Specificity phenomena in mycorrhizal symbioses: community-ecological consequenses and practical implications. In: Allen MF (ed) Mycorrhizal functioning – an integrative plant-fungal process. Chapman & Hall, New York, pp 357–423

Moser M (1953) Die Röhrlinge und Blätterpilze. Bd IIb/2, 1st edn. G Fischer, Stuttgart

Mousain D, Bousquet N, Polard C (1988) Comparison of phosphatase activities in ectomycorrhizal homobasidiomycetes cultured in vitro. Eur J For Pathol 18:299–309

Natarjan K, Mohan V, Ingleby K (1992) Correlation between basidiomata production and ectomycorrhizal formation in *Pinus patula* plantation. Soil Biol Biochem 24:279–280

Niini SS, Raudaskoski M (1993) Response of ectomycorrhizal fungi to benomyl and nocodazole: growth inhibition and microtubule depolymerization. Mycorrhiza 3:83–91

Niini SS, Raudaskoski M (1997) Growth patterns in non-mycorrhizal and mycorrhizal short roots of *Pinus sylvestris*. Symbiosis 25:101–114

Niini SS, Tarkka M, Raudaskoski M (1996) Tubulin and actin protein patterns in Scots pine (*Pinus sylvestris*) roots and developing ectomycorrhiza with *Suillus bovinus*. Physiol Plant 96:186–192

Nurmiaho-Lassila EL, Timonen S, Haahtela K, Sen R (1997) Bacterial colonization patterns of intact *Pinus sylvestris* mycorrhizospheres in dry pine forest soil: an electron microscopy study. Can J Microbiol 43:1017–1035

Ohenoja E (1993) Effect of weather conditions on the larger fungi at different forest sites in northern Finland in 1976–1988. PhD Thesis, Oulu University, Oulu

Olsson PA, Chalot M, Baath E, Finlay RD, Söderström B (1996) Ectomycorrhizal mycelia reduce bacterial activity in a sandy soil. FEMS Microbiol Ecol 21:77–86

Pantidou ME, Groves JW (1966) Cultural studies of Boletaceae, some species of *Suillus* and *Fuscoboletinus*. Can J Bot 44:1371–1392

Parladé J, Álvarez IF, Pera J (1996) Ability of native ectomycorrhizal fungi from northern Spain to colonize Douglas-fir and other introduced conifers. Mycorrhiza 6:51–55

Pegler DN, Young TWK (1981) A natural arrangement of the Boletales with reference to spore morphology. Trans Br Mycol Soc 76:103–146

Pomerleau R, Smith AH (1962) *Fuscoboletinus*, a new genus of Boletales. Brittonia 14:156–172

Price RA, Olsen-Stojkovich J, Lowenstein JM (1987) Relationships amoung the genera of Pinaceae: an immunological comparison. Syst Bot 12:91–97

Read DJ, Boyd R (1986) Water relations of mycorrhizal fungi and their host plants. In: Ayers PG, Boddy L (eds) Water, fungi and plants. Cambridge University Press, Cambridge, pp 287–304

Repac I (1996) Inoculation of *Picea abies* (L.) Karst. seedlings with vegetative inocula of ectomycorrhizal fungi *Suillus bovinus* (L. Fr.) O. Kuntze and *Inocybe lacera* (Fr.) Kumm. New For 12:41–54

Riffle JW (1975) Two *Aphelenchoides* species suppress formation of *Suillus granulatus* ectomycorrhizae with *Pinus ponderosa* seedlings. Plant Dis Rep 59:951–955

Rigou L, Nignard E, Plassard C, Arvieu J-C, Remy J-C (1995) Influence of ectomycorrhizal infection on the rhizosphere pH around roots of maritime pine (*Pinus pinaster* Soland in Ait.) New Phytol 130:141–147

Roldán A, Albaladejo J (1994) Effect of mycorrhizal inoculation and soil restoration on the growth of *Pinus halepensis* seedlings in a semi arid soil. Biol Fertil Soil 18:143–149

Romell LG (1939) The ecological problem of mycotrophy. Ecology 20:163–167

Rudawska M, Kieliszewska-Rokika B, Debaud J-C, Lewandowski A, Gay G (1994) Enzymes of ammonium metabolism in ectendomycorrhizal and ectomycorrhizal symbionts of pine. Physiol Plant 92:279–285

Rühling Å, Söderström B (1990) Changes in fruitbody production of mycorrhizal and litter decomposing macromycetes in heavy metal polluted coniferous forests in north Sweden. Water Air Soil Pollut 49:375–387

Sarand I, Timonen S, Nurmiaho-Lassila E-L, Koivula T, Haahtela K, Romantschuk M, Sen R (1998) Microbial biofilms and catabolic plasmid harbouring degradative fluorescent pseudomonads in Scots pine mycorrhizospheres developed on petroleum contaminated soil. FEMS Microbiol Ecol 27:115–126

Sarjala T (1990) Effect of nitrate and ammonium concentration on nitrate reductase activity in five species of mycorrhizal fungi. Physiol Plant 79:65–70

Sarjala T (1991) Effect of mycorrhiza and nitrate nutrition on nitrate reductase activity in Scots pine seedlings. Physiol Plant 81:89–94

Sarjala T (1993) Effect of ammonium on glutamate synthetase activity in ectomycorrhizal fungi and in mycorrhizal and non-mycorrhizal Scots pine seedlings. Tree Physiol 12:93–100

Schelkle M, Peterson RL (1996) Suppression of common root pathogens by helper bacteria and ectomycorrhizal fungi in vitro. Mycorrhiza 6:481–485

Sen R (1990a) Intraspecific variation in two species of *Suillus* from Scots pine *Pinus sylvestris* L. forests based on somatic incompatibility and isozyme analyses. New Phytol 114:607–616

Sen R (1990b) Isozymic identification of individual ectomycorrhizas synthesized between Scots pine *Pinus sylvestris* L. and isolates of two species of *Suillus*. New Phytol 114:617–626

Shaw TM, Dighton J, Sanders FE (1995) Interactions between ectomycorrhizal and saprotrophic fungi on agar and in association with seedlings of lodgepole pine (*Pinus contorta*). Mycol Res 99:159–165

Shishido M, Massicotte HB, Chanway CP (1996a) Effect of plant growth promoting *Bacillus* strains on pine and spruce seedling growth and mycorrhizal infection. Ann Bot 77:433–441

Shishido M, Petersen DJ, Massicotte HB, Chanway CP (1996b) Pine and spruce seedling growth and mycorrhizal infection after inoculation with plant growth promoting *Pseudomonas* strains. FEMS Microbiol Ecol 21:109–119

Singer R (1949) The genus *Gomphidius* in North America. Mycologia 41:462–489

Singer R (1951) Agaricales (mushrooms) in modern taxonomy, 1st edn. Lilloa 22:1–832

Singer R (1986) Agaricales in modern taxonomy, 4th edn. Koeltz, Königstein

Söderström B, Read DJ (1987) Respiratory activity of intact and excised ectomycorrhizal mycelial systems growing in unsterilised soil. Soil Biol Biochem19:231–236

Söderström B, Finlay RD, Read DJ (1988) The structure and function of the vegetative mycelium of ectomycorrhizal plants. IV. Qualitative analysis of carbohydrate contents of mycelium interconnecting host plants. New Phytol 109:163–166

Stenström E (1991) The effects of flooding on the formation of ectomycorrhizae in *Pinus sylvestris* seedlings. Plant Soil 131:247–250

Stenström E, Ek M (1990) Field growth of *Pinus sylvestris* following nursery inoculation with mycorrhizal fungi. Can J For Res 20:914–918

Surtiningsih T, Leyval C, Berthelin J (1992) Dissolution and immobilization of phosphorus and cadmium from rock phosphates by ectomycorrhizal fungi. In: Read DJ, Lewis DH, Fitter AH, Alexander IJ (eds) Mycorrhiza in ecosystems. CAB International, Wallingford, pp 403–404

Sutherland JR, Fortin JA (1968) Effect of the nematode *Aphelencus avenae* on some ectotrophic mycorrhizal fungi and on a red pine mycorrhizal relationship. Phytopathology 58:519–523

Termorshuizen AJ, Ket PC (1991) Effects of ammonium and nitrate on mycorrhizal seedlings of *Pinus sylvestris*. Eur J For Pathol 21:404–413

Theodorou C, Bowen GD (1970) Mycorrhizal responses of radiata pine in experiments with different fungi. Aust For 34:183–191

Theodorou C, Bowen GD (1987) Germination of basidiospores of mycorrhizal fungi in the rhizosphere of *Pinus radiata* D. Don. New Phytol 106:217–223

Thompson GW, Medve RJ (1984) Effects of aluminum and manganese on the growth of ectomycorrhizal fungi. Appl Environ Microbiol 48:556–560

Timonen S, Sen R (1998) Heterogeneity of fungal and plant enzyme expression in intact Scots pine–*Suillus bovinus* and –*Paxillus involutus* mycorrhizospheres developed in natural forest humus. New Phytol 138:355–366

Timonen S, Finlay RD, Söderström B, Raudaskoski M (1993) Identification of cytoskeletal components in pine ectomycorrhizas. New Phytol 124:83–92

Timonen S, Finlay RD, Olsson S, Söderström B (1996) Dynamics of phosphorus translocation in intact ectomycorrhizal systems: non-destructive monitoring using a beta-scanner. FEMS Microbiol Ecol 19:171–180

Timonen S, Tammi H, Sen R (1997a) Outcome of interactions between genets of two *Suillus* spp. and different *Pinus sylvestris* genotype combinations: identity and distribution of ecto-mycorrhizas and effects on early seedling growth in N-limited nursery soil. New Phytol 137:691–702

Timonen S, Tammi H, Sen R (1997b) Characterization of the host genotype and fungal diversity in Scots pine mycorrhiza from natural humus using isoenzyme and PCR-RFLP analyses. New Phytol 135:313–323

Todd NK, Rayner ADM (1980) Fungal individualism. Sci Prog Oxf 66:331–354

Trappe JM (1963) Fungus associates of ectotrophic mycorrhiza. Bot Rev 28:538–606

Treu R, Miller OK (1993) Nuclear status of two *Suillus* species. Mycologia 85:46–50

Unestam T, Sun Yu-P (1995) Extramatrical structures of hydrophobic and hydrophilic ectomycorrhizal fungi. Mycorrhiza 5:301–311

Väre H (1990) Aluminum polyphosphate in the ectomycorrhizal fungus *Suillus variegatus* Fr. O. Kunze as revealed by energy dispersive spectrometry. New Phytol 116:663–668

Väre H, Ohenoja E, Ohtonen R (1996) Macrofungi of oligotrophic Scots pine forest in northern Finland. Karstenia 36:1–18

Vares GC, Portinaro S, Trotta A, Scannerini S, Luppi-Mosca A-M, Martinotti MG (1996) Bacteria associated with *Suillus grevillei* sporocarps and ectomycorrhizae and their effects on in vitro growth of the mycobiont. Symbiosis 21:129–147

Visser S (1995) Ectomycorrhizal fungal succession in jack pine stands following wildfire. New Phytol 129:389–401

Wallander H, Nylund J-E (1992) Effects of excess nitrogen and phosphorus starvation on the extramatrical mycelium of ectomycorrhizas of *Pinus sylvestris* L. New Phytol 120:495–503

Wallander H, Wickman T, Jacks G (1997) Apatite as a P source in mycorrhizal and non-mycorrhizal *Pinus sylvestris* seedlings. Plant Soil 196:123–131

Watteau F, Berthelin J (1990) Iron solubilization by mycorrhizal fungi producing siderophores. Symbiosis 9:59–68

Watteau F, Berthelin J (1994) Microbial dissolution of iron and aluminium from soil minerals: efficiency and specificity of hydroxamate siderophores compared to aliphatic acids. Eur J Soil Biol 30:1–9

Wilkins DA (1991) The influence of sheathing ectomycorrhizas of trees on the uptake and tox-icity of metals. Agric Ecosyst Environ 35:245–260

Yakovlev EB (1987) Insect infestation of edible mushrooms in Soviet South Karelia and bioeco-logical characteristics of the pests. Acta Bot Fenn 136:99–103

Zhu H, Higginbotham KO, Danick BP, Navratil S (1988) Intraspecific genetic variability of isoen-zymes in the ectomycorrhizal fungus *Suillus tomentosum*. Can J Bot 66:588–594

Chapter 3

Laccaria

B. R. Kropp[1] and G. M. Mueller[2]

3.1
Introduction

Laccaria is a cosmopolitan and common genus of mushrooms (Agaricales). The genus has been reported from ectotrophic plant communities in every continent except Antarctica, although it is not known from native forests of Amazonian South America or Africa south of the Sahara. It is also frequently associated with planted pines and eucalypts throughout the world. Species of *Laccaria* have been reported to form ectomycorrhizal (ECM) associations with numerous tree species, including many that are of major economic importance (Table 3.1).

Both the systematics and biogeography of *Laccaria* species have been well studied, particularly in North America (Mueller 1992). Some species of *Laccaria*, notably *L. bicolor* (Maire) Orton, *L. laccata* (Scop.: Fr.) Cooke, and *L. proxima* (Boud.) Pat., are easy to manipulate under experimental conditions. Because of this, they have been used extensively in both basic and applied research on ECM associations and ECM fungi (Fries 1983; Fries and Mueller 1984; Kropp and Fortin 1988; Armstrong et al. 1989; Barrett et al. 1989, 1990; Doudrick and Anderson 1989; Gardes et al. 1990a, 1991a,b; Mueller 1991; Mueller and Gardes 1991; Mueller and Ammirati 1993; Mueller et al. 1993; Albee et al. 1996; Kropp 1997). A large number of workers have included one or more *Laccaria* species in their studies; thus it would be difficult to incorporate into this chapter every article in which an isolate of *Laccaria* was used. Rather, our emphasis is on the research done in areas where *Laccaria* species have played a key role in advancing our understanding of ECM fungi.

[1] Department of Biology, Utah State University, Logan, Utah 84322-5303, USA
[2] Department of Botany, The Field Museum, Chicago, Illinois 60605-2496, USA

Table 3.1. Genera and families of plants forming ECM with species of *Laccaria*

Pinaceae	Dipterocarpaceae	Myrtaceae
Pinus	*Dipterocarpus*	*Eucalyptus*
Larix	Fagaceae	*Leptospermum*
Picea	*Betula*	Tiliaceae
Pseudotsuga	*Fagus*	*Tilia*
Tsuga	*Nothofagus*	Salicaceae
Abies	*Quercus*	*Salix*

3.2
Taxonomy and Ecology

3.2.1
Taxonomy

Laccaria is placed within the Tricholomataceae and most modern systematists consider it to be an autonomous genus that is easily separated from other members of the family (Singer 1986). However, *Hydnangium* and *Podohydnangium* are very closely related genera that need to be included in a discussion of *Laccaria* systematics. While these are both sequestrate fungi (false truffles) that lack a stipe and lamellae and do not forcibly discharge their basidiospores, they share a distinctive type of basidiospore ornamentation. This echinulate basidiospore ornamentation is only known from *Laccaria*, *Hydnangium*, and *Podohydnangium* and indicates a close phylogenetic relationship between the three genera. Preliminary cladistic analysis of sequence data from the internal transcribed spacer (ITS) region of the rDNA repeat and the 5′ end of the large subunit rDNA supports the hypothesis that the three genera form a monophyletic group (G. M. Mueller and E. M. Pine, unpubl. data; Fig. 3.1). In fact, this analysis indicates that *Hydnangium* and *Podohydnangium* are derived from within *Laccaria*, but we feel that more data and further analyses are necessary to confirm this.

Because many *Laccaria* taxa appear similar upon superficial examination, delimitation of infrageneric taxa is often difficult. Although nearly 100 species epithets have been used for *Laccaria* worldwide, Singer (1986) recognized only 18 clearly defined species while McNabb (1972) indicated that there may be as many as 43 species worldwide. Mueller (1992) recognized 19 species for North America, with a tentative world total of 40 species. Mueller's (1992) North American monograph of the genus provides descriptions, keys, illustrations, discussion, phylogenetic information, and a classification for the genus based on cladistic analysis. While the focus of the work is on the North American taxa, it includes a tentative key to worldwide taxa and descriptions of all available type specimens. An electronic version of the North American treatment, complete with color photographs, is now available as well (Mueller

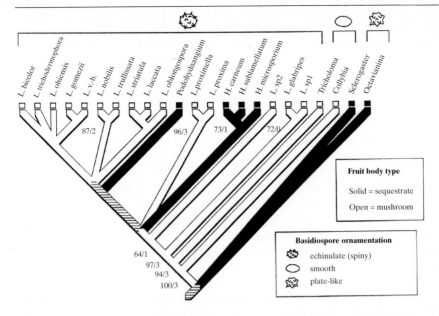

Fig. 3.1. Hypothetical relationships among species of *Hydnangium*, *Laccaria*, and *Podohyd-nangium* based on analysis of ITS and partial large subunit rDNA sequences (G. M. Mueller and E. M. Pine, unpubl. data). Numbers indicate bootstrap values and decay index, respectively. *Laccaria* sp. *1* and *2* are from Australia. Drs. J. Trappe and N. Malajczuk provided specimens of most of the Australasian taxa. While these analyses indicate that the three genera are monophyletic and of Australasian origin, inclusion of more taxa in the analyses is needed to test these hypotheses

1997). This monograph also includes references to numerous published myco-logical surveys and revisions which include descriptions and illustrations of *Laccaria* in most areas of the world.

3.2.2
Ecology and Distribution

Until the mid 1980s, most species of *Laccaria* were considered to be nearly cosmopolitan. However, detailed comparative study of macro- and micro-morphology along with interfertility tests and analysis of genetic divergence (Gardes et al. 1990a, 1991a; Mueller 1991, 1992; Mueller and Gardes 1991) have documented that species of *Laccaria* often display limited geographic ranges. Few species, except circumboreal taxa and those introduced with exotic species of *Pinus* and *Eucalyptus*, can be found in more than one or two continents, and many have much narrower distribution ranges (Mueller 1992).

Some degree of host specificity occurs within *Laccaria*. For example, *L. amethystina* Cooke and *L. ochropurpurea* (Berk.) Peck appear to be associated solely with temperate hardwoods, especially species of *Quercus* and *Fagus grandifolia*, while *L. vinaceobrunnea* G. M. Mueller has only been found under live oak, *Quercus virginiana*, in southeastern North America. *L. ohiensis* (Mont.) Singer is found associated with *Quercus* in temperate and tropical zones and with *Nothofagus* in south temperate regions. There is a suite of *Laccaria* species that are restricted to either *Eucalyptus* or *Nothofagus* in Australasia, and two of the four species reported from temperate South America are Chilean and Argentine endemics associated only with species of *Nothofagus*. Many other species of *Laccaria* show similar levels of specificity. *Hydnangium* and *Podohydnangium* are restricted to species of *Eucalyptus*, and only *H. carneum* Wallroth apud Klotzsche is routinely found outside of Australia (associated with planted *Eucalyptus*).

The three members of the *L. bicolor* complex (*L. bicolor*, *L. nobilis* Sm. apud G. M. Mueller and *L. trichodermophora* G. M. Mueller) have fairly well defined distributions. *L. trichodermophora* is found from the southeastern United States through Mexico to Colombia. *L. nobilis* is distributed in western North America and in the Great Lakes region in North America, while *L. bicolor* occurs both in Europe and North America (Mueller 1992). All three species are associated with the Pinaceae in North America, and *L. bicolor* is also associated with the Pinaceae in Europe. However, *L. trichodermophora* has been found with tropical *Quercus* in Costa Rica and Colombia and can apparently shift hosts in the absence of Pinaceae (Mueller and Strack 1992).

There are a few exceptions to the generalized statement on host specificity in the genus. *L. proxima* and *L. laccata* var. *pallidifolia* (Peck) Peck are commonly associated with both Fagaceae and Pinaceae. *L. montana* Singer and *L. pumila* Fayod are commonly collected in arctic and alpine habitats where they appear to be associated with Pinaceae (especially *Pinus*), *Salix* and *Betula*. *L. amethysteo-occidentalis* G. M. Mueller is a species from western North America that is often found associated with conifers, especially *Pseudotsuga menziesii*; however, it is also associated with *Quercus* in California.

While many *Laccaria* are encountered growing among mosses, only *L. laccata* var-*moelleri* Singer, *L. longipes* G. M. Mueller, *L. bicolor* and *L. proxima* are found commonly in *Sphagnum* bogs. *L. proximella* Singer appears to be restricted to very poor rocky soil in south temperate South America. Two species, *L. trullissata* (Ellis) Peck and *L. maritima* (Teodorowicz) Singer ex Huhtinen, occur only in sand dunes or other very sandy areas. *L. trullissata* is found in the eastern portion of North America associated with species of *Pinus*, while *L. maritima* occurs primarily in northern Europe, although there are also a few reports from Canada.

3.2.3
Population Biology

Populations of *Laccaria* species have a structure similar to that reported for a number of other Basidiomycota. They are composed of genetically distinct mycelial individuals or genets that may persist in a given area for a lengthy period of time. The *L. bicolor* genets studied by Baar et al. (1994), located in a 17-year-old *Pinus sylvestris* stand, were estimated to be between 13 and 31 years old and, although they varied in size, their maximum width was 12.5 m. Genets of *L. bicolor* occurring in a 4-year-old plantation were also studied by de la Bastide et al. (1994). The spatial distribution of the basidiomata they produced was followed for several years and used to estimate the growth rate and size of the genets. During the study, they grew at a rate of about 17–20 cm year^{-1}. Although it was not determined when they became established, they had attained a maximum size of around 2 m across by the time the plantation was 7 years old.

Some taxa of *Laccaria* are reported to act as pioneer species (Singer and Moser 1965; Watling 1977; Mason et al. 1982). Basidiomata of certain *Laccaria* species are frequently found in disturbed sites and in association with young stands, but not in mature forests (Danielson 1984; Dighton and Mason 1985; Dighton et al. 1986). Thus, *Laccaria*, at least in some situations, may play an important role in primary and secondary succession. Because of this, *Laccaria* genets are probably initiated by basidiospores. It is likely that a relatively large number of genets will begin the colonization process and that the genets best adapted to the site will gradually dominate. This pattern was observed for *Suillus bovinus* (L.: Fr.) O. Kuntze (Dahlberg and Stenlid 1990) and may be typical of pioneering ECM fungi. Although the studies of de la Bastide et al. (1994) and Baar et al. (1994) described above were carried out under entirely different circumstances, the two studies taken together indicate that *L. bicolor* follows this pattern. The work of de la Bastide et al. (1994) was done in a young plantation in which several small *L. bicolor* genets occurred. The older stand in which Baar et al. (1994) worked contained fewer and larger genets, implying that genets adapted to that site were becoming dominant.

Because species of *Laccaria* are often pioneer species existing under a wide range of conditions, they may need a mechanism for incorporating variation into established genets. Genets of ECM fungi might be able to accept genetic input from sexually or somatically compatible mycelia that vary at loci not involved in somatic interactions. Controlled studies have shown that inoculation of a single root system with basidiospores of *L. bicolor* may result in colonization of the roots by more than one fungal strain. Even when the roots have been colonized by a single strain, it is still possible for new strains to become established by means of basidiospores (de la Bastide et al. 1995b). If dikaryon–monokaryon matings happened between dikaryotic mycelia established on tree roots and germinating basidiospores, genets would have the ability to incorporate new genetic information. In one study, dikaryon–monokaryon matings between compatible *L. bicolor* strains did not occur on root systems (de la Bastide et al. 1995b). However, in another study, dikaryon–monokaryon matings readily took place in Petri dishes and occurred on the roots of inoculated seedlings (de la Bastide et al. 1995c). Gardes et al. (1990b) also found that when roots were inoculated with compatible dikaryotic and monokaryotic mycelia, new dikaryotic genotypes appeared. This demonstrates that dikaryon–monokaryon matings can occur on root systems.

The degree to which this occurs under natural conditions is unknown. However, evidence that it does occur in the field comes from a study involving genets of *L. bicolor* in a *Picea abies* plantation (de la Bastide et al. 1995a). The results of this study showed that, based on sexual incompatibility tests, two adjoining genets in the plantation shared a nucleus. This could have come about if two independent monokaryon–monokaryon pairings occurred adjacent to one another and one of the monokaryons in each pairing had identical mating type alleles. However, the most likely explanation for this is that these genets arose from a dikaryon–monokaryon pairing between an established dikaryotic mycelium and a monokaryotic mycelium initiated by a basidiospore. Considering this, genets of *L. bicolor* appear to be capable of incorporating new genetic information as they develop under field conditions.

3.2.4
Markers for Population Studies

Work on the population structure of ECM fungi or on following strains that have been released into the field requires reliable markers that allow unambiguous strain recognition. A useful marker must be relatively quick and easy to use. It should also have a degree of specificity and, depending on the work being done, the markers might need to be able to recognize fungi at the genus, species, or even strain level. For studies of ECM fungi, it should also be possible to work directly with root samples. Because *Laccaria* species are useful experimental subjects for nursery inoculations or genetic studies, a number of markers have been developed for use with members of the genus.

Somatic and sexual incompatibility genes have both been used as markers in population studies of *Laccaria* species. Since incompatibility tests are the primary means for defining genets, other methods such as isoenzyme profiles or DNA-based techniques should be tied to incompatibility tests. Baar et al. (1994) used somatic incompatibility tests to define genets of *Laccaria bicolor* in their studies on litter and humus removal. The tests are inexpensive and easy to perform, but they have the disadvantage of requiring that the fungus being studied is obtained in pure culture. The sexual incompatibility test has also been used to define genets of *Laccaria* species (de la Bastide et al. 1994, 1995a). This test is reliable and has the advantage of allowing individual nuclei to be recognized. However, like somatic incompatibility tests, it requires pure cultures of the fungus. In addition, monokaryotic cultures must be obtained by spore germination and their mating types determined. *Laccaria* spores germinate readily (Fries 1983), but the process of germinating them and determining their mating types is time consuming.

Both types of incompatibility tests allow identification of individual genets and are thus useful in studies of population structure. However, they reflect differences or similarities only in the loci that code for mating type or somatic interactions. Work using randomly amplified polymorphic DNA markers has shown that somatic compatibility between strains of *Suillus* does not necessarily indicate that they are genetically identical (Jacobsen et al. 1993). In addition, the work of Raffle et al. (1995) demonstrates that populations of *L. bicolor* defined by sexual compatibility do not always correspond to patterns revealed by randomly amplified polymorphic DNA (RAPD) markers.

Isozyme patterns permit detailed studies to be done on ECM fungi and are reliable markers for strain identification. Isozymes have been used successfully as markers for *L. bicolor*. Gardes et al. (1990b) were able to use leucine amino peptidase patterns to distinguish between strains of *L. bicolor* isolated from roots of pine seedlings grown under laboratory conditions. The patterns proved sensitive enough to distinguish between different monokaryons used in the study. They could therefore be used to study the occurrence of dikaryon–monokaryon matings and dikaryotization of monokaryotic strains associated with roots. Leucine amino peptidase patterns were also useful in studies exploring the interactions between different *L. bicolor* strains after they were inoculated onto roots of pine (de la Bastide et al. 1995b). Enzymes other than leucine amino peptidase could probably also be employed in studies of *Laccaria*. One disadvantage that limits the usefulness of isozymes is the need to grow pure cultures for protein extraction.

Markers based on DNA sequences have also been employed successfully in *Laccaria* species. Restriction fragment length polymorphisms, using either ribosomal or mitochondrial DNA as a probe, can distinguish between *Laccaria* species and often between strains within a species (Armstrong et al. 1989; Gardes et al. 1991a). A repeated DNA sequence has also been found that, when used as a probe, can discriminate between *Laccaria* and other fungal genera. It has the added advantage of being useful to discriminate between both species and strains of *Laccaria* (Sweeney et al. 1996). The main disadvantage of using either of these methods is that they require isolation of DNA from pure cultures of the fungi followed by hybridization with a labelled DNA probe. As a result, these steps are time-consuming and cannot be used to identify fungi directly from root samples.

Markers that take advantage of DNA sequence variation offer much flexibility. Both conserved and variable DNA regions can be useful in marker development. The conserved regions are useful for developing primers for amplification of DNA using the polymerase chain reaction while the variable regions allow differentiation between fungi. Ribosomal DNA sequences have been used most often to create markers for *Laccaria* species because of their high copy number and the availability of both conserved and variable regions.

Gardes et al. (1991b) were the first to amplify and identify fungal DNA directly from ECM roots. Soon afterwards, Henrion et al. (1992) found that it was possible to amplify the entire ribosomal repeat unit and both spacer regions using the polymerase chain reaction. When the amplification products were digested with restriction enzymes, the resulting band pattern allowed them to distinguish between *Laccaria* species; however, this technique could not unambiguously identify all strains within *Laccaria* species. These workers found that when the small intergenic spacer was amplified and digested separately it could be used as a marker to follow a strain of *Laccaria* introduced into the field. Especially useful was the fact that it could be used to identify the strain directly from root samples rather than relying on pure cultures (Henrion et al. 1994). The large intergenic spacer of both *L. proxima* and *L. bicolor* also has potential for use as a marker and contains enough variation to clearly distinguish between strains of these species (Selosse et al. 1996). Albee et al. (1996) showed that the large intergenic spacer region of *L. proxima* could be amplified directly from ECM.

Both of these spacer regions are known to vary in size and for their sequences. Selosse et al. (1996) demonstrated that recombination may occur in the large intergenic spacer region by crossing over during meiosis. They also found that the size of the small intergenic spacer may vary between monokaryotic strains. This is due to a deletion that sometimes occurs in a series of TAACCC repeats located within the small intergenic spacer. Although meiotic recombination occurs in the large intergenic spacer, neither Selosse et al. (1996) nor Albee et al. (1996) found evidence of somatic recombination. Thus, the spacer sequences appear to remain stable as long as meiosis does not occur, and these regions appear to have good potential for use in population genetics of *Laccaria* species. If the haplotypes in a population were well characterized, it would be possible to study interbreeding in the field. By following specific polymorphisms in the large intergenic spacer or perhaps the TAACCC deletion in the small intergenic spacer, it should also be possible to learn whether introduced strains mix with indigenous populations.

RAPD markers have proven useful in population studies of *Laccaria* species (de la Bastide et al. 1994). RAPD markers take advantage of many more loci than the ribosomal repeats and are able to separate closely related strains. In addition, they are quick and relatively easy to generate. Two disadvantages of using RAPD markers are that they cannot be used to identify strains directly from root samples and that they may not always be reproducible. However, Tommerup et al. (1995) showed that if the components and temperature profile of the PCR reaction were carefully standardized, RAPD markers could be reliably and repeatedly used as markers for *Laccaria*.

3.3
ECM Formation

3.3.1
Host–Fungus Specificity

Mueller (1992) discusses results of in vitro ECM synthesis studies employing the growth pouch technique of Fortin et al. (1983). These studies were done to characterize particular species of *Laccaria* and to document their ability to form ECM with select tree species under laboratory conditions. ECM were synthesized between several *Laccaria* species [*L. amethysteo-occidentalis, L. bicolor, L. laccata* var. *pallidifolia, L. proxima,* and *L. striatula* (Peck) Peck] and the following North American trees: *Picea sitchensis, Pinus ponderosa, Pinus resinosa,* and *Pseudotsuga menziesii.* ECM were also synthesized between *L. trullissata* and *P. resinosa* as well as several South American isolates of *L. ohiensis* and seedlings of the southern beech, *Nothofagus obliqua.* This work, along with a survey of reports in the literature, indicates that *Laccaria* species form ECM with a wide range of hosts (Table 3.1). The ECM formed with different tree species varied in their macromorphology (e.g. degree of branching), but their micromorphology was similar. All of the ECM that were studied exhibited a well-defined mantle that was 15–60 μm thick and consisted of clamped, morphologically undifferentiated hyphae that were tightly interwoven. The Hartig net was consistently well developed and extended through 70–100% of the cortex (Mueller 1992).

3.3.2
Production of Sporocarps

Not only do the commonly studied species of *Laccaria* grow well on artificial media, some fruit spontaneously after inoculation onto suitable hosts (Molina and Chamard 1983; Gagnon et al. 1987). Fruiting of *L. bicolor* in association with a host may also be controlled artificially by manipulating physical parameters. Godbout and Fortin (1990) found that lowering P and N levels and decreasing the photoperiod enhanced fruiting of *L. bicolor.* However, lowering the temperature from 24 °C day/18 °C night to 18 °C day/12 °C night decreased fruiting. Another factor that affects fruiting of *L. bicolor* is the rate of photosynthesis of the host plant. Lamhamedi et al. (1994) demonstrated that a clear correlation exists between the basidiome biomass produced and net photosynthesis of the host seedlings.

The host that the fungus is associated with may also affect basidiome production. Association with *Pinus monticola* resulted in more fruiting and faster fruiting in *L. bicolor* than association with white spruce or *Pinus taeda* (Godbout and Fortin 1990). Molina (1982) reported that fruiting of *L. laccata* occurred more rapidly in association with *P. ponderosa* and *P. menziesii* than with *P. sitchensis* and *Tsuga heterophylla.* Even differences between clones or

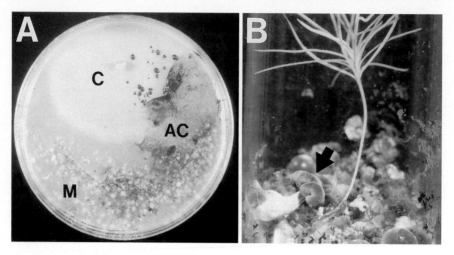

Fig. 3.2. A Method for germination of basidiospores of *Laccaria bicolor* showing an actively growing *L. bicolor* colony (*C*), activated charcoal (*AC*) scattered on the agar surface, and monokaryotic colonies (*M*) growing from germinating basidiospores. **B** Sterile basidiome (*arrow*) of *L. bicolor* produced by a monokaryotic culture

provenances of the same species have been shown to affect fruiting of other ECM fungi (Last et al. 1984).

3.3.3
Sexual Incompatibility and Behavior of Monokaryotic Strains

Laccaria species have been used as model organisms for work on the genetics of ECM formation because they grow readily and their sexual stage can be produced. In addition, haploid cultures can be readily obtained by germinating basidiospores. The method described by Fries (1983) has proven effective in isolating monokaryons from germinating basidiospores of some *Laccaria* species. To germinate *Laccaria* basidiospores using this method, a clean spore print is obtained and spores are spread out onto an agar plate containing activated charcoal and an actively growing *Laccaria* colony (Fig. 3.2a). Alternatively, an aseptically germinated conifer seedling may be placed on the agar to stimulate spore germination. Monokaryons are isolated by subculturing hyphae produced from germinating spores that lack clamp connections.

The sexual incompatibility system of *L. bicolor* and *L. laccata* is bifactorial and multiallelic (Fries and Mueller 1984; Kropp and Fortin 1988; Doudrick and Anderson 1989). Having a multiallelic incompatibility system enhances the potential for outbreeding and thus helps maintain populations with genetic diversity. Estimates were made for the number of alleles present in *L. laccata* populations across north-central North America by Doudrick et al. (1990). They estimated that 17 alleles are present for the A factor and that 18 alleles are present for the B factor. The population of *L. laccata* in the region that they studied appeared to be freely interbreeding and was calculated to have an outbreeding potential of 88.6%.

Similar studies have been done on the mating system of *L. bicolor*, but estimates of the number of alleles vary. Doudrick and Anderson (1989) predicted an infinite number of alleles for the B factor and 23 alleles for the A factor. The study of de la Bastide et al. (1995a) was limited to the province of Quebec, but they estimated 10 alleles for the A factor and 9 for the B factor. Raffle et al. (1995) used a larger sample to predict 45 alleles for the A factor and 24 for the B factor. The estimates for the outbreeding potential of *L. bicolor* vary from 93 to 85.7% (de la Bastide et al. 1995a; Raffle et al. 1995).

Monokaryotic isolates of *Laccaria* species vary greatly in their properties. Wei et al. (1995) found that *L. bicolor* monokaryons differed in their mycelial growth on artificial medium. They also found that monokaryons of *L. bicolor* varied in their ability to produce indole-3-acetic acid (IAA) and noted that, on average, monokaryons produced much less IAA than the parental dikaryon. Monokaryotic strains of *L. bicolor* also differ in ECM formation and root colonization, and sibling monokaryotic strains vary in a number of ways. Some sibling monokaryons colonize roots as well as their parental dikaryon while others appear to be incapable of ECM formation (Kropp et al. 1987). Different isolates produced mantles which vary in thickness and density. The ECM also vary for the number of cells penetrated by the Hartig net (Wong et al. 1989). Root colonization by dikaryotic strains, produced by crossing compatible monokaryons, differs in the same ways as monokaryotic strains and different strains vary in the amount of time it takes for mantle development to occur (Wong et al. 1990b).

Although a great deal of variation exists in the behavior of *Laccaria* monokaryons, they generally grow and form ECM normally. Interestingly, monokaryons of *L. bicolor* appear to carry the genetic information necessary for initiating basidiomes. Kropp and Fortin (1988) report that one monokaryon formed primordia when in association with a host plant. In one case, this strain produced a basidiome with a stipe and pileus that lacked hymenial structures (Fig. 3.2b). Monokaryotic fruiting has been reported for other basidiomycota and, in the case of *Agrocybe aegerita* (Briganti) Singer, two genes have been shown to control initiation and differentiation of basidiomes. Depending on the segregation of these genes, monokaryons of this species can produce basidiomes with two-spored basidia, produce only primordia, or remain mycelial (Esser and Meinhardt 1977). The genetic control of fruiting in *L. bicolor* is unknown, but the production of sterile basidiomes by a monokaryotic strain indicates that similar mechanisms may be involved.

3.3.4
Genetics of ECM Formation

Because fruiting of *L. bicolor* can be achieved, it is possible to investigate the inheritance of characteristics relevant to ECM formation and function. The evidence currently available indicates that the ability to form ECM is poly-genically controlled (Kropp 1997). This evidence comes from a study that was carried out using both haploid and dikaryotic progeny of crosses between compatible monokaryons that differed in their ability for ECM colonization. Although a relatively small portion of the variation in ECM colonization was under genetic control, ECM formation for both the dikaryotic and monokary-otic progeny was continuously distributed. This indicates that ECM coloniza-tion is under polygenic control.

It is logical that ECM colonization would be polygenically modulated. Very heavy colonization would divert too many carbohydrates to the fungus and has been shown to result in decreased plant size (Kropp and Fortin 1988; Colpaert et al. 1992). In this case, the fungal partner could act more like a par-asite than a symbiont. On the other hand, poor colonization would provide little benefit to the host. Thus, stabilizing selection may keep the relationship balanced at an optimal level that would provide benefits to both partners but prevent the fungus from draining host carbohydrates (Fig. 3.3). ECM colo-nization has also been shown to be under polygenic control by *Pisolithus tinctorius* (Pers.) Coker and Couch. (Rosado et al. 1994), and a number of additional factors important to ECM functioning have been shown to be under polygenic control. For example, production of indole-3-acetic acid and glutamate dehydrogenase is polygenically controlled in *Hebeloma cylin-drosporum* Romagnesi (Gay and Debaud 1987; Wagner et al. 1988). In the case of *L. bicolor*, acid phosphatase activity appears to be under polygenic control (Kropp 1990b).

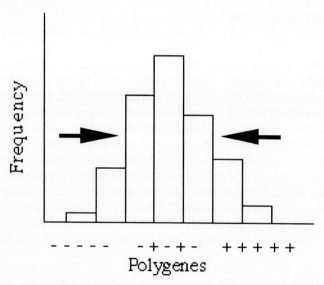

Fig. 3.3. Hypothetical distribution of polygenes that represent an increase (+) or decrease (−) in mycorrhizal colonization. The net interaction between + and − polygenes would balance the relationship (as indicated by *arrows*) at a beneficial level. (After Kropp and Anderson 1994)

Cytoplasmic factors do not appear to be involved in ECM colonization by *L. bicolor*. In the one instance where this has been examined, a cross between an ECM monokaryon of *L. bicolor* and one that was essentially non-ECM resulted in a non-ECM dikaryon (Kropp 1990a). To determine whether cytoplasmic factors were involved, the two component nuclei were re-isolated from this dikaryon in protoplasts and regenerated to produce monokaryotic cultures. The nuclei in the monokaryons were identified by their mating type alleles and the resulting monokaryons were tested for their ability to form ECM. The monokaryons containing ECM nuclei regained their ability to form ECM while those that contained the poorly ECM nuclei did not. Since these nuclei had been isolated from the same cytoplasm, cytoplasmic factors did not appear to influence ECM colonization by these strains. When these new monokaryons were re-crossed with one another, the resulting dikaryons were again non-ECM, indicating that the non-ECM state resulted from interactions between the nuclei.

Future research might benefit from identifying the locations of genes that regulate characters important to ECM formation and functioning. A linkage map has been constructed for *L. bicolor* using RAPD markers by Doudrick et al. (1995). Their markers covered roughly half of the genome of this species and were distributed among 15 linkage groups. Although it has not yet been determined whether the linkage groups correspond to chromosomes, an interesting finding of this study is that a linkage exists between the mating type loci in *L. bicolor*. This might explain the unequal distribution of mating types in germinated spores of *L. bicolor* reported by some authors (Fries and Mueller 1984; Kropp et al. 1987) as opposed to the equal proportion that would result from random segregation.

3.3.5
Molecular Genetics

Our understanding of the ECM symbiosis could be greatly enhanced by manipulating the genes that regulate the symbiosis. ECM fungi are relatively difficult to manipulate at a molecular level, but enormous advances have been made in understanding the biology of other organisms using molecular techniques. Progress is now being made with several ECM fungi. Introduction of genetic material into fungi may be accomplished by transformation using protoplasts. Protoplasts made from monokaryotic cultures of ECM fungi could serve as an unlimited supply of uninucleate propagules that might be used in mutagenesis. They may also be used in cell fusion experiments or in studies of cytoplasmic factors. Methods are available for preparing and regenerating protoplasts of a number of ECM fungi, but species of *Hebeloma* and *Laccaria* appear to be best suited for research involving protoplasts (Barrett et al. 1989). Protoplast release from mycelia of *Laccaria* species is strongly affected by the age of the mycelium and by the temperature and length of incubation. In addition, different strains of *L. bicolor* and *L. laccata* appear to vary significantly in their ability to produce protoplasts (Kropp and Fortin 1986; Barrett et al. 1989).

Protoplasts of *L. laccata* have been successfully transformed with a gene that confers resistance to hygromycin B and the transforming DNA was shown to be integrated into the *Laccaria* genome (Barrett et al. 1990). In this case, a promotor and transcription terminator from *Aspergillus nidulans* (Eidam) Winter were used, indicating that these are also functional in members of the basidiomycota such as *Laccaria*. Bills et al. (1995) report that particle bombardment can also be used to transform both *Paxillus involutus* (Batsch) Fr. and *L. bicolor*. The ability to transform ECM fungi creates possibilities for studying the genetic regulation of ECM symbioses by gene disruption, site-directed mutagenesis, or introduction of functional genes.

No research has yet been published on gene expression during ECM formation by *Laccaria* species. However, research has been done on changes in protein synthesis that occur during ECM formation by other species of fungi (e.g. Hilbert and Martin 1988). This work indicates that, at least with some fungus–host combinations, enormous alterations in patterns of protein synthesis occur when ECM form. Symbiosis-related proteins have been shown to exist and work towards understanding their function has been initiated (Martin and Tagu 1995). Similar studies could readily be done using species of *Laccaria*. This, in combination with the ability to study inheritance and to transform *Laccaria* species, sets the stage for rapid development in our understanding of how ECM symbioses formed by members of the genus are regulated.

3.3.6
ECM Development

Work on other fungi has suggested that ECM species begin colonization with attachment to the root surface by means of extracellular material (Piché et al. 1983; Bonfante-Fasalo 1987). The work carried out thus far on *Laccaria* species supports this. Wong et al. (1990a) observed that the hyphae of *L. bicolor* initially became embedded into the root mucilage but that the fungus did not appear to degrade the mucilage. Lectins may aid in the attachment process by recognizing extracellular sugar residues in polysaccharides or glycoproteins. Lei et al. (1991) found that concanavalin A binding sites were abundant on the extracellular materials at the hypha–root interface formed by an aggressive ECM strain of *L. bicolor*, but that labelling was weak in the case of a poor ECM strain. The extracellular material appears to be fibrillar in some cases, but this appears to be variable since fibrils were either not seen or rarely observed by Kottke (1997) and Wong et al. (1990a).

An ultrastructural study of ECM colonization by *L. amethystea* (Bull.) Murrill (= *L. amethystina*) showed that the hyphal attachment was nonspecific (Kottke 1997). Hyphae attached to root hairs, cortical cells, and root cap cells. Adhesive pads were formed at the tips of hyphae contacting the root surface, followed by enzymatic digestion of the cuticle in the contact zone (Kottke 1997). After penetration, the hyphae of *L. amethystea* branched and grew between the cortical cells to form a typical Hartig net.

Relatively little is known about initial signalling processes or the molecular events involved in ECM formation by *Laccaria* species. We know that changes in protein expression occur during ECM formation by other fungi (Martin and Tagu 1995), but the details are not yet clear. Tagu and Martin (1996) reported increased levels of certain symbiosis-related proteins, including hydrophobins, during ECM formation and proposed that they may be involved in both adhesion of the fungus to root surfaces and subsequent colonization of roots. A study by Münzenberger et al. (1995) found that ECM formed between *Laccaria amethystea* and *Larix decidua* have much lower amounts of soluble phenolic compounds. These authors proposed that, since phenolics are often inhibitory to pathogenic fungi (Jorgensen 1961; Sylvia and Sinclair 1983a), reductions in phenolics may be a prerequisite for ECM development by this fungus. The underlying mechanism for this is unclear, but altered protein expression like that reported by Hilbert and Martin (1988) might be involved since the activity of enzymes involved in altering the structure of phenolic compounds is changed in ECM rootlets (Münzenberger et al. 1997). These enzymatic changes could result in the decreased levels of soluble phenolics that these workers found in ECM roots. On the other hand, a histochemical study of ECM formed by *L. laccata* showed an overall increase in accumulated phenolic materials compared to ECM roots (Sylvia and Sinclair 1983a). Thus, it might be that such changes depend on the interactions in a specific host–fungus combination or the stage of ECM development.

3.4
Host Plant Growth Responses and Fungus-Derived Benefits

3.4.1
Nursery and Field Application

Species of *Laccaria* have been used in a number of studies designed to evaluate the survival and growth of outplanted ECM tree seedlings. The response of outplanted seedlings to inoculation may be positive, neutral, or negative depending on environmental factors or on the *Laccaria* species or host plant used. Le Tacon et al. (1988) reported that outplanted tree seedlings often respond positively to inoculation with *L. bicolor* and *L. laccata* but that the response depends on the site and may be best when indigenous fungi are either depleted or not optimal for the host. Browning and Whitney (1992a) found that growth of *Pinus banksiana* planted on a stony loam soil was increased by inoculation with both *L. bicolor* and *L. proxima*. However, on a sandy site, only *L. proxima* improved growth. For *Picea mariana*, inoculation with *L. proxima*, but not *L. bicolor*, sometimes increased growth. In fact, inoculation with *L. bicolor* decreased shoot dry weight by the end of the second growing season on one site (Browning and Whitney 1992a). In a similar study, inoculation with *L. proxima* and *L. bicolor* had no effect on growth of either *P. banksiana* or *P. mariana*, but it did improve the percent survival of *P. banksiana* planted on sandy soil (Browning and Whitney 1993). Shaw et al. (1987) found that the survival and growth of *P. sitchensis* was not improved by inoculation with *L. laccata*.

At times, seedling response to inoculation with *Laccaria* species can be striking. Inoculation of *Pseudotsuga menziesii* seedlings with *L. bicolor* gave impressive growth increases 2 years after outplanting (Villeneuve et al. 1991). In this study, shoot dry weight increased by 239% and the number of root tips and root dry weight also increased by 142 and 160% respectively. Garbaye and Churin (1996) noted that inoculation with *L. laccata* also stimulated the growth of *Tilia tomentosa* used as ornamentals in an urban setting. Even though the trees were 8 years old when planted and were not inoculated until outplanting, shoot growth was improved by about 44 cm over a 4-year period by inoculation. Another benefit was that *L. laccata* sometimes prolonged leaf retention during the autumn. The diameter growth of *T. tomentosa* was improved by inoculation with other ECM fungi but not by *L. laccata*.

Although rather large increases in seedling growth due to inoculation with *Laccaria* were noted by Villeneuve et al. (1991), the increases reported by Browning and Whitney (1992a) were relatively small. The largest growth increase due to inoculation that they observed was only 3.4 cm compared to their controls. If measurements were made over longer periods, greater differences might have been observed between treatments. In this context, even small gains in seedling response, especially survival, due to inoculation may make it economically feasible to inoculate nurseries with ECM fungi (Kropp and Langlois 1990).

Responses of seedlings in the nursery to inoculation with *Laccaria* species also vary. As often happens after outplanting, seedlings in the nursery sometimes respond to inoculation by increased growth. At other times, growth may be unaffected or even decreased after inoculation. The growth of container-grown *P. banksiana*, *Quercus rubra*, *P. menziesii*, and *P. ponderosa* seedlings either decreased or showed no change as a result of inoculation with *Laccaria* species (Molina 1982; Molina and Chamard 1983; Gagnon et al. 1987, 1991). Browning and Whitney (1992a, 1993) reported that *P. banksiana* and *P. mariana* either showed no growth response or minor changes due to inoculation with *L. bicolor* and *L. proxima*. Le Tacon et al. (1988) showed that several tree species inoculated with *L. laccata* grew better under nursery conditions. *Pseudotsuga menziesii*, *Pinus sylvestris* and *Picea abies* were larger than uninoculated seedlings after 2 years. However, when *Larix* seedlings were inoculated, the growth stimulation lasted only the first year. Inoculation of nursery beds with *L. proxima* can greatly increase emergence of *P. sitchensis* seedlings (Ingleby et al. 1994). Ingleby et al. (1994) applied inoculum as a layer just under the surface of the nursery beds before the seed were sown. Application of inoculum in both peat-vermiculite and alginate beads increased emergence. In some cases, however, when fertilizer was applied in addition to inoculum emergence was decreased.

The level of fertilization also affected ECM formation and seedling response in other studies. Browning and Whitney (1992b) observed that ECM colonization of *P. banksiana* and *P. mariana* was sharply reduced at high P levels. In their study, both tree species produced greater shoot dry weights in response to inoculation with *L. bicolor*, but their responses differed depending on fertility treatments. Inoculated *P. mariana* seedlings showed increased dry weight only at lower levels of P, while *P. banksiana* weights increased at both lower and higher P levels. Molina and Chamard (1983) found that different fertility treatments had no effect on ECM colonization by *L. laccata*. However, increased N levels dramatically reduced ECM formation by *L. bicolor*. At N levels of 28.7 mg per seedling per season, *L. bicolor* only colonized about 5% of the available roots but seedling shoot growth increased. Work has also been done to learn how *Laccaria* species affect host interactions with other environmental factors. In one study, *L. laccata* was found to have a minimal effect on drought tolerance of both *P. banksiana* and *P. mariana* (Boyle and Hellenbrand 1991). In another, it improved the growth of *Eucalyptus globulus* as pH increased. *L. laccata* also appears to provide some protection against certain organic components of landfill leachate (Tosh et al. 1993; Thomson et al. 1996).

Although responses vary, most of the work measuring the effect of inoculation with *Laccaria* species on seedlings under nursery conditions indicates that inoculation usually has either a negative or neutral impact on seedling growth. Most of the studies involving outplanted seedlings indicate that increased growth or survival can be expected after inoculation with *Laccaria*.

Perhaps because they are early-stage fungi, *Laccaria* species are competitive once they become associated with seedlings. Browning and Whitney (1993) noted that when either *L. proxima* or *L. bicolor* were inoculated onto seedlings grown under nursery conditions, colonization by indigenous *Thelephora terrestris* Ehrh.: Fr. was almost eliminated. The presence of *Laccaria* species also appears to inhibit root colonization by indigenous fungi after outplanting. When inoculated seedlings are planted in the field, they remain associated with *Laccaria* while indigenous fungi readily colonize uninoculated seedlings, although the degree to which *Laccaria* inhibits colonization by other fungi appears to depend on site characteristics (Browning and Whitney 1993). Work using either incompatibility tests or RAPD markers indicates that a given strain of *L. bicolor* can remain associated with the roots of inoculated seedlings for at least 2.5 years (Buschena et al. 1992; Henrion et al. 1994).

3.4.2
Interactions with Other Organisms

One of the benefits attributed to ECM fungi is an increased resistance to root diseases (Smith and Read 1997). For example, a number of ECM fungi have been shown to either inhibit pathogen growth in pure culture or reduce the incidence of root disease on pine seedlings (Marx 1969; Duchesne et al. 1988). Species of *Laccaria* have also been shown to be effective in reducing the incidence of root disease. Stack and Sinclair (1975) were the first to report that *L. laccata* provided protection against *Fusarium* root rot on *P. menziesii* seedlings in a controlled environment. Further work by Sinclair et al. (1975) showed that *L. laccata* exhibited some protective influence in nursery beds as well as under laboratory conditions. However, protection was only provided by the fungus when the beds were fumigated using methyl bromide and then only for the first 7 weeks after planting. Interestingly, ECM formation did not appear to be necessary in either of these studies for the protective influence to occur.

In order to explain why *L. laccata* provided protection against *Fusarium oxysporum* Schlechtend.: Fr., Sylvia and Sinclair (1983b) examined the interaction between the two fungi in pure culture. Under certain conditions, *L. laccata* slowed mycelial growth of *Fusarium* on agar medium and culture filtrates delayed germination of *Fusarium* spores. These workers also found that ECM formed by *L. laccata* on *P. menziesii* seedlings had higher accumulations of phenolics and that less *Fusarium* infection occurred in roots with increased phenolic contents. Thus, protection against pathogens by *L. laccata* appears to be at least partially due to chemical activity.

Interactions between *Laccaria* species and soil-inhabiting bacteria can be either positive or negative, but the overwhelming majority of reports are of positive interactions. Certain bacteria act synergistically with *Laccaria* species to inhibit the growth of *F. oxysporum*, while others do not (Schelkle and Peterson 1996). One recent study showed that bacterial activity, as measured by thymidine and leucine incorporation, could be decreased by mycelium of *L. proxima* in the absence of ECM roots. However, bacterial counts were not significantly affected and when roots ECM with *L. bicolor* were assayed, both bacterial activity and counts could increase (Olsson et al. 1996).

Inoculation of seedlings with *Laccaria* along with *Pseudomonas* species isolated from ECM often results in large increases in percent ECM colonization and in seedling growth (Duponnois and Garbaye 1991; Duponnois et al. 1993; Garbaye 1994). *P. menziesii* seedlings inoculated with *L. laccata* and selected bacteria, called mycorrhization helper bacteria (MHB), showed better ECM formation and seedling growth than those inoculated with *Laccaria* alone or uninoculated controls (Garbaye et al. 1990; Duponnois et al. 1993). In fact, since this was done in non-fumigated nursery soil, these authors suggest that some strains of *Pseudomonas fluorescens* have the potential to serve as an alternative to soil fumigation.

Some MHB exhibit fungus specificity, increasing ECM formation by *L. laccata* and decreasing it for *H. cylindrosporum* (Duponnois et al. 1993). They do not appear to exhibit host specificity since both *Quercus* and *P. menziesii* seedlings inoculated with the same strains of *L. laccata* and *Pseudomonas* showed improved ECM formation (Garbaye and Duponnois 1992; Garbaye et al. 1992; Duponnois et al. 1993). Frey et al. (1997) provided evidence suggesting that *L. bicolor* exerts some selection pressure for strains of bacteria that metabolize trehalose.

MHB are effective in very low concentrations and populations may decline below the limit of detection as quickly as 19 weeks after treatment. The bacteria do not appear to be endophytic or true rhizobacteria, but they may attach to the hyphal wall of *L. bicolor* and occur in the mantle of ECM (Garbaye 1994; Frey-Klett et al. 1997). The mechanisms by which MHB function are not understood, but Garbaye (1994) proposed that a number of mechanisms could underlie the effect. Thus, MHB might facilitate root-fungus recognition, make roots more receptive to the fungus, stimulate fungal growth and propagule germination, or modify rhizosphere soil properties. In addition, Li and Castellano (1987) reported the isolation of an N-fixing *Azospirillum* species from sporocarps of *L. laccata* and posed the question of whether N fixed by these bacteria contributes to the nutritional requirements of the fungus.

References

Albee SR, Mueller GM, Kropp BR (1996) Polymorphisms in the large intergenic spacer of the nuclear ribosomal repeat identify *Laccaria proxima* strains. Mycologia 88:970–976

Armstrong JL, Fowles NL, Rygiewicz PT (1989) Restriction fragment length polymorphisms distinguish ectomycorrhizal fungi. Plant Soil 116:1–7

Baar J, Ozinga WA, Kuyper ThW (1994) Spatial distribution of *Laccaria bicolor* genets reflected by sporocarps after removal of litter and humus layers in a *Pinus sylvestris* forest. Mycol Res 98:726–728

Barrett V, Lemke PA, Dixon RK (1989) Protoplast formation from selected species of ectomycorrhizal fungi. Appl Microbiol Biotechnol 30:381–387

Barrett V, Dixon RK, Lemke PA (1990) Genetic transformation of a mycorrhizal fungus. Appl Microbiol Biotechnol 33:313–316

Bills SN, Richter DL, Podila GK (1995) Genetic transformation of the ectomycorrhizal fungus *Paxillus involutus* by particle bombardment. Mycol Res 99:557–561

Bonfante-Fasolo P, Perotto S, Testa B, Faccio A (1987) Ultrastructural localization of cell surface sugar residues in ericoid mycorrhizal fungi by gold-labeled lectins. Protoplasma 139:25–35

Boyle D, Hellenbrand KE (1991) Assessment of the effect of mycorrhizal fungi on drought tolerance of conifer seedlings. Can J Bot 69:1764–1771

Browning MHR, Whitney RD (1992a) Field performance of black spruce and jack pine inoculated with selected species of ectomycorrhizal fungi. Can J For Res 22:1974–1982

Browning MHR, Whitney RD (1992b) The influence of phosphorus concentration and frequency of fertilization on ectomycorrhizal development in containerized black spruce and jack pine seedlings. Can J For Res 22:1263–1270

Browning MHR, Whitney RD (1993) Infection of containerized jack pine and black spruce by *Laccaria* species and *Thelephora terrestris* and seedling survival and growth after outplanting. Can J For Res 23:330–333

Buschena CA, Doudrick RL, Anderson NA (1992) Persistence of *Laccaria* spp. as ectomycorrhizal symbionts of container-grown black spruce. Can J For Res 22:1883–1887

Colpaert JV, Van Assche JA, Luijtens K (1992) The growth of the extramatrical mycelium of ectomycorrhizal fungi and the growth response of *Pinus sylvestris* L. New Phytol 120:127–135

Dahlberg A, Stenlid J (1990) Population structure and dynamics in *Suillus bovinus* as indicated by spatial distribution of fungal clones. New Phytol 115:487–493

Danielson RM (1984) Ectomycorrhizal associations in jack pine stands in northeastern Alberta. Can J Bot 62:932–939

De La Bastide PY, Kropp BR, Piché Y (1994) Spatial distribution and temporal persistence of discrete genotypes of the ectomycorrhizal fungus *Laccaria bicolor* (Maire) Orton. New Phytol 127:547–556

De La Bastide PY, Kropp BR, Piché Y (1995a) Population structure and mycelial phenotypic variability of the ectomycorrhizal basidiomycete *Laccaria bicolor* (Maire) Orton. Mycorrhiza 5:371–379

De La Bastide PY, Kropp BR, Piché Y (1995b) Mechanisms for the development of genetically variable mycorrhizal mycelia in the ectomycorrhizal fungus *Laccaria bicolor*. Appl Environ Microbiol 61:3609–3616

De La Bastide Y, Kropp BR, Piché Y (1995c) Vegetative interactions among mycelia of *Laccaria bicolor* in pure culture and in symbiosis with *Pinus banksiana*. Can J Bot 73:1768–1779

Dighton J, Mason PA (1985) Mycorrhizal dynamics during forest tree development. In: Moore D, Casselton IA, Wood DA, Frankland JC (eds) Development biology of higher fungi. Cambridge University Press, Cambridge, pp 117–139

Dighton J, Poskitt JM, Howard DM (1986) Changes in occurrence of basidiomycete fruit bodies during forest stand development with specific reference to mycorrhizal species. Trans Br Mycol Soc 87:163–171

Doudrick RL, Anderson NA (1989) Incompatibility factors and mating competence of two *Laccaria* spp. (Agaricales) associated with black spruce in northern Minnesota. Phytopathology 79:694–700

Doudrick RL, Furnier GR, Anderson NA (1990) The number and distribution of incompatibility alleles in *Laccaria laccata* var. *moelleri*. Phytopathology 80:869–872

Doudrick RL, Raffle VL, Nelson CD, Furnier GR (1995) Genetic analysis of homokaryons from a basidiome of *Laccaria bicolor* using random amplified polymorphic DNA (RAPD) markers. Mycol Res 99:1361–1366

Duchesne LC, Peterson RL, Ellis BE (1988) Interaction between the ectomycorrhizal fungus *Paxillus involutus* and *Pinus resinosa* induces resistance to *Fusarium oxysporum*. Can J Bot 66:558–562

Duponnois R, Garbaye J (1991) Effect of dual inoculation of Douglas-fir with the ectomycorrhizal fungus *Laccaria laccata* and mycorrhization helper bacteria (MHB) in two bare-root nurseries. Plant Soil 138:169–176

Duponnois R, Garbaye J, Bouchard D, Churin JL (1993) The fungus-specificity of mycorrhization helper bacteria (MHBs) used as an alternative to soil fumigation for ectomycorrhizal inoculation of bare-root Douglas-fir planting stocks with *Laccaria laccata*. Plant Soil 157:257–262

Esser K, Meinhardt F (1977) A common genetic control of dikaryotic and monokaryotic fruiting in the basidiomycete *Agrocybe aegerita*. Mol Genet 155:113–115

Fortin JA, Piché Y, Godbout C (1983) Methods for synthesizing ectomycorrhizas and their effect on mycorrhizal development. Plant Soil 71:275–284

Frey P, Frey-Klett P, Garbaye J, Berge O, Heulin T (1997) Metabolic and genotypic fingerprinting of fluorescent pseudomonads associate with the Douglas-fir–*Laccaria bicolor* mycorrhizosphere. Appl Environ Microbiol 63:1852–1860

Frey-Klett P, Pierrat JC, Garbaye J (1997) Location and survival of mycorrhiza helper *Pseudomonas fluorescens* during establishment of ectomycorrhizal symbiosis between *Laccaria bicolor* and Douglas fir. Appl Environ Microbiol 63:139–144

Fries N (1983) Spore germination, homing reaction, and intersterility groups in *Laccaria laccata* (Agaricales). Mycologia 75:221–227

Fries N, Mueller GM (1984) Incompatibility systems, cultural features and species circumscriptions in the ectomycorrhizal genus *Laccaria* (Agaricales). Mycologia 76:633–664

Gagnon J, Langlois CG, Fortin JA (1987) Growth of containerized jack pine seedlings inoculated with different ectomycorrhizal fungi under a controlled fertilization schedule. Can J For Res 17:840–845

Gagnon J, Langlois CG, Garbaye J (1991) Growth and ectomycorrhiza formation of container-grown red oak seedlings as a function of nitrogen fertilization and inoculum type of *Laccaria bicolor*. Can J For Res 21:966–973

Garbaye J (1994) Tansley review, no. 76. Helper bacteria: a new dimension to the mycorrhizal symbiosis. New Phytol 128:197–210

Garbaye J, Churin JL (1996) Effect of ectomycorrhizal inoculation at planting on growth and foliage quality of *Tilia tomentosa*. J Arboriculture 22:29–34

Garbaye J, Duponnois R (1992) Specificity and function of mycorrhization helper bacteria (MHB) associated with the *Pseudotsuga menziesii–Laccaria laccata* symbiosis. Symbiosis 14:335–344

Garbaye J, Duponnois R, Wahl JL (1990) The bacteria associated with *Laccaria laccata* ectomycorrhizas or sporocarps: effect on symbiosis establishment on Douglas fir. Symbiosis 9:267–273

Garbaye J, Churin JL, Duponnois R (1992) Effects of substrate sterilization, fungicide treatment, and mycorrhization helper bacteria on ectomycorrhizal formation of pedunculate oak (*Quercus robur*) inoculated with *Laccaria laccata* in two peat bare-root nurseries. Biol Fertil Soil 13:55–57

Gardes M, Fortin JA, Mueller GM, Kropp BR (1990a) Restriction fragment length polymorphisms in the nuclear ribosomal DNA of four *Laccaria* spp.: *L. bicolor, L. laccata, L. proxima,* and *L. amethystina*. Phytopathology 80:1312–1317

Gardes M, Wong KKY, Fortin JA (1990b) Interactions between monokaryotic and dikaryotic isolates of *Laccaria bicolor* on roots of *Pinus banksiana*. Symbiosis 8:233–250

Gardes M, Mueller GM, Fortin JA, Kropp BR (1991a) Mitochondrial DNA polymorphisms in *Laccaria bicolor, L. laccata, L. proxima* and *L. amethystina*. Mycol Res 95:206–216

Gardes M, White TJ, Fortin JA, Bruns TD, Taylor JW (1991b) Identification of indigenous and introduced symbiotic fungi in ECM by amplification of nuclear and mitochondrial ribosomal DNA. Can J Bot 69:180–190

Gay G, Debaud JC (1987) Genetic study on indole-3-acetic acid production by ectomycorrhizal *Hebeloma* species: inter- and intraspecific variability in homo- and dikaryotic mycelia. Appl Microbiol Biotechnol 26:141–146

Godbout C, Fortin JA (1990) Cultural control of basidiome formation in *Laccaria bicolor* with container-grown white pine seedlings. Mycol Res 94:1051–1058

Henrion B, Di Battista C, Bouchard D, Vairelles D, Thomson BD, Le Tacon F, Martin F (1994) Monitoring the persistence of *Laccaria bicolor* as an ectomycorrhizal symbiont of nursery-grown Douglas fir by PCR of the rDNA intergenic spacer. Mol Ecol 3:571–580

Henrion F, Le Tacon F, Martin F (1992) Rapid identification of genetic variation of ectomycorrhizal fungi by amplification of ribosomal RNA genes. New Phytol 122:289–298

Hilbert JL, Martin F (1988) Regulation of gene expression in ectomycorrhizas. I. Protein changes and the presence of ectomycorrhiza specific polypeptides in the *Pisolithus-Eucalyptus* symbiosis. New Phytol 110:339–346

Hilbert JL, Costa G, Martin F (1991) Ectomycorrhizin synthesis and polypeptide changes during the early stage of eucalypt mycorrhiza development. Plant Physiol 97:977–984

Ingleby K, Wilson J, Mason PA, Munro RC, Walker C, Mason WL (1994) Effects of mycorrhizal inoculation and fertilizer regime on emergence of Sitka spruce seedlings in bare-root nursery seedbeds. Can J For Res 24:618–623

Jacobson KM, Miller OK, Turner BJ (1993) Randomly amplified polymorphic DNA markers are superior to somatic incompatibility tests for discriminating genotypes in natural populations of the ectomycorrhizal fungus *Suillus granulatus*. Proc Natl Acad Sci USA 90:9159–9163

Jorgensen E (1961) The formation of pinosylvin and its monomethyl ether in the sapwood of *Pinus resinosa* Ait. Can J Bot 39:1765–1772

Kottke I (1997) Fungal adhesion pad formation and penetration of root cuticle in early stage mycorrhizas of *Picea abies* and *Laccaria amethystea*. Protoplasma 196:55–64

Kropp BR (1990a) Variable interactions between non-mycorrhizal and ectomycorrhizal strains of the basidiomycete *Laccaria bicolor*. Mycol Res 94:412–415

Kropp BR (1990b) Variation in acid phosphatase activity among progeny from controlled crosses in the ectomycorrhizal fungus *Laccaria bicolor*. Can J Bot 68:864–866

Kropp BR (1997) Inheritance of the ability for ectomycorrhizal colonization of *Pinus strobus* by *Laccaria bicolor*. Mycologia 89:578–585

Kropp BR, Anderson AJ (1994) Molecular and genetic approaches to understanding variability in mycorrhizal formation and functioning. In: Pleger FL, Linderman RG (eds) Mycorrhizae and plant health. APS Press, St Paul, pp 309–336

Kropp BR, Fortin JA (1986) Formation and regeneration of protoplasts from the ectomycorrhizal basidiomycete *Laccaria bicolor*. Can J Bot 64:1224–1226

Kropp BR, Fortin JA (1988) The incompatibility system and relative ectomycorrhizal performance of monokaryons and reconstituted dikaryons of *Laccaria bicolor*. Can J Bot 66:289–294

Kropp BR, Langlois CG (1990) Ectomycorrhizae in reforestation. Can J For Res 20:438–451

Kropp BR, McAffee BJ, Fortin JA (1987) Variable loss of ectomycorrhizal ability in monokaryotic and dikaryotic cultures of *Laccaria bicolor*. Can J Bot 65:500–504

Lamhamedi MS, Godbout C, Fortin JA (1994) Dependence of *Laccaria bicolor* basidiome development on current photosynthesis of *Pinus strobus* seedlings. Can J For Res 24:1797–1804

Last FT, Mason PA, Pelham J, Ingleby K (1984) Fruitbody production by sheathing mycorrhizal fungi: effects of "host" genotypes and propagating soils. For Ecol Manage 9:221–227

Lei J, Wong KKY, Piché Y (1991) Extracellular concanavalin A-binding sites during early interactions between *Pinus banksiana* and two closely related genotypes of the ectomycorrhizal basidiomycete *Laccaria bicolor*. Mycol Res 95:357–363

Le Tacon F, Garbaye J, Bouchard D, Chevalier G, Olivier JM, Guimberteau J, Poitou N, Frochot H (1988) Field results from ectomycorrhizal inoculation in France. In: Lalonde M, Piché Y (eds) Canadian workshop on mycorrhizae in forestry, CRBF, Université Laval, Ste-Foy, Quebec, pp 51–74

Li CY, Castellano MA (1987) *Azospirillum* isolated from within sporocarps of the mycorrhizal fungi *Hebeloma crustuliniforme*, *Laccaria laccata* and *Rhizopogon vinicolor*. Trans Br Mycol Soc 88:563–564

Martin F, Tagu D (1995) Ectomycorrhiza development: a molecular perspective. In: Varma A, Hock B (eds) Mycorrhiza structure, function, molecular biology and biotechnology. Springer, Berlin Heidelberg New York, pp 29–58

Marx DH (1969) The influence of ectotrophic ectomycorrhizal fungi on the resistance of pine roots to pathogenic infections. I. Antagonism of mycorrhizal fungi to pathogenic fungi and soil bacteria. Phytopathology 59:153–163

Mason PA, Last FT, Pelham J, Inglebey K (1982) Ecology of some fungi associated with an ageing stand of birches (*Betula pendula* and *B. pubescens*). For Ecol Manage 4:19–39

McNabb RFR (1972) The Tricholomataceae of New Zealand. I. *Laccaria* Berk. & Br. N Z J Bot 10:461–484

Molina R (1982) Use of the ectomycorrhizal fungus *Laccaria laccata* in forestry. I. Consistency between isolates in effective colonization of containerized conifer seedlings. Can J For Res 12:469–473

Molina R, Chamard J (1983) Use of the ectomycorrhizal fungus *Laccaria laccata* in forestry. II. Effects of fertilizer forms and levels on ectomycorrhizal development and growth of container-grown Douglas-fir and ponderosa pine seedlings. Can J For Res 13:89–95

Mueller GJ, Mueller GM, Shih LH, Ammirati JF (1993) Cytological studies in *Laccaria* (Agaricales). I. Meiosis and postmeiotic mitosis. Am J Bot 80:316–321

Mueller GM (1991) *Laccaria laccata* complex in North America and Sweden: intercollection pairing and morphometric analyses. Mycologia 83:578–594

Mueller GM (1992) Systematics of *Laccaria* (Agaricales) in the continental United States and Canada, with discussions on extralimital taxa and descriptions of extant types. Fieldiana Bot NS 30:1–158

Mueller GM (1997) The mushroom genus *Laccaria* in North America. World Wide Web site at URL: <http://www.fmnh.org/candr/botany/ botany_sites>

Mueller GM, Ammirati JF (1993) Cytological studies in *Laccaria* (Agaricales). II. Assessing phylogenetic relationships among *Laccaria*, *Hydnangium*, and other Agaricales. Am J Bot 80:322–329

Mueller GM, Gardes M (1991) Intra- and interspecific relations within *Laccaria bicolor sensu lato*. Mycol Res 95:592–601

Mueller GM, Strack BA (1992) Evidence for a mycorrhizal host shift during migration of *Laccaria trichodermophora* and other agarics into neotropical oak forests. Mycotaxon 45:249–256

Münzenberger B, Kottke I, Oberwinkler F (1995) Reduction of phenolics in mycorrhizas of *Larix decidua* Mill. Tree Physiol 15:191–196

Münzenberger B, Otter T, Wüstrich D, Polle A (1997) Peroxidase and laccase activities in mycorrhizal and nonmycorrhizal fine roots of Norway spruce (*Picea abies*) and larch (*Larix decidua*). Can J Bot 75:932–938

Olsson PA, Chalot M, Bååth E, Finlay RD, Söderström B (1996) Ectomycorrhizal mycelia reduce bacterial activity in a sandy soil. FEMS Microbiol Ecol 21:77–86

Piché Y, Peterson RL, Howarth MJ, Fortin JA (1983) A structural study of the interaction between the ectomycorrhizal fungus *Pisolithus tinctorius* and *Pinus strobus* roots. Can J Bot 61:1185–1193

Raffle VL, Anderson NA, Furnier GR, Doudrick RL (1995) Variation in mating competence and random amplified polymorphic DNA in *Laccaria bicolor* (Agaricales) associated with three tree host species. Can J Bot 73:884–890

Rosado SCS, Kropp BR, Piché Y (1994) Genetics of ectomycorrhizal symbiosis. II. Fungal variability and heritability of ectomycorrhizal traits. New Phytol 126:111–117

Schelkle M, Peterson RL (1996) Suppression of common root pathogens by helper bacteria and ectomycorrhizal fungi in vitro. Mycorrhiza 6:481–485

Selosse MA, Costa G, Di Battista C, Le Tacon F, Martin F (1996) Meiotic segregation and recombination of the intergenic spacer of the ribosomal DNA in the ectomycorrhizal basidiomycete *Laccaria bicolor*. Curr Genet 30:332–337

Sen R (1990) Intraspecific variation in two species of *Suillus* from Scots pine (*Pinus sylvestris* L.) forests based on somatic incompatibility and isozyme analyses. New Phytol 114:607–616

Shaw CG, Sidle RC, Harris AS (1987) Evaluation of planting sites common to a southeast Alaska clearcut. III. Effects of microsite type and ectomycorrhizal inoculation on growth and survival of Sitka spruce seedlings. Can J For Res 17:334–339

Sinclair WA, Cowles DP, Hee SM (1975) *Fusarium* root rot of Douglas-fir seedlings: suppression by soil fumigation, fertility management, and inoculation with spores of the fungal symbiont *Laccaria laccata*. For Sci 21:390–399

Singer R (1986) The Agaricales in modern taxonomy 4th edn. Koeltz Scientific Books, Koenigstein, 981 pp

Singer R, Moser M (1965) Forest mycology and forest communities in South America. I. The early fall aspect of the mycoflora of the Cordillera Pelada (Chile), with mycogeographic analysis and conclusions regarding the heterogeneity of the Valdivean flora district. Mycopathol Mycol Appl 26:129–191

Smith SE, Read DJ (1997) Mycorrhizal symbiosis 2nd edn. Academic Press, San Diego

Stack RW, Sinclair WA (1975) Protection of Douglas-fir seedlings against *Fusarium* root rot by a mycorrhizal fungus in the absence of mycorrhiza formation. Phytopathology 65:468–472

Sweeney M, Harmey MA, Mitchell DT (1996) Detection and identification of *Laccaria* species using a repeated DNA sequence from *Laccaria proxima*. Mycol Res 100:1515–1521

Sylvia DM, Sinclair DM (1983a) Phenolic compounds and resistance to fungal pathogens induced in primary roots of Douglas-fir seedlings by the ectomycorrhizal fungus *Laccaria laccata*. Phytopathology 73:390–397

Sylvia DM, Sinclair WA (1983b) Suppressive influence of *Laccaria laccata* on *Fusarium oxysporum* and on Douglas-fir seedlings. Phytopathology 73:384–389

Tagu D, Martin F (1996) Molecular analysis of cell wall proteins expressed during the early steps of ectomycorrhiza development. New Phytol 133:73–85

Thomson BD, Grove TS, Malajczuk N, Hardy GEStJ (1996) The effect of soil pH on the ability of ectomycorrhizal fungi to increase the growth of *Eucalyptus globulus* Labill. Plant Soil 178:209–214

Tommerup IC, Barton JE, O'Brien PA (1995) Reliability of RAPD fingerprinting of three basidiomycete fungi, *Laccaria*, *Hydnangium* and *Rhizoctonia*. Mycol Res 99:179–186

Tosh JE, Senior E, Smith JE, Watson-Craik IA (1993) Ectomycorrhizal seedling response to selected components of landfill leachate. Mycol Res 97:129–135

Villeneuve N, Le Tacon F, Bouchard D (1991) Survival of inoculated *Laccaria bicolor* in competition with native ectomycorrhizal fungi and effects on the growth of outplanted Douglas-fir seedlings. Plant Soil 135:95–107

Wagner F, Gay G, Debaud JC (1988) Genetical variability of glutamate dehydrogenase activity in the monokaryotic and dikaryotic mycelia of the ectomycorrhizal fungus *Hebeloma cylindrosporum*. Appl Microbiol Biotechnol 28:566–571

Watling R (1977) Relationships between the development of higher plants and fungal communities. Abstr 2nd Int Mycological Congr, University of South Florida, Tampa, 718 pp

Wei Y, Le Tacon F, Lapeyrie F (1995) Auxin accumulation by homocaryotic and dicaryotic mycelia of the ectomycorrhizal fungus *Laccaria bicolor*. In: Mycorrhizas for plantation forestry in Asia. Proc Int Symp and Worksh. Australian Centre for International Agricultural Research, Kaiping, Guangdong Province, PR China, pp 86–90

Wong KKY, Piché Y, Montpetit D, Kropp BR (1989) Differences in the colonization of *Pinus banksiana* roots by sib-monokaryotic and dikaryotic strains of ectomycorrhizal *Laccaria bicolor*. Can J Bot 67:1717–1726

Wong KKY, Montpetit D, Piché Y, Lei J (1990a) Root colonization by four closely related genotypes of the ectomycorrhizal basidiomycete *Laccaria bicolor* (Maire) Orton – comparative studies using electron microscopy. New Phytol 116:669–679

Wong KKY, Piché Y, Fortin JA (1990b) Differential development of root colonization among four closely related genotypes of ectomycorrhizal *Laccaria bicolor*. Mycol Res 94:876–884

Chapter 4

Hebeloma

R. Marmeisse, H. Gryta, P. Jargeat, L. Fraissinet-Tachet, G. Gay and J.-C. Debaud

4.1
Introduction

During the period 1987–1997, approximately 70 species names of *Hebeloma* were quoted in the index list of the *Abstracts of Mycology*. Amongst a total of 301 publications referenced, 193 dealt with one of the following two species: *Hebeloma crustuliniforme* (Bull.: St-Amans) Quél., 132 references, and *H. cylindrosporum* Romagn., 61 references. Nine other species were less frequently studied: *H. sinapizans* (Paul.: Fr.) Gillet, 11 references; *H. radicosum* (Bull.: Fr.) Ricken, 9; *H. westraliense* Bougher, Tommerup & Malajczuk; 8; *H. arenosa* Burdsall, MacFall & Albers, 8; *H. sacchariolens* Quél., 6; *H. hiemale* Bresadola, 6; *H. longicaudum* (Pers.: Fr.) Kumm., 6; *H. vinosophyllum* Hongo, 5; *H. mesophaeum* (Pers.: Fr.) Quél., 5. Fourty-four other species were only quoted once, mainly in taxonomic surveys. These statistics certainly do not reflect either the biological diversity within the genus *Hebeloma* or the position and function of *Hebeloma* species in the various ecosystems where they occur. However, they justify why this chapter is based largely on the two species *H. crustuliniforme* and *H. cylindrosporum*.

 H. crustuliniforme and *H. cylindrosporum* are two ectomycorrhizal (ECM) species which have been studied from many points of view: ecology, biology, biochemistry and physiology as well as genetics and molecular biology. *H. cylindrosporum* is a species for which geographic range is very restricted and its use in various studies has been dictated essentially because it can easily be manipulated in the laboratory. Conversely, *H. crustuliniforme* is a well-studied species because of its very wide host range and potential interest as a commercial inoculum for forestry purposes.

Laboratoire d'Ecologie Microbienne du Sol (UMR CNRS 5557), Bât. 405, 43 Bd. du 11 Novembre 1918, 69622 Villeurbanne Cedex, France

4.2
Taxonomy and Ecology

4.2.1
Taxonomy

The genus *Hebeloma* belongs to the family Cortinariaceae (Basidiomycotina, Homobasidiomycetes, Agaricomycetideae, Cortinariales). Since the 1960s different authors have made important taxonomic contributions on this genus: Favre (1960), Romagnesi (1965, 1983), Bruchet (1970), Moser (1970, 1985) and Vesterholt (1995) for European species, and Smith et al. (1983) for the North American species.

Hebeloma is divided into three sections: two of these – Indusiata and Hebeloma (= Denudata) – were defined by Fries; the third section (= sub-genus) – Myxocybe – was defined by Singer. The complete definition and characteristics of the genus are detailed in Singer (1986) and Moser and Jülich (1987). As for most fungal genera, the number of described *Hebeloma* species is difficult to give because of synonymy. In France, where temperate, arctico-alpine and mediterranean species can be collected, it is estimated to contain ca. 75, plus ca. 10 infraspecific, taxa. In the USA, recent studies have yielded remarkably high numbers of new species, for example 95 (= 85%) of the 112 veiled species of *Hebeloma* presently described (Smith et al. 1983). The total world number of *Hebeloma* species is presently estimated at 250–600 species.

The precise taxonomic position of many species either remains uncertain or has only recently been clarified, as for example the *Hebeloma* species of northern European countries belonging to the section Indusiata (Vesterholt 1989). This is the case for *H. cylindrosporum*, first placed by Romagnesi (1965) in the section Indusiata, moved by Bruchet (1980) to the section Denudata (= section Hebeloma) and now placed by Vesterholt (1989) in section Myxocybe. This is likewise the case for *H. crustuliniforme*, which is a member of a species complex [together with, for example, *H. alpinum* (Favre) Bruchet] within which species are difficult to identify (Vesterholt 1995).

4.2.2
Saprophytic and ECM *Hebeloma* Species

There is no comprehensive survey of the life histories of *Hebeloma* species. Some are known to be ECM species from the results of in vitro syntheses performed in the laboratory; others are presumed to be ECM since sporocarps of these species have only been observed in the close vicinity of ECM plants. Only a few *Hebeloma* species appear saprophytic, or more precisely do not need the association with a plant to survive during their mycelial stage in soil and to differentiate sporocarps. This is the case for *H. truncatum* (Schaeff.: Fr.) Karst. and *H. calyptrosporum* Bruchet living in grassland ecosystems (meadows) or for *H. anthracophilum* R. Maire which fruits on burnt ground. Mycelium of *H. calyptrosporum*, however, forms ECM with a clear Hartig net on *Pinus virginiana* (Hacskaylo and Bruchet 1972) and both species form ECM on birch (Giltrap 1982). These two species can therefore be considered as facultative symbionts. There are also reports of *H. aminophilum* Hilton & Miller as a "sarcophilous fungus" found in the vicinity of decaying animal matter (Sagara 1992).

Some *Hebeloma* species fruit after soil disturbance. Among these species some appear as late-stage fungi on burnt ground. Others, such as *H. spoliatum* (Fr.) Gillet, *H. vinosophyllum* and *H. radicosoides* (in Sagara 1992), have been classified as "ammonia fungi" (species fruiting after urea or ammonia treatment of the soil) (Sagara 1992; Sagara et al. 1985). These species are, however, known to form ECM with birch trees in culture (Giltrap 1982) and are able to grow and fruit saprophytically in axenic culture (Giltrap 1982; Sagara 1992).

Most *Hebeloma* species are considered to be ECM since their sporocarps occur only in the close vicinity of ECM plants (Bruchet 1970). Furthermore, of the 27 species studied by Hacskaylo and Bruchet (1972), 17 formed ECM with *P. virginiana*, five others penetrated only the first layer of the root cortex and only five did not interact at all with the root of this plant species. In a similar study by Giltrap (1982), of the 22 *Hebeloma* species tested, 21 formed ECM with both *Betula pendula* and *B. pubescens*. Some species such as *H. crustuliniforme* were also demonstrated, along with other ECM fungi, to be able to form arbutoid mycorrhizas with *Arbutus menziesii* and *Arctostaphylos uva-ursi* (both Ericales) (Zak 1976a,b). These mycorrhizas differ from the typical ECM by the penetration of fungal hyphae in root epidermal cells. In this case it is clear that the plant, not the fungus, controls the development of the mycorrhizas.

4.2.3
Host Range

The definition of the host range of ECM fungal species is largely based upon observed associations between host plants and sporocarps in the field. Such observations include not only the host range of the fungal species but also the geographic distribution of both partners of the symbiosis and their adaptations to particular climatic and edaphic conditions. Another definition of the host range of an ECM species can be deduced from laboratory or greenhouse experiments where the fungus is confronted with various ECM plants which spontaneously grow either in the same ecosystem or in other ecosystems. These experiments can eventually serve to establish links between host specificity and plant taxonomy or phylogeny. However, the relevance of some associations obtained under laboratory conditions has to be interpreted cautiously as the substrates used are usually quite artificial and often supplemented with carbohydrates, and, as such, do not reflect natural conditions.

For some *Hebeloma* species the host range can be qualified as broad, with little ambiguity, as these species occur naturally associated in the field with various gymnosperms and dicots. For example, this is the case for *H. crustuliniforme*, *H. edurum* Métrod, *H. sarcophyllum* Peck and *H. sinapizans*. In the case of *H. crustuliniforme*, which is widely distributed in the temperate northern hemisphere, the ability to form ECM with numerous plant species is further substantiated by laboratory and glasshouse experiments (Table 4.1). Examples of narrow host specificities as observed in the field can be illustrated with arctico/alpine *Hebeloma* species (Bruchet 1974). Of the seven species known to occur in this ecosystem, five have only been observed associated with *Salix herbacea* (*H. kuehneri* Bruchet, *H. nigellum* Bruchet, *H. repandum* Bruchet, *H. minus* Bruchet and *H. subconcolor* Bruchet) whereas two [*H. alpinum* and *H. marginatulum* (Favre) Bruchet] present in the same ecosystems can be found under either *Salix* or *Dryas octopetala*.

The lack of correlation between ecological specificity and host range defined in the laboratory can be illustrated with *H. cylindrosporum* and *Hebeloma* species from the alpine/arctic ecosystem. *H. cylindrosporum* is a European species which occurs in sandy soils, essentially in dune ecosystems along the Atlantic coast but also in Italy. In these ecosystems it is found associated with *Pinus* species, such as *P. pinaster* in south-west France or *P. sylvestris* in the Netherlands (Jansen 1982). In Sardinia, it has been reported to fruit in pure *Cistus* stands (Contu 1991). In the laboratory it has been successfully induced to form ECM with plant species as diverse as American gymnosperm species such as *Pinus banksiana* and *Larix laricina* (McAfee and Fortin 1988) and also with various species from different dicots families such as the North American arctic *Dryas integrifolia* (Melville et al. 1987a,b, 1988), *Betula* species (Giltrap 1982) and *Castanea sativa* (Valjalo 1979; see Table 4.1). In the case of the European alpine/arctic species, *H. alpinum* and *H. marginatulum* are almost exclusively associated in the field with the rosaceous alpine *Dryas octopetala* or the alpine shrubs *Salix retusa* and *S. reticulata*. They can also form ECM in vitro with the American pine *P. virginiana* (Hacskaylo and Bruchet 1972). For these different fungus–plant combinations it is doubtful that either one or the other partner would survive long after transplantation to the ecosystem where either the plant or the fungus naturally occurs. Such data suggest that specificity for numerous *Hebeloma* species in the field is due more to environmental conditions than to direct host–fungus interactions.

4.2.4
Geographical Distribution

As for many genera of Homobasidiomycetes, most studies on *Hebeloma* have considered species which naturally occur in the northern hemisphere mediterranean, temperate or arctico/alpine ecosystems and, indeed, numerous species belonging to this genus occur in any of these ecosystems. Furthermore, *Hebeloma* species are well represented in Europe, Asia and

Table 4.1. A non-exhaustive list of plant species reported to form mycorrhizas with either *H. crustuliniforme* or *H. cylindrosporum*. The references combine field observations and laboratory mycorrhizal syntheses

Plant species	*H. crustuliniforme*	*H. cylindrosporum*
Angiosperms		
Alnus rubra	Miller et al. (1991)[a]	nd
Arbutus menziesii	Zak (1976)	nd
Arctostaphylos uva-ursi	Zak (1976)	nd
Betula pendula	Giltrap (1982)	Giltrap (1982)
B. pubescens	Giltrap (1982)	Giltrap (1982)
Castanea sativa	nd	Valjalo (1979)
Cistus sp.	Bruchet et al. (1985)[a]	Contu (1991)[a,b]
Dryas integrifolia	nd	Melville et al. (1987b)
Quercus acutissima	nd	Oh et al. (1995)
Q. serrata	nd	Oh et al. (1995)
Q. robur	ShemakLanova (1967)	nd
Q. robur (vitroplants)	Lei and Dexheimer (1987)	nd
Q. suber	Romano et al. (1994)	nd
Gymnosperms		
Larix x eurolepis	nd	Piola et al. (1995)
L. laricina	nd	Wong and Fortin (1989)
Picea abies	Brunner et al. (1991)	Le Tacon et al. (1985); Brunner et al. (1991)
P. mariana	Abuzinadah and Read (1989)	Cote and Thibault (1988); Browning and Withney (1992)
P. sitchensis	Chu-Chou (1979)[a] Abuzinadah and Read (1989)	nd nd
Pinus banksiana	nd	McAfee and Fortin (1988); Wong and Fortin (1989); Browning and Withney (1992)
P. contorta	Chu-Chou (1979)[a]	nd
P. halepensis	nd	Delmas (1978)
P. pinaster	nd	Debaud and Gay (1987)[b]
P. ponderosa	Perry et al. (1989)	nd
P. radiata	Chu-Chou and Grace (1985)	nd
P. strobus	nd	Piché and Fortin (1982)
P. sylvestris	nd	Jansen (1982)[a,b]
Pseudotsuga douglasii	nd	Le Tacon et al. (1985)
Tsuga heterophylla	Rygiewicz et al. (1984a,b)	nd

nd, A fungus–plant association which has not been studied.
[a] Field observations.
[b] Natural host species.

North America. In Australia, according to Bougher et al. (1991), of the six *Hebeloma* species described, three may be introduced species as were first described in the northern hemisphere and are mostly associated with introduced trees. Among the other three species, *H. westraliense* is associated with *Eucalyptus* trees, and could be native to Australia. The genus *Hebeloma* seems to be poorly represented, if not absent, from tropical–intertropical regions. Some *Hebeloma* species, however, have a very wide distribution, as illustrated by the case of *H. crustuliniforme* which occurs in Europe and North America, and which may have disseminated in Australia. As discussed above, this species has a very broad host range under natural conditions (Table 4.1) and its potential hosts are widely distributed in the areas considered.

According to Bruchet (1973), who studied *Hebeloma* species collected in France, three groups of species with different geographic distributions can be distinguished: (1) species having a wide distribution in temperate forests (e.g. *H. crustuliniforme*); (2) artico/alpine species with different host specificities in the field (e.g. *H. kuehneri*, *H. minus* and *H. subconcolor*, which are almost exclusively associated with the Salicacae, whereas species such as *H. alpinum* and *H. marginatulum* form ECM with both alpine Rosaceae and Salicacae); (3) species limited to mediterranean (and sub-mediterranean) areas (e.g. *H. sarcophyllum*, associated with *Quercus ilex* or *Pinus halepensis*, other species that are preferentially, if not exclusively, associated with *Cistus* or *Helianthemum* plants, *H. cistophilum* R. Maire, *H. album* Peck and *H. hiemale*). This, incomplete, classification illustrates that *Hebeloma* species have colonized a wide range of different temperate forest ecosystems.

4.2.5
Hebeloma Species as Successional Pioneers

Some *Hebeloma* species appear in the early stages of fungal succession and may represent pioneer species. Such species colonize seedlings efficiently, are dominant species on young trees and preferentially occupy the periphery of the root system of older trees. During the succession of sheathing ECM fungi these pioneer species are progressively replaced by late-stage fungi which ultimately dominate the ECM community when the forest canopy closes.

Evidence of this behaviour has been reported by Last et al. (1983) who observed that sporocarps of *Hebeloma* spp. can appear in the first years after planting of *Betula* spp. trees and in the following years are produced further away from the tree trunk as the root system expands in the soil. Similar observations were reported by Fleming et al. (1984) showing that *Hebeloma* species were able to form ECM on <1-year-old birch seedlings; other species such as *Lactarius pubescens* could not. The critical influence of the age of the tree and of its root system was demonstrated by planting seedlings in soil cores sampled beneath old trees. These seedlings were rapidly colonized by early-stage species even for soil cores sampled in areas where mostly late-stage fungi were observed (Deacon et al. 1983). Furthermore, when *H. crustuliniforme* and *L. pubescens* were co-inoculated on the root system of birch trees, the former could form ECM, regardless of the position of the short roots, whereas the latter species preferentially formed ECM on short roots which emerged from the older parts of the root system (Gibson and Deacon 1988). Such observations may explain the occurrence of *Hebeloma* spp. ECM in newly afforested sites, whereas *L. pubescens* appears as one of the first of the late-stage species. Moreover, species of early-stage fungi such as *Hebeloma* species are able to establish ECM from spores on *Betula* and *Pinus* seedlings. This could prevent ECM formation by other fungi occurring naturally in the rooting media.

A possible physiological difference separating early- from late-stage fungi was presented by Gibson and Deacon (1990) who compared the effect of added glucose on the establishment of ECM by fungal species representing each of these two classes. Typically, mycelia of early-stage fungi (such as *H. crustuliniforme*) grew even in the absence of glucose in the culture medium whereas all the studied late-stage fungi needed glucose for hyphal extension. Furthermore, late-stage species also required glucose to form ECM with *Betula pubescens* as opposed to early-stage fungi.

Field observations also classify *H. cylindrosporum* as an early-stage fungal species in the fungal succession which characterizes *P. pinaster* forests established on the back of coastal sand dunes (Gryta et al. 1997; Guinberteau 1997). This species is one of the dominant species around isolated pine trees planted just behind the unstable dune. It progressively disappears in the older parts of the forest where a humus layer accumulates on the soil surface, and is completely absent further inland except in places such as the borders of forest-free corridors constructed to prevent forest fires.

4.3
Life Cycle and Breeding Systems

The life cycles of many homobasidiomycete species have been characterized and it is clear that, within one genus, different species have adopted different life styles. While most species are heterothallic with one or two mating type genes, others have evolved towards homothallism. Knowledge of the life cycle of a fungal species is critical for interpretation of data in the field of population genetics. The ability to obtain and reproduce in the laboratory the different stages of a life cycle is also necessary to unravel the genetic and molecular basis of ECM differentiation and functioning.

4.3.1
Basidiospore Germination

The control of basidiospore germination is a critical step for analysis of both the life cycle and breeding system of a basidiomycete species. For heterothallic species it allows the recovery of haploid homokaryotic mycelia and allows genetic analyses to be performed. In different *Hebeloma* species, it has been shown that different biotic factors can influence basidiospore germination and the ecological significance of these factors has been discussed. However, in the results described below none of the germination rates reported is higher than 50%, and in most cases they do not exceed 1–2%. It is still not clear whether these results indicate that most basidiospores are naturally non-viable.

Basidiospore germination on culture media has been reported for several species of *Hebeloma* without any particular treatment. This was the case for 19 of the 36 species studied by Bruchet (1973). In this study, all species belonging to the section Denudata produced some viable basidiospores which germinated within 2 weeks of plating. For most other species, germination was delayed and monosporous isolates were recovered after 4–5 weeks. Germination rates were always low, around 1%. This is the case for *H. cylindrosporum* for which the germination rate is 2–4% and decreases rapidly upon storage of the spores at 4°C (unpubl. results from the authors' laboratory). This rapid decrease, reported for many different species, and which is generally interpreted as a loss of viability, could also reflect the entry of spores into a quiescent stage. For species like *H. cylindrosporum* which produce sporocarps at the end of autumn, it could be of some advantage for their spores to germinate after winter.

Stimulation of basidiospore germination by the roots of pine or birch seedlings has been reported by Fries (1984) for *H. mesophaeum* and *H. crustuliniforme*. Other biotic factors such as mycelia of the same species or the soil yeast *Rhodotorula* had no effect on germination. Ali and Jackson (1988) demonstrated that birch root was the most stimulatory in the case of *H. crustuliniforme*, with germination rates up to 30% for spores at 0–1 mm from the root edge. Fries and Birraux (1980) demonstrated that *Pinus sylvestris* roots improved the germination of four of seven spore collections belonging to different strains of *H. crustuliniforme*, *H. ingratum* Bruchet and *H. mesophaeum*, the germination rate being at least 10 times higher than in the absence of roots. The degree of specificity of this interaction has not been fully elucidated. Fries and Swedjemark (1986) showed that all five species of trees tested, including gymnosperms and dicots, exerted a strong effect on the germination of *H. mesophaeum*, whereas only carrot roots showed comparable effect among 16 herbaceous plants tested. Basidiospore germination is certainly of great importance for installation of early-stage fungi on the root system of young trees. In a comparative study, Fox (1986) demonstrated that *H. crustuliniforme*, *H. leucosarx* P. D. Orton and *H. sacchariolens* were able to establish ECM from basidiospores in unsterile soils, whereas late-stage fungi were not.

4.3.2
Nuclear Status of Mycelia

To our knowledge, all species of *Hebeloma* studied so far are heterothallic. Basidiospore germination gives rise to haploid homokaryotic mycelia. Mating between two compatible homokaryotic mycelia gives rise to a heterokaryotic mycelium which will ultimately form the sporocarps. Two species, *H. cylindrosporum* (Debaud et al. 1988) and *H. westraliense* (Bougher et al. 1991), have been studied in detail. *H. cylindrosporum* produces binucleate basidiospore containing two copies of the same haploid nucleus. The homokaryotic mycelia resulting from their germination contain one nucleus per cell, except for the apical one which contains from three to ten nuclei. These monokaryotic mycelia do not produce asexual spores and have never been isolated as such in the field. Mating between two compatible monokaryons gives rise to dikaryotic mycelia which are easily identified by the presence of clamp connections. Each cell of the dikaryon contains two unfused haploid nuclei. Dikaryotic mycelia represent the permanent (durable) stage of the fungus in the field. Both mono- and dikaryons are able to produce ECM in the laboratory, but only dikaryons form ECM and sporocarps in the field. Karyogamy of the two haploid nuclei occurs in the basidia and is immediately followed by meiosis. The resulting recombinant haploid nuclei migrate in each of the four basidiospores. For *H. westraliense*, cells in both juvenile and mature tissues of the sporocarps are dikaryotic with clamp connections, the spores having two nuclei. In developing basidia, a post-meiotic nucleus migrates into each of the four spores where a post-meiotic mitosis occurs. The mycelium originating from a single spore germination is monokaryotic and lacks clamp connections.

4.3.3
Breeding Systems

The breeding system of an ECM fungus is of special significance for colonization of forest sites. Airborne basidiospores are agents of long-distance dispersal. For homothallic or pseudohomothallic species, a single spore is sufficient to establish a new, sexually reproducing mycelial individual. In contrast, a single spore of a heterothallic species would result in a homokaryotic mycelium, which would probably be unable to survive without mating with a second compatible homokaryon. This could be of particular importance for the establishment of pioneer *Hebeloma* species.

All *Hebeloma* species studied so far appear to be heterothallic, and to have a bifactorial mating system. This is the case for *H. circinans* Qué, *H. edurum*, *H. sinapizans*, *H. subsaponaceum* Karst. and *H. trucatum* (Bruchet 1973), for *H. cylindrosporum* (Debaud et al. 1986) and for *H. crustuliniforme* (D. Aanen, pers. comm.). In bifactorial heterothallic species, four different mating types segregate in the homokaryotic progeny of a single dikaryon and only 25% of all the possible crosses result in the formation of a dikaryon. Mating types are determined by two different unlinked mating type factors called A and B with multiple allelic forms. Two homokaryons can mate and give a dikaryon only if they possess different alleles at both the A and B factors. Since more than two alleles exist at both the A and B factors, matings between homokaryon progeny of dikaryons that have different alleles at both A and B are all compatible. This breeding system therefore favours outcrossing and limits inbreeding. Multiple alleles at the A and B loci have been demonstrated in the case of *H. cylindrosporum* after pairing homokaryon progeny of six different dikaryotic wild strains collected along the south-west Atlantic coast of France (Debaud et al. 1986).

The bifactorial breeding system is the predominant breeding system among homobasidiomycete species and it is thus not surprising that the studies conducted so far on a limited number of species within the genus *Hebeloma* have identified this mating system only. These data should not be generalized to the whole *Hebeloma* genus since studies carried out in different genera such as *Laccaria* (Tommerup et al. 1991) show that different breeding systems can co-exist within one genus.

4.3.4
Sporocarp Formation

For several of the well-studied ECM fungal species such as *Laccaria* spp., *Pisolithus arhizus* (Scop.: Pers.) Rauschert, *Suillus collinitus*, *Thelephora terrestris* (Ehrb.) Fr. and *Tuber maculatum* Vitt., sporocarp production has occasionally been reported to occur under glasshouse conditions, during symbiotic association with a host plant grown in containers (see Godbout and Fortin 1990). This is also the case for several *Hebeloma* species. This genus seems to show a greater number of fruiting species than other ECM genera. In most cases the species which have been reported to fruit are species which produce small sporocarps, a situation which may be favourable, the host plant being the main source of carbohydrate for the fungus.

H. sarcophyllum differentiates sporocarps in the glasshouse when associated with one of its natural host plants, *Pinus halepensis* (Fig. 4.1b; Gilles Gay, unpubl. data), but also with *P. virginiana*, a non-habitual host (Hacskaylo and Bruchet 1972). This species, as well as *H. spoliatum*, has also been observed to fruit in the absence of a host plant (Bruchet 1973). Similarly, *H. crustuliniforme* was reported to fruit in the nursery after association with different host plants such as *Pinus radiata* (Chu-Chou 1979) or *Pseudotsuga menziesii* (Chu-Chou and Grace 1987). Another report of sporocarp formation under non-sterile conditions concerns the two arctico/alpine species *H. alpinum* and *H. marginatulum* associated with one of their habitual hosts, the alpine Rosaceae *D. octopetala* (Fig. 4.1c,d; Debaud et al. 1981a,b). As for almost all other ECM species, in most of these reports the presence of a host plant appears necessary for sporocarp production.

Fig. 4.1. Laboratory fruiting of four different *Hebeloma* species: **A** *H. cylindrosporum* associated with *P. pinaster* under sterile conditions (after Debaud and Gay 1987); **B** *H. sarcophyllum* associated with a young seedling of *P. halepensis* under sterile conditions (G. Gay, unpubl. data); **C** *H. alpinum* associated with *Dryas octopetala* under greenhouse conditions (Debaud et al. 1981a); **D** *H. marginatulum* associated with *D. octopetala* under greenhouse conditions (Debaud et al. 1981a)

Although, for some of these fungal species fruiting has been reported with different host plants, no study on the effect of the nature of the host plant on sporocarp production has been published and it is not clear whether an habitual host is more "efficient" than a non-habitual one. Using clones of *Betula* species, Last et al. (1984), however, demonstrated that the production of *Hebeloma* sporocarps was dependent upon host plant genotype, some genotypes being more productive than others. The soil composition also had a strong effect, one soil favouring fruiting of *Laccaria tortilis* (Fr.: Bolt.) Boud. whereas another soil stimulating fruiting of a *Hebeloma* sp.

The reproducible formation of sporocarps under axenic conditions has only been reported in the case of *H. cylindrosporum* (Debaud and Gay 1987). Sporocarps of this species differentiate after a dikaryotic mycelium has formed ECM with *P. pinaster* seedlings, an habitual host plant of this species (Fig. 4.1a). Fruiting is obtained on a sterile vermiculite substrate watered with a synthetic nutrient solution, under controlled light and temperature conditions. The ability of a mycelium to fruit is a very stable attribute, and does not decline for up to 28 years in the case of the HC1 strain isolated in 1970 and preserved at 4°C. Moreover, many different wild dikaryons of this species have been induced to fruit as well as dikaryons resulting from laboratory crosses between wild type, mutant or transformed homokaryons (Debaud and Gay 1987; Durand et al. 1992; unpubl. results from the authors' laboratory). This species is also able to fruit in pot cultures under glasshouse conditions after association with *Quercus acutissima* (Oh et al. 1995), a non-habitual host species. However, only primordia or abnormal immature sporocarps have very rarely been obtained in the absence of the host plant (Debaud and Gay 1987).

4.4
Quantitative Variation

Hebeloma is a taxonomic entity, but can it also be considered as a functional entity? It was suggested above that, in different ecosystems, *Hebeloma* species preferentially occur in the earlier stages of the vegetation succession or on pioneer sites. A possible interpretation of these observations is that, in these ecosystems, different *Hebeloma* species carry out similar functions and therefore present similar physiological potentials. This argument, at the interpecific level, should also apply to different strains of the same species and is often consciously, or not, followed in many research papers where a single strain of different ECM species is compared with discussion and conclusion where the results obtained are generalized to the whole species.

Quantitative variation between species or strains of the same species can result not only from intrinsic genetic differences but also from environmental effects to which different genotypes will respond differentially. The study of quantitative variation is necessary not only to select strains with increased ECM efficiency but also to appreciate the role of individual species or strains in an ecosystem. Among *Hebeloma* species, *H. cylindrosporum* has been extensively used to study different aspects of intraspecific variation and to elucidate the genetic control of this variation as different sets of strains resulting from laboratory crosses between known genotypes are available. Variation within *Hebeloma* has been studied and quantified for different biochemical and physiological traits linked to N and P metabolism: NADP-glutamate dehydrogenase (GDH) and nitrate reductase (NR) activities, implicated in NH_4^+ and NO_3^- assimilation respectively, and acid phosphatase (APase) activity. Auxin production has also been studied, as this hormone, produced by the fungus, may be implicated in ECM differentiation.

4.4.1
Interspecific Variation

Interspecific variation has been studied only for auxin (IAA) production using a set of 12 *Hebeloma* species occurring in different ecosystems, using only one heterokaryotic isolate for each species studied. All species produced IAA in axenic culture, with production ranging from 50.7 to 354.4 nmol IAA synthesized $h^{-1} mg^{-1}$ protein. This variability is of the same order of magnitude as intergeneric or interspecific variability reported for various physiological properties studied in other ECM fungi (Wong and Fortin 1990; Smith and Read 1997). No correlation could be detected between the IAA-synthesizing activity of the mycelia and their taxonomic position within the genus *Hebeloma*, their geographic origin or the host plant under which they were collected.

4.4.2
Intraspecific Variation

Studies in which genetically different wild-type mycelia of a single species have been compared indicate wide intraspecific variability in ECM fungi (Zhu et al. 1988; Debaud et al. 1995). In particular, intraspecific variability in ECM-forming ability has been demonstrated within *H. crustuliniforme* (Giltrap 1982; Gibson and Deacon 1990) and within *H. sinapizans* and *H. truncatum* (Giltrap 1982) associated with *Betula pendula* and *B. pubescens*.

In *H. cylindrosporum*, intraspecific variability of different physiological properties has been studied at two levels: between wild-type dikaryons isolated from different sporocarps collected in the field and within mono- and dikaryotic progenies of a single fruiting dikaryon. In 11 wild-type dikaryotic strains of *H. cylindrosporum* isolated under *P. pinaster* in "Les Landes" forest, South-West France, variability in IAA-synthesizing activity of these strains was of the same magnitude as that recorded at the interspecific level. Similarly, variation in GDH (Wagner et al. 1988), NR (Wagner et al. 1989), or APase (Meysselle et al. 1991) activities of these dikaryons was comparable with variation recorded for IAA production (Table 4.2).

In order to understand the origin and to study the genetic control of variation of a character under polygenic control, it is necessary to study

Table 4.2. Range of intraspecific variation of different biochemical properties of *H. cylindrosporum* and estimates of environmental and genetic components of total phenotypic variance in the case of the 50 synthesized dikaryons. (Data from Gay and Debaud, 1987; Wagner et al, 1988, 1989; and Meysselle et al. 1991)

Character	11 Wild-type dikaryons		20 Sib-homokaryons		50 Synthesized dikaryons		Environmental variance (%)	Genetic variance (%)
	Range of variation	Average	Range of variation	Average	Range of variation	Average		
GDH activity (nkat mg^{-1} protein)	1.5–11.6	7.8	0.19–9.0	2.8	4.0–19.8	8.5	6.0–14.1	94–85.9
Nitrate reductase activity (nmol NO$_2$ mg^{-1} protein h^{-1})	201–700	345	51–510	211	72–689	344	13.3	86.7
Acid phosphatase activity (nmol p-nitrophenyl phosphate hydrolysed mg^{-1} protein min^{-1})	49–676	238	85–791	352	186–756	454	–	–
IAA synthesizing activity (nmol IAA mg^{-1} protein h^{-1})	85–321	222	108–561	220	94–437	241	14.3–57.8	42.2–85.7

GDH, NADPH-specific glutamate dehydrogenase; IAA, indole-3-acetic acid.

variation of this character within groups of individuals resulting from controlled crosses between known parental lines. This was done for *H. cylindrosporum* using 20 sib-homokaryons randomly chosen in the progeny of a single dikaryon. Variation for GDH (Wagner et al. 1988), NR (Wagner et al. 1989), APase activities (Meysselle et al. 1991) and IAA production (Gay and Debaud 1987) was studied within this group of homokaryons as well as within a group of 50 dikaryons resulting from compatible laboratory crosses between these homokaryons. For each of these four quantitative characters, variations between sib-homokaryons as well as between synthesized dikaryons were in the same range as variations recorded between unrelated wild-type dikaryons of this species (Table 4.2).

From the results obtained with the 50 synthesized dikaryons, the different components of the total phenotypic variation could be estimated. The relative parts played by the environmental and genetic components respectively, and the parental and interactive components within the genetic effects, were far from identical for the different quantitative characters studied. In all cases, these characters had a genetic basis and the environmental effects accounted for <15% of the variation observed, except in the case of auxin production which represented 55% of the total variation (Table 4.2). This result was surprising as experiments conducted under identical conditions in the laboratory tend to minimize environmental effects. The rest of the variation resulted from genetic effects which were further partitioned into parental (additive) and interactive effects. GDH and IAA synthesizing activities were similar, whereas the parental one was almost negligible in the case of the NR activity. When the interactive component predominates, the selection of improved strains is made difficult as the outcome of a cross cannot be deduced from the values of the two parental lines because of unpredictable interactions between the parental genomes.

These data, along with other results of growth rates on different media (Gay et al. 1993), demonstrate that quantitative variation within a single species and even within the progeny of a single strain can be similar to, if not greater than, the variation observed between different species. It therefore appears very difficult, at least for the type of physiological characters which have been studied, to extrapolate the results obtained on one strain to the whole species. Furthermore, these results demonstrate that the genetic control of different activities linked to ECM functioning can be very different with respect to the relative parts taken by the parental and interactive components. In selection practice, this suggests that it is extremely difficult to simultaneously improve the different characters in a single selection scheme.

4.5
Population Structure and Dynamics

To understand the consequences of inter- and intraspecific variation of ECM fungi on the functioning of the symbiosis in natural ecosystems, it is necessary to have some knowledge of the structures and dynamics of the communities and populations of these fungal species. At the community level, different studies based on the observation of sporocarps have clearly established that (1) different fungal species can co-exist within small areas, suggesting that the corresponding below-ground mycelia overlap and can be associated with the root system of a single tree, and (2) the qualitative and quantitative compositions of such fungal communities change during the evolution and ageing of the forest ecosystem, suggesting that the root system of a single tree will have to deal with different fungi during its life span. In studies of fungal communities, *Hebeloma* species have frequently been shown to be more frequent in the earliest stages of the evolution of the forest ecosystem (see above).

Until now, the study of populations of homobasidiomycete species has almost exclusively been based on the sampling of sporocarps which are the only macroscopic structures these species produce during their life cycles. Sporocarps collected in one forest stand can emerge from one or several below-ground mycelia. For fungal species which do not form vegetative spores which can disseminate genets over long distances, two opposite and non-exclusive models of population dynamics can be envisaged and tested by sampling basidiocarps. The first model postulates that vegetative mycelia are long lived, that they survive and expand in the soil year after year and that the essential role of sexual reproduction is for the colonization of new sites through meiospore dispersal. If this model is true, then different basidiocarps collected at some distance in a forest site should be of the same genotype, the density of genotypes should be low and the same genotypes should be identified in the same places during many years. The second model postulates that sexual reproduction plays a major role in the dynamics of the populations. In this case, new genets appear every year in a given forest stand, replacing the previous ones whose mycelia do not expand in the soil. If this model is true, few sporocarps should be of the same genotype and always located very close to each other. Moreover, the same genets should be rarely sampled during consecutive fruiting seasons. These two models were tested with *H. cylindrosporum* in three different sites of "Les Landes" forest located along the Atlantic coast in south-west France (Gryta et al. 1997). These sites were chosen as they represent three different stages in the stabilization of the sand dune system and in the evolution of a monospecific *P. pinaster* forest. At each of these sites, ranging in size from 60 to 500 m^2, sporocarps were precisely mapped, collected and used to establish pure cultures of the corresponding dikaryotic mycelia.

The unambiguous characterization of any fungal genotype can be performed by looking at polymorphisms in different unlinked DNA regions. Different DNA sequences from the genome of *H. cylindrosporum* were shown to be highly polymorphic and suitable for the identification of even closely related genotypes of this species. The studied sequences included the ribosomal intergenic spacer (IGS) region which can easily be amplified by PCR using conserved oligonucleotide primers (Gryta et al. 1996), a monolocus probe (Marmeisse et al. 1992b) and a multilocus probe both of nuclear origin, as well as the mitochondrial genome. In the study of the different *H. cylindrosporum* populations, each of these molecular methods were implemented on each isolate. When two isolates were characterized by identical polymorphisms with each of the four probes it was considered that the corresponding sporocarps were of the same genotype and therefore were formed by the same mycelium.

One of the sampled sites was located behind the coastal sand dune (the extreme limit of pine tree survival in such an unstable environment). At this site, the diversity of ECM species was low, with *H. cylindrosporum* along with *Rhizopogon luteolus* Fr. & Nordh. and *T. terrestris* constituting the dominant species associated with 10- to 20-year-old trees. At this site, which was sampled in the course of a single fruiting season, some of the genets identified enclosed more than 20 sporocarps and the greatest distance found between two sporocarps of the same genotype was 3.6 m. Furthermore, genetic diversity at this site was low, with only two mitochondrial and two IGS types identified. The data for this site are compatible with the survival of the genets for several years and with a very limited number of genets being at the origin of this population (Fig. 4.2, GC site). The two other sites were more stable environments, with 20- to 60-year-old trees. The soil contained a higher percentage of organic matter and more ECM fungal species were present. These two sites were both sampled twice at a 3-year interval. In contrast to the previous site where sporocarps were approximately evenly distributed, at the two latter sites basidiocarps were clustered in groups of more than ten on patches of ground of a few decimetres square. With very few exceptions, the location of patches of sporocarps changed between the two sampling seasons.

The main characteristics of these two sites were that almost all sporocarps were of a unique genotype, even when they almost touched each other. No more than two sporocarps were of the same genotype and, when this was the case, the two sporocarps were always <15 cm apart. None of the genets identified during the first year were re-identified 3 years later. Furthermore, based on the number of mitochondrial types which are inherited from a single parent in such heterothallic homobasidiomycete species, it was clear that several founding events were at the origin of these micropopulations, and that they are probably not isolated from adjacent sites colonized by *H. cylindrosporum*. At the two latter sites it is clear that sexual reproduction played a major role in the short-term evolution of the corresponding populations and that fungal genets were short lived and very rapidly replaced by new ones. From the spatial distribution of the genets it can be deduced that a very small portion of the root system of a single tree can be associated with several genets of the same ECM species (Fig. 4.2).

This study demonstrates that at different sites, the structures of the populations of *H. cylindrosporum* can differ with respect to the survival of the mycelia in the soil and the turnover rate of the genets. These differences raise

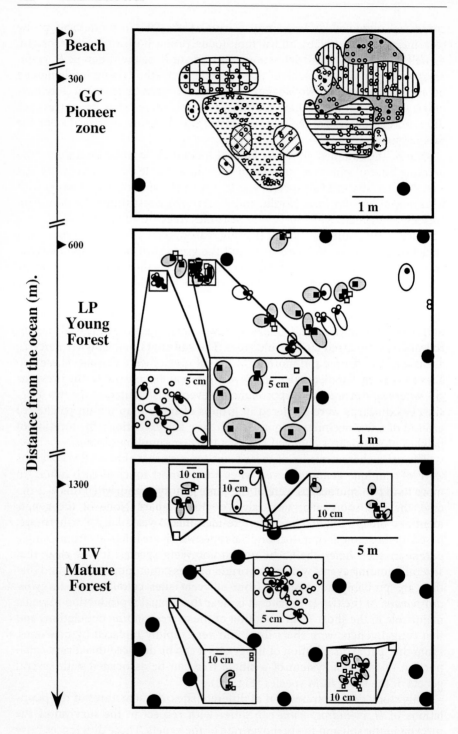

Distance from the ocean (m).

0
Beach

300
GC
Pioneer
zone

1 m

600
LP
Young
Forest

5 cm

5 cm

1 m

1300
TV
Mature
Forest

10 cm

10 cm

10 cm

5 m

5 cm

10 cm

10 cm

the question of their origins. They could result from the intrinsic nature of the sampled sites, with the survival of the mycelia in the soil being favoured at the pioneer sites where fewer competing symbiotic species seem to exist. This could represent an adaptation to poor sandy soils by *H. cylindrosporum*. This study also demonstrates that in natural ecosystems, a plant root system may be associated with a diverse set of fungal symbionts which can change in qualitative and quantitative composition very rapidly. A future challenge that research on ECM symbiosis will have to face is to appreciate the consequences of this diversity on the overall physiology of the host plant under natural conditions.

4.6
Genetics and Molecular Biology

To gain further insights into the developmental processes and potential changes in gene regulation which take place during differentiation of ECM, it is of interest to select mutants of either of the two partners of the symbiosis altered in their symbiotic abilities. The search for such mutants using the fungal partner seems more reasonable than the use of woody plants which have a long generation time. Some species of *Hebeloma* are good candidates for developing a genetic approach to the study of ECM. Among the most interesting features are the possibility of growing haploid mycelia (which can differentiate ECM) in pure culture, the possibility of making crosses and, in the case of *H. cylindrosporum*, the possibility of obtaining fruit bodies producing viable spores which allow genetic analyses to be performed.

4.6.1
Protoplasts and Mutagenesis

For fungal species which do not, or rarely, differentiate asexual spores, the possibility of obtaining large numbers of viable protoplasts may be useful for genetic transformation and to induce mutations on a genetically simple and uniform biological material. Hébraud and Fèvre (1988) and Barrett et al. (1989) both screened a number of different ECM fungal species, including *Hebeloma*, for the formation of viable protoplasts. For the *Hebeloma* species studied [*H. circinans* (Barrett et al. 1989), *H. cylindrosporum* (Hébraud and Fèvre 1988; Barrett et al. 1989), *H. edurum* and *H. sinapizans* (Hébraud and

◀―――

Fig. 4.2. Spacial distribution of genets of *H. cylindrosporum* in three *P. pinaster* forest stands located along the south-west Atlantic coast of France. *Large filled circles* represent *P. pinaster trees*. Sporocarps of *H. cylindrosporum* were collected in 1990 (*small squares*) and in 1993 (*small circles*) from sites at Grand Crohot (*GC*), Le Porge (*LP*) and Truc Vert (*TV*). *Filled circles* and *squares* represent sporocarps whose mycelia were compared for DNA polymorphisms. Areas around sporocarps enclose those of identical genotype. For the *LP* and *TV* sites, enlarged views of small patches of ground with many sporocarps are shown. (After Gryta et al. 1997)

Fèvre 1988)], protoplasts were obtained with yields varying between 0.5×10^6 (*H. edurum*) and $5 \times 10^6 \mathrm{ml}^{-1}$ or more (*H. cylindrosporum*). Protoplast regeneration was only reported for *H. circinans* and *H. cylindrosporum*, with frequencies varying between 1.6 and 18% for the latter species, depending on the strain used.

Among ECM fungi, only mutants of *H. cylindrosporum* have been reported and all mutations were induced by UV irradiation of protoplasts prepared from homokaryotic strains. The first mutants selected were fungicide-resistant mutants (Hébraud and Fèvre 1988). These mutations conferred resistance to either benomyl, benodanil, carboxine or oxycarboxine. Further analyses of these strains (R. Marmeisse, unpubl. data) showed that the resistant phenotypes were dominant in dikaryons heterozygous for these mutations and that mutants selected for resistance toward either benodanil, carboxine or oxycarboxine showed cross-resistance to the other two fungicides.

Among the mutants obtained with *H. cylindrosporum* were mutants affected in metabolic pathways, which could be used to study the nutritional interactions between the fungus and the plant. One class of such mutants are auxotrophs that require either an amino acid (e.g. lysine or methionine) or adenine for growth (R. Marmeisse, unpubl. data). Selection of such mutants is rather tedious as it requires the manual transfer of mycelia from a complete medium on which the protoplasts were plated and irradiated to a minimal medium. Another class of such mutants are those unable to use NO_3^- as the sole N source as a result of mutations affecting nitrate reductase activity (R. Marmeisse et al., unpubl. data). A final class of mutants are those affected in ECM morphogenesis, as illustrated by the auxin-overproducing strains described above.

4.6.2
Gene Cloning and Genetic Transformation

Only four protein sequences from three *Hebeloma* species are presently registered in GenBank. Two of these sequences represent a class II chitin synthase from *H. crustuliniforme* and *H. mesophaeum* (Mehmann et al. 1994). The two protein sequences are highly similar, with only two amino-acid differences. The other two protein sequences are from a linear plasmid which occurs in strains of *H. circinans*; one sequence is of a DNA-polymerase, the other of an RNA-polymerase (Bai et al. 1997). Genetic transformation is a very powerful technique as it allows introduction of very specific modifications in a metabolic or developmental pathway. These modifications can be the overexpression of a cloned gene, the introduction of a novel gene in the fungal genome or the specific inactivation of a known gene. Of the three ECM species transformed so far, one is *H. cylindrosporum* (Marmeisse et al. 1992b), the two others being *Laccaria laccata* (Scop.: Fr.) Berk & Br. (Barrett et al. 1990) and *Paxillus involutus* (Batsch.) Fr. (Bills et al. 1995). In *H.*

cylindrosporum, transformation is performed on protoplasts treated with polyethylene glycol and $CaCl_2$ and a plasmid conferring resistance to the antibiotic hygromycin B (Marmeisse et al. 1992b). The pAN7.1 plasmid used contained the resistance gene fused to regulatory sequences from the ascomycete *Aspergillus nidulans* (Eidam) Wint. (Punt et al. 1987) and transformants of *H. cylindrosporum* were recovered at a frequency of about $5\,\mu g^{-1}$ plasmid and 10^7 protoplasts. In all cases transformation resulted in the stable integration of several copies of the plasmid in the fungal genome (Marmeisse et al. 1992b). All transformants tested retained their ability to form ECM on *P. pinaster* plants. Introduction of additional genes was performed by co-transformation, where the plasmid containing the resistance gene was mixed with a second plasmid containing the gene of interest, which when present in the genome did not confer a selectable phenotype. As for many other fungal species, co-transformation of *H. cylindrosporum* resulted in a high (<70%) yield of transformed hygromycin-resistant mycelia which had integrated the second plasmid. The additional genes introduced by co-transformation were an NADP-glutamate dehydrogenase gene which can participate in NH_4^+ assimilation by the fungus and genes of the tryptophan biosynthetic pathway, this amino acid being the precursor of auxin. These results showed that transformation has the potential to modify functions linked to the ECM ability of the fungus.

4.7
ECM Formation

4.7.1
Developmental Aspects

Hebeloma ECM, either collected in the field or obtained in the laboratory, have been extensively described using light or electron microscopy. Most of the early studies concerned alpine *Hebeloma* species. For instance, Debaud et al. (1981a,b) obtained synthetic ECM by inoculating *D. octopetala* with two of its natural symbionts, *H. alpinum* and *H. marginatulum*. The ECM showed a well-developed mantle and a Hartig net which completely surrounded host cortical cells which were not radially elongated. Polyphosphate granules and glycogen were detected in the mantle and Hartig net hyphae. An American *Dryas*, *D. integrifolia*, has also been used as a host plant to obtain synthetic ECM in association with *H. cylindrosporum* (Fig. 4.3c; Melville et al. 1988). Although *D. integrifolia* is not the natural host of *H. cylindrosporum*, ECM had the same characteristics as those described by Debaud et al. (1981b). Melville et al. (1987a,b) showed that the ontogeny of early stages of *H. cylindrosporum–D. integrifolia* ECM are similar to those described for other ECM. The alpine species *H. repandum* is generally associated with the dwarf willow, *Salix herbacea*. Synthetic ECM were obtained between this fungus and seedlings or cuttings of *S. herbacea* (Graf and Brunner 1996). In comparison to other

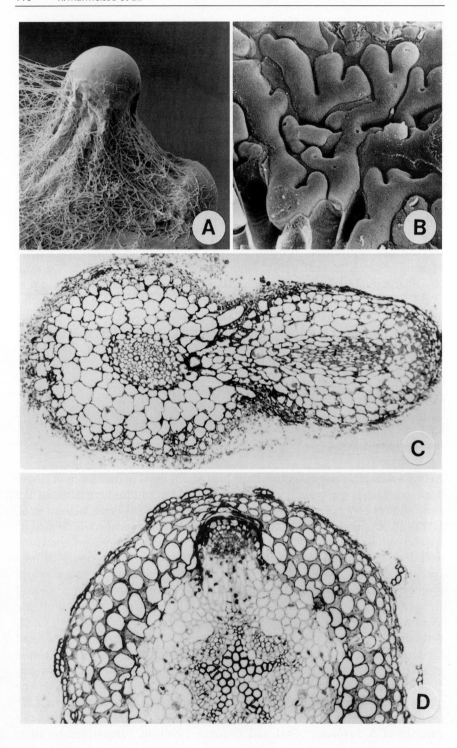

natural associates of this plant, *H. repandum* was less infective and formed ECM with a poorly developed mantle and a Hartig net restricted to the radial walls of epidermal cells.

Numerous *Hebeloma* ECM formed with forest trees have also been described at both structural and ultrastructural levels. Interestingly, descriptions of ECM formed by different *Hebeloma* species associated with various host plants raise questions regarding the use of ECM morphology for the identification of the fungal associates. For example, *H. crustuliniforme* and *H. cylindrosporum* associated with *Picea abies* induced ECM with the same aspect (Brunner 1991). *H. cylindrosporum* which forms a well-developed mantle when associated with *P. abies*, a non-natural host plant, forms a very diffuse one with *P. pinaster*, one of its natural host plants. This, together with the fact that morphological and anatomical characteristics of ECM can be affected by environmental factors (Brunner et al. 1991), complicates the identification of the species of the fungal partner using only ECM morphology, at least for this species.

Under field conditions, ECM are generally formed by dikaryotic mycelia, but monokaryotic strains are also able to form typical ECM. Debaud et al. (1988) compared synthetic ECM formed by mono- and dikaryotic strains of *H. cylindrosporum* associated with *P. pinaster*. Monokaryons were able to form typical ECM identical to those induced by their parental dikaryon. These ECM were functional as demonstrated by an ultrastructural localization of acid phosphatase activity. However, as reported with other species, monokaryons were generally less infective than dikaryons.

Tran Van and Sotta (pers. comm.) studied the dynamics of *P. pinaster* root system colonization and ECM formation by wild-type and IAA-overproducing mutant strains of *H. cylindrosporum*. Long roots (either tap roots or first-order laterals) could be colonized by the fungus, with the formation of a typical Hartig net. Short roots which formed in such colonized long roots were colonized by the hyphae present in the mother root cortex immediately after the long root endodermis broke under the pressure of the growing primordium (Fig. 4.3d). As a consequence of this endogenous colonization, short root primordia had a typical, well-developed Hartig net

Fig. 4.3. Structural aspects of different *Hebeloma* ECM: **A** SEM view of a short dichotomous *H. cylindrosporum–P. pinaster* ECM (Debaud et al. 1997). **B** SEM view of a tangential fracture in the Hartig net of an *H. crustuliniforme–P. abies* mycorrhiza showing branched fingerlike hyphae (Scheidegger and Brunner 1993). **C** Cross section of a lateral root emerging from a first-order lateral root of *D. integrifolia* colonized by *H. cylindrosporum* (Melville et al. 1987b). The first-order root and basal part of the emerging root are surrounded by a fully developed hyphal mantle. The apical part of the emerging root remains uncolonized. **D** Cross section of a tap root of *P. pinaster* colonized by an auxin-overproducer mutant of *H. cylindrosporum*. The Hartig net which reaches the endodermis is comprised of several layers of fungal hyphae. A lateral root is growing through the fully colonized cortex of the main root (H. Tran Van, G. Gay and B. Sotta, unpubl. data)

and could be considered as ECM before they emerged from the mother root. In contrast, a short root formed on a mother root that was devoid of a Hartig net emerged as a non-ECM root and was subsequently colonized by extra-matrical hyphae. This example indicates that for a given fungus–plant combination, different sequences of events can operate during ECM development, depending on the ECM status of mother roots.

These observations are in line with those reported by Scheidegger and Brunner (1993) and Brunner and Scheidegger (1992) who described the ontogeny of *P. abies–H. crustuliniforme* ECM. For both *H. crustuliniforme* and *H. cylindrosporum*, the Hartig net starts to develop in basal parts of the developing ECM and develops towards the apex. Ontogeny of *Dryas integrifolia–H. cylindrosporum* ECM was also studied by Melville et al. (1987b). The only difference between *D. integrifolia* and *P. pinaster* was that lateral root colonization by hyphae emanating from the Hartig net in a primary root was much more rapid with *P. pinaster* than with *D. integrifolia*.

In addition to forming ECM, *Hebeloma* species can interact with in vitro propagated microcuttings or calli. This is the case for *H. cylindrosporum* which has been used to enhance in vitro rooting of micropropagated cuttings of ECM trees such as *P. pinaster* and *P. sylvestris* (Normand et al. 1996), as well as of non-ECM plants such as *Prunus avium* and *Prunus cerasus* (Grange et al. 1997). This fungus has also been used to obtain synthetic ECM in association with somatic embryo-derived *Larix* plantlets (Piola et al. 1995). Similarly, hyphae of *H. crustuliniforme* colonized spruce calli, and fungal structures resembling the fungal sheath and Hartig net were observed (Sirrenberg et al. 1995).

4.7.2
Role of Fungal Auxin

The mechanisms involved in the different morphogenetic steps of ECM formation are still poorly understood. One of the best known, and most controversial, hypotheses is the hormonal theory of ECM formation (Slankis 1973). According to this theory, fungal auxin is responsible for the typical morphology of ECM (Fig. 4.3a) and is necessary for the establishment and functioning of the symbiosis. This theory has frequently been discussed but, as yet, has never been experimentally tested. To assess the role of fungal auxin, IAA-overproducing mutants of *H. cylindrosporum* have been obtained and studied. These mutants were selected using a two-step method based on results which demonstrated that the main limiting factor for fungal IAA production in ECM was tryptophan availability in root exudates (Rouillon et al. 1986; Gay et al. 1989). Therefore, Durand et al. (1992) used UV-irradiated monokaryotic protoplasts to isolate IAA-overproducing mutants which were able to metabolize endogenous tryptophan into IAA.

All monokaryotic mutants were able to form ECM with *P. pinaster*, and they formed three to five times more ECM than did the corresponding wild types

(Durand et al. 1992; Gay et al. 1994). Synthesized dikaryons heterozygous for an IAA$^+$ mutation as well as synthesized dikaryons containing two different IAA$^+$ mutations had ECM activity ca. equal to that of the corresponding wild-type dikaryon. This indicates that the selected mutations are recessive. For this reason, dikaryons homozygous for an IAA$^+$ mutation were synthesized. These dikaryons are more infective than the wild types. All together, these results confirm Slankis's hypothesis that IAA released by ECM fungi is one of the determinants of their infectivity. By increasing plant rhizogenesis, fungal auxin could facilitate the establishment of the symbiosis (Gay et al. 1995).

Striking differences occurred in the extent of Hartig net development by mutants compared to wild-type mycelia (Gay et al. 1994; Gea et al. 1994). In ECM formed by wild-type monokaryons, the Hartig net is generally uniseriate, and restricted to the outer root cortex. By contrast, IAA-overproducing mutants form a hypertrophic pluriseriate Hartig net which frequently reaches the endodermis (Fig. 4.3d). This seems to indicate that fungal IAA may facilitate Hartig net development by affecting host cell wall metabolism and structure. This hypothesis is supported by cytometrical studies which show that both wild-type and mutant monokaryons affect growth polarity of host cortical cells and induce a change of shape of the cells (unpubl. results from the authors' laboratory).

Another unexpected result was the occurrence of mutant hyphae penetrating living host cortical cells (Gea et al. 1994). No necrotic or hypersensitive response by the host tissue was observed and the mutants stimulated the growth of host plants to the same extent as the wild types. Thus, although very invasive, IAA$^+$ mutants did not show pathogen-like behaviour and it was concluded that IAA overproduction does not modify the symbiotic status of the fungus. Instead, it only increases Hartig net formation and allows it to overcome barriers which normally prevent ECM fungi from developing inside living host cortical cells. By reference to the acid growth theory, Gay et al. (1995) hypothesized that fungal auxin may induce a loosening of host cortical cell wall, favourable for fungal intercellular penetration and subsequent Hartig net formation.

The hormonal theory of ECM formation speculates that the absence of ECM frequently observed at high external N concentrations may be ascribed to the inhibition of fungal auxin production. This was investigated by Brunner and co-workers using *H. crustuliniforme-P. abies* ECM as a model. They documented a variety of structural modifications at high levels of NH_4^+ in the culture medium. The most noticeable modification was the absence of Hartig net formation (Brunner and Scheidegger 1995), accompanied by an increase in Ca concentration in the host cortical cell wall (Frey et al. 1997), and callose deposition in the cell wall of epidermal and cortical cells (Brunner and Schneider 1996). Since *H. crustuliniforme* can synthesize auxin in pure culture (Ek et al. 1983; Gay and Debaud 1987), but not at high N concentrations (Rudawska 1983), Brunner and Scheidegger (1995) hypothesized that the absence of the Hartig net in ECM formed at high N level may be ascribed to the inhibition of fungal IAA production. Unfortunately, fungal auxin production within the roots was not studied and no definitive conclusion can thus be drawn.

4.8
Nutritional Interactions

4.8.1
Effect on Host Plant Growth

Different species and strains of *Hebeloma* have been shown to improve tree growth under nursery and field conditions and have been used as commercial inocula. This was the case for *Eucalyptus globulus* associated with *H. westraliense* under nursery conditions (Thomson et al. 1994). In this study, improved growth of the plants was correlated with increased P uptake. Similarly, *H. cylindrosporum* significantly improved growth and survival of different tree species in nurseries (Le Tacon and Valdenaire 1980; Le Tacon et al. 1986). For this fungal species, entrapment of the inoculum in polymeric gels of sodium alginate, which results in better protection of the hyphae (Le Tacon et al. 1985), significantly improved seedling growth compared to other inocula. Long-term beneficial effects of *H. cylindrosporum* could not be demonstrated, with even a depressing effect on plants by this fungus after 10 years compared to uninoculated plants (Le Tacon et al. 1997). This negetive effect may reflect the ill-adaptation of this fungus to the edaphic and climatic conditions of the plantation site. Commercial inoculum of *H. crustuliniforme* has also been produced and used in large-scale inoculation of pine plantations in the USA (Marx 1991).

In the field, plants can form ECM with different fungal partners simultaneously. Consequently, the functioning of such associations under natural conditions is, in most cases, the result of a "multipartner" symbiosis in which the different fungal strains or species may play different and complementary roles, as demonstrated with *P. radiata* inoculated simultaneously with *H. crustuliniforme* and *Rhizopogon* spp. (Chu-Chou and Grace 1985). As a result, the beneficial effect of one fungal species is very difficult if not impossible to quantify. It has also been demonstrated that different host plants can be interconnected by a common mycelium, leading to development of plant communities in which there is evidence of hyphal transport of nutrients between plants and subsequently a competition mediated by ECM. This is the case between *P. menziesii* and *P. ponderosa* co-inoculated with *H. crustuliniforme* and other ECM fungi (Perry et al. 1989).

4.8.2
Influence on Host Carbon Economy

In terms of C metabolism, it is usually considered that ECM fungi represent a sink for plant photosynthates and ECM association may modify the whole plant carbohydrate metabolism. When associated with *H. cylindrosporum*, *P. pinaster* plants showed an increase in net photosynthesis and root respiration rates compared with non-ECM control plants (Conjeaud et al. 1996). In conditions of low P supply, this could result in growth depression of the plant as the fungal partner may represent a stronger sink. Addition of P to non-ECM plants increased their growth but did not result in increased photosynthesis. Similarly, Dosskey et al. (1991) showed that *H. crustuliniforme* tended to enhance net photosynthesis rate of *P. menziesii* plants. The increase in respiration rate of ECM roots, which was quantified as 23% in the *H. crustuliniforme*–*P. contorta* association, was largely thought to result from respiration in the fungal partner (Rygiewicz and Andersen 1994). Durall et al. (1994), studying photosynthate allocation during ageing of *P. ponderosa* ECM, demonstrated that the host plant continues to allocate C to morphologically old roots, with differences existing between ECM species: allocation to old *Laccaria* ECM was significantly stronger than and different to allocation to *H. crustuliniforme* ECM or non-ECM old roots. Given that sucrose is the predominant sugar molecule present in the phloem sap, ECM fungi may depend upon plant cell wall-bound invertase for its assimilation. It was demonstrated that *H. crustuliniforme* lacked this enzyme system and as a consequence could not use sucrose as a C source when it could easily use glucose or fructose (Salzer and Hager 1991). Plant-derived carbohydrates may serve for hyphal growth but also accumulate in the form of, for example, lipids or glycogen in the Hartig net and fungal sheath, as seen in ECM formed between *H. alpinum* or *H. marginatulum* and *D. octopetala* (Debaud et al. 1981b). As suggested by the authors, these important accumulations during fruiting may serve to sustain the rapid differentiation of the sporocarps.

Using ^{14}C-labelled protein added to the growth substrate, Abuzinadah and Read (1989c) demonstrated that mycelia of *H. crustuliniforme*, which are capable of degrading proteins, can transfer a significant amount of C derived from these proteins to *Betula pendula* plants. After a 55-day growth period it was quantified that up to 9% of the plant C came from protein degradation and assimilation by the fungal partner, thus demonstrating that under certain circumstances C allocation can be a bi-directional process.

4.8.3
Phosphorus Metabolism

Hebeloma species are able to improve the P nutrition and subsequently the growth of their host plants, as demonstrated by MacFall et al. (1992), with *P. resinosa* associated with *H. arenosa* in soils with varying levels of P. Mousain et al. (1978) also showed that *P. pinaster* seedlings inoculated with *H. cylindrosporum* in sandy soils have an increased P content. The beneficial effect appears to be greater in soils with low P content and could result from the ability of these fungi to produce abundant extracellular and cell wall-bound acid phosphatases (Ho and Zak 1979) which solubilize insoluble organic or mineral P forms, as demonstrated with *H. crustuliniforme* and

H. cylindrosporum by Lapeyrie et al. (1990) and by MacFall et al. (1991a,b) for *H. arenosa* associated with red pine. Furthermore, acid phosphatase production increases at low P concentrations, as demonstrated with *H. edurum* by Calleja et al. (1980). In contrast, high levels of P can also inhibit ECM formation, as in the case of *H. arenosa* (MacFall et al. 1992).

Mycelia of *H. cylindrosporum* can utilize different forms of P substrates, including Na-phytate and Na-polyphosphate in axenic culture. Phytase activity represents ca. 2% of the para-nitrophenyl phosphatase (PNPPase) activity (Mousain et al. 1988) and ca. 52% of the polyphosphatase activity (Deransart et al. 1990). These authors demonstrated a positive correlation between the secreted PNPase and polyphosphatase activities. A phosphatase secreted by a dikaryotic strain of *H. cylindrosporum* has been purified and used to prepare polyclonal antibodies which could be used to study the regulation of the fungal phosphatase activities during symbiotic association (Deransart et al. 1990). In pure culture, mycelia of *H. cylindrosporum* store little polyphosphate (4–9% of total P), and most of the P in P-starved mycelia is stored in the form of orthophosphate (Rolin et al. 1984). These results confirm those reported by Martin et al. (1983) using ^{31}P nuclear magnetic resonance to study polyphosphate metabolism in intact *H. crustuliniforme* and *H. cylindrosporum* mycelia. In contrast, in the mycelium of *H. crustuliniforme*, Martin et al. (1985) found a large fraction of soluble polyphosphates. This is also the case in the *H. arenosa–P. resinosa* symbiosis, with polyphosphate being present only in ECM roots.

Meysselle et al. (1991) demonstrated a wide range of variation for acid phosphatase activity between unrelated dikaryotic wild strains and within mono- and dikaryotic progenies of a dikaryotic strain of *H. cylindrosporum* (Table 4.2). If a correlation between phosphatase activity of a mycelium and its ability to improve its host-plant P nutrition could be established, this activity could be used as a criterion for the selection of strains to be used in forestry for the plantation of trees in sandy soils where P is a limiting element and where *H. cylindrosporum* occurs naturally. The ecological importance of variability for acid phosphate activity of ECM fungi has been suggested by Tibbett et al. (1998) who compared strains of *Hebeloma* sp. collected in either temperate or arctic ecosystems. Although arctic strains grew more slowly at low temperatures (<12°C) than temperate strains, arctic strains produced significantly more acid phosphatases at these temperatures. The authors suggested that this pattern of enzyme production could represent an adaptation to cold climate by allowing soil organic P to be hydrolysed in frozen soils.

4.8.4
Nitrogen Metabolism

In soils, both mineral (NH_4^+ and NO_3^-) and organic (proteins, peptides, amino acids) forms of N can be found. Organic forms may account for a significant proportion of the total N pool, especially in soils with a low rate of mineralization. Studies on the N nutrition of ECM plants have shown that the contribution of the fungal partner can result from both its ability to assimilate and transfer to the plants N sources located at some distance from the root and also its ability to assimilate organic N sources which cannot be or cannot significantly be assimilated directly by the plant. These different aspects have been well studied with *Hebeloma* species: data exist on the range of N sources which they can use either in pure culture or in association with different host plants, and also on the regulation of some of the metabolic pathways implicated.

H. crustuliniforme has been studied extensively with respect to its use of organic forms of N. In pure culture, mycelia of this species grow well with proteins as the sole N source (Abuzinadah and Read 1986, 1989a,b; Langdale and Read 1989; Read et al. 1989). Extracellular protease activities are detected in the culture filtrates (El-Badaoui and Botton 1989; Zhu et al. 1990; Finlay et al. 1992) and an acid protease has been purified (Zhu et al. 1990). This enzyme, which actively functions between pH 2 and 5, can hydrolyse several proteins such as albumin and casein. Protease production is regulated by the presence of proteins in the medium which act as inducers (El-Badaoui and Botton 1989) and by NH_4^+ which, when present at high concentrations, acts as a repressor (El-Badaoui and Botton 1989; Zhu et al. 1994). The presence of glucose even at high concentrations does not prevent protease production (Zhu et al. 1994), suggesting that in this fungal species this enzyme activity is not subjected to C-mediated catabolic repression. Protein degradation leads to the production of oligopeptides and of amino acids which are ultimately assimilated by the hyphae. *H. crustuliniforme* can use several but not all amino acids as N sources (Abuzinadah and Read 1988). Among the amino acids which are not metabolized are the aromatic amino acids or the S-containing ones; however, these amino acids are usually not present in significant amounts in soils as opposed to molecules such as aspartate, glutamate and arginine which are readily assimilated by the fungus. Protein degradation and the use of amino acids by the fungus may be relevant for the nutrition of the host plants under natural conditions, as suggested in experiments by Abuzinadah and Read (1986), who showed that *B. pendula*, *P. sitchensis* and *P. contorta* plants were dependent upon ECM colonization for growth on subtrates where proteins were available as the sole N source.

Hebeloma species can use NO_3^- as well as NH_4^+ as inorganic N sources. In pure culture, for both *H. crustuliniforme* (Littke et al. 1984; Finlay et al. 1992; Quoreshi et al. 1995) and *H. cylindrosporum* (Wagner et al. 1988, 1989) intraspecific differences exist for the use of these two ions, with, for instance, some strains of *H. cylindrosporum* achieving a higher biomass on NO_3^- than on NH_4^+ (Mention and Plassard 1983; Plassard et al. 1986). In *H. cylindrosporum*, as opposed to model ascomycetes such as *Aspergillus* or *Penicillium*, a significant in vitro nitrate reductase activity can be detected in mycelia growing on NH_4^+, or glutamine-containing media, which suggests that this molecule does not act as a repressor for the transcription of the corresponding gene (Scheromm et al. 1990a,b). A strong increase in nitrate reductase activity was observed when mycelia of this species were starved of N. This enzyme was characterized by Plassard et al. (1984a): it has a strict requirement for NADPH and its affinities (K_m) for NO_3^- (0.15 µM), NADPH (0.18 µM) and FAD (22.7 µM) are similar to the values reported for plant nitrate reductases. Similarly, the *H. cylindrosporum* nitrite reductase needs NADPH and does not seem to be a rate-limiting enzyme for NO_3^- assimilation as NO_2^- does not accumulate in fungal hyphae (Plassard et al. 1984b).

In fungi the two main metabolic pathways which can contribute to the assimilation of NH_4^+ are the glutamate dehydrogenase (GDH) and the glutamine synthetase (GS)/glutamate synthase (GOGAT) pathways. In *Hebeloma* species studied so far, the role of NADP-GDH appears predominant, NADP-GDH being involved only in the catabolic breakdown of glutamate. Both NADP-GDH and GS activities have been detected in extracts of *H. crustuliniforme* (Quoreshi et al. 1995) and of *H. cylindrosporum* (Wagner et al. 1988; Chalot et al. 1991a). For the GS enzyme, one single isoform was found in *H. crustuliniforme*. It remains unclear through which pathway most of the assimilated NH_4^+ is channelled. Downstream of primary NH_4^+ assimilation, by either GDH or GS, is the synthesis of other amino acids by aminotransferases. Alanine- and aspartate-aminotransferase activities, which catalyse the synthesis of two amino acids which accumulate in mycelia, have been measured in *H. cylindrosporum* (Chalot et al. 1990) and *H. crustuliniforme* (Quoreshi et al. 1995). These two enzyme systems, whose activities increase during growth on NH_4^+, have been purified from the latter species. Aspartate-aminotransferase is characterized by one major cationic isoform and a minor anionic one, whereas alanine-aminotransferase occurs as two anionic isoforms (Quoreshi et al. 1995). Interestingly, some of these enzyme systems are differentially regulated in the symbiosis depending on the host plant. The use of ^{15}N-labelled NH_3^+ has clearly demonstrated that in the association *Hebeloma* sp.–*Picea excelsa*, both the fungal GDH and the plant GS pathways are functional for the assimilation of NH_4^+ (Chalot et al. 1991b). Immunolabelling using fungal GDH-specific antibodies showed in this case that the concentration of this GDH decreases from the most external cells of the fungal sheath to the Hartig net hyphae (Botton et al. 1989; Dell et al. 1989). In the case of the association between *Fagus sylvatica* and *H. crustuliniforme*, the fungal GDH activity is extremely low, suggesting that NH_4^+ is primarily assimilated by the GS/GOGAT pathway (Botton et al. 1989; Dell et al. 1989; Chalot et al. 1991b). Similar host-plant-dependent regulations seem to affect fungal aminotransferase activities (Chalot et al. 1990, 1991b). The regulation of fungal enzyme activities, however, is not always under the control of the host plant, as demonstrated by Botton and Dell (1994), who found in same *Eucalyptus* species a strong repression of the GDH activity of *L. laccata* but not of the GDH activity of *H. westraliense*.

When considering the whole plant, *H. crustuliniforme* has been shown to stimulate NH_4^+ assimilation by different host plants (Rygiewicz et al. 1984a).

The same also applies for NO_3^- assimilation by the association *H. cylindrosporum–Pinus pinaster* (Plassard et al. 1994) and also *H. crustuliniforme–P. menziesii* (Rygiewicz et al. 1984b), whereas no stimulation of NO_3^- assimilation was observed when this latter species was associated with *T. heterophylla* or *P. sitchensis*.

4.9
Non-nutritional Interactions

Although the nutritional interactions between plants and ECM fungi have received much attention, the beneficial effect of fungal symbionts on their host plant fitness can also result under some circumstances from non-nutritional effects. *H. crustuliniforme* has, for instance, been shown to reduce radionucleide, present in soils as a result of atmospheric nuclear tests and after the Chernobyl nuclear plant accident, accumulation in *Picea abies* plants. This was studied for ^{134}Cs (Brunner et al. 1996) and for ^{85}Sr (Riesen and Brunner 1996), with inoculated plants containing lower concentrations of the elements compared to non-inoculated controls. Both Cs and K are localized mainly in the vacuoles of the Hartig net hyphae, whereas Sr occurs mainly in electron-opaque and P-rich granules (Frey et al. 1997). Kasuya et al. (1990) demonstrated different sensitivities of ECM fungi to Al (in the form of $AlCl_3$). *H. cylindrosporum* showed partial tolerance and *H. crustuliniforme* a good tolerance to Al. Mycorrhization by such fungi thus has the potential to reduce Al damage to plant roots. Mycelium of *H. crustuliniforme* is also tolerant in vitro to water stress (Coleman et al. 1989) and inoculation of *P. mariana* with *H. longicaudum* resulted in better drought tolerance than in uninoculated controls (Boyle and Hellenbrand 1991).

In the context of protection towards soilborne fungal pathogens, *H. crustuliniforme* was shown to protect beech seedlings against *Pythium ultimum* Trow (Perrin and Garbaye 1983). This species is also resistant (not sensitive) to the deleterious effect of *Verticillium bulbillosum*, a rhizospheric fungus which inhibits ECM formation (Marchetti and Varese 1996).

4.10
Conclusions

ECM fungi belonging to the genus *Hebeloma* and more particularly *H. crustuliniforme*, *H. cylindrosporum* and *H. westraliense* appear as suitable model species for the study of several aspects of the ECM symbiosis. These species can easily be manipulated in the laboratory and so aspects of their physiology, biochemistry and genetics have been studied. Furthermore, for species such as *H. cylindrosporum*, all stages of their life cycles can be obtained under laboratory conditions, from basidiospore germination to sporocarp differentiation, and the mating system has been characterized. These features allow the quantitative analysis of intraspecific variation in characters involved in the ECM association. The availability of viable protoplasts has also allowed, in the case of *H. cylindrosporum*, production of mutant and transgenic strains altered in the differentiation (role of the fungal auxin) and in the functioning (GDH overproducing transformants) of the symbiotic association. These preliminary results should soon be followed by the cloning of fungal genes involved in these pathways and the study of their expression in response to ECM formation. These fungal species are also suitable for studies on ecology and population genetics. Several *Hebeloma* species are found in pioneer environments and differentiate sporocarps a few years after the establishment of woody plants in the ecosystem. Study of these fungal species could serve to establish a link between fungal properties, soil composition, host plant physiology and the evolution of ECM communities as influenced by ageing of the forest ecosystem. *Hebeloma* species whose life cycles have been characterized and for which molecular methods to identify genotypes have or can easily be established could be used as model species to study, in different ecosystems, the structure of ECM fungal populations and colonization strategies in relation to the evolution and diversity of the host plant populations.

Acknowledgements. We would like to thank J. Guinberteau and D. Aanen for their help with the taxonomy of *Hebeloma*; I. Brunner, R. Hugueney, R. L. Peterson, H. Tran Van and B. Sotta for contributing to the illustrations; P. Audenis, C. Raffier and M. C. Verner for printing the pictures and typing the text; and S. Hitchin for correcting the manuscript.

References

Abuzinadah RA, Read DJ (1986) The role of proteins in the nutrition of ectomycorrhizal plants. III. Protein utilization by *Betula*, *Picea* and *Pinus* in mycorrhizal association with *Hebeloma crustuliniforme*. New Phytol 103:507–514

Abuzinadah RA, Read DJ (1988) Amino acids as nitrogen sources for ectomycorrhizal fungi: utilization of individual amino acids. Trans Br Mycol Soc 91:473–479

Abuzinadah RA, Read DJ (1989a) The role of proteins in the nutrition of ectomycorrhizal plants. IV. The utilization of peptides by birch (*Betula pendula* L.) infected with different mycorrhizal fungi. New Phytol 112:55–60

Abuzinadah RA, Read DJ (1989b) The role of proteins in the nutrition of ectomycorrhizal plants. V. Nitrogen transfer in birch (*Betula pendula*) grown in association with mycorrhizal and non-mycorrhizal fungi. New Phytol 112:61–68

Abuzinadah RA, Read DJ (1989c) Carbon transfer associated with assimilation of organic nitrogen sources by silver birch (*Betula pendula* Roth). Trees 3:17–23

Ali NA, Jackson RM (1988) Effects of plant roots and their exudates on germination of spores of ectomycorrhizal fungi. Trans Br Mycol Soc 91:253–260

Bai X, Debaud JC, Schruender J, Meinhardt F (1997) The ectomycorrhizal basidiomycete *Hebeloma circinans* harbours a typical linear plasmid encoding a viral RNA- and DNA-polymerase. J Gen Appl Microbiol 43:273–279

Barrett V, Lemke PA, Dixon RK (1989) Protoplast formation from selected species of ectomycorrhizal fungi. Appl Microbiol Biotechnol 30:381–387

Barrett V, Dixon RK, Lemke PA (1990) Genetic transformation of a mycorrhizal fungus. Appl Microbiol Biotechnol 33:313–316

Bills SN, Richter DL, Podila GK (1995) Genetic transformation of the ectomycorrhizal fungus *Paxillus involutus* by particle bombardment. Mycol Res 99:557–561

Botton B, Dell B (1994) Expression of glutamate dehydrogenase and aspartate aminotransferase in eucalypt ectomycorrhizas. New Phytol 126:249–257

Botton B, Chalot M, Dell B (1989) Changing electrophoretic patterns of glutamate dehydrogenases and aspartate aminotransferases in a few tree species under the influence of ectomycorrhization. Ann Sci For 46:718–720

Bougher NL, Tommerup IC, Malajczuk N (1991) Nuclear behaviour in the basidiomes and ectomycorrhizas of *Hebeloma westraliense* sp. nov. Mycol Res 95:683–688

Boyle CD, Hellenbrand KE (1991) Assessment of the effect of mycorrhizal fungi on drought tolerance of conifer seedlings. Can J Bot 69:1764–1771

Browning MHR, Whitney RD (1992) Field performance of black spruce and jack pine inoculated with selected species of ectomycorrhizal fungi. Can J For Res 12:1974–1982

Bruchet G (1970) Contribution à l'étude du genre *Hebeloma* (Fr.) Kummer. Partie spéciale. Bull Soc Linn Lyon Fr 6:1–132

Bruchet G (1973) Contribution à l'étude du genre *Hebeloma* (Fr.) Kumm. (Basidiomycètes-Agaricales). Essai taxinomique et ecologique. PhD Thesis, University of Lyon, Lyon

Bruchet G (1974) Recherches sur l'écologie des *Hebeloma* arctico-alpins (Basidiomycètes-Agaricales). Etude spéciale de l'aptitude ectomycorrhizogène des espèces. Numéro spécial, Bull Soc Linn Lyon Fr, pp 85–96

Bruchet G (1980) *Hebeloma cylindrosporum* Romagnési. Rev For Fr 3:294–295

Bruchet G, Debaud JC, Gay G (1985) Genetic variations in the physiology of *Hebeloma*. In: Gianinazzi-Pearson V, Gianinazzi S (eds) Physiological and genetical aspects of mycorrhizae. INRA Publ, Paris, pp 121–131

Brunner I (1991) Comparative studies on ectomycorrhizae synthesized with various in vitro techniques using *Picea abies* and two *Hebeloma* species. Trees 5:90–94

Brunner I, Scheidegger C (1992) Ontogeny of synthesized *Picea abies* (L.) Karst.–*Hebeloma crustuliniforme* (Bull. ex St. Amans) Quél. ectomycorrhizas. New Phytol 120:359–369

Brunner I, Scheidegger C (1995) Effects of high nitrogen concentrations on ectomycorrhizal structure and growth of seedlings of *Picea abies* (L.) Karst. New Phytol 129:83–95

Brunner I, Schneider B (1996) Callose in ectomycorrhizas grown at high nitrogen concentrations. In: Azcon-Aguilar C, Barea JM (eds) Mycorrhizas in integrated systems. European Commission, Brussels, pp 178–181

Brunner I, Amiet R, Schneider B (1991) Characterization of naturally grown and in vitro synthesized ectomycorrhizas of *Hebeloma crustuliniforme* and *Picea abies*. Mycol Res 95:1407–1413

Brunner I, Frey B, Riesen TK (1996) Influence of ectomycorrhization and cesium/potassium ratio on uptake and localization of cesium in Norway spruce seedlings. Tree Physiol 16:705–711

Calleja M, Mousain D, Lecouvreur B, d'Auzac J (1980) Influence de la carence phosphatée sur les activités phosphatases acides de trois champignons mycorhiziens: *Hebeloma edurum* Metrod, *Suillus granulatus* (L. and Fr.) O. Kuntze et *Pisolithus tinctorius* (Pers.) Coker and Couch. Physiol Vég 18:489–504

Chalot M, Brun A, Khalid A, Dell B, Rohr R, Botton B (1990) Occurrence and distribution of aspartate aminotransferase in spruce and beech ectomycorrhizas. Can J Bot 68:1756–1762

Chalot M, Brun A, Debaud JC, Botton B (1991a) Ammonium assimilating enzymes and their regulation in wild and NADP-glutamate dehydrogenase-deficient strains of the ectomycorrhizal fungus *Hebeloma cylindrosporum*. Physiol Plant 83:122–128

Chalot M, Stewart GR, Brun A, Martin F, Botton B (1991b) Ammonium assimilation by spruce-*Hebeloma* sp. ectomycorrhizas. New Phytol 119:541–550

Chu-Chou M (1979) Mycorrhizal fungi of *Pinus radiata* in New Zealand. Soil Biol Biochem 11:557–562

Chu-Chou M, Grace LJ (1985) Comparative efficiency of the mycorrhizal fungi *Laccaria laccata*, *Hebeloma crustuliniforme* and *Rhizopogon* species on growth of radiata pine seedlings. NZ J Bot 23:417–424

Chu-Chou M, Grace LJ (1987) Mycorrhizal fungi of *Pseudotsuga menziesii* in the South Island of New Zealand. Soil Biol Biochem 19:243–246

Coleman MD, Bledsoe C, Lopushinsky W (1989) Pure culture response of ectomycorrhizal fungi to imposed water stress. Can J Bot 67:29–39

Conjeaud C, Scheromm P, Mousain D (1996) Effects of phosphorus and ectomycorrhiza on maritime pine seedlings (*Pinus pinaster*). New Phytol 133:345–351

Contu M (1991) Contributo allo studio del genere *Hebeloma* (Basidiomycetes, Cortinariaceae) in Sardegna (Italia). I. Rev Iberoam Micol 8:38–42

Cote JF, Thibault JR (1988) Allelopathic potential of raspberry foliar leachates on growth of ectomycorrhizal fungi associated with black spruce. Am J Bot 75:966–970

Deacon JW, Donaldson SJ, Last FT (1983) Sequences and interactions of mycorrhizal fungi on birch. Plant Soil 71:257–262

Debaud JC (1987) Ecophysiological studies on alpine macromycetes: saprophytic *Clitocybe* and mycorrhizal *Hebeloma* associated with *Dryas octopetala*. In: Laursen GA, Ammirati JF, Redhead SA (eds) Arctic and alpine mycology II. Plenum Press, London, pp 47–60

Debaud JC, Gay G (1987) In vitro fruiting under controlled conditions of the ectomycorrhizal fungus *Hebeloma cylindrosporum* associated with *Pinus pinaster*. New Phytol 105:429–435

Debaud JC, Pepin R, Bruchet G (1981a) Étude des ectomycorrhizes de *Dryas octopetala*. Obtention de synthèses mycorrhiziennes et de carpophores d'*Hebeloma alpinum* et *H. marginatulum*. Can J Bot 59:1014–1020

Debaud JC, Pepin R, Bruchet G (1981b) Ultrastructure des ectomycorhizes synthétiques à *Hebeloma alpinum* et *Hebeloma marginatulum* de *Dryas octopetala*. Can J Bot 59:2160–2166

Debaud JC, Gay G, Bruchet G (1986) Intraspecific variability in an ectomycorrhizal fungus: *Hebeloma cylindrosporum*. I. Preliminary studies on in vitro fruiting, spore germination and sexual comportment. In: Gianinazzi-Pearson V, Gianinazzi S (eds) Physiological and genetical aspects of mycorrhizae. INRA Publ, Paris, pp 581–588

Debaud JC, Gay G, Prevost A, Lei J, Dexheimer J (1988) Ectomycorrhizal ability of genetically different homokaryotic and dikaryotic mycelia of *Hebeloma cylindrosporum*. New Phytol 108:323–328

Debaud JC, Marmeisse R, Gay G (1995) Intraspecific genetic variation in ectomycorrhizal fungi. In: Varma AK, Hock B (eds) Mycorrhiza: structure, molecular biology and function. Springer, Berlin Heidelberg New York, pp 79–113

Debaud JC, Marmeisse R, Gay G (1997) Genetic and molecular biology of the fungal partner in the ectomycorrhizal symbiosis *Hebeloma cylindrosporum* × *Pinus pinaster*. In: Carroll GC, Tudzynski P (eds) The Mycota, part B. Springer, Berlin Heidelberg New York, pp 95–115

Dell B, Botton B, Martin F, Le Tacon F (1989) Glutamate dehydrogenases in ectomycorrhizas of spruce (*Picea excelsa* L.) and beech (*Fagus sylvatica*). New Phytol 111:683–692

Delmas J (1978) The potential cultivation of various edible fungi. In: Chan ST, Hayes WA (eds) The biology and cultivation of edible mushrooms. Academic Press, New York, pp 699–724

Deransart C, Chaumat E, Cleyet-Marel JC, Mousain D, Labarère J (1990) Purification assay of phosphatases secreted by *Hebeloma cylindrosporum* and preparation of polyclonal antibodies. Symbiosis 9:185–194

Dosskey MG, Boersma L, Linderman RG (1991) Role for the photosynthate demand of ectomycorrhizas in the response of Douglas fir seedlings to drying soil. New Phytol 117:327–334

Durall DM, Marshall JD, Jones MD, Crawford R, Trappe JM (1994) Morphological changes and photosynthate allocation in ageing *Hebeloma crustuliniforme* (Bull.) Quel. and *Laccaria bicolor* (Maire) Orton mycorrhizas of *Pinus ponderosa* Dougl. ex Laws. New Phytol 127:719–724

Durand N, Debaud JC, Casselton LA, Gay G (1992) Isolation and preliminary characterization of 5-fluoroindole-resistant and IAA-overproducer mutants of the ectomycorrhizal fungus *Hebeloma cylindrosporum* Romagnesi. New Phytol 121:545–553

Ek M, Jungquistl PO, Stenström E (1983) Indole-3-acetic acid production by mycorrhizal fungi determined by gas chromatography-mass spectrometry. New Phytol 94:401–407

El-Badaoui K, Botton B (1989) Production and characterization of exocellular proteases in ectomycorrhizal fungi. Ann Sci For 46:728-730

Favre J (1960) Catalogue descriptif des champignons superieurs de la zone alpine du parc national Suisse. Liestal, Lüdin

Finlay RD, Frostegard A, Sonnerfeldt AM (1992) Utilization of organic and inorganic nitrogen sources by ectomycorrhizal fungi in pure culture and in symbiosis with *Pinus contorta* Dougl. ex Loud. New Phytol 120:105-115

Fleming LV, Deacon JW, Last FT, Donaldson SJ (1984) Influence of propagating soil on the mycorrhizal succession of birch seedlings transplanted to a field site. Trans Br Mycol Soc 82:707-711

Fox FM (1986) Groupings of ectomycorrhizal fungi of birch and pine, based on establishment of mycorrhizas on seedlings from spores in unsterile soils. Trans Br Mycol Soc 87:371-380

Frey B, Brunner I, Walther P, Scheidegger C, Zierold K (1997) Element localization in ultrathin cryosections of high-pressure frozen ectomycorrhizal spruce roots. Plant Cell Environ 20:929-937

Fries N (1984) Spore germination in the higher Basidiomycetes. Proc Indian Acad Sci 3:205-222

Fries N, Birraux D (1980) Spore germination in *Hebeloma* stimulated by living plant roots. Experientia 36:1056-1057

Fries N, Swedjemark G (1986) Specific effects of tree roots on spore germination in the ectomycorrhizal fungus, *Hebeloma mesophaeum* (Agaricales). In: Gianinazzi-Pearson V, Gianinazzi S (eds) Physiological and genetical aspects of mycorrhizae. INRA Publ, Paris, pp 725-730

Gay G, Debaud JC (1987) Genetic study on indole-3-acetic acid production by ectomycorrhizal *Hebeloma* species: inter- and intraspecific variability in homo- and dikaryotic mycelia. Appl Microbiol Biotechmol 26:141-146

Gay G, Rouillon R, Bernillon J, Favre-Bonvin J (1989) IAA biosynthesis by the ectomycorrhizal fungus *Hebeloma hiemale* as affected by different precursors. Can J Bot 67:2235-2239

Gay G, Bernillon J, Debaud JC (1993) Comparative analysis of IAA production in ecto-mycorrhizal, ericoid and saprophytic fungi in pure culture. In: Read DJ, Lewis DH, Fitter AH, Alexander IJ (eds) Mycorrhizas in ecosystems. CAB International, Cambridge, pp 356-366

Gay G, Normand L, Marmeisse R, Sotta B, Debaud JC (1994) Auxin overproducer mutants of *Hebeloma cylindrosprum* Romagnesi have increased mycorrhizal activity. New Phytol 128:645-657

Gay G, Sotta B, Tran-Van H, Gea L, Vian B (1995) Fungal auxin is involved in ectomycorrhiza formation: genetical, biochemical and ultrastructural studies with IAA-overproducer mutants of *Hebeloma cylindrosporum*. In: Sandermann H, Bonnet-Masimbert M (eds) Eurosilva contribution to forest tree physiology. INRA Publ, Paris, pp 215-231

Gea L, Normand L, Vian B, Gay G (1994) Structural aspects of ectomycorrhiza of *Pinus pinaster* (Ait.) Sol. formed by an IAA-overproducer mutant of *Hebeloma cylindrosporum* Romagnesi. New Phytol 128:659-670

Gibson F, Deacon JW (1988) Experimental study of establishment of ectomycorrhizas in different regions of birch root systems. Trans Br Mycol Soc 91:239-251

Gibson F, Deacon JW (1990) Establishment of ectomycorrhizas in aseptic culture: effects of glucose, nitrogen and phosphorus in relation to successions. Mycol Res 94:166-172

Giltrap NJ (1982) *Hebeloma* spp. as mycorrhizal associates of birch. Trans Br Mycol Soc 79:157-160

Godbout C, Fortin JA (1990) Cultural control of basidiome formation in *Laccaria bicolor* with container-grown white pine seedlings. Mycol Res 94:1051-1058

Graf F, Brunner I (1996) Natural and synthesised ectomycorrhizas of the alpine dwarf willow *Salix herbacea*. Mycorrhiza 6:227-235

Grange O, Bärtschi H, Gay G (1997) Effect of the ectomycorrhizal fungus *Hebeloma cylindrosporum* on in vitro rooting of micropropagated cuttings of arbuscular-forming *Prunus avium* and *Prunus cerasus*. Trees 12:49-56

Grove TS, Le Tacon F (1993) Mycorrhiza in plantation forestry. Adv Plant Pathol 23:191-227

Gryta H, Marmeisse R, Gay G, Guinberteau J, Debaud JC (1996) Molecular characterization of *Hebeloma cylindrosporum*: an ectomycorrhizal basidiomycete associated with *Pinus pinaster* in coastal sand dunes. In: Azcon-Aguilar C, Barea JM (eds) Mycorrhizas in integrated systems. European Commission, Brussels, pp 35–38

Gryta H, Debaud JC, Effosse A, Gay G, Marmeisse R (1997) Fine-scale structure of populations of the ectomycorrhizal fungus *Hebeloma cylindrosporum* in coastal sand dune forest ecosystem. Mol Ecol 6:353–364

Guinberteau J (1997) La mycoflore des écosystèmes dunaires du Bas-Médoc (Gironde, France). Approche écodynamique des mycocoenoses dunaires Atlantiques. In: Favennec J, Barrère P (eds) Biodiversité et protection dunaire. Lavoisier Tec and Doc, Paris, pp 242–261

Hacskaylo E, Bruchet G (1972) Hebelomas as mycorrhizal fungi. Bull Torrey Bot Club 99:17–20

Hawksworth DL (1991) The fungal dimension of biodiversity: magnitude, significance and conservation. Mycol Res 95:641–655

Hébraud M, Fèvre M (1988) Protoplast production and regeneration from mycorrhizal fungi and their use for isolation of mutants. Can J Microbiol 34:157–161

Ho I, Zak B (1979) Acid phosphatase activity of six ectomycorrhizal fungi. Can J Bot 79:1203–1205

Jansen AE (1982) *Lactarius hysginus* en *Hebeloma cylindrosporum* in Nederland. Coolia 25:62–67

Kasuya MCM, Muchovej RMC, Muchovej JJ (1990) Influence of aluminium on in vitro formation of *Pinus caribaea* mycorrhizae. Plant Soil 124:73–78

Langdale AR, Read DJ (1989) Substrate decomposition and product release by ericoid and ectomycorrhizal fungi grown on protein. Agric Ecol Environ 28:285–291

Lapeyrie F, Ranger J, Vairelles D (1990) Phosphate solubilizing activity of ectomycorrhizal fungi in vitro. Can J Bot 69:342–346

Last FT, Mason PA, Wilson J, Deacon JW (1983) Fine roots and sheathing mycorrhizas: their formation, function and dynamics. Plant Soil 71:9–21

Last FT, Mason PA, Pelham J, Ingleby K (1984) Fruitbody production by sheathing mycorrhizal fungi: effects of "host" genotypes and propagating soils. For Ecol Manage 9:221–227

Le Tacon F, Valdenaire JM (1980) La mycorhization contrôlée en pépinière, premiers résultats obtenus à la pépinière du Fond Forestier national de Peyrat-le-Chateau sur Epicéa et Douglas. Rev For Fr 3:281–292

Le Tacon F, Jung G, Mugnier J, Michelot P, Mauperin C (1985) Efficiency in a forest nursery of an ectomycorrhizal fungus inoculum produced in a fermentor and entrapped in polymeric gels. Can J Bot 9:1664–1668

Le Tacon F, Bouchard D, Perrin R (1986) Effects of soil fumigation and inoculation with pure culture of *Hebeloma cylindrosporum* on survival, growth and ectomycorrhizal development of Norway spruce and Douglas fir seedlings. Eur J For Pathol 16:257–265

Le Tacon F, Mousain D, Garbaye J, Bouchard D, Churin JL, Argillier C, Amirault JM, Généré B (1997) Mycorhizes, pépinières et plantations forestières en France. Rev For Fr Spec Issue (Champignons et mycorhizes en forêt):131–154

Lei J, Dexheimer J (1987) Preliminary results concerning the controlled mycorrhization of oak (*Quercus robur* L.) vitroplants. Ann Sci For 44:315–324

Littke WR, Bledsoe CS, Edmonds RL (1984) Nitrogen uptake and growth in vitro by *Hebeloma crustuliniforme* and other Pacific Northwest mycorrhizal fungi. Can J Bot 62:647–652

MacFall J, Slack SA, Iyer J (1991a) Effects of *Hebeloma arenosa* and phosphorus fertility on growth of red pine (*Pinus resinosa*) seedlings. Can J Bot 69:372–379

MacFall J, Slack SA, Iyer J (1991b) Effects of *Hebeloma arenosa* and phosphorus fertility on root acid phosphatase activity of red pine (*Pinus resinosa*) seedlings. Can J Bot 69:380–383

MacFall J, Slack SA, Wehrli S (1992) Phosphorus distribution in red pine roots and the ectomycorrhizal fungus *Hebeloma arenosa*. Plant Physiol 100:713–717

Marchetti M, Varese GC (1996) Influence of *Vertilcilium bulbillosum* on in vitro formation of mycorrhizae by *Laccaria laccata* and *Hebeloma crustuliniforme* with *Picea abies*. Allionia 34:45–54

Marmeisse R, Debaud JC, Casselton LA (1992a) DNA probes for species and strain identification in the ectomycorrhizal fungus *Hebeloma*. Mycol Res 96:161–165

Marmeisse R, Gay G, Debaud JC, Casselton LA (1992b) Genetic transformation of the symbiotic basidiomycete fungus *Hebeloma cylindrosporum*. Curr Genet 22:41–45

Martin F, Canet D, Marchal JP, Larher F (1983) Phosphorus 31 nuclear magnetic resonance study of phosphate metabolism in intact ectomycorrhizal fungi. Plant Soil 71:469–476

Martin F, Marchal JP, Timinska A, Canet D (1985) The metabolism and physical state of polyphosphates in ectomycorrhizal fungi. A ^{31}P nuclear magnetic resonance study. New Phytol 101:275–290

Marx DH (1991) The practical significance of ectomycorrhizae in forest establishment. In: Ecophysiology of ectomycorrhizae of forest trees. Marcus Wallenberg Foundation Symp Proc, vol 7. Stockholm, Sweden, pp 54–90

McAfee BJ, Fortin JA (1988) Comparative effects of the soil microflora on ectomycorrhizal inoculation of conifer seedlings. New Phytol 108:443–449

Mehmann B, Brunner I, Braus GH (1994) Nucleotide sequence variation of chitin synthase genes among ectomycorrhizal fungi and its potential use in taxonomy. Appl Environ Microbiol 60:3105–3111

Melville LH, Massicotte HB, Peterson RL (1987a) Morphological variations in developing mycorrhizae of *Dryas integrifolia* and five fungal species. Scanning Microsc 1:1455–1464

Melville LH, Massicotte HB, Peterson RL (1987b) Ontogeny of early stages of ectomycorrhizae synthesized between *Dryas integrifolia* and *Hebeloma cylindrosporum*. Bot Gaz 148:332–341

Melville LH, Massicotte HB, Ackerley CA, Peterson RL (1988) An ultrastructural study of modifications in *Dryas integrifolia* and *Hebeloma cylindrosporum* during ectomycorrhiza formation. Bot Gaz 149:408–418

Mention M, Plassard C (1983) Comparaison de la nutrition nitrique et ammoniacale de quatre espèces de Basidiomyètes ectomycorhiziens. CR Acad Sci Fr 297:489–492

Meysselle JP, Gay G, Debaud JC (1991) Intraspecific genetic variation of acid phosphatase activity in monokaryotic and dikaryotic populations of the ectomycorrhizal fungus *Hebeloma cylindrosporum*. Can J Bot 69:808–813

Miller SL, Koo CD, Molina R (1991) Characterization of red alder ectomycorrhizae: a preface to monitoring belowground ecological responses. Can J Bot 69:516–531

Moser M (1970) Beiträge zur Kenntnis der Gattung *Hebeloma*. Z Pilzkd 36:61–75

Moser M (1985) Beiträge zur Kenntnis der Gattung *Hebeloma*. II. Sydowia 38:171–177

Moser M, Jülich W (1987) Farbatlas der Basidiomyceten. Gustav Fischer, Stuttgart

Mousain D, Poitou N, Delmas J (1978) La symbiose mycorhizienne: résultats obtenus avec *Hebeloma cylindrosporum* et *Pisolithus tinctorius* et les perspectives d'application agronomique. In: Delmas J (ed) Mushroom science X part I. Center for Agricultural Publications and Documentation, Wageningen, pp 949–956

Mousain D, Bousquet N, Polard C (1988) Comparaison des activités phosphatases d'homobasidiomycètes ectomycorhiziens en culture in vitro. J Eur Pathol For 18:299–309

Normand L, Bärtschi H, Debaud JC, Gay G (1996) Rooting and acclimatization of micropropagated cuttings of *Pinus pinaster* and *Pinus sylvestris* are enhanced by the ectomycorrhizal fungus *Hebeloma cylindrosporum*. Physiol Plant 98:759–766

Oh KI, Melville LH, Peterson RL (1995) Comparative structural study of *Quercus serrata* and *Q. acutissima* formed by *Pisolithus tinctorius* and *Hebeloma cylindrosporum*. Trees 3:171–179

Perrin R, Garbaye J (1983) Influence of ectomycorrhizae on infectivity of *Pythium*-infested soils and substrates. Plant Soil 71:345–351

Perry DA, Margolis H, Choquette C, Molina R, Trappe JM (1989) Ectomycorrhizal mediation of competition between coniferous tree species. New Phytol 112:501–512

Piché Y, Fortin JA (1982) Development of mycorrhizae, extramatrical mycelium and sclerotia on *Pinus strobus* seedlings. New Phytol 91:211–220

Piola F, Rohr R, von Aderkas P (1995) Controlled mycorrhizal initiation as a means to improve root development in somatic embryo plantlets of hybrid larch (*Larix* x *eurolepis*). Physiol Plant 95:575–580

Plassard P, Mousain D, Salsac L (1984a) Mesure in vitro de l'activité nitrate réductase dans les thalles de *Hebeloma cylindrosporum*, champignon basidiomycète. Physiol Vég 22:67–74

Plassard P, Mousain D, Salsac L (1984b) Mesure in vitro et in vivo de l'activité nitrite réductase dans les thalles de *Hebeloma cylindrosporum*, champignon basidiomycète. Physiol Vég 22:147–154

Plassard C, Martin F, Mousain D, Salsac L (1986) Physiology of N assimilation by mycorrhizas. In: Gianinazzi-Pearson V, Gianinazzi S (eds) Physiological and genetical aspects of mycorrhizae. INRA Publ, Paris, pp 11–21

Plassard C, Barry D, Eltrop L, Mousain D (1994) Nitrate uptake in maritime pine (*Pinus pinaster*) and the ectomycorrhizal fungus *Hebeloma cylindrosporum*: effect of ectomycorrhizal symbiosis. Can J Bot 72:189–197

Punt PJ, Oliver RP, Dingemanse MA, Pouwels PH, van den Hondel CAMJJ (1987) Transformation of *Aspergillus* based on the hygromycin B resistance marker from *Escherichia coli*. Gene 56:117–124

Quoreshi AM, Ahmad I, Malloch D, Hellebust JA (1995) Nitrogen metabolism in the ectomycorrhizal fungus *Hebeloma crustuliniforme*. New Phytol 131:263–271

Read DJ, Leake JR, Langdale AR (1989) The nitrogen nutrition of mycorrhizal fungi and their host plants. In: Boddy L, Marchant R, Read DJ (eds) Nitrogen, phosphorus and sulphur utilization by fungi. Cambridge University Press, Cambridge, pp 181–204

Riesen TK, Brunner I (1996) Effect of ectomycorrhizae and ammonium on ^{134}CS and ^{85}SR uptake into *Picea abies* seedlings. Environ Pollut 93:1–8

Rolin D, Le Tacon F, Larher F (1984) Characterization of the different forms of phosphorus in the mycelium of the ectomycorrhizal fungus *Hebeloma cylindrosporum* cultivated in pure culture. New Phytol 98:335–343

Romagnesi H (1965) Étude sur le genre *Hebeloma*. Bull Soc Mycol Fr 81:321–344

Romagnesi H (1983) Étude sur le genre *Hebeloma* II. Sydowia 36:255–268

Romano A, Martins-Loucao MA (1994) Mycorrhization of cork oak (*Quercus suber* L.). Rev Biol 15:1–4

Rouillon R, Gay G, Bernillon J, Favre Bonvin J, Bruchet G (1986) Analysis by HPLC-mass spectrometry of the indole compounds released by the ectomycorrhizal fungus *Hebeloma hiemale* in pure culture. Can J Bot 64:1893–1897

Rudawska M (1983) The effect of nitrogen and phosphorus on auxin and cytokinin production by mycorrhizal fungi. Arbor Kornickie 28:219–236

Rygiewicz PT, Andersen CP (1994) Mycorrhizae alter quality and quantity of carbon allocated below ground. Nature 369:58–60

Rygiewicz PT, Bledsoe CJ, Zasoski RJ (1984a) Effects of ectomycorrhizae and solution pH on (^{15}N) ammonium uptake by coniferous seedlings. Can J For Res 14:885–892

Rygiewicz PT, Bledsoe CJ, Zasoski RJ (1984b) Effects of ectomycorrhizae and solution pH on (^{15}N) nitrate uptake by coniferous seedlings. Can J For Res 14:893–899

Sagara N (1992) Experimental disturbances and epigeous fungi. In: Carroll GC, Wicklow DT (eds) The fungal community, its organization and role in the ecosystem. Marcel Dekker, New York, pp 427–454

Sagara N, Kitamoto Y, Nishio R, Yoshimi S (1985) Association of two *Hebeloma* species with decomposed nests of vespine wasps. Trans Br Mycol Soc 84:349–352

Salzer P, Hager A (1991) Sucrose utilization of the ectomycorrhizal fungi *Amanita muscaria* and *Hebeloma crustuliniforme* depends on the cell wall-bound invertase activity of their host *Picea abies*. Bot Acta 104:439–445

Scheidegger C, Brunner I (1993) Freeze-fracturing for low-temperature scanning electron microscopy of Hartig net in synthesized *Picea abies–Hebeloma crustuliniforme* and– *Tricholoma vaccinum* ectomycorrhizas. New Phytol 123:123–132

Scheromm P, Plassard C, Salsac L (1990a) Effect of nitrate and ammonium nutrition on the metabolism of the ectomycorrhizal basidiomycete *Hebeloma cylindrosporum* Romagn. New Phytol 114:227–234

Scheromm P, Plassard C, Salsac L (1990b) Regulation of nitrate reductase in the ectomycorrhizal basidiomycete *Hebeloma cylindrosporum* Romagn. cultured on nitrate or ammonium. New Phytol 114:441–447

Shemakhanova NM (1967) Mycotrophy of woody plants. Israel Program for Scientific Translation Ltd, Jerusalem, Israel

Singer R (1986) The agaricales in modern taxonomy 4th edn. Sven Koeltz Scientific Books, Koenigstein

Sirrenberg A, Salzer P, Hager A (1995) Induction of mycorrhiza-like structures and defence reactions in dual cultures of spruce callus and ectomycorrhizal fungi. New Phytol 130:149–156

Slankis V (1973) Hormonal relationships in mycorrhizal development. In: Marks GC, Koslowski TT (eds) Ectomycorrhizae, their ecology and physiology. Academic Press, New York, pp 231–298

Smith AH, Everson VS, Mitchell DH (1983) The veiled species of *Hebeloma* in the western United States. University of Michigan Press, Ann Arbor

Smith SE, Read DJ (1997) Mycorrhizal symbiosis, 2nd edn. Academic Press, London

Thomson BD, Grove TS, Malajczuk N, Hardy GEStJ (1994) The effectiveness of ectomycorrhizal fungi in increasing the growth of *Eucalyptus globulus* Labill. in relation to root colonization and hyphal development in soil. New Phytol 126:517–524

Tibbett M, Sanders FE, Cairney JWG (1998) The effect of temperature and inorganic phosphorus supply on growth and acid phosphatase production in arctic and temperate strains of ectomycorrhizal *Hebeloma* spp. in axenic culture. Mycol Res 102:129–135

Tommerup IC, Bougher NL, Malajczuk N (1991) *Laccaria fraterna*, a common ectomycorrhizal fungus with mono- and bi-sporic basidia and multinucleate spores: comparison with the quadristerigmate, binucleate spored *L. laccata* and the hypogeous relative *Hydnangium carneum*. Mycol Res 95:689–698

Valjalo J (1979) Étude de la nature et du rôle de l'association ectomycorhizienne chez le châtaignier: cas de l'*Hebeloma cylindrosporum* (Romagnési). In: Delmas J (ed) Mushroom science X, part I. Center for Agricultural Publications and Documentation, Wageningen, pp 903–918

Vesterholt J (1989) A revision of *Hebeloma* sect. *Indusiata* in the Nordic countries. Nord J Bot 9:289–319

Vesterholt J (1995) *Hebeloma crustuliniforme* and related taxa – notes on some characters of taxonomic importance. Symb Bot Ups 3:129–137

Wagner F, Gay G, Debaud JC (1988) Genetical variability of the glutamate dehydrogenase activity in monokaryotic and dikaryotic mycelia of the ectomycorrhizal fungus *Hebeloma cylindrosporum*. Appl Microbiol Biotechnol 28:566–571

Wagner F, Gay G, Debaud JC (1989) Genetical variation of nitrate reductase activity in mono- and dikaryotic populations of the ectomycorrhizal fungus *Hebeloma cylindrosporum* Romagnesi. New Phytol 113:259–264

Wong KKY, Fortin JA (1989) A Petri dish technique for the aseptic synthesis of ectomycorrhizae. Can J Bot 67:1713–1716

Wong KKY, Fortin JA (1990) Root colonization and intraspecific mycobiont variation in ectomycorrhiza. Symbiosis 8:197–231

Zak B (1976a) Pure culture synthesis of bearberry mycorrhizae. Can J Bot 54:1297–1305

Zak B (1976b) Pure culture synthesis of Pacific madrone ectendomycorrhizae. Mycologia 68:362–369

Zhu H, Higginbotham KO, Dancik BP (1988) Intraspecific genetic variability of isoenzymes in the ectomycorrhizal fungus *Suillus tomentosus*. Can J Bot 66:588–594

Zhu H, Guo DC, Dancik BP (1990) Purification and characterisation of an extracellular acid proteinase from the ectomycorrhizal fungus *Hebeloma crustuliniforme*. Appl Environ Microbiol 56:837–843

Zhu H, Dancik BP, Higginbotham KO (1994) Regulation of extracellular proteinase production in an ectomycorrhizal fungus *Hebeloma crustuliniforme*. Mycologia 86:227–234

Chapter 5

Rhizopogon

R. Molina[1], J. M. Trappe[2], L. C. Grubisha[3] and J. W. Spatafora[3]

5.1
Introduction

Rhizopogon is the largest genus of hypogeous Basidiomycota, with worldwide distribution among Pinaceae. Several *Rhizopogon* species are important members of ectomycorrhizal (ECM) fungal communities, contributing significantly to sporocarp productivity and ECM dominance. They occur in young and old forest stands alike and in diverse habitats. This ecological amplitude was recognized early in the twentieth century when *Rhizopogon* species were observed as dominant ECM fungi on *Pinus* in exotic plantations. Consequently, *Rhizopogon* has been the focus of considerable application research in forestry. The ease of culturing from sporocarps, manipulation of pure cultures of *Rhizopogon* and practical use of spore inoculation has made *Rhizopogon* a model genus to explore morphological, physiological, ecological, and symbiotic mutualisms of ECM. Nearly 200 papers have been published on *Rhizopogon* taxonomy, host range and specificity, ECM morphology, distribution, ecology, physiology, and applications in forestry.

5.2
Taxonomy

Elias Fries described the genus and species *R. luteolus* in 1817. The quaintly descriptive generic name literally translates to "root beard" in reference to the clustered, root-like rhizomorphs "bearding" sporocarps of *R. luteolus*.[4]

---

[1] USDA Forest Service, Pacific Northwest Research Station, Forestry Sciences Laboratory, 3200 Jefferson Way, Corvallis, Oregon 97331, USA
[2] Department of Forest Science, Oregon State University, Corvallis, Oregon 97331, USA
[3] Department of Botany and Plant Pathology, Oregon State University, Corvallis, Oregon 97331, USA
[4] The authority for the names *Rhizopogon* and *luteolus* is often incorrectly cited as "Fries and Nordholm." Although Johan Nordholm's name appears on the 1817 paper, he was a student of Fries, the sole author. Svengunnar Ryman, Curator of the Mycological Herbarium of the University of Uppsala, explains why (pers. comm.): "According to the constitution of Uppsala University from 1655 and Lund University from 1666, every professor must write and officially discuss a thesis at least once a year. The thesis was typically defended not by the faculty author

Fries (1823) described three more species, and others (Vittadini 1831; Tulasne and Tulasne 1844, 1851; Zobel 1854) added to the understanding and misunderstanding of the genus (see review by Martín 1996).

Zeller and Dodge (1918) were the first to attempt a worldwide monograph of *Rhizopogon*, recognizing 12 species in all. Coker and Couch (1928) did a later account for eastern US and Canadian species, and Zeller (1939, 1941) described several additional North American species. Alexander H. Smith (Smith and Zeller 1966) introduced new characters for describing *Rhizopogon* species. He expanded the number of named species to well over a hundred and established sections and subsections within the genus. Trappe (1975) transferred Smith's section *Rhizopogonella* to the genus *Alpova*, and molecular methods have provided new insights on relationships between species and groups of species within *Rhizopogon*.

Smith and Zeller (1966) enabled later workers to evaluate new finds of *Rhizopogon* in various parts of the world: the United States (Smith 1964, 1966, 1968; Harrison and Smith 1968; Hosford 1972, 1975; Miller 1986), Mexico (Trappe and Guzman 1971; Hosford and Trappe 1980; Cázares et al. 1992), Japan (Hosford and Trappe 1988), China (Liu 1985), Italy (Pacione 1984a), and North Africa (Pacione 1984b). Martín (1996) discovered new species in Europe and reviewed other work on the genus in detail (Martín 1996).

Martín (1996) classified different peridial types for *Rhizopogon* spp. and demonstrated by scanning electron microscopy that some species have a minute spore ornamentation. Martín and Sánchez (1996) tested thin layer chromatography as an additional and often useful tool to explore differences between closely related species. Martín and Högberg (1996) applied molecular methods to explore phylogeny and speciation of the genus *Rhizopogon*. They used a relatively conserved approach that limits interpretations about infrageneric relationships. Grubisha (1998) used the ITS region for study of generic and infrageneric relationships of *Rhizopogon*. She tested sectional relationships through maximum parsimony analysis of nuclear ribosomal DNA internal transcribed spacer (ITS) sequences of 16 *Rhizopogon* spp., 7 *Suillus* spp., *Gomphidius glutinosus* Fr. and *Truncocolumella citrina* Zeller (Table 5.1), resulting in a single most parsimonious tree (Fig. 5.1). *Rhizopogon* and *Suillus* were well-supported monophyletic groups as evidenced by the high bootstrap values of 99 and 86, respectively. The sister-group status of *Suillus* and *Rhizopogon* supports their close relationship; both are monophyletic (Bruns et al. 1989; Bruns and Szaro 1992; Kretzer et al. 1996).

Section *Rhizopogon*, the most taxonomically diverse of the four sections, forms two well-supported sister clades. Its most basal clade is the most

---

but by the author's student, a doctoral candidate, who also had to pay the printing costs. This curious Swedish custom was the only way for the faculty to publish without cost to themselves; it persisted until 1852, when it was changed so that the candidate must both write and defend the thesis."

Table 5.1. Taxa, sectional affiliations, voucher number, and ITS GenBank accession numbers of species examined for phylogenetic relationships within *Rhizopogon*. (Grubisha 1998)

Taxa	*Rhizopogon* section	Voucher number[a]	GenBank accession number
Gomphidius glutinosus (Schaeff. ex Fr.) Fr.			L54114
Rhizopogon burlinghamii Smith	*Rhizopogon*	JMT 17882	AF058303
R. ellenae Smith	*Amylopogon*	JMT 17476	AF058311
R. evadens Smith	*Rhizopogon*	JMT 16402	AF058312
R. fuscorubens Smith	*Rhizopogon*	JMT 17446	AF058313
R. gilkeyae Smith	*Villosuli*	JMT 19383	AF058304
R. occidentalis Zeller & Dodge	*Rhizopogon*	JMT 17564	AF058305
R. ochraceisporus Smith	*Fulviglebae*	JMT 17944	AF058306
R. parksii Smith	*Villosuli*	JMT 19446	AF058314
R. rubescens (Tul.) Tulasne	*Rhizopogon*	JMT 8227	AF058315
R. semireticulatus Smith	*Amylopogon*	JMT 7899	AF058307
R. subcaerulescens Smith	*Amylopogon*		M91613
R. subpurpurascens Smith	*Amylopogon*	JMT 19168	AF058308
R. villescens Smith	*Villosuli*	JMT 17681	AF058309
R. villosulus Zeller	*Villosuli*	JMT 19466	AF058310
R. vinicolor Smith	*Fulviglebae*	JMT 17899	AF058316
Suillus americanus (Peck) Snell ex Slipp & Snell			L54103
S. caerulescens Smith & Thiers			L54096
S. cavipes (Opat.) Smith & Thiers			L54085
S. lakei (Murrill) Smith & Thiers			L54086
S. luteus (Fries) Gray			L54100
S. sinuspaulianus (Pomerleau & Smith) Dick & Snell			L54078
Truncocolumella citrina Zeller			L54097

[a] All vouchers are those of James M. Trappe (JMT) and are deposited in the Oregon State University Herbarium (OSC).

dissimilar at the ITS sequence level as compared to the other three clades and sections on the phylogram (Fig. 5.1). Its more derived clade relates more closely to section *Villosuli* than to the most basal clade in section *Rhizopogon*. The two species sampled from section *Fulviglebae*, *R. vinicolor* Smith and *R. ochraceisporus* Smith, share affinities with section *Villosuli*. Morphologically and ecologically they also relate to section *Villosuli*. Section *Amylopogon* forms the most supported section in this study and is also held together by phylogenetically informative morphological characters.

The morphological, molecular and ecological data accumulating on species of *Rhizopogon* will permit broad revision of the genus within the next few years. Sections can be tightened into monophyletic subgenera, species synonyms can be tended, the array of taxonomically useful characters can be strengthened, and new species can be added. All this will fulfill the prediction of A. H. Smith (Smith and Zeller 1966), whose pioneering work laid the foundation for later progress.

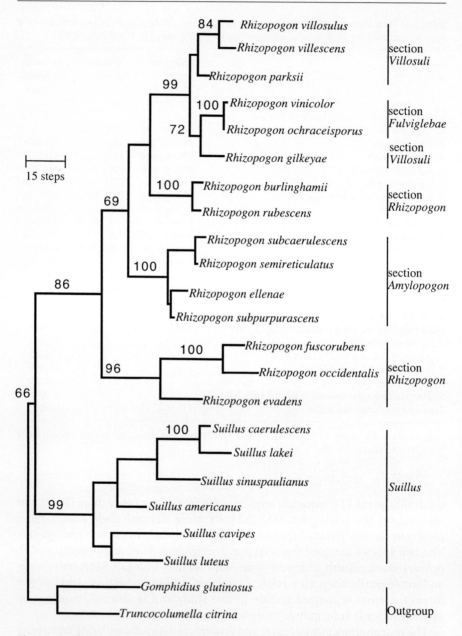

Fig. 5.1. Single most parsimonius tree obtained from maximum parsimony analysis of ITS1, ITS2, and 5.8S subunit nrDNA sequences. Bootstrap values are shown above the respective internode. CI (consistency index) = 0.592, RI (retention index) = 0.693, RC (rescale consistency index) = 0.411

5.3
Morphology and Anatomy of *Rhizopogon* ECM

Given the diversity of *Rhizopogon* species, it is not surprising that they also show a diversity in ECM morphology and anatomy. Several species, however, do share common features, particularly within sections (Molina and Trappe 1994). Common features include tightly clustered to coralloid or tuberculate structure and presence of rhizomorphs, often strongly differentiated in structure and with crystal deposits. Many early descriptions of *Rhizopogon* ECM are from in vitro syntheses which may contain artifacts, particularly if glucose is present in the rooting substrate (Duddridge 1986a,b); many of those descriptions were rudimentary and primarily noted color, branching, size, and presence of Hartig net development. A few detailed descriptions have been published for *Pseudotsuga menziesii*, various pine species, and *Tsuga heterophylla*.

5.3.1
Pseudotsuga menziesii ECM

Molina and Trappe (1994) note similarity in overall appearance in ECM synthesized between *Pseudotsuga menziesii* and section *Villosuli* species. Most synthesized and field-collected ECM form pinnate, often dense clusters that develop patches of dark pigmented surface hyphae overlaying typically white or pale interior mantles (Trappe 1967; Molina and Trappe 1982b, 1994; Chu-Chou and Grace 1983a; Massicotte et al. 1994, unpubl. data). The density and coverage of the dark surface hyphae are most pronounced on older ECM and described for *P. menziesii–R. parksii* Smith as a moderate reddish brown to greyish reddish brown fibrillose epicutus (Massicotte et al. 1994).

P. *menziesii–R. vinicolor* ECM have received the most detailed anatomical study. They often develop as dense coralloid or tuberculate clusters and the tubercles are encased in a rind of thick, darkly pigmented hyphae (Trappe 1965; Zak 1971; Chu-Chou and Grace 1983a; Massicotte et al. 1992; Goodman et al. 1996; Figs. 5.2–5.4). Trappe (1965) believed this rind to be of phycomycetous origin, but Zak (1971) later showed the rind to be differentiated aseptate, amber, thick-walled hyphae belonging to *R. vinicolor*. ECM synthesized between *R. vinicolor* and *P. menziesii* by Zak (1971), Molina and Trappe (1982b), and Massicotte et al. (1994) did not fully develop into tubercles but did show a patchy structure of the differentiated, pigmented surface hyphae as noted later by Molina and Trappe (1994) and Massicotte et al. (1994, unpubl. data) for several species of section *Villosuli* on *P. menziesii*. The presence of this unique epicutus of dark hyphae is diagnostic for distinguishing *P. menziesii* + section *Villosuli* ECM in the field or on seedlings grown in soil bioassays. Mantle structure for *R. vinicolor* and other section *Villosuli* species is typically multilayered, often with a loose outer layer and compact inner layer. Rhizomorphs are usually abundant and highly

Fig. 5.2. ECM of *R. vinicolor–P. menziesii*. Several tubercles (*T*) and non-tuberculate lateral roots (*arrow*). (Massicotte et al. 1992)

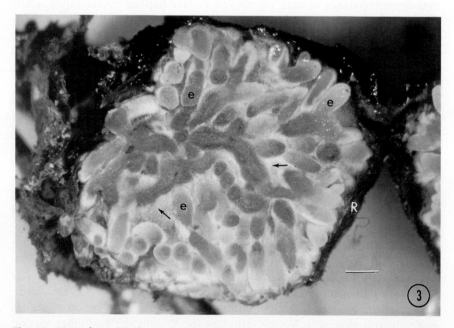

Fig. 5.3. ECM of *R. vinicolor–P. menziesii*. Hand section of fresh tubercle showing cluster of ECM root tips (*e*), surrounding mycelium (*arrows*) and peridium-like rind (*R*). *Bar* 1 mm. (Massicotte et al. 1992)

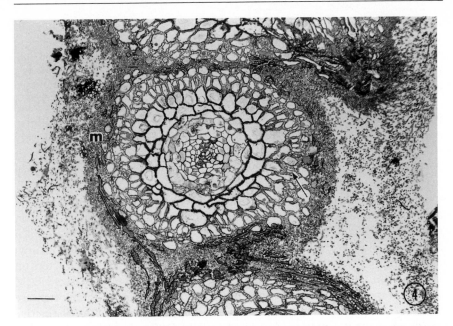

Fig. 5.4. ECM of *R. vinicolor–P. menziesii*. Transverse sections of adjacent roots. Each root has developed a hyphal mantle (*m*) and a Hartig net (*arrow*). *Bar* 0.1 mm. (Massicotte et al. 1992)

differentiated (Zak 1971). Massicotte et al. (1992, unpubl. data) also noted the presence of crystal deposits in the mantle of *R. vinicolor* ECM.

5.3.2
Pinus spp. ECM

R. luteolus and *R. roseolus* Corda *sensu* Smith (*rubescens* Tul.) have received the most anatomical attention, particular in early in vitro synthesis studies (Modess 1941; Björkman 1942; Fontana and Centrella 1967; Pachlewski and Pachlewska 1974; Froidevaux and Amiet 1975; Molina and Trappe 1982b, 1994; Mohan et al. 1993; Massicotte et al. 1994; Molina et al. 1997). Uhl (1988) provides the first comprehensive description of *R. luteolus* on pine and discusses similarities between *R. luteolus* and *R. roseolus* (*rubescens*) on *Pinus sylvestris* (Fontana and Centrella 1967; Pachlewski and Pachlewska 1974). He describes *P. sylvestris–R. luteolus* as typically coralloid, yellowish to brownish, with a two-layered mantle and strongly differentiated rhizomorphs. Pachlewski and Pachlewska (1974) describe this ECM as a creamy white to rusty brown type but also with a two-layered mantle. Fontana and Centrella (1967) note a two-layered mantle for *R. roseolus* (*rubescens*) + Scots pine. In comparing *P. sylvestris–R. luteolus* ECM to other descriptions of *Rhizopogon* ECM, Uhl (1988) notes three features shared in common: (1) a plectenchymatic, two-layered mantle; (2) highly differentiated rhizomorphs; and (3) incrustations (crystalline deposits) at least on some hyphae. *Pinus patula–R. luteolus* synthesized via spore inoculation (Mohan et al. 1993) were similar to those described by Uhl (1988), but Mohan et al. (1993) noted colors as whitish to brownish yellow, a distinct three-layered mantle, and abundant but undifferentiated rhizomorphs.

Molina and Trappe (1994) and Massicotte et al. (1994, 1999) describe ECM formed by several *Rhizopogon* species in sections *Amylopogon* and *Rhizopogon* on *Pinus contorta* and *Pinus ponderosa*, respectively. These ECM were usually white when young but then either turned greyish to light brown or developed some gradient of yellow to reddish coloration with age. Sporocarp peridium of several of these species also shows the same color changes as they mature or upon bruising. Nearly all ECM were compact coralloid structures and formed rhizomorphs; rhizomorph colors were similar to those seen on mantles. Massicotte et al. (1999) recently performed detailed anatomical studies on many of the ECM synthesized in the Massicotte et al. (1994) study; they examined *Rhizopogon* species from three sections for ECM characters on *P. ponderosa*. They found that even though several of the synthesized ECM share common mantle features and Hartig net development, there often are subtle or unique differences. Some species, for example, show the distinctive two-layered mantle pattern described by Uhl (1988), but others have thin mantles with no layering. They also found that most species had birefringent deposits in the mantle, possibly calcium oxalate, as noted in previously described *Rhizopogon* ECM: *R. subcaerulescens* Smith (section *Amylopogon*) contains peg-like appendages in the outer mantle; and *R. truncatus* Linder (*R. Cokeri*) (section *Fulviglebae*) develops uniquely bright lemon yellow subtuberculate ECM and, similar to *P. menziesii–R. vinicolor* ECM, a rind-like layer over the thick mantle of adjacent ECM.

5.3.3
Tsuga heterophylla ECM and Arbutoid Mycorrhizas

Agerer et al. (1996) provide detailed descriptions of *T. heterophylla–R. subcaerulescens*. These ECM develop into often dense clusters of coralloid systems up to 40 mm in diameter. They are similar in color to those formed by *R. subcaerulescens* on *Pinus* and *Pseudotsuga menziesii* (Massicotte et al. 1994) with readily apparent cystidia on the mantle surface. The mantle is two-layered and shows a strong amyloid reaction. Rhizomorphs are infrequent but highly differentiated and with dense crystal deposits.

Molina and Trappe (1982a) and Molina et al. (1997) synthesized several *Rhizopogon* species on the ericaceous hosts *Arbutus menziesii* and *Arctostaphylos uva-ursi*. All developed the typical cruciform (trilobate) branching pattern, often strongly coralloid; mantle colors and textures were similar to those synthesized on *Pinus* and *Pseudotsuga menziesii*. A primary difference, however, was that Hartig net developed only paraepidermally and hyphae grew intracellularly in the epidermal cells, typical of arbutoid mycorrhizae.

5.4
Ecology

5.4.1
Geographical Distribution and Host Specificity

Rhizopogon sporocarps have been collected on all continents except Antarctica. The worldwide distribution of *Rhizopogon* sharply coincides with distribution of Pinaceae, particularly the genus *Pinus*, owing to strong specificity with Pinaceae (Molina and Trappe 1994). Where Pinaceae are not native (e.g. Australia, New Zealand, large areas of South America, and numerous islands), *Rhizopogon* species have followed their introduced Pinaceae hosts into these exotic locations, often becoming dominant members of the introduced ECM flora. In New Zealand, for example, Chu-Chou (1979) reports that 32% of her 7444 isolation attempts from ECM of Pinaceae yielded *Rhizopogon* spp. Chu-Chou and Grace (1981, 1983a,c) later discovered *R. vinicolor* and *R. parksii* to be dominant ECM fungi of introduced *P. menziesii* seedlings in nurseries and plantations in New Zealand.

By far the greatest concentration and diversity of *Rhizopogon* is in the Pacific northwestern United States (Smith and Zeller 1966), reflecting the strong specificity of *Rhizopogon* with Pinaceae. This region contains a rich diversity of Pinaceae: six genera of ECM conifers, including 20 species in Oregon and Washington alone. Apparently the Pacific Northwest has been a major center of evolution, diversification and speciation of *Rhizopogon* and their conifer hosts. Many *Rhizopogon* species show strong specificity to either the genus *Pinus* or *Pseudotsuga*. About 20 *Pinus* species occur in the major forest types of western North America with more than 40 additional species in Central America. *Pseudotsuga menziesii* dominates large tracts of wet temperate (Coast, Cascade, and Sierra Nevada ranges) and interior montane (Rocky Mountain) forests from British Columbia south into central Mexico. Thus, these two genera contribute strongly to the overall distribution and diversification of *Rhizopogon* in North America.

By comparison, only 21 *Rhizopogon* species are reported from Europe (Martín 1996), and at least two of these have been introduced with *P. menziesii*. This may be due to lower conifer diversity, extinctions related to severe geologic and climatic changes during past epochs, or geographic isolation from major population centers of *Rhizopogon*. Few species are reported from Asia, but this may be due to a lack of collecting for hypogeous fungi on the Asian mainland. The distribution of *Rhizopogon* in exotic plantations reflects interesting patterns of origin. Exotic pine plantations in the Caribbean and South America host *R. nigrescens* Coker & Couch and *R. subaustralis* Smith, both native to the southwestern United States; pine plantations in Australia, New Zealand, and South Africa are populated primarily by *R. luteolus* and *R. roseolus* (*rubescens*) of European origin (J. Trappe and M. Castellano, unpubl. data). *R. luteolus* occurs rarely if at all in North America.

Molina and Trappe (1994) provided a comprehensive list of known or suggested ECM hosts of *Rhizopogon* species: that list includes field observations, experimental inoculations, and pure-culture syntheses. With few exceptions, field observations and pure-culture syntheses confirm the specificity of *Rhizopogon* spp. for Pinaceae (Molina and Trappe 1994). The only experimentally confirmed exceptions (pure-culture syntheses) are the ericaceous host genera *Arbutus* and *Arctostaphylos*, which seem receptive to many otherwise conifer-host-specific ECM fungi (Zak 1976; Molina and Trappe 1982a). Molina et al. (1997) also report the ability of three *Rhizopogon* species to form arbutoid mycorrhizae with *Arbutus* and *Arctostaphylos* following spore inoculation, but only when grown in dual culture with *Pinus ponderosa*. The *R. subcaerulescens* species group has also been confirmed with molecular based polymerase chain reaction (PCR) techniques to be the primary fungal symbiont on the achlorophyllous ericaceous monotrope *Pterospora andromedea* (Cullings et al. 1996). *Pterospora andromedea* occurs exclusively in Pinaceae forests where *Rhizopogon* species are common. Such a strongly connected guild structure (Pinaceae overstorey–Pinaceae ECM fungus specialist–understorey Monotropoideae) emphasizes the important role ECM specificity can play in forest community organization (Molina et al. 1992; Cullings et al. 1996).

Other exceptions to Pinaceae association noted by Molina and Trappe (1994) are from field observations, which can be erroneous particularly in mixed forests (Trappe 1962). For example, Thoen (1974) notes the close proximity of *R. luteolus* with four *Eucalyptus* species; he also notes the presence of many *Pinus* species in the same plantation and considers the matter of eucalypt association with *Rhizopogon* as unresolved. The unsuccessful pure-culture syntheses of Chilvers (1973) between *R. luteolus* and *Eucalyptus st.-johnii* and those of Malajczuk et al. (1982) involving three *Rhizopogon* species on 11 *Eucalyptus* spp. make a *Eucalyptus–Rhizopogon* association seem unlikely. Similarly, while some Fagaceae and Betulaceae are listed as possible hosts, Molina et al. (1997) and Massicotte et al. (1994) report that *Castanopsis chrysophylla*, *Lithocarpus densiflorus* and *Alnus rubra* do not form ECM with *Rhizopogon* when grown in dual culture with *Pinus* as shown for *Arctostaphylos* and *Arbutus*. Simarly, when *Alnus rubra* or *Betula papyrifera* are grown in dual culture with *Pseudotsuga menziesii* in field soil bioassays, the dominant, host-specific *Rhizopogon* section *Villosuli*-type on *P. menziesii* does not form ECM on the co-cultured angiosperm (Miller et al. 1992; Simard et al. 1997a). Associations with angiosperms should not be entirely discounted, however. Allen et al. (1999) recently reported the association of *R. mengii* M. Allen & Trappe with *Adenostoma fasiculatum* (Rosaceae); careful search revealed no Pinaceae closer than ca. 1 km.

Smith (1964) and Smith and Zeller's (1966) taxonomic treatises on *Rhizopogon* provide the first data to indicate patterns of host specificity at the subgeneric fungus level. They found some species such as section *Villosuli* in constant association with *P. menziesii* and others found with pines. They also report, however, that many occur in "mixed conifer" forests; a mixture could include various combinations of *Pseudotsuga, Pinus, Abies, Tsuga, Larix,* and *Picea.* Molina and Trappe (1994) conducted an extensive pure culture synthesis experiment with 20 species of *Rhizopogon* (29 isolates representing the four sections) on *Pseudotsuga menziesii, Pinus contorta,* and *T. heterophylla* to test the sectional level of specificity. They recorded three general responses: strong specificity to *P. menziesii,* specificity or strongest development on *P. contorta,* and an intermediate response where ECM formed on two or three of the hosts. Section *Villosuli* isolates showed strong specificity for *P. menziesii* and developed similar morphology and mantle colors. *R. vinicolor,* although a member of section *Fulviglebae* based on truncate spores, also strongly associated with *P. menziesii.* As noted previously, Molina and Trappe (1994) and results from Grubisha (1998) support transferring *R. vinicolor* to section *Villosuli.* Section *Rhizopogon* shows the strongest affinity for pines, and indeed this section houses species such as *R. luteolus,* which associate with pines in Europe and exotic pine plantations worldwide. Section *Amylopogon* isolates perform equally well on *Pseudotsuga menziesii* and *Pinus contorta* and are often found in mixed forests.

Although synthesis results supported the sectional affinities delimited by Smith and Zeller (1966) and the patterns of sporocarp-host associations in the field, there remained potential artifacts in the results due to the presence of glucose in the synthesis medium (Duddridge 1986a,b). Massicotte et al. (1994) designed a follow-up synthesis experiment in the greenhouse that used spore inoculation to address this concern. They tested 13 species (16 isolates) from the four *Rhizopogon* sections on *P. menziesii, Pinus ponderosa, T. heterophylla, Picea sitchensis, Abies grandis* and *A. rubra.* Spore inoculations were performed on each host species grown singly and in dual culture with either *P. menziesii* or *P. ponderosa* to examine effects of neighboring host species on ECM development. Overall, the results support the patterns shown in the pure culture synthesis experiment by Molina and Trappe (1994) and emphasized the importance of *P. menziesii* and *Pinus.* In single host inoculations, section *Villosuli* isolates (including *R. vinicolor*) developed ECM only with *P. menziesii* and abundantly so; the remaining species developed ECM only with *P. ponderosa* in single host culture. In dual host culture, section *Villosuli* isolates continued to develop ECM with *P. menziesii* even though hyphae and rhizomorphs were seen intertwined on the co-cultured host. Several of the isolates in the remaining sections, however, did develop ECM on the "secondary" hosts, such as *A. grandis, T. heterophylla* and *P. sitchensis,* in dual culture with *P. ponderosa.* Frequency of occurrence on seedlings and degree of ECM development were typically less on *A. grandis, T. heterophylla* and *P. sitchensis* than on *P. ponderosa,* but ECM were of typical anatomy (H. B. Massicotte, unpubl. data). Section *Rhizopogon* isolates showed the strongest specificity and development on pine, and section *Amylopogon* showed a predominantly intermediate response. None of the *Rhizopogon* species developed ECM with *A. rubra.* Massicotte et al. (1994) hypothesized that the various

Rhizopogon species first germinate and develop ECM with their preferred host (pine) and then spread to the "secondary" host.

Molina et al. (1997) provide additional evidence on the influence of *P. menziesii* and *Pinus* on host specificity and ECM development by *Rhizopogon*. They used a spore inoculation and dual host design similar to Massicotte et al. (1994) to examine effects of *P. menziesii* and *P. ponderosa* on *Rhizopogon* ECM development with the ericaceous hosts *A. menziesii* and *A. uva-ursi*. None of the tested isolates developed arbutoid mycorrhizae on the ericaceous hosts in single host culture. In dual culture with *P. ponderosa*, however, *R. occidentalis* Zeller & Dodge (section *Rhizopogon*) and *R. ellenae* Smith and *R. subcaerulescens* (section *Amylopogon*) formed well-developed arbutoid mycorrhizae with both ericaceous hosts. External ECM characters were similar to those seen with pine, but intracellular hyphal proliferation occurred in the epidermal cells typical of arbutoid mycorrhizae. In dual culture with *P. menziesii*, none of the section *Villosuli* isolates developed mycorrhizae with either of the ericaceous hosts, even though abundant mycelium was seen on root surfaces. Thus, section *Villosuli* remained strongly specific to *P. menziesii*, even in the presence of these "receptive" (Molina and Trappe 1982a) arbutoid hosts. Whether the influence on host specificity by dominant hosts such as *P. menziesii* and pine occurs in field situations remains to be tested. The advent of powerful molecular-based PCR tools, however, provides the technology to examine in vivo whether the intermingled roots of different tree species are developing ECM with common fungal species. Cullings et al. (1996), for example, used molecular tools to show the development and specificity of the *R. subcaerulescens* group on the achlorophyllos ericad *P. andromedea*. Such studies will shed considerable light on the role of specificity phenomena on fungal community dynamics and plant interactions.

5.4.2
Habitats and Community Dynamics

Given the diversity of *Rhizopogon* species and their occurrence wherever Pinaceae are found, it is difficult to generalize about the habitats that characterize *Rhizopogon* species. Known habitats range from moist, cool coastal forests to harsh, cold, and often dry interior forests; they are found in highly organic soils but also in pumice and sandy (including sand dune) substrates. Many species show broad ecological amplitude and are found in a variety of habitats. Although detailed habitat attributes are lacking for most species, we have observed some cases of habitat specialization in the Pacific Northwest. In Oregon, for example, *R. fuscorubens* Smith is known only from the coastal sand dune–*P. contorta* forest community; but it is also reported under pines in the southeastern United States (Miller 1986). *R. occidentalis* is the most common hypogeous fungus in the Oregon dune–*P. contorta* communities, but it is also widespread in interior pine forests. *R. subsalmoneus* Smith is typically found at high elevations under *Abies* spp. *R. flavovibrillosus* Smith is known primarily from serpentine soils. The tuberculate ECM of *R. vinicolor* and *R. truncatus* (with brilliant yellow ECM and sporocarps) are commonly found in buried, decomposing logs and highly organic soils. In Europe, Pachlewski and Pachlewska (1968) note that *R. luteolus* is common with pine in peatlands and dunes, but *R. luteolus* is also one of the more widespread *Rhizopogon* species in Europe (Martín 1996) and in exotic plantations.

Although there are few studies on ECM hypogeous sporocarp fungal communities, several *Rhizopogon* species are known from young to old forests. We suspect that many *Rhizopogon* species fall into the "multistage" classification of Visser (1995). J. E. Smith (unpubl. data) found similar frequency and abundance of *R. vinicolor* and *R. parksii* (two of the more common *P. menziesii*-specific associates) in 40- to 60–80, and >400-year-old *P. menziesii* forests in the Oregon Cascades. These two species did not respond to a moisture gradient in similar *P. menziesii* forests (Luoma 1988; Luoma et al. 1991). Visser (1995) reports *R. roseolus (rubescens)* to be equally frequent in a 6 as in a 65-year-old *Pinus banksiana* forest. We have noted *R. vulgaris* (Vitt.) M. Lange fruiting prolifically with 2-year-old *P. ponderosa* seedlings in an Oregon bareroot nursery, yet this species is one of the more common pine *Rhizopogon* in the western United States (J. M. Trappe, unpubl. data). Similarly, *R. luteolus* and *R. roseolus* are present on *Pinus radiata* from the nursery stage through forest rotation age in Australia (Molina and Trappe 1994) and South America (Garrido 1986).

Examination of ECM morphotypes, either in vivo or with greenhouse seedling bioassays of forest soils, also emphasizes the "multistage" aspect and response to disturbance by several *Rhizopogon* species. *Rhizopogon*-type ECM are among the more common types found on *P. menziesii* and pine seedlings growing particularly in disturbed soil in the Pacific Northwest. Studies by Alvarez and Linderman (1983), Pilz and Perry (1983), Castellano and Trappe (1985), Castellano (1987), Amaranthus and Perry (1989), Roth and Berch (1992) and E. Cázares, R. Molina, and J. E. Smith (unpubl. data) have shown the prevalence of *Rhizopogon* on *P. menziesii* seedlings in a variety of sites. Roth and Berch (1992), for example, found that *R. vinicolor*-like ECM were the second most abundant type on *P. menziesii* outplanted on recently clearcut forest sites on Vancouver Island, British Columbia. In southern British Columbia, Simard et al. (1997b,c) found that *R. vinicolor*-like ECM were among the dominant types on *P. menziesii* seedlings planted into mixed *P. menziesii*-*B. papyrifera* forests; trenching around the seedling plots drastically reduced the amount of *R. vinicolor*-like ECM on seedlings. Similarly, greenhouse bioassays of soils from variously disturbed sites (e.g. thinned, clearcut, burned) and undisturbed forest sites with *P. menziesii* seedlings consistently yield a preponderance of *R. vinicolor*-like ECM (Borchers and Perry 1990; Miller et al. 1992; Smith et al. 1995; Simard et al. 1997b). Based on our extensive synthesis records of *Rhizopogon* ECM (Molina and Trappe

1982b, 1994; Massicotte et al. 1994), we refer to this group of *R. vinicolor*-like ECM as section *Villosuli*-type because of the similarity of *Rhizopogon* ECM within section *Villosulii*.

In the most comprehensive examination of ECM morphotypes in Oregon, D. L. Luoma and J. Eberhart (unpubl. data) found that the section *Villosuli*-type ECM ranked second in frequency among 192 morphotypes found in 90-year-old *P. menziesii* forests. This morphotype occurred in 74% of the 189 soil cores and was among the five most abundant types, with a relative frequency in soil cores ranging from 14 to 29% depending on the specific site location (D. L. Luoma, Oregon State University, pers. comm.). On Vancouver Island, Goodman and Trofymow (1998) found that of 17 500 root tips examined in old-growth and mature *P. menziesii* stands, *R. vinicolor*-like ECM types were the second most abundant and third most frequent among 69 morphotypes; in soil cores from old-growth stands, *R. vinicolor*-like types averaged 9% of the total abundance and 11.7% frequency of total types encountered. In New Zealand, Chu-Chou and Grace (1981) isolated the section *Villosuli/R. vinicolor*-type fungus from 14% of over 2000 root tips sampled, a proportion greater than all other ECM fungi combined. They also report that this ECM type was isolated from trees of all age classes examined.

Two reproductive factors likely contribute to the widespread and often abundant nature of *Rhizopogon* in diverse conifer forests and in quickly developing ECM on tree seedlings following disturbance. Many *Rhizopogon* species produce abundant sporocarps and the basidiospores show high viability and inoculum effectiveness.

5.4.3
Sporocarp Productivity

The abundant productivity of *Rhizopogon* sporocarps was first reported in pine nurseries. Kessel (1927) reported that *R. luteolus* are the first fruiting bodies found in association with scattered "healthy" pine patches in Australian nurseries. Baxter (1928) found 370 sporocarps of *R. roseolus* (*rubescens*) in *P. sylvestris* seedling beds in Michigan. J. M. Trappe and R. Molina (pers. observ.) found *R. vulgarus* fruiting prolifically in *P. ponderosa* seedling beds in an Oregon nursery.

Actual data on *Rhizopogon* sporocarp productivity in natural forests are primarily limited to ECM fungal community studies in *P. menziesii* forests of Oregon and Washington. Investigators often use different sampling schemes and analyses to describe annual sporocarp productivity or standing crop; readers should closely examine the published methods and analyses to make comparisons between studies (Vogt et al. 1992). In the first study of hypogeous fungus (truffle) productivity in *P. menziesii* forests, Fogel (1976) reports that in a coastal forest habitat *R. parksii*, *R. villosulus* Zeller and *R. vinicolor* together yielded from 122 to 291 g ha^{-1} year^{-1}, and a combined total of 433 sporocarps ha^{-1} year^{-1}. Over 2 years in a nearby coastal *P. menziesii* forest, Hunt and Trappe

(1987) collected about 2000 sporocarps of *R. parksii* and *R. vinicolor* on a south-facing slope; total biomass for *R. parksii* ranged from 116 to 463 g ha^{-1} year^{-1} and for *R. vinicolor* from 10–116 g ha^{-1} year^{-1}. In *P. menziesii* forests of the central Cascade Range in Oregon, Luoma et al. (1991) found that three *Rhizopogon* species were dominant among all truffle fungi and yielded a minimum estimate of 974 g ha^{-1} year^{-1} biomass. *R. parksii* was one of five truffle fungi accounting for >70% of the total biomass; during the autumn fruiting season, *R. parksii* produced >50% of the total truffle biomass and had an 88% frequency of occurrence. In a similarly located study in the central Cascade Range of Oregon, J. E. Smith and colleagues (unpubl. data) examined differences in truffle productivity in three forest age classes (40, 60–80, and >400 years old). Over 3.5 years (four autumn and three spring samplings) they found ten *Rhizopogon* species that yielded a minimum estimate of 694 g ha^{-1} year^{-1} over all sites; *Rhizopogon* biomass was fairly evenly distributed across the three age classes. *R. vinicolor* and *R. parksii* were most abundant and yielded 342 and 241 g ha^{-1} year^{-1} across all sites respectively. In low-elevation *P. menziesii* forests of west-central Washington, Colgan (1997) found that *R. vinicolor* and *R. hawkeri* Smith produced 35% of the total number of hypogeous sporocarps and yielded 126 g ha^{-1} year^{-1} in biomass. These same two species produced 39% of the winter standing crop of truffle biomass, a significant winter food resource for small mammals. From a study in *P. menziesii*-dominated forests of the Oregon and Washington Cascade ranges, D. L. Luoma and J. Eberhart (unpubl. data, pers. comm.) found *R. parksii* mean standing crop biomass to range from 937 to 1158 g ha^{-1}. Clearly, these *P. menziesii*-specific *Rhizopogon* species are important fungal components in these temperate forest ecosytems. Even in young managed *T. heterophylla* forest stands in northern Washington with only 25% cover by *P. menziesii*, *R. parksii* and *R. subcaerulescens* were among the four most abundant truffle species (North et al. 1997); *R. vinicolor* yielded 18.5 g ha^{-1} year^{-1} and *R. subcaerulescens* 12.3 g ha^{-1} year^{-1}.

In the only systematic study of truffle sporocarps in pine forests, States and Gaud (1997) found that over 3 years in *P. ponderosa* stands in northern Arizona four *Rhizopogon* species collectively accounted for the greatest portion of the annual sporocarps in total numbers and biomass. *R. evadens* Smith and *R. subcaerulescens* were two of five species that accounted for 87% of the total annual biomass; *R. evadens* yielded 1232–1604 g ha^{-1} year^{-1} and *R. subcaerulescens* 181–386 g ha^{-1} year^{-1}. *R. evadens* was the most important of all truffle species, producing 39% of the total biomass, and was present during all four seasons.

5.4.4
Spore Biology

The effectiveness of *Rhizopogon* basidiospores was first apparent in their use to inoculate nursery seedlings. Lamb and Richards (1974a) showed that *R. luteolus* basidiospores have high survival potential and later (Lamb and

Richards 1978) used them to effectively inoculate pine seedlings. Several researchers have successfully used spore inoculation with *Rhizopogon*; Castellano and Molina (1989) summarize these techniques for use in nursery inoculation programs. Spore viability studies by Miller et al. (1993) and Torres and Honrubia (1994a) report that *Rhizopogon* species show high viability (activity with vital stains) and germinability; activity was nearly 100% for *R. roseolus*. Although Torres and Honrubia (1994a) found that spore viability of *R. roseolus* declines after storage or freezing, Castellano and Molina (1989) found that refrigerated slurries of several *Rhizopogon* species remain effective as inoculum after several years.

In vivo studies of basidiospore abundance and activity are difficult so data are scarce. Miller et al. (1994) followed the persistence of *Rhizopogon* basidiospores in natural soils under pine by noting the location of sporocarps in the autumn and returning the following spring to retrieve spores from the same location. They found that spores of *R. subcaerulescens* and *R. rubescens* are present in high numbers in the lowermost soil fraction after snow melt. They hypothesize that although hypogeous fungi are primarily dispersed long distances by small mammals, in situ persistence or short-distance dispersal in soil solution may be significant by maintaining spores in a soil "seed bank" for several years. Our experience with soil bioassays of variably disturbed soils that yield *Rhizopogon* morphotypes supports this hypothesis. In a recent study of naturally regenerating pine seedlings following fire, Horton et al. (1998) have provided strong molecular evidence to support this hypothesis. They used random fragment length polymorphism (RFLP) techniques to identify ECM morphotypes on *Pinus muricata* seedlings the first year following catastrophic fire in both a formerly *P. muricata* stand and adjacent scrub-brush, vesicular-arbuscular mycorrhizal (VAM) community with no *P. muricata* component. *R. subcaerulescens* and *R. ochraceorubens* Smith were two of three dominant fungi found on seedlings growing in the burned VAM scrub community. Because this was the first year following fire, they assumed that there was little opportunity for dispersal of fresh basidiospores into the site with all nearby pines killed. They hypothesized that *Rhizopogon* spores dispersed in previous years, probably by small mammals, persisted in these non-ECM communities, and remained viable and effective as inoculum following the catastrophic disturbance. The ability of ECM fungi to persist in vegetative forms cannot be discounted, however. Theodorou and Bowen (1971) found that *R. luteolus* could grow in the presence of non-ECM (grass) host roots. The ability to grow vegetatively in non-host root rhizosphere needs to be further investigated in the field.

5.4.5
Mycophagy of Sporocarps

A thorough discussion of the ecosystem importance of *Rhizopogon* would not be complete without noting their ecological role in forest food webs.

Consumption of hypogeous fungi by small mammals (mycophagy) is well known (Fogel and Trappe 1978; Maser et al. 1978). Given the sporocarp abundance and widespread nature of *Rhizopogon* in coniferous forests, it is not surprising to find them as dominant components in mammal diets. The following studies report dietary results after examination of stomach or fecal material. Maser et al. (1978) found *Rhizopogon* in five classes of rodents, averaging 22% frequency of occurrence, the highest of all fungal genera encountered. Maser et al. (1985) examined 91 trapped flying squirrels and found *Rhizopogon* most frequent among all fungi. In a study of the yellow pine chipmunk in northeastern Oregon, Z. Maser and C. Maser (1987) found that *Rhizopogon* occurred with 90% frequency and occupied as much as 96% of total stomach contents. Four species of mice in the genus *Peromyscus* showed a 25% sample frequency of *Rhizopogon* species (C. Maser and Z. Maser 1987). In a study of eight squirrel species (five genera) distributed throughout Oregon forests, Maser and Maser (1988) report *Rhizopogon* species as the dominant fungal dietary component. Colgan (1997) conducted a comprehensive study of truffle production and fecal pellets taken from two species of live-trapped small mammals occupying the same forest stands; the stands had received silvicultural treatments of variable thinning or no thinning (controls). He found that *Rhizopogon* species were the dominant fungal species for both the northern flying squirrel and Townsends chipmunk in all stand types and for all seasons. *Rhizopogon* sporocarps were also the most frequently encountered and abundant in the stands. Cázares et al. (1999) found *Rhizopogon* spores present with a 99% frequency in fecal pellets of western red-backed voles, northern flying squirrels, and Siskiyou chipmunks. Because *Rhizopogon* sporocarps can be abundant over winter (Colgan 1997; States and Gaud 1997), they are significant food resources during periods of food scarcity. Of course, many large mammals prey on these small mammals as do birds such as the endangered northern spotted owl. Thus, *Rhizopogon* and other truffle genera are critical links in forest food webs.

5.4.6
Interactions with Other Micro-organisms

Rhizopogon spp. exhibit complex interactions with other soil microorganisms. Among these are various antagonisms toward pathogens. Zak (1971) reports strong inhibition of the growth of the root pathogens *Phytophthora cinnamomi* Rands, *Pythium debaryanum* Hesse and *Pythium sylvaticum* Cambell & Hendrix in vitro by *R. vinicolor*. *Heterobasidion (Fomes) annosus* (Fr.) Bref. and *Phellinus weirii* (Murr.) Gilb. were moderately inhibited. *R. roseolus* is known to be antagonistic toward *H. annosus* (Hyppel 1968). Sasek and Musilek (1967) found that *R. roseolus* exhibits strong antibiotic activity. In a study of the interaction of bacteria and ECM fungi, Bowen and Theodorou (1979) found *R. luteolus* superior to other ECM fungi in colonizing root surfaces in both the presence and absence of bacteria and suggest that this is why it is so common as an ECM fungus on *P. radiata* in Australia. Some bacterial associates may be beneficial to *Rhizopogon* ECM. Massicotte et al. (1992) report bacterial associates along the hyphae within the outer rind or on the surface of the tubercle of *R. vinicolor* ECM. Tilak et al. (1988) discovered free-living, nitrogen-fixing bacteria in the genus *Azospirillum* in sporocarps of *R. vinicolor*. Li et al. (1992) isolated a nitrogen-fixing, spore-forming *Bacillus* sp. from *R. vinicolor* tubercles. Some *Rhizopogon* spp. can also influence the development of other ECM fungi. For example, Agerer (1990, 1991) and Agerer et al. (1996) describe the close association of *Chroogomphus* and *Gomphidius* spp. within ECM formed by *Rhizopogon* spp.

5.5
Physiology

5.5.1
Growth in Culture and Physiological Traits

Rhizopogon species are relatively easy to isolate from sporocarp tissue. Fresh sporocarps without obvious holes, cracks or larvae are broken in half and a small piece of gleba tissue aseptically removed and placed on nutrient agar (Molina and Palmer 1982). If successful, growth usually commences within 1 week. We have successfully isolated about 40 *Rhizopogon* species and maintain working cultures on modified Melin-Norkrans medium (MMN) (Marx 1969), although several other synthetic and natural media (e.g. potato dextrose agar, PDA) also work well (see Molina and Palmer 1982). Hacskaylo et al. (1965) found that *R. roseolus* grew optimally at 18°C for 24 days and 13°C for 48 days; growth also occurs at 32°C. Working cultures are typically maintained at 20°C in most growth experiments. Agar discs of *Rhizopogon* cultures can be stored in sterile water at 4–5°C for at least 2 years (Marx and Daniel 1976; Smith et al. 1994) and many species also store well in sterile water at 18°C (Smith et al. 1994).

Modess (1941) reports that an isolate of *Rhizopogon* could grow in pure cultures at pHs ranging from 3.5 to 6.8. He found it to fruit in nature in an even broader range of soil pHs, but it was most abundant in soil pHs of 4.5–6.4. The optimum was pH 5.9. Hung and Trappe (1983) found that an isolate of *R. vinicolor* would grow well over a span of 4 pH units, a broader range than most fungi they studied. This adaptability reflects the rather wide habitat adaptability observed for *R. vinicolor* in its native range.

In both agar and liquid substrates, many *Rhizopogon* species grow rapidly compared to other ECM fungi. Zak (1971) reports, for example, that several cultures of *R. vinicolor* grow an average of 6.6 cm on MMN to 7.5 cm in diameter on PDA after 20 days at 20°C, nearly covering a 9-cm diameter Petri dish. Molina (1980b) examined 25 species (59 isolates) and found strong variability in growth rate after 4 weeks at 20°C; *R. vinicolor* grew at rates similar to those noted by Zak (1971) as did cultures of *R. subcinnamo-*

meus Smith, *R. clavitisporous* Smith, *R. villescens* Smith, *R. villosulus, R. fuscorubens, R. roseolus,* and *R. vulgaris.* Many of the remaining species grew more slowly, averaging ca. 3–5 cm in diameter.

Zak (1971) provides detailed cultural descriptions for *R. vinicolor,* and several, albeit less detailed, descriptions are available for the often studied *R. luteolus* and *R. roseolus (rubescens)* (Modess 1941). These latter two species typically produce a white to cream or yellowish buff-colored culture, a characteristic also noted in field and laboratory descriptions of their ECM. Chu-Chou and Grace (1984) note that *R. rubescens* also has a pinkish to salmon pink coloration; they provide a key to differentiate *R. luteolus* from *R. rubescens* cultures. Molina (1980b) and R. Molina, J. M. Trappe, D. R. Hosford, and A. Jumpponen (unpubl. data) studied cultures of about 30 species and 70 isolates in the four sections. They found remarkable similarity within sections for several species, thus providing further support for several of Smith's (Smith and Zeller 1966) original sectional designations of species. Those cultural descriptions are too numerous to note in entirety here, so only a few general findings are reported.

Often mycelial characters noted in ECM and sporocarps were also seen in axenic cultures. Aerial hyphae were often analogous to the peridial epicutus of sporocarps. For example, representatives of section *Amylopogon* have a peridial epicutis of fine, interwoven rhizomorphs and produce pink to blue pigments in response to KOH; cultures of all but one of the species we studied had similar rhizomorphs, and all produced the pink to blue pigments. The sporocarp epicutis of species in section *Villosuli* is characterized as having hyphae with brown and often thickened walls, and many produce flagellate setal or inflated versiform cells. Similar phenomena occurred in the aerial mycelium of the isolates we studied. ECM mantles of section *Villosuli* isolates also display this epicutus of thickened, dark brown pigmented hyphae in synthesized ECM and those found in the field (see Sect. 5.3.1).

Clamp connections have not been observed on sporocarps of *Rhizopogon* species except for *R. abietis* Smith, in which they are extremely rare (Smith and Zeller 1966). Zak (1973), however, reports clamp connections in cultures of *R. vinicolor,* and we found them on hyphae in cultures of 15 of the species studied: eight in section *Villosuli,* two in section *Amylopogon,* two in section *fulviglebae,* and three in section *Rhizopogon.* Except for section *Villosuli,* no pattern was discernible in the relation of clamp connections to taxa above the species level.

Rhizopogon species, as most ECM fungi, grow best on glucose as a C source. Hacskaylo (1973) found that *R. roseolus* grows on 18 C sources, but only mannose, cellobiose, trehalose, and pectin were slightly better than glucose. Similarly, Lamb (1974) grew *R. roseolus* and *R. luteolus* on 21 C sources; growth was best on glucose and mannose, but they also grew well on trehalose, rafinose, mannitol, xylose, and pectin. Lamb (1974) also found that both *Rhizopogon* species grew better on the poorer of the substrates when a small amount of starter glucose was incorporated into the medium.

As for many ECM fungi, cultures of *Rhizopogon* grow poorly on plant polymers such as cellulose and lignin. Lundeberg (1970) found low cellulase activity in *R. luteolus* and *R. roseolus*, and Haselwandter et al. (1990) reported little growth of *R. roseolus* on lignin. When in ECM symbiosis, however, *Rhizopogon* and other species can show greater capacity to decompose organic C than in pure culture. Trojanowski et al. (1984) found that ECM *R. luteolus* and *R. roseolus* are able to decompose ^{14}C-labeled plant lignin, ^{14}C-lignocellulose, and ^{14}C-DHP-lignin, although still at lower rates than the white rot fungi *H. annosum* and *Sporotrichium pulvervulentum*. Durall et al. (1994) found that *R. vinicolor* in association with *P. menziesii* releases about 23% of the ^{14}C in hemicellulose and small amounts of ^{14}C-labeled needles in the substrate. Thus, ability to attack organic C while in natural ECM associations may be greater than pure culture experiments indicate.

A similar history of N studies also leads to the conclusion that some ECM fungi, including *Rhizopogon*, can decompose organic N. Studies by Lundeberg (1970) show that *R. luteolus* and *R. roseolus* prefer NH_4^+-N to most amino acids, although in his tests both species grew reasonably well on L-asparagine; *R. luteolus* grew well on KNO_3 and *R. roseolus* did not. Rudawska (1982) found that the amino acids glutamic acid, glutamine, aspartic acid, arginine and urea increase growth of *R. luteolus*, but alanine, glycine, asparagine, and proline decrease growth. Actual use of protein was examined by Abuzinadah and Read (1986a). In addition to using peptides, *R. roseolus* was able to use protein (bovine serum albumin) with growth yields similar to those with NH_4^+ as the sole N source. Abuzinadah and Read (1986a) noted that these results were contrary to those of Lundeberg (1970), who found little protease activity. They cite the likely occurrence of catabolic repression of enzyme synthesis in Lundeberg's experiments, an unknown metabolic process at the time. In a follow-up experiment, Abuzinadah and Read (1986b) found that in ECM association with *P. contorta* pine, *R. roseolus* grows as well on protein as on a sole source of NH_4^+. They placed *R. roseolus* in a category of "protein fungi" and discussed the significance of ECM fungi in using organic N in nutrient cycling and in fungal succession phenomena.

5.5.2
Nutrient Translocation

Radiotracer experiments have examined direct translocation of elements between *Rhizopogon* and pine hosts. Melin and Nilsson (1957) conducted the first ECM studies to conclusively demonstrate that photosynthetic carbon moves from host needles to ECM fungi; they used *R. roseolus* and *Suillus variegatus* (Fr.) O. Kuntze for these experiments. They found that "considerable amounts" of $^{14}CO_2$-C moved from labeled pine to *R. roseolus* and *S. variegatus* ECM. Other studies show the movement of soil elements or nutrients from *Rhizopogon* to hosts. Melin et al. (1958) showed that large amounts of ^{22}Na (a surrogate for K) move from *R. roseolus* mycelium to *Pinus virgini-*

ana seedlings. Skinner and Bowen (1974b) report rapid uptake of ^{32}P-orthophosphate by *R. luteolus* ECM and translocation to *P. radiata*, although the amount of uptake differed among fungal strains. Finlay et al. (1988) used intact ECM microcosms of *Pinus–R. roseolus* seedlings to examine uptake and movement of ^{15}N-labeled NH_4^+ and conversion to amino acids in plant tissue. They found high levels of uptake and conversion to ^{15}N-labeled glutamate/glutamine, aspartate/asparagine, and alanine in plant tissue; lower levels were recorded for other amino acids.

5.5.3
Enzyme and Hormone Activities

In addition to production of cellulase and hydrolytic enzymes as needed in the above experiments to degrade organic C and N sources, other enzymes are reported in growth studies of *Rhizopogon* species. Lundeberg (1970) reports barely perceptible pectinase and proteinase and no laccase activity in *R. luteolus* and *R. roseolus*. Ho and Trappe (1980) report that *R. vinicolor* produces high levels of nitrate reductase compared to other ECM fungi. Bowen and Theodorou (1967) found that soluble P occurs in cultures of *R. roseolus* containing rock phosphate, and Theodorou (1968) indirectly showed indication of phosphatase activity in *R. roseolus* by growing the fungus with phytate as the sole P source. Ho and Zak (1979) report that *R. vinicolor* produces low amounts of acid phosphatase, and later Ho and Trappe (1987) found that six *Rhizopogon* species from three sections produced acid and alkaline phosphatases as well as nitrate reductase. They also report that all species produced cytokinins and IAA. Crafts and Miller (1974) found that *R. ochraceorubens* produces cytokinins but four other *Rhizopogon* species do not. Cytokinins or their precursors also are reported in cultures of *R. roseolus* and *R. luteolus* (Laloue and Hall 1973; Miura and Hall 1973; Rudawska 1982). Rudawska (1982, 1983) found that the amount and type of N and P in the growth media affect auxin and cytokinin production by *R. luteolus*.

5.6
Applications in Forestry

The importance of *Rhizopogon* spp. as ECM symbionts was first seen in the successful introduction of exotic pines into the southern hemisphere. In 1927, Kessel recognized *R. luteolus* as being among the first fungi to fruit in association with scattered "healthy" *P. radiata* in Australian nurseries. By 1937, Birch had listed *R. roseolus* (*rubescens*) as an important fungus associated with *Pinus radiata* and *Pinus caribaea* in New Zealand plantations. Chu-Chou (1979) re-emphasized the importance of *Rhizopogon* in exotic plantations and nurseries in New Zealand; 32% of her 7444 isolation attempts from ECM of Pinaceae yielded *Rhizopogon* spp. Chu-Chou and Grace (1981, 1983a,c) later discovered *R. vinicolor* and *R. parksii* to be dominant ECM fungi of introduced *P. menziesii* seedlings in nurseries and plantations. In Nigeria, Momoh (1976) has found *R. luteolus* associated with introduced pines. *Rhizopogon* ECM also have been associated with the establishment of exotic conifers in Africa (Donald 1975; Fogel 1980; Ivory 1980), Puerto Rico (Volkart 1964), Europe (Levisohn 1956, 1965; Gross et al. 1980; Jansen and de Vries 1989; Alvarez et al. 1993; Parlade and Alvarez 1993; Parlade et al. 1996), New Zealand (Birch 1937; Chu-Chou and Grace 1981, 1983a,c), South America (Garrido 1986), and the United States (Baxter 1928).

5.6.1
Seedling Inoculation

Extensive research has been conducted on methods to artificially inoculate seedlings with selected *Rhizopogon* spp. Chopped sporocarps and basidiospore suspensions mixed into soil have frequently been used to inoculate both bareroot and container-grown seedlings (Rayner 1938; Levisohn 1956; Theodorou 1971, 1980, 1984; Theodorou and Bowen 1973, 1987; Lamb and Richards 1974a,b,c, 1978; Skinner and Bowen 1974a; Donald 1975; Theodorou and Skinner 1976; Trappe 1977; Hodgson 1979; Lamb 1979; Marx 1980; Chu-Chou and Grace 1981; Ivory and Munga 1983; Theodorou and Benson 1983; Castellano 1985, 1987; Castellano and Trappe 1985; Castellano et al. 1985; Chu-Chou and Grace 1985, Castellano and Molina 1989; Torres and Honrubia 1994b; Parladé et al. 1996). Coating seeds of *P. radiata* with *R. luteolus* basidiospores has also been effective (Theodorou 1971, 1980; Theodorou and Bowen 1973; Theodorou and Skinner 1976). Given the high survival potential of *Rhizopogon* basidiospores shown by Lamb and Richards (1974c), Miller et al. (1993), and Torres and Honrubia (1994a), and the abundance of sporocarps reported in some nurseries and plantations (Baxter 1928; Young 1937; Adams 1951; Purnell 1957; Chu-Chou 1979; Chu-Chou and Grace 1983b; Croghan 1984; R. Molina and J. M. Trappe, pers. observ.), spore inoculation remains a practical method for introducing *Rhizopogon*. For example, Castellano (1987) and Castellano and Molina (1989) report the successful inoculation of 6 million container-grown *P. menziesii* seedlings simply by incorporating a spore suspension of *R. vinicolor* into a fertilizer injector system and mist-irrigating the spores onto the seedlings. They further report successful spore inoculation with several previously untested *Rhizopogon* species. Massicotte et al. (1994) and Molina et al. (1997) used spore inoculation of several *Rhizopogon* spp. to investigate host specificity.

Because *Rhizopogon* species are easily isolated into pure culture from sporocarp tissue and grow comparatively rapidly, they are prime candidates for mycelial inoculations and several researchers report success with this approach, albeit at experimental scales (Volkart 1964; Fontana and Centrella 1967; Theodorou 1967, 1978, 1980; Theodorou and Bowen 1970; Lamb and Richards 1971; Odeyinde and Ekwebelam 1974; Momoh 1976; Trappe 1977; Sands and Theodorou 1978; Ekwebelam 1980; Cline and Reid 1982; Ford et al.

1985; Jones et al. 1986). Work in our laboratory, however, has shown mycelial inoculum of several *Rhizopogon* species grown in a vermiculite carrier to be ineffective (Molina 1980a, and unpubl. data). Current interest by researchers, industry, and foresters in production and use of ECM inoculum in forestry programs emphasizes the need to find efficient isolates of *Rhizopogon* and to enhance inoculum effectiveness for this purpose.

5.6.2
Benefits to Hosts

In his review of inoculation success of numerous ECM fungi used in forestry and subsequent host effects, Castellano (1996) reports that overall *R. parksii* and *R. vinicolor* nearly always stimulate growth of *P. menziesii* seedlings, but *R. luteolus, R. nigrescens,* and *R. roseolus* yield variable results, often not affecting the growth of different pine species. Nevertheless, growth enhancement of seedlings in nurseries and experiments has been repeatedly reported after inoculation with *Rhizopogon* species (Levisohn 1956; Volkart 1964; Henderson and Stone 1970; Theodorou and Bowen 1970; Lamb and Richards 1971, 1974c; Theodorou 1971; Donald 1975; Momoh 1976; Theodorou and Skinner 1976; Ekwebelam 1980; Castellano and Trappe 1985; Chu-Chou 1985; Chu-Chou and Grace 1985; Ekwebelam and Odeyinde 1985; Parladé et al. 1996). Outplanting performance of inoculated seedlings has also been improved (Volkart 1964; Theodorou and Bowen 1970; Theodorou 1971; Momoh 1976; Castellano and Trappe 1985; Ekwebelam and Odeyinde 1985). Significantly increased uptakes of P (Theodorou and Bowen 1970; Lamb and Richards 1971, 1974a; Skinner and Bowen 1974a,b; Chu-Chou 1979; Chu-Chou and Grace 1985), K (Theodorou and Bowen 1970; Lamb and Richards 1971), Na (Melin et al. 1958) and N (Chu-Chou and Grace 1985; Finlay et al. 1988) by pine seedlings inoculated with either *R. luteolus* or *R. roseolus (rubescens)* have been demonstrated.

Many *Rhizopogon* spp. produce prolific rhizomorphs in soil; these have been noted and described in synthesized ECM (Trappe 1967; Zak 1971; Molina and Trappe 1982a, 1994; Uhl 1988; Mohan et al. 1993; Massicotte et al. 1994; Agerer et al. 1996). Bowen (1968) suggests that the extensive rhizomorph network of *R. luteolus* is largely responsible for enhanced nutrient uptake and seedling growth. Rhizomorphs also play an important role in water uptake and movement in ECM systems (Duddridge et al. 1980; Brownlee et al. 1983; Read and Boyd 1986). Parke et al. (1983) and Dosskey et al. (1990) demonstrate enhanced tolerance to drought stress of *P. menziesii* seedlings inoculated with *R. vinicolor* and attribute this enhancement in part to rhizomorph production and function in water transport. Rhizomorphs also link plant roots and provide conduits for interplant exchange of nutrients (Read 1984; Read et al. 1985; Simard et al. 1997a).

A series of field and greenhouse soil bioassays emphasize the ecological importance of *Rhizopogon* species, particularly in disturbed forest habitats of the Pacific Northwest and California (Alvarez and Linderman 1983; Pilz and Perry 1983; Castellano 1987; Borchers and Perry 1990; Miller et al. 1992; Simard et al. 1997b; Horton et al. 1998). *Rhizopogon*-type ECM are among the most common and often dominant types found on *P. menziesii* and pine seedlings grown in disturbed forest soils. Studies by Alvarez and Linderman (1983), Castellano and Trappe (1985), and Castellano (1987) show the prevalence of *Rhizopogon* ECM colonizing outplanted *P. menziesii* and pine seedlings on a variety of reforestation sites. Castellano (1987) further reports that *P. menziesii* seedlings inoculated with *R. vinicolor* have higher survival and better growth after outplanting than uninoculated seedlings, and that *R. vinicolor* rapidly forms ECM on new, emerging feeder roots of planted seedlings, thus maintaining a position of dominance on the root system. Horton et al. (1999) found two *Rhizopogon* species to be dominant components on root systems of *P. muricata* naturally regenerating in scrub communities after fire in coastal California. *Rhizopogon* ECM are also often found on seedlings grown in soils taken from undisturbed, mature forests, indicating their ability to function on root systems at various stages of forest community development. The widespread occurrence of *Rhizopogon* spp., their ecological adaptation to disturbed and undisturbed habitats, competitive interactions with other ECM fungi, ability to provide drought tolerance, and ease of spore inoculation confirm them as prime candidates for seedling inoculation programs in forestry practice.

Acknowledgements. We thank Dan Luoma, Jane Smith, Annette Kretzer, and Tom Horton for reviews of early drafts, and Lew Melville for copies of the *Rhizopogon vinicolor* + Douglas-fir ECM.

References

Abuzinadah RA, Read DJ (1986a) The role of proteins in the nitrogen nutrition of ectomycorrhizal plants. I. Utilization of peptides and proteins by ectomycorrhizal fungi. New Phytol 103:481–493

Abuzinadah RA, Read DJ (1986b) The role of proteins in the nitrogen nutrition of ectomycorrhizal plants. I. Utilization of protein by mycorrhizal plants of *Pinus contorta*. New Phytol 103:495–506

Adams AJS (1951) A forest nursery for *Pinus radiata* at Mt Burr in the southeast of South Australia. Aust For 15:47–56

Agerer R (1990) Studies on ectomycorrhizae XXIV. Ectomycorrhizae of *Chroogomphus helveticus* and *C. rutilus* (Gomphidiaceae, Basidiomycetes) and their relationship to those of *Suillus* and *Rhizopogon*. Nova Hedwigia 50:1–63

Agerer R (1991) Studies on ectomycorrhizae XXXIV. Mycorrhizae of *Gomphidius glutinosus* and of *G. roseus* with some remarks on Gomphidiaceae (Basidiomycetes). Nova Hedwigia 53:127–170

Agerer R, Müller WR, Bahnweg G (1996) Ectomycorrhizae of *Rhizopogon subcaerulescens* on *Tsuga heterophylla*. Nova Hedwigia 63:397–415

Albertini IB, Schweiniz LD (1805) Conspectus fungorum in Lusatiae superioris. Sumtibus Kummerianis, Leipzig, Germany

Allen MF, Trappe JM, Horton TR (1999) Nats truffle and truffle-like fungi 8: *Rhizopogon mengei* sp. nov. (Boletaceae, Basidiomycota). Mycotaxon (in press)

Alvarez IF, Linderman RG (1983) Effects of ethylene and fungicide dips during cold storage on root regeneration and survival of western conifers and their mycorrhizal fungi. Can J For Res 13:962–971

Alvarez IF, Parladé J, Trappe JM, Castellano MA (1993) Hypogeous mycorrhizal fungi of Spain. Mycotaxon 47:201–217

Amaranthus MP, Perry DA (1989) Interaction effects of vegetation type and Pacific madrone soil inocula on survival, growth, and mycorrhiza formation of Douglas-fir. Can J For Res 19:550–556

Baxter DV (1928) Mycorrhiza and Scotch pine in the University of Michigan forest nursery. Mich Acad Arts Sci Lett Pap 9:509–516

Birch TTC (1937) A synopsis of forest fungi of significance in New Zealand. N Z J For 4:109–125

Björkman E (1942) Über die Bedingungen der Mykorrhizabildung bei Kiefer und Fichte. Symb Bot Ups 6:1–190

Borchers SL, Perry DA (1990) Growth and ectomycorrhiza formation of Douglas-fir seedlings grown in soils collected at different distances from pioneering hardwoods in southwestern Oregon clearcuts. Can J For Res 20:712–721

Bowen GD (1968) Phosphate uptake by mycorrhizas and uninfected roots of *Pinus radiata* in relation to root distribution. Trans 9th Congr Soil Sci 2:219–228

Bowen GD, Theodorou C (1967) Studies on phosphate uptake by mycorrhizas. 14th IUFRO Congr (Munich) 5:116–138

Bowen GD, Theodorou C (1979) Interactions between bacteria and ectomycorrhizal fungi. Soil Biol Biochem 11:119–126

Brownlee CJ, Duddridge A, Malibari A, Read DJ (1983) The structure and function of mycelial systems of ectomycorrhizal roots with special reference to their role in forming inter-plant connections and providing pathways for assimilate and water transport. Plant Soil 71:433–443

Bruns TD, Szaro TM (1992) Rate and mode differences between nuclear and mitochondrial small-subunit rRNA genes in mushrooms. Mol Biol Evol 9:836–855

Bruns TD, Fogel RD, White TJ, Palmer J (1989) Accelerated evolution of a false truffle from a mushroom ancestor. Nature 339:140–142

Castellano MA (1985) Basidiospores of *Rhizopogon vinicolor* and *Rhizopogon colossus* as ectomycorrhizal inoculum. In: Molina R (ed) 6th North American Conf on Mycorrhiza. Forest Research Laboratory, Oregon State University, Corvallis, 11 pp

Castellano MA (1987) Ectomycorrhizal inoculum production and utilization in the Pacific Northwestern US - a glimpse at the past, a look to the future. In: Sylvia DM, Hung LL, Graham JH (eds) 7th North American Conf on Mycorrhizae. Institute of Food and Agricultural Sciences, University of Florida, Gainsville, pp 290–291

Castellano MA (1996) Outplanting performance of mycorrhizal inoculated seedlings. In: Mukerji KG (ed) Concepts in mycorrhizal research. Kluwer Dordrecht, pp 223–301

Castellano MA, Molina R (1989) Mycorrhizae. In: Landis TD, Tinus RW, McDonald SE, Barnett JP (eds) The container tree nursery manual, vol 5, Agric Handb 674. US Department of Agriculture, Forest Service, Washington, DC, pp 101–167

Castellano MA, Trappe JM (1985) Ectomycorrhizal formation and plantation performance of *P. menziesii* nursery stock inoculated with *Rhizopogon* spores. Can J For Res 15:613–617

Castellano MA, Trappe JM, Molina R (1985) Inoculation of container-grown *P. menziesii* seedlings with basidiospores of *Rhizopogon vinicolor* and *R. colossus*: effects of fertility and spore application rate. Can J For Res 15:10–13

Cázares E, García J, Castillo J, Trappe JM (1992) Hypogeous fungi from northern Mexico. Mycologia 84:341–359

Cázares E, Luoma DL, Amaranthus MP, Chambers CL, Lehmkuhl JF (1999) Interaction of fungal sporocarp production with small mammal abundance and diet in Douglas-fir stands of the southern Cascade Range. Northwest Sci 73:64–76

Chilvers GA (1973) Host range of some eucalypt mycorrhizal fungi. Aust J Bot 21:103–111

Chu-Chou M (1979) Mycorrhizal fungi of *Pinus radiata* in New Zealand. Soil Biol Biochem 11:557–562

Chu-Chou M (1985) Effect of different mycorrhizal fungi on *Pinus radiata* seedling growth. In: Molina R (ed) 6th North American Conf on Mycorrhiza. Forest Research Laboratory, Oregon State University, Corvallis, 208 pp

Chu-Chou M, Grace LJ (1981) Mycorrhizal fungi of *Pseudotsuga menziesii* in the North Island of New Zealand. Soil Biol Biochem 13:247–249

Chu-Chou M, Grace LJ (1983a) Characterization and identification of mycorrhizas of Douglas-fir in New Zealand. Eur J For Pathol 13:251–260

Chu-Chou M, Grace LJ (1983b) Characterization and identification of mycorrhizas of radiata pine in New Zealand. Aust For Res 13:121–132

Chu-Chou M, Grace LJ (1983c) Hypogeous fungi associated with some forest trees in New Zealand. N Z J Bot 21:183–190

Chu-Chou M, Grace LJ (1984) Cultural characteristics of *Rhizopogon* spp. associated with *Pinus radiata* seedlings. N Z J Bot 22:35–41

Chu-Chou M, Grace LJ (1985) Comparative efficiency of the mycorrhizal fungi *Laccaria laccata*, *Hebeloma crustuliniforme*, and *Rhizopogon* species on growth of radiata pine seedlings. N Z J Bot 23:417–424

Cline ML, Reid PP (1982) Seed source and mycorrhizal fungus effects on growth of containerized *Pinus contorta* and *Pinus ponderosa* seedlings. For Sci 28:237–250

Coker WC, Couch JN (1928) The Gasteromycetes of the eastern United States and Canada. University of North Carolina Press, Chapel Hill

Colgan W III (1997) Diversity, productivity, and mycophagy of hypogeous mycorrhizal fungi in a variably thinned Douglas-fir forest. PhD Thesis, Oregon State University, Corvallis

Crafts CB, Miller CO (1974) Detection and identification of cytokinins produced by mycorrhizal fungi. Plant Physiol 54:586–588

Croghan CG (1984) Survey for mycorrhizal fungi in lake states tree nurseries. Mycologia 76:951–953

Cullings KW, Szaro TM, Bruns TD (1996) Evolution of extreme specialization within a lineage of ectomycorrhizal epiparasites. Nature 379:63–66

Donald DGM (1975) Mycorrhizal inoculation for pines. S Afr For J 92:27–29

Dosskey M, Boersma L, Linderman RG (1990) Role for photosynthate demand by ectomycorrhizas in the response of Douglas-fir seedlings to drying soil. New Phytol 117:327–324

Duddridge JA (1986a) The development and ultrastructure of ectomycorrhizas. III. Compatible and incompatible interactions between *Suillus grevillei* (Klotzsch) Sing. and 11 species of ectomycorrhizal hosts in vitro in the absence of exogenous carbohydrate. New Phytol 103:457–464

Duddridge JA (1986b) The development and ultrastructure of ectomycorrhizas. IV. Compatible and incompatible interactions between *Suillus grevillei* (Klotzsch) Sing. and a number of ectomycorrhizal hosts in vitro in the presence of exogenous carbohydrate. New Phytol 103:465–471

Duddridge JA, Malibari A, Read DJ (1980) Structure and function of mycorrhizal rhizomorphs with special reference to their role in water transport. Nature 287:834–836

Durall DM, Todd AW, Trappe JM (1994) Decomposition of ^{14}C-labeled substrates by ectomycorrhizal fungi in association with Douglas fir. New Phytol 127:725–729

Ekwebelam SA (1980) Effect of mycorrhizal fungi on the growth and yield of *Pinus oocarpa* and *Pinus caribaea* var *bahamensis* seedlings. East Afr Agric For J 45:290–295

Ekwebelam SA, Odeyinde MA (1985) Field response of *Pinus* species inoculated with ectomycorrhizal fungi in Nigeria. In: Molina R (ed) 6th North American Conf on Mycorrhiza. Forest Research Laboratory, Oregon State University, Corvallis, 220 pp

Finlay RD, Ek H, Odham G, Soderstrom B (1988) Mycelial uptake, translocation and assimilation of nitrogen from ^{15}N-labeled ammonium by *Pinus sylvestris* plants infected with four different ectomycorrhizal fungi. New Phytol 110:59–66

Fogel R (1976) Ecological studies of hypogeous fungi. II. Sporocarp phenology in a western Oregon Douglas fir stand. Can J Bot 54:1152–1162

Fogel R (1980) Additions to the hypogeous mycoflora of the Canary Islands and Madeira. Contrib Univ Mich Herb 14:75–82

Fogel R, Trappe JM (1978) Fungus consumption (mycophagy) by small animals. Northwest Sci 52:1–31

Fontana A, Centrella E (1967) Ectomicorrize prodotte da funghi ipogei. Allionia 13:149–176

Ford VL, Torbert JL, Burger JA, Miller OK (1985) Comparative effects of four mycorrhizal fungi on loblolly pine seedlings growing in a greenhouse in a Piedmont soil. Plant Soil 83:215–221

Fries EM (1823) Systema mycologicum vol II. Greifswald, Lund

Froidevaux PL, Amiet R (1975) Synthèse en culture pure de l'association mycorrhizienne *Pinus silvestris* L. + *Rhizopogon rubescens* Tul. Eur J For Pathol 5:53–57

Garrido N (1986) Survey of ectomycorrhizal fungi associated with exotic forest trees in Chile. Nova Hedwigia 43:423–442

Goodman DM, Trofymow JA (1998) Comparison of communities of ectomycorrhizal fungi in old-growth and mature stands of Douglas-fir at two sites on southern Vancouver Island. Can J For Res 28:574–581

Goodman DM, Durral DM, Trofymow JA, Berch SM (eds) (1996) A manual of concise descriptions of North American ectomycorrhizae. Mycologue Publications, Sydney, British Columbia, Canada

Gross G, Runge A, Winterhoff W (1980) Bauchpilze (Gasteromycetes S L) in der Bundesrepublik Deutschland und Westberlin. Z Mykol Beih 2:1–220

Grubisha L (1998) Systematics of the genus *Rhizopogon* inferred from nuclear ribosomal DNA large subunit and internal transcribed spacer sequences. Masters Thesis, Oregon State University, Corvallis

Hacskaylo E (1973) Carbohydrate physiology of ectomycorrhizae. In: Marks GC, Kozlowski TT (eds) Ectomycorrhizae, their ecology and physiology. Academic Press, New York, pp 207–230

Hacskaylo E, Palmer JG, Vozzo JA (1965) Effect of temperature on growth and respiration of ectotrophic mycorrhizal fungi. Mycologia 57:748–756

Harrison KA, Smith AH (1968) Some new species and distribution records of *Rhizopogon* in North America. Can J Bot 46:881–889

Haselwandter K, Ortwin B, Read DJ (1990) Degradation of ^{14}C-labeled lignin and dehydropolymer of coniferyl alcohol by ericoid and ectomycorrhizal fungi. Arch Microbiol 153:352–354

Henderson GS, Stone EL (1970) Growth of mycorrhizal Monterey pine supplied with phosphorus fixed on perlite. In: Youngberg CT, Davey DB (eds) Tree growth and forest soils. Oregon State University Press, Corvallis, pp 171–180

Ho I, Trappe JM (1980) Nitrate reductase activity of nonmycorrhizal Douglas-fir rootlets and of some associated mycorrhizal fungi. Plant Soil 54:395–398

Ho I, Trappe JM (1987) Enzymes and growth substances of *Rhizopogon* species in relation to mycorrhizal hosts and infrageneric taxonomy. Mycologia 79:553–558

Ho I, Zak B (1979) Acid phosphatase activity of six ectomycorrhizal fungi. Can J Bot 57:1203–1205

Hodgson TJ (1979) Basidiospore inoculation of soil: the effect of application timing on *Pinus elliottii* seedling development. S Afr For J 108:10–15

Horton TR, Cázares E, Bruns TD (1998) Ectomycorrhizal, vesicular-arbuscular and dark septate fungal colonization of Bishop pine (*Pinus muricata*) seedlings in the first 5 months of growth after wildfire. Mycorrhiza 8:11–18

Hosford DR (1972) *Rhizopogon* of the northwestern United States. PhD Thesis, University of Washington, Seattle

Hosford DR (1975) Taxonomic studies on the genus *Rhizopogon*. I. Two new species from the Pacific Northwest. Beih Nova Hedwigia Kryptogamenkd 6:163–169

Hosford DR, Trappe JM (1980) Taxonomic studies on the genus *Rhizopogon*. II. Notes and new records of species from Mexico and Caribbean countries. Bol Soc Mex Micol 14:3–15

Hosford DR, Trappe JM (1988) A preliminary survey of Japanese species of *Rhizopogon*. Trans Mycol Soc Jpn 29:63–72

Hung LL, Trappe JM (1983) Growth variation between and within species of ectomycorrhizal fungi in response to pH in vitro. Mycologia 75:234–241

Hunt GA, Trappe JM (1987) Seasonal hypogeous sporocarp production in a western Oregon Douglas-fir stand. Can J Bot 65:438–445

Hyppel A (1968) Antagonistic effects of some soil fungi on *Fomes annosus* in laboratory experiments. Stud For Suec 64:1–18

Ivory MH (1980) Ectomycorrhizal fungi of lowland tropical pines in natural forests and exotic plantations. In: Mikola P (ed) Tropical mycorrhiza research. Oxford University Press, New York, pp 110–117

Ivory MH, Munga FM (1983) Growth and survival of container-grown *Pinus caribaea* infected with various ectomycorrhizal fungi. Plant Soil 71:339–344

Jansen AE, de Vries FW (1989) Mycorrhizas on Douglas fir in the Netherlands. Agric Ecosyst Environ 28:197–200

Jones MD, Browning MHR, Hutchinson TC (1986) The influence of mycorrhizal associations on paper birch and jack pine seedlings when exposed to elevated copper, nickel or aluminum. Water Air Soil Pollut 31:441–448

Kessel SL (1927) Soil organisms. The dependence of certain pine species on a biological soil factor. Emp For J 6:70–74

Kretzer A, Li Y, Szaro T, Bruns TD (1996) Internal transcribed spacer sequences from 38 recognized species of *Suillus sensu lato*: phylogenetic and taxonomic implications. Mycologia 88:776–785

Laloue M, Hall RH (1973) Cytokinins in *Rhizopogon roseolus*: secretion of N-[9-(B-D-ribofuranosyl-9H) purin-6-ylcarbamoyl]threonin into the culture medium. Plant Physiol 51:559–562

Lamb RJ (1974) Effect of D-glucose on utilization of single carbon sources by ectomycorrhizal fungi. Trans Br Mycol Soc 63:295–306

Lamb RJ (1979) Factors responsible for the distribution of mycorrhizal fungi of *Pinus* in eastern Australia. Aust For Res 9:25–34

Lamb RJ, Richards BN (1971) Effect of mycorrhizal fungi on the growth and nutrient status of slash and radiata pine seedlings. Aust For 35:1–7

Lamb RJ, Richards BN (1974a) Inoculation of pines with mycorrhizal fungi in natural soils. I. Effects of density and time of application of inoculum and phosphorus amendments on mycorrhizal infection. Soil Biol Biochem 6:167–171

Lamb RJ, Richards BN (1974b) Inoculation of pines with mycorrhizal fungi in natural soils. II. Effects of density and time of application of inoculum and phosphorus amendment on seedling yield. Soil Biol Biochem 6:173–177

Lamb RJ, Richards BN (1974c) Survival potential of sexual and asexual spores of ectomycorrhizal fungi. Trans Br Mycol Soc 62:181–191

Lamb RJ, Richards BN (1978) Inoculation of pines with mycorrhizal fungi in natural soils. III. Effects of soil fumigation on rate of infection and response to inoculum density. Soil Biol Biochem 10:273–276

Levisohn I (1956) Growth stimulation of forest tree seedlings by the activity of free living mycorrhizal mycelia. Forestry 29:53–59

Levisohn I (1965) Nutritional problems in forest nurseries – mycorrhizal investigations. G B For Comm Bull 37:228–235

Li CY, Massicote HB, Moore LVH (1992) Nitrogen-fixing *Bacillus* sp. associated with Douglas-fir tuberculate ectomycorrhiza. Plant Soil 140:35–40

Liu B (1985) New species and new records of hypogeous fungi from China. Acta Mycol Sin 4:84–89

Lundeberg G (1970) Utilisation of various nitrogen sources, in particular bound soil nitrogen, by mycorrhizal fungi. Stud For Suec 79:1–95

Luoma DL (1988) Biomass and community structure of sporocarps formed by hypogeous ectomycorrhizal fungi within selected forest habitats of the HJ Andrews Experimental Forest, Oregon. PhD Thesis, Oregon State University, Corvallis

Luoma DL, Frenkel RE, Trappe JM (1991) Fruiting of hypogeous fungi in Oregon Douglas-fir forests: seasonal and habitat variation. Mycologia 83:335–353

Malajczuk N, Molina R, Trappe JM (1982) Ectomycorrhiza formation in *Eucalyptus*. I. Pure culture synthesis, host specificity and mycorrhizal compatibility with *Pinus radiata*. New Phytol 91:467–482

Martín MP (1996) The genus *Rhizopogon* in Europe. Societat Catalana de Micologia, Barcelona

Martín MP, Högberg N (1996) Molecular analysis confirms morphological reclassification of the genus *Rhizopogon*. Br Mycol Soc Centenary Symp (Abstr)

Martín MP, Sánchez A (1996) Thin layer chromatography patterns of *Rhizopogon* species and their possible use as a taxonomic criterion. Rev Catalan Micol 19:91–98

Marx DH (1969) The influence of ectotrophic mycorrhizal fungi on the resistance of pine roots to pathogenic infections. I. Antagonism of mycorrhizal fungi and soil bacteria. Phytopathology 59:153–163

Marx DH (1980) Ectomycorrhizal fungus inoculations: a tool for improving forestation practices. In: Mikola P (ed) Tropical mycorrhiza research. Clarendon Press, Oxford, pp 11–71

Marx DH, Daniel WJ (1976) Maintaining cultures of ectomycorrhizal and plant pathogenic fungi in sterile water cold storage. Can J Microbiol 22:338–341

Maser C, Maser Z (1987) Notes on mycophagy in four species of mice in the genus *Peromyscus*. Great Basin Nat 47:308–313

Maser C, Maser Z (1988) Interactions among squirrels, mycorrhizal fungi, and coniferous forests in Oregon. Great Basin Nat 48:358–369

Maser C, Trappe JM, Nussbaum RA (1978) Fungal–small mammal interrelationships with emphasis on Oregon coniferous forests. Ecology 59:799–809

Maser Z, Maser C (1987) Notes on mycophagy of the yellow-pine chipmunk (*Eutamias amoenus*) in northeastern Oregon. Murrelet 68:24–27

Maser Z, Maser C, Trappe JM (1985) Food habits of the northern flying squirrel (*Glaucomys sabrinus*) in Oregon. Can J Zool 63:1084–1088

Massicotte HB, Melville LH, Li CY, Peterson RL (1992) Structural aspects of Douglas-fir [*Pseudotsuga menziesii* (Mirb.) Franco] tuberculate ectomycorrhizae. Trees 6:137–146

Massicotte HB, Molina R, Luoma DL, Smith JE (1994) Biology of the ectomycorrhizal genus *Rhizopogon*. II. Patterns of host-fungus specificity following spore inoculation of diverse hosts grown in mono- and dual cultures. New Phytol 126:677–690

Massicotte HB, Melville LH, Peterson RL, Molina R (1999) Biology of the ectomycorrhizal genus *Rhizopogon*. IV. Comparative morphology and anatomy of ectomycorrhizas synthesized between several *Rhizopogon* species on ponderosa pine (*Pinus ponderosa*). New Phytol (In press)

Melin E, Nilsson H (1957) Transport of C^{14}-labeled photosynthate to the fungal associate of pine mycorrhiza. Sven Bot Tidskr 51:1–166–186

Melin E, Nilsson H, Hacskaylo E (1958) Translocation of cations to seedlings of *Pinus virginiana* through mycorrhizal mycelium. Bot Gaz 119:243–246

Miller OK (1983) Ectomycorrhizae in the Agaricales and Gasteromycetes. Can J Bot 61:909–916

Miller SL (1986) Hypogeous fungi from the southeastern United States. I. The genus *Rhizopogon*. Mycotaxon 27:193–218

Miller SL, Koo CD, Molina R (1992) Early colonization of red alder and Douglas-fir by ectomycorrhizal fungi and *Frankia* in soils from the Oregon Coast Range. Mycorrhiza 2:53–61

Miller SL, Torres P, McClean TM (1993) Basidiospore viability and germination in ectomycorrhizal and saprotrophic basidiomycetes. Mycol Res 97:141–149

Miller SL, Torres P, McClean TM (1994) Persistence of basidiospores and sclerotia of ectomycorrhizal fungi and *Morchella* in soil. Mycologia 86:89–95

Miura G, Hall RH (1973) *trans*-Ribosylzeatin: its biosynthesis in *Zea mays* endosperm and the mycorrhizal fungus, *Rhizopogon roseolus*. Plant Physiol 51:563–569

Modess O (1941) Zur Kenntnis der Mycorrhizabildner von Kiefer und Fichte. Symb Bot Ups 1:1–146

Mohan V, Natarajan K, Ingleby K (1993) Anatomical studies on ectomycorrhizas. III. The ectomycorrhizas produced by *Rhizopogon luteolus* and *Scleroderma citrinum* on *Pinus patula*. Mycorrhiza 3:51–56

Molina R (1980a) Ectomycorrhizal inoculation of containerized western conifer seedlings. Research Note PNW-357, Pacific Northwest Forest and Range Experiment Station, US Department of Agriculture, Forest Service, Corvallis

Molina R (1980b) Patterns of ectomycorrhizal host-fungus specificity in the Pacific Northwest. PhD Thesis, Oregon State University, Corvallis

Molina R, Palmer JG (1982) Isolation, maintenance, and pure culture manipulation of ectomycorrhizal fungi. In: Schenck NC (ed) Methods and principles of mycorrhizal research. American Phytopathological Press, St Paul, pp 115–129

Molina R, Trappe JM (1982a) Lack of mycorrhizal specificity by the ericaceous hosts *Arbutus menziesii* and *Arctostaphylos uva-ursi*. New Phytol 90:495–509

Molina R, Trappe JM (1982b) Patterns of ectomycorrhizal host specificity and potential among Pacific Northwest conifers and fungi. For Sci 28:423–458

Molina R, Trappe JM (1994) Biology of the ectomycorrhizal genus *Rhizopogon*. I. Host associations, host-specificity and pure culture syntheses. New Phytol 125:653–675

Molina R, Massicotte H, Trappe JM (1992) Specificity phenomena in mycorrhizal symbioses: community-ecological consequences and practical implications. In: Allen MF (ed) Mycorrhizal functioning: an integrative plant–fungal process. Chapman and Hall, New York, pp 357–423

Molina R, Smith JE, McKay D, Melville LH (1997) Biology of the ectomycorrhizal genus, *Rhizopogon*. III. Influence of co-cultured conifer species on mycorrhizal specificity with the arbutoid hosts *Arctostaphylos uva-ursi* and *Arbutus menziesii*. New Phytol 137:519–528

Momoh ZO (1976) Synthesis of mycorrhiza on *Pinus oocarpa*. Ann Appl Biol 82:221–226

North M, Trappe J, Franklin J (1997) Standing crop and animal consumption of fungal sporocarps in Pacific Northwest forests. Ecology 78:1543–1554

Odeyinde MA, Ekwebelam SA (1974) In search of a suitable pine mycorrhiza fungus in the high forest zones of Nigeria. Nig J For 4:93–97

Pachlewski R, Pachlewska J (1968) *Rhizopogon luteolus* Fr. in a synthesis with pine (*Pinus silvestris* L.) in pure culture in agar. Inst Bad Lesn 346:77–95

Pachlewski R, Pachlewska J (1974) Studies on symbiotic properties of mycorrhizal fungi of pine (*Pinus silvestris* L.) with the aid of the method of mycorrhizal synthesis in pure cultures on agar. Forest Research Institute, Warsaw, Poland

Pacioni G (1984a) Champignons hypogés nouveaux pour l'Afrique du Nord. Bul Soc Mycol Fr 100:111–124

Pacioni G (1984b) Un nuovo fungo ipogeo raccolto in Sardegna: *Rhizopogon sardöus* nov. sp. Micol Ital 2:45–47

Parke JL, Linderman RG, Black CH (1983) The role of ectomycorrhizas in drought tolerance of Douglas-fir seedlings. New Phytol 95:83–95

Parladé J, Alvarez IF (1993) Coinoculation of aseptically grown Douglas fir with pairs of ectomycorrhizal fungi. Mycorrhiza 3:93–96

Parladé J, Pera J, Alvarez IF (1996) Inoculation of containerized *Pseudotsuga menziesii* and *Pinus pinaster* seedlings with spores of five species of ectomycorrhizal fungi. Mycorrhiza 6:236–245

Pilz P, Perry DA (1983) Impact of clearcutting and slash burning on ectomycorrhizal associations of Douglas-fir seedlings. Can J For Res 14:94–100

Purnell H (1957) Notes on fungi found in Victorian plantations. III. The mycorrhizal fungi. Plantat Tech Pap For Comm Victoria 3:9–13

Rayner MC (1938) The use of soil or humus inocula in nurseries and plantations. Emp For J 17:236–243

Read DJ (1984) The structure and function of vegetative mycelium of mycorrhizal roots. In: Jennings DH, Rayner ADM (eds) Ecology and physiology of the fungal mycelium. Cambridge University Press, Cambridge, pp 215–240

Read DJ, Boyd R (1986) Water relations of mycorrhizal fungi and their host plants. In: Ayres PG, Boddy L (eds) Water, fungi, and plants. Cambridge University Press, Cambridge, pp 287–303

Read DJ, Francis R, Finlay RD (1985) Mycorrhizal mycelia and nutrient cycling in plant communities. In: Fitter AH, Atkinson D, Read DA, Usher MB (eds) Ecological interactions in soil. Blackwell, Palo Alto, pp 193–217

Roth AL, Berch SM (1992) Ectomycorrhizae of Douglas-fir and western hemlock seedlings outplanted on eastern Vancouver Island. Can J For Res 22:1646–1655

Rudawska M (1982) Effect of various organic sources of nitrogen on the growth of mycelium and content of auxin and cytokinin in cultures of some mycorrhizal fungi. Acta Physiol Plant 4:11–20

Rudawska M (1983) The effect of nitrogen and phosphorus on auxin and cytokinin production by mycorrhizal fungi. Arbor Kornickie 28:219–236

Sands R, Theodorou C (1978) Water uptake by mycorrhizal roots of radiata pine seedlings. Aust J Plant Physiol 5:301–309

Sasek V, Musilek V (1967) Cultivation and antibiotic activity of mycorrhizal basidiomycetes. Folia Microbiol 12:515–523

Simard SW, Molina R, Smith JE, Perry, Jones MD (1997a) Shared compatibility of ectomycorrhizae on *Pseudotsuga menziesii* and *Betula papyrifera* seedlings grown in mixture in soils from southern British Columbia. Can J For Res 27:331–342

Simard SW, Perry DA, Jones MD, Myrold DD, Durall DM, Molina R (1997b) Net transfer of carbon between ectomycorrhizal tree species in the field. Nature 388:579–582

Simard SW, Perry DA, Smith JE, Molina R (1997c) Effects of soil trenching on occurrence of ectomycorrhizas on *Pseudotsuga menziesii* seedlings grown in mature forests of *Betula papyrifera* and *Pseudotsuga menziesii*. New Phytol 136:327–340

Skinner MF, Bowen GD (1974a) The penetration of soil by mycelial strands of ectomycorrhizal fungi. Soil Biol Biochem 6:57–61

Skinner MF, Bowen GD (1974b) The uptake and translocation of phosphate by mycelial strands of pine mycorrhizas. Soil Biol Biochem 6:53–56

Smith AH (1964) *Rhizopogon*, a curious genus of false truffle. Mich Bot 3:13–19

Smith AH (1966) New and noteworthy higher fungi from Michigan. Mich Bot 5:18–25

Smith AH (1968) Further studies on *Rhizopogon*. I. J Elisha Mitchell Sci Soc 84:274–280

Smith AH, Zeller SM (1966) A preliminary account of the North American species of *Rhizopogon*. Mem N Y Bot Gard 14:1–178

Smith JE, McKay D, Molina R (1994) Survival of mycorrhizal fungal isolates stored in sterile water at two temperatures and retrieved on solid and liquid nutrient media. Can J Microbiol 40:736–742

Smith JE, Molina R, Perry DA (1995) Occurrence of ectomycorrhizas on ericaceous and coniferous seedlings grown in soils from the Oregon Coast Range. New Phytol 129:73–81

States JS, Gaud WS (1997) Ecology of hypogeous fungi associated with ponderosa pine. I. Patterns of distribution and sporocarp production in some Arizona forests. Mycologia 89:712–721

Theodorou C (1967) Inoculation with pure cultures of mycorrhizal fungi of radiata pine growing in partially sterilized soil. Aust For 31:303–309

Theodorou C (1968) Inositol phosphate in needles of *Pinus radiata* D. Don and the phytase activity of mycorrhizal fungi. Proc 9th Int Congr Soil Sci 3:483–493

Theodorou C (1971) Introduction of mycorrhizal fungi into soil by spore inoculation of seed. Aust For 35:17–22

Theodorou C (1978) Soil moisture and the mycorrhizal association of *Pinus radiata* D. Don. Soil Biol Biochem 10:33–37

Theodorou C (1980) The sequence of mycorrhizal infection of *Pinus radiata* D. Don following inoculation with *Rhizopogon luteolus* Fr. and Nordh. Aust For Res 10:381–387

Theodorou C (1984) Mycorrhizal inoculation of pine nurseries by spraying basidiospores onto soil prior to sowing. Aust For 47:76–78

Theodorou C, Benson AD (1983) Operational mycorrhizal inoculation of nursery beds with seed-borne fungal spores. Aust For 46:43–47

Theodorou C, Bowen GD (1970) Mycorrhizal responses of radiata pine in experiments with different fungi. Aust For 34:183–191

Theodorou C, Bowen GD (1971) Effects of non-host plants on growth of mycorrhizal fungi of radiata pine. Aust For 35:17–22

Theodorou C, Bowen GD (1973) Inoculation of seeds and soil with basidiospores of mycorrhizal fungi. Soil Biol Biochem 5:765–771

Theodorou C, Bowen GD (1987) Germination of basidiospores of mycorrhizal fungi in the rhizosphere of *Pinus radiata* D. Don. New Phytol 106:217–223

Theodorou C, Skinner M (1976) Effects of fungicides on seed inocula of basidiospores of mycorrhizal fungi. Aust For Res 7:53–58

Thoen D (1974) Premières indications sur les mycorrhizes et les champignons mycorrhiziques des plantations d'exotiques du Hant-Shaba (République du Zaire). Bull Rech Agron Gembloux 9:215–227

Tilak KVBR, Li CY, Trappe JM (1988) Characterization of nitrogen-fixing *Azospirillum* within sporocarps of ectomycorrhizal fungi associated with Douglas-fir (*Pseudotsuga menziesii* (Mirb.) Franco). Ind J Microbiol 28:315–319

Torres P, Honrubia M (1994a) Basidiospore viability in stored slurries. Mycol Res 98:527–530

Torres P, Honrubia M (1994b) Inoculation of containerized *Pinus halepensis* (Miller) seedlings with basidiospores of *Pisolithus arhizus* (Pers.) Rauschert, *Rhizopogon roseolus* (Corda) and *Suillus collinitus* (Fr.) O. Kuntze. Ann Sci For 51:521–528

Trappe JM (1962) Fungus associates of ectotrophic mycorrhizae. Bot Rev 28:538–606

Trappe JM (1965) Tuberculate mycorrhizae of Douglas-fir. For Sci 11:27–32

Trappe JM (1967) Pure culture synthesis of Douglas-fir mycorrhizae with species of *Hebeloma, Suillus, Rhizopogon,* and *Astraeus.* For Sci 13:121–130

Trappe JM (1975) A revision of the genus *Alpova* with notes on *Rhizopogon* and the Melanogastraceae. Beih Nova Hedwigia 51:279–309

Trappe JM (1977) Selection of fungi for ectomycorrhizal inoculation in nurseries. Annu Rev Phytopathol 15:203–222

Trappe JM, Guzman G (1971) Notes on some hypogeous fungi from Mexico. Mycologia 63:317–332

Trojanowski J, Halder K, Hüttermann A (1984) Decomposition of ^{14}C-labeled lignin, holocellulose and lignocellulose by mycorrhizal fungi. Arch Microbiol 139:202–206

Tulasne L-R, Tulasne C (1844) Fungii hypogaei nonnulli, novi vel minus cogniti. G Bot Ital 2:56–63

Tulasne L-R, Tulasne C (1851) Fungi hypogaei. Friedrich Klincksieck, Paris

Uhl M (1988) Studies on ectomycorrhizae. XV. Mycorrhizae formed by *Rhizopogon luteolus* on *Pinus sylvestris.* Persoonia 13:449–458

Visser S (1995) Ectomycorrhizal fungal succession in jack pine stands following wildfire. New Phytol 129:389–401

Vittadini C (1831) Monographia Tuberacearum. Felicis Rusconi, Milan

Vogt KA, Bloomfield J, Ammirati JF, Ammirati SR (1992) Sporocarp production by basidiomycetes, with emphasis on forest ecosystems. In: Carroll GC, Wicklow DT (eds) The fungal community, its organization and role in the ecosystem. Marcel Dekker, New York, pp 563–581

Volkart CM (1964) Formacion de micorrizas en pinos centro-americanos bajo condiciones controladas. Turrialba 14:203–205

Young HE (1937) *Rhizopogon luteolus,* a mycorrhizal fungus of *Pinus.* Forestry 11:30–31

Zak B (1971) Characterization and classification of mycorrhizae of Douglas-fir. II. *Pseudotsuga menziesii + Rhizopogon vinicolor*. Can J Bot 49:1079–1084

Zak B (1973) Classification of ectomycorrhizae. In: Marks GC, Kozlowski TT (eds) Ectomycorrhizae, their ecology and physiology. Academic Press, New York, pp 43–78

Zak B (1976) Pure culture synthesis of bearberry mycorrhizae. Can J Bot 54:1297–1305

Zeller SM (1939) New and noteworthy Gasteromycetes. Mycologia 31:1–32

Zeller SM (1941) Further notes on fungi. Mycologia 33:196–214

Zeller SM, Dodge CW (1918) *Rhizopogon* in North America. Ann Mo Bot Gard 5:1–36

Zobel JB (1854) Iconum fungorum hucusque cognitorum. VI. Friderici Ehrlich, Prague

Chapter 6

Tuber

G. Pacioni and O. Comandini

6.1
Introduction

Our knowledge of ectomycorrhizas (ECM) first stemmed from the truffles. Thus, Frank (1885) identified the symbiotic partnership between plants and a truffle (*Tuber aestivum*) that we now know as ECM. Since Frank's discovery, truffle-farming has become widespread, and truffle orchards have been established over many thousands of hectares. Truffles, along with the connected nursery activities, are a significant source of income in southern Europe; however, despite extensive literature, our knowledge of the biology of *Tuber* species remains relatively poor. This is partly explained by acknowledged difficulties in growing *Tuber* mycelia under axenic conditions. The first reliable isolations of *Tuber* mycelia from young sporocarps and ECM were reported in the 1970s. Unfortunately these mycelia were often short-lived and most were lost after being studied, their potentially high economic value meaning that the studies were not disseminated widely in the scientific community (Fontana 1971; Chevalier 1972; Palenzona et al. 1972). In this chapter we review what is currently known regarding *Tuber* species, placing emphasis on their interactions with plants in ECM associations.

6.2
Taxonomy and Life Cycle

Tuber Mich. ex F. H. Wigg.: Fr. (Ascomycotina, Pezizales) is an ECM genus producing hypogeous cleistothecial ascomata with spiny or reticulate ascospores (ochraceous to black in colour) that are produced in globose pedunculate asci and spread by animals. The almost 100 known species have a natural distribution in the temperate zones of the northern hemisphere, with at least three differentiation areas occurring in Europe, Southeast Asia and North America. Following forestry and truffle farming plantation activities, however, several species have now spread worldwide.

Dipartimento di Scienze Ambientali, Università, 67100 L'Aquila, Italy

Sporocarps of *Tuber* are mainly characterised by a combination of peridial and sporal features. They are generally distinct in black and white truffles, the peridium being warty and black in the former, whitish and smooth in the latter. While a wide variability exists between the two extreme forms, the distinction is helpful in separating the species with economical interest. Isoenzyme analysis and population genetics indicate that all *Tuber* taxa are fixed genetically in homozygosity (Pacioni and Pomponi 1989, 1991; Pacioni et al. 1993; Bullini et al. 1994), suggesting that the species concept in *Tuber* should be considered a set of clones with a common reference morphotype. Some morphotypes exhibit wide variability at the morphological, biochemical or molecular levels (e.g. *T. aestivum* Vittad.), while others display limited polymorphism (e.g. *T. melanosporum* Vittad. and *T. magnatum* Pico: Fr.) (Gandeboeuf et al. 1995; Lanfranco et al. 1995; Paolocci et al. 1995a; Guillemaud et al. 1996).

The life cycle of *Tuber* appears to be highly complex, requiring the succession of several steps that are unusual for an ascomycete, being reminiscent of ECM basidiomycetes, with a permanent secondary dikariotic mycelium (Grente and Delmas 1974). Sporocarps require several months to grow and mature, and appear to absorb nutrients from soil via mycelial tufts that sprout from the peridium. This saprophytic phase was confirmed using ^{32}P-, ^{3}H$_2$O- and ^{14}C-mannose-labelling (Barry et al. 1994, 1995). It is further clear that *Tuber* sporocarps can absorb Ca from the external environment and accumulate high concentrations of Cl, Co, Cu, K, Mg, S, Sr and Zn (Pirazzi and Di Gregorio 1987; Bencivenga and Granetti 1989; Beuchat et al. 1993; Pirazzi 1993). Sulphate, which can be present at concentrations of up to 6800 ppm (Bencivenga and Granetti 1989; Beuchat et al. 1993), plays important physiological (detoxification of methanol and formaldehyde) and ecological (insect attraction) roles in sporocarp biology via its reduction to methylsulphides (Pacioni 1992).

Sporocarp development is correlated with the expression of tyrosinase genes and may be controlled by them, as previously shown in *Neurospora crassa* Shear & Dodge (Miranda et al. 1992; Ragnelli et al. 1992). Sporocarp development is also characterised by high L-dopa oxidase and tyrosinase hydroxylase activities that mark a layer of cells from which peridium, veins and asci originate. These activities are implicated in the formation of allomelanins, which are important wall components in peridial cells and spores. Some volatile compounds (alcohols and aldehydes with 2–5 C molecules) are produced before spore formation, while phenolic and S compounds along with ketones and esters appear later, accompanied by a steady decrease in tyrosine (Pacioni et al. 1995). Metabolites in unripe sporocarps are normal fungal metabolites produced via the pyruvate pathway, and the same substances are probably produced by the mycelium. Some of these are ecologically important (see below), as is dimethyl sulphide, which is produced at maturity and attracts truffle-eating (hydnophagous) animals and insects which are the agents of spore dispersal (Talou et al. 1990; Pacioni et al. 1991).

The physical factors involved in truffle production, such as temperature, moisture and soil permeability, have also been described (Bencivenga et al. 1992; Callot and Jaillard 1996), as has the influence of host plant growth (Grente and Delmas 1974; Shaw et al. 1996)

6.3
Ecology

6.3.1
General Ecology

Tuber generally grows in calcareous soils with a pH close to 7–8 or higher, although a few taxa, such as *T. mesentericum*, are adapted to slightly acidic soils (pH 6–7) (Napoliello et al. 1990). Different pH requirements appear to exist between mycelial and sporocarp metabolism in some species, such as those linked to coniferous trees. The environmental requirements of *Tuber* have, however, been studied mainly in economically important species such as *T. melanosporum*, *T. aestivum*, *T. uncinatum* Chatin and *T. magnatum* (Pacioni 1985; Chevalier and Frochot 1990; Chevalier and Poitou 1990). These data indicate that truffles generally grow in poor soils of temperate zones, with low levels of N, P and Fe, but rich in Ca and with a good quantity of K and S. Black truffles prefer well-drained stony soils, while white ones grow in deeper, fresh, sandy-clay soils (Bencivenga and Granetti 1990a). Slight changes in water potential, pH, nutrient and/or ion concentration can favour the ECM of one species and repress those of another, especially in black truffle species (Chevalier et al. 1982; Mamoun et al. 1985).

Some black truffle species (e.g. *T. melanosporum* and *T. aestivum*) produce a drastic change in the rhizosphere of their host trees, creating what is referred to as the "hydnosphere" (Pacioni 1991). This effect is typified by an area around the tree, known as a "brûlé" or "pianello", that is almost circular in shape and appears as if it has been burnt (Fig. 6.1). Several investigations have reported negative effects on the growth of grasses (Montacchini and Caramiello-Lomagno 1977) and micro-organisms in these areas (Chalvignac et al. 1959; Luppi-Mosca et al. 1970; Luppi-Mosca 1972; Luppi-Mosca and Fontana 1977). Brûlé formation was attributed to an allelopathic effect; however, a chemical explanation has also been hypothesised (Fasolo-Bonfante et al. 1971; Papa et al. 1992a). This is supported by the fact that Pacioni (1991), assaying the volatile substances produced by sporocarps, found that three aldehydes (2-methyl propanal, 2-methyl butanal and 3-methyl butanal) and two alcohols (2-methyl butanol and 3-methyl butanol) had a negative effect on soil fungi and plants respectively. A strain of *Pseudomonas* isolated from a truffle surface, however, tolerated higher concentrations of these substances. In addition, Plattner and Hall (1995) demonstrated that *T. melanosporum* mycelium can behave as a minor pathogen, producing abundant and extensive lesions on the roots of plant weeds. This pathogenic effect was demon-

Fig. 6.1. Truffle habitat (*T. melanosporum* associated with *Q. pubescens*) showing an extensive "burnt" area

strated in New Zealand weeds that were exposed to introduced truffle mycelium in a recently established truffle farm. Interestingly, the European grass *Festuca ovina*, when co-cultured with the ECM hazels, was not affected or only contacted at the root level by *T. melanosporum* mycelium (Mamoun and Olivier 1997).

6.3.2
Tuber ECM and Bacteria

Tuber sporocarps typically contain large numbers of bacteria (Parguey-Leduc et al. 1987; Pacioni 1990). Some isolated bacteria appear to improve mycelial growth in culture (Fontana 1971, Pirazzi 1990a) or to be N-fixing (Li and Castellano 1985). *Pseudomonas fluorescens* Migula is commonly isolated from *Tuber* sporocarps, accompanied, according to the truffle host, by other species of *Pseudomonas*, *Micrococcus*, *Staphylococcus* and rarely other genera (Citterio et al. 1995). The fluorescent *Pseudomonas* populations show seasonal variations in nature and in relation to truffle-farming practices. Black truffle ECM, for example, decrease in autumn and winter, while *Pseudomonas* numbers in ECM rhizoplane rise (Olivier and Mamoun 1988; Mamoun and Oliver 1990). Furthermore, fluorescent *Pseudomonas* species [*P. fluorescens* and *P. putida* (Trevisan) Migula] seem to dominate the *Tuber* ECM mantle and are the most powerful with regard to their Fe-chelating capability (Mamoun and Olivier 1989). Glasshouse experiments indicate that irrigation promotes root colonisation by *T. melanosporum* in favour of other competitive ECM fungi, an effect perhaps resulting from a modification of the $Fe^{2+}:Fe^{3+}$ ratio in the soil as a result of water potential variations (Lindsay 1974) and altered competition towards the available Fe. The addition of *P. fluorescens* and *P. putida* strains seemed to provide a permanent inhibition of basidiomycete and ascomycete competitors, promoting the spread of *Tuber* ECM (Mamoun and Olivier 1992). Following the addition of the bacteria, however, *Tuber* ECM numbers declined markedly, suggesting an antagonistic effect of pseudomonads on ECM. Antagonistic effects of *Pseudomonas* on the growth of *T. melanosporum* and *T. brumale* Vittad. mycelia have also been shown in vitro (Mamoun et al. 1985).

The relationships between *Pseudomonas* and *Tuber* ECM are further modified by the available Fe concentration in the soil. Although in vitro growth of *T. melanosporum* mycelium is not affected by Fe in pot experiments (supplied as ferric ethylenediaminedi-phenyl-acetic acid [Fe-EDDHA] 6 mg kg^{-1}) (Mamoun and Olivier 1991), Fe amendment significantly depresses *T. melanosporum* infection and the antagonistic activity of fluorescent *Pseudomonas*, favoring root colonisation by the competing ECM fungus *Pulvinula globifera* (Berk. & Curt. apud Berk.) Le Gal.

6.3.3
Spore Germination and Mycelial Growth

Germination of spores from black and white truffles near host plant roots has been observed and documented (Sappa 1940; Rouquerol and Payre 1974–1975; Zambonelli and Govi 1990). The conditions required for such germination are poorly understood; however, germination in vitro has been claimed and questioned by several researchers for some time. Sappa (1940) maintained that he achieved spore germination in *T. magnatum* after 14 months in water. Germination of *T. melanosporum* spores was reported by Guiochon (1959) and later under axenic conditions by Grente et al. (1972), while a clear picture of a spore with emerging mycelium was provided by Chevalier and Dupré (1990).

The walls of *Tuber* spores are composed mainly of proteins linked to allomelanins of polyketide origin (quinonoid and polyphenolic biopolymers) (Fonvieille et al. 1990; De Angelis et al. 1996), and protease and laccase activities are probably involved in the first stages of germination. Indeed, Chevalier and Desmas (1977) noted that, prior to germination, *T. melanosporum* spores become translucent, suggesting decomposition of melanin, probably via laccase activity.

Truffle spores seem to be activated by passage through the intestine of hydnophagous animals (Fogel and Trappe 1978; Pacioni 1989). In the laboratory, spores of *T. melanosporum* can be activated by storing sporocarps in wet sterile sand at 5°C for 4 months. During this treatment, endogenous glycogen is degraded to 1,5-D-anhydrofructose and is then converted to an unusual sugar, ascopyroneT (Deffieux et al. 1994). The means of this conversion, however, remains unclear. If spores are then treated with the lytic enzyme helicase, and repeatedly rinsed with antibiotic and disinfectant solutions, germination can occur after 2–3 days in either water or Ringer's solution. The percentage of germination can reach 70%, which is high compared with the percentages obtained with spores of other ECM fungi, such as *Terfezia* spp., where germination rate is generally <20% (Awameh and Alsheikh 1980a,b). Germination of *Tuber* ascospores has unfortunately not been confirmed by other groups and a reproducible method is still required. Our own unpublished data indicate that small germ tubes with few ramifications can be obtained from sterile spores on MMN, to which a cold concentrated water extract from uninfected oak roots (sterilised by filtration) is added.

The necessity of a secondary mycelium (either dikariotic or heterokariotic) for successful ECM establishment is, as with many ECM basidiomycetes, unclear. Grente et al. (1972) found that a mycelium obtained from a germinated ascospore did not produce ECM. Fasolo-Bonfante and Brunel (1972) demonstrated that such mycelium is uninucleate, whereas mycelia from ECM or sporocarp tissue, that are able to produce ECM, contain two or more nuclei per cell. However, a number of nuclei produced by several mitoses can co-exist inside spores (Marchisio 1964) and all examined species appear to be homozygous, despite the presence of two or more nuclei in their hyphae (Pacioni and Pomponi 1989, 1991; Pacioni et al. 1993). A more accurate study of the kariology of germinating spores and germ tube is therefore required. Establishment of *T. magnatum* ECM using conventional spore suspension inocula thus gives poor infection; however, direct contact between ECM and uninfected roots represents an alternative system for ECM propagation (Gregori and De Paoli 1992).

Apical growth of *Tuber* hyphae has been demonstrated to be under the control of the enzyme protein kinase (PKC). The PKC-encoding gene was isolated in *T. magnatum* and *T. borchii* Vittad. and DNA sequence analysis has shown a close similarity to that of *N. crassa*. As with the latter, the *Tuber* PKC is inhibited by chelelitrin and permanently activated by the elimination of a portion of the terminal aminic region (Stocchi 1997; G. Mancino, G. Arpaia and R. Ambra, unpubl. data).

Since chitin (β-1-4-linked N-acetylglucosamine) is the main component of fungal walls (Bulawa 1993) and it has been located in the walls of symbiotic and non-symbiotic hyphae of truffles (Balestrini et al. 1996), the genes for chitin biosynthesis (CHS) have been extensively studied in several truffle species (Mehmann et al. 1994; Lanfranco et al. 1995). Five to six groups of CHS genes (constituting a genetic family) have now been identified in *Tuber* (Stocchi 1997), and these genes appear to be expressed differently according to the growth phase. The role of these genes in ECM formation, however, remains to be demonstrated.

6.4
ECM Formation

6.4.1
Host–Fungus Specificity

Tuber species, such as *T. melanosporum*, seem to have different abilities to penetrate the root cortical layers according to the host plant. With usual host trees, such as *Quercus pubescens* or *Corylus avellana*, the Hartig net develops within a five to seven cell layer. With other suitable, but not usual, hosts, such as *Pinus sylvestris*, the intracellular hyphae appear more aggressive, penetrating six to ten cells deep, often reaching the endodermis (Palenzona et al. 1972). In addition, depth of penetration can vary in different genotypes of a single host species. This variability has led to the selection of clonal lines of host trees for truffle ECM, such as the clone H219 of *C. avellana* (Mamoun and Olivier 1996) and others of *Tilia platyphyllos* and *Populus alba* (Zambonelli et al. 1989; Gregori and De Paoli 1992; Pirazzi and Lubrano 1994). Clone H219, for example, is said to be strongly receptive towards *T. melanosporum*, producing enhanced resistance to competing ECM fungi.

Since the first observations of Frank (1885), many studies have been devoted to the identification of host plants for different *Tuber* species (Trappe 1962; Ceruti 1965; Chevalier et al. 1975; Fontana and Giovannetti 1979; Yun 1990; Granetti 1994, 1995; Maia et al. 1996), largely with the goal of enhancing truffle production (Ceruti 1965, 1968; Scannerini and Palenzona 1967; Bencivenga and Granetti 1990b,c; Chevalier and Frochot 1990). A number of natural and/or artificial host–*Tuber* combinations have been demonstrated in the laboratory or proposed based on field observations. Compatible host–*Tuber* associations so far confirmed by laboratory synthesis experiments are summarised in Table 6.1. Host specificity in nature seems to be linked largely to geographical factors rather than to genetic specificity, since all host plant genera so far tested have formed ECM with *Tuber*, although in nature symbiosis between them is unlikely because of different habitat preferences. Spores of truffles, as with other hypogeous fungi, are spread by animals which are confined within their natural habitat, often true "forest island" refugia (Fogel and Trappe 1978; Pacioni 1989; Fogel 1992). There is thus little possibility of contact with other ECM plant genera living in different habitats, such as acid soils, or in disconnected habitats, such as the southern temperate forests of *Nothofagus* or *Eucalyptus*.

Like most ECM fungi, truffles do not fruit in axenic culture and can currently be grown commercially only by inoculating suitable host trees. Because of the difficulties associated with isolation and growth of *Tuber* mycelia, the best method for producing ECM in the laboratory seems to be the use of spore suspensions, or pounded sporocarp fragments as inoculum (Fassi and Fontana 1967; Fontana and Bonfante-Fasolo 1971; Pacioni 1985). These methods have been used successfully with seedlings from sterilised

Table 6.1. Host genera with which *Tuber* spp. isolates have been confirmed as forming ECM in mycorrhizal synthesis experiments

Host genus	Reference
Abies	Palenzona et al. (1972)
Alnus	Gregori and Tocci (1985)
Carpinus	Palenzona et al. (1972)
Carya	Pirazzi (1994)
Castanea	Chevalier et al. (1973)
Cedrus	Pirazzi (1990b)
Cistus	Fontana and Giovannetti (1979)
Corylus	Palenzona (1969)
Fagus	Palenzona et al. (1972)
Fumana	Chevalier et al. (1975)
Helianthemum	Chevalier et al. (1975)
Ostrya	Bencivenga and Granetti (1990a)
Picea	Blaschke (1987)
Pinus	Fassi and Fontana (1967)
Populus	Fontana and Palenzona (1969)
Pseudotsuga	Chu-Chou and Grace (1983)
Quercus	Palenzona et al. (1972)
Salix	Granetti (1987)
Tilia	Palenzona et al. (1972)

seeds, cuttings and sterile micropropagated seedlings (Zambonelli et al. 1989; Bencivenga and Granetti 1990c; Pirazzi and Lubrano 1994). Nevertheless, despite the promising results from in vitro mycorrhiza synthesis, planting inoculated seedlings remains problematic. Such difficulties probably reflect the particular edaphic and soil conditions required to maintain *Tuber* ECM (Bencivenga et al. 1991; Granetti 1994). Moreover, the host mycorrhizal efficiency in vitro may differ from that under more natural conditions and competition from other fungi will also vary according to soil conditions (Zambonelli and Branzanti 1987; Zambonelli and Govi 1991; Donnini and Bencivenga 1995).

6.4.2
Developmental Aspects of the Fungus–Host Interaction

Two-dimensional protein electrophoresis of extracts from *T. borchii* mycelium, *T. platyphyllos–T. borchii* ECM and non-ECM roots revealed nine ECM-specific polypeptides (Stocchi 1997). While the significance of these polypeptides remains unclear, it is known that, in *Cistus incanus–T. melanosporum* ECM at least, the mantle begins to form at the periphery of the root cap (Fusconi 1983). Hyphal growth in the mantle occurs in the inner apical area, where the youngest hyphae are present and contain dense cytoplasm with few vacuoles. Hyphae penetrate inwards through the remains of root cap cells which are trapped in the inner sheath. Growth of hyphae into dead layers of root cap cells then leads to the formation and development of the mantle. Older elements are progressively observed from the apex to the base, and from the inner to the outer part of the mantle. Differences in both structure and function of the different mantle layers were noted by Scannerini (1968), who observed an abundance of hyphal cells rich in organelles (nuclei, mitochondria, endoplasmatic reticulum) in the inner layer of the mantle, while in the outer part, cells with vacuolate cytoplasm prevailed. The stratification of the mantle was further revealed by combining ultrastructural cytochemistry with the selective extraction of polysaccharides by various solvents or lytic enzymes (Pargney 1990). Mantles of young *C. avellana–T. macrosporum* Vittad. ECM are characterised by active hyphae with simple septa, Woronin bodies, polyphosphate granules and glycogen particles. In this host–fungus combination, polyphosphate granules were shown to be more abundant in the outer hyphae than in the inner ones (Scannerini and Bonfante-Fasolo 1983).

In *Tuber* ECM as well as in those of other ascomycetes, the inner layer hyphae are cemented by electron-dense material (Scannerini 1968; Strullu 1976; Scannerini and Bonfante-Fasolo 1983). Some authors have suggested that this matrix is composed of polysaccharides and proteins (Nylund 1987; Pargney et al. 1988). However, immunological experiments using a monoclonal antibody (JIM 5), specific for unesterified pectin, clearly demonstrated that the cementing material in ECM of *C. avellana–T. magnatum* does not contain this pectin epitope (Balestrini et al. 1996). The cementing material also showed no labelling by antibodies specific for components of the host wall, or to chitin and β-1,3-glucans, typical wall components of ascomycetes (Bartniki-Garcia 1968). The presence of mannan and glycoproteins in the cementing material has been confirmed and, since these molecules also exist on hyphal surfaces, it may be that the cementing substance originates from material sloughed off from the fungal surface (Balestrini et al. 1996).

In all ECM studied so far, the fungal sheath and tannin layer are contiguous. During the establishment of *Tuber* ECM, differentiation of root cap cells occurs rapidly (Pargney and Brimont 1995). Vacuoles and cytoplasm of these cells accumulate phenols and die. Dead cells are then degraded and a layer of polyphenols remains. This tannin layer is penetrated by *Tuber* hyphae and polyphenols are incorporated into the fungal mantle. Below the fungal sheath and tannin layer, the Hartig net, formed by profusely branched hyphae (Nylund and Unestam 1982), surrounds the cortical cells, penetrating throughout the middle lamella. The hyphae of the Hartig net have similar characteristics to those of the fungal sheath and both are covered by the same cementing material (Scannerini 1968; Nylund 1980; Strullu and Gourret 1980; Scannerini and Bonfante-Fasolo 1983).

During fungal infection, host walls become swollen, less compact, and sometimes degraded (Gea et al. 1994). The symbiosis does not, however, lead to a qualitative change in wall components. Regarding this, Balestrini and coworkers conducted a detailed study on cell wall components in ECM formed by *C. avellana* and *T. magnatum* using immunocytochemistry and enzyme/lectin-gold techniques (Balestrini et al. 1996). In particular, they wanted to understand the extent to which morphological changes during fungal infection correspond to modifications in cell wall components. This study provides an accurate map of cell wall components in a hazel root and reveals that *T. magnatum* causes only subtle changes in host cell walls (Balestrini et al. 1996).

Parenchyma cells surrounded by the Hartig net show important ultrastructural modifications (Scannerini 1968). In *C. avellana–T. uncinatum* and *Quercus robur–T. magnatum* ECM, differentiation of cortical cells in contact with the Hartig net starts from lateral meristematic cells. Three simultaneous processes occur during this differentiation: (1) radial cell growth, probably induced by the production of fungal hormones; (2) an increase in vacuolar volume; (3) cytoplasmic changes, including a reduction in mitochondria, plastids, dictyosomes and an increase in endoplasmic reticulum (Pargney and Brimont 1993).

Studies on Ca localisation in walls and matrix components of several *Tuber* species have led to the suggestion that this element may be involved in hyphal attachment (Pargney and Brimont 1993; Genet and Pargney 1994; Pargney and Kottke 1994). In *Tuber* ECM, as well as in mycelia of other ascomycetes, Ca concentration in younger hyphae has been shown to be higher than in older hyphae (Botton et al. 1980; Pargney and Kottke 1994). The resulting decrease in Ca concentration in the outer part of the hyphal mantle has been taken as evidence that in *Tuber*, Ca may be mobilised from the outer hyphae and transported to the inner living ones (Pargney and Kottke 1994).

6.4.3
Identification of ECM

In general, *Tuber* ECM are monopodial (simple or ramified) in angiosperms and gymnosperms, but are dichotomous in the Pinaceae. Mantle colour can be light amber, ochre, yellowish brown, light brown or red brown, depending on the plant host or the age of the ECM. Mantle structure is pseudo-parenchymatous, with polygonal, angular or more or less lobed hyphal cells. ECM dimensions are variable within the same ECM type and vary further in different *Tuber* species. Some ECM have cystidia on the surface, the characteristics of which can be useful for a rapid identification of the fungal symbiont in some instances. However, different features can sometimes be observed even within the same *Tuber* species. For example, in *T. melanosporum*, two ECM types can develop with ageing. The first type (young) are clavate, single or dichotomous ECM, while the second type (aged) are strictly branched ECM with a thick mantle or a slimy envelope of extramatrical mycelium (glomerulus type). In fact, the latter was noted and described from oak roots producing black truffles (Condamy 1876) some years before the discovery of the ECM symbiosis by Frank (1885). The presence of this type of ECM is thought to be indicative of sporocarp-producing ability (Grente and Delmas 1974).

Several studies have been conducted in order to distinguish the ECM formed by economically important *Tuber* species from the less important, but similar, species. In some instances, the mantle pattern can be a useful diagnostic character to distinguish between even very similar *Tuber* ECM, such as those of *T. magnatum* and *T. albidum* Pico on *Quercus pubescens* (Zambonelli et al. 1993b). Unfortunately, this does not hold true in all cases, since Zambonelli et al. (1995) were unable to distinguish ECM formed by several *Tuber* species on *Pinus pinea* on this basis. While the form, length, colour and cystidial characteristics seem to be the most useful diagnostic elements, they may vary according to season and are not present in all *Tuber* species (Granetti et al. 1995a,b,c; Rauscher et al. 1995; Zambonelli et al. 1995; Comandini and Pacioni 1997, and references therein). The combination of both macroscopic, cystidial and mantle characteristics is thus the most reliable morphological method for discriminating different *Tuber* ECM. Some diagnostic features for the common *Tuber* species are reported in Fig. 6.2.

Discrimination of *Tuber* ECM, however, remains a difficult task and, in the past decade, other approaches have been used to address the problem. Biochemical markers such as protein patterns and isoenzyme polymorphisms have been shown to be useful in clarifying some aspects of *Tuber* taxonomy and ecology (Pacioni and Pomponi 1989, 1991; Palenzona et al. 1990; Dupré and Chevalier 1991; Dupré et al. 1992; Pacioni et al. 1993). Nevertheless, these tools are only species-specific and their sensitivity is low, since up to 100 ECM are necessary for successful analysis (Henrion et al. 1994). ELISA has also been successfully used to distinguish *T. albidum* ECM from *T. magnatum*, *T. aestivum*, *T. brumale* and an unidentified ECM on *Q. pubescens* and *P. pinea* (Zambonelli and Poggi Pollini 1991; Zambonelli et al. 1993a). The same method was further successful in discriminating between *T. albidum* and *T. magnatum* ECM (Plattner et al. 1991; Papa et al. 1992b), but was unable to separate *T. melanosporum* from *T. magnatum* ECM (Corocher et al. 1992).

More recently, molecular approaches, including polymerase chain reaction (PCR) amplification of variable regions of ribosomal genes [internal transcribed spacer (ITS) or intergenic spacer (IGS) regions], followed by restriction fragment length polymorphism (RFLP), allele-specific hybridization, direct sequencing, or single-strand conformation polymorphism analyses have been successfully employed in typing and distinguish sporocarps and ECM of *Tuber* species. Such methods were used successfully to distinguish between *T. brumale*, *T. ferrugineum* Vittad. and *T. melanosporum* (Henrion et al. 1994), and to separate *T. melanosporum* from *T. borchii* (Mello et al. 1996), but several other species remained difficult to distinguish using these approaches. Random amplified polymorphic DNA (RAPD) markers and comparisons of ITS-PCR products allowed rapid identification of several *Tuber* species sporocarps and, with more difficulty, their respective ECM

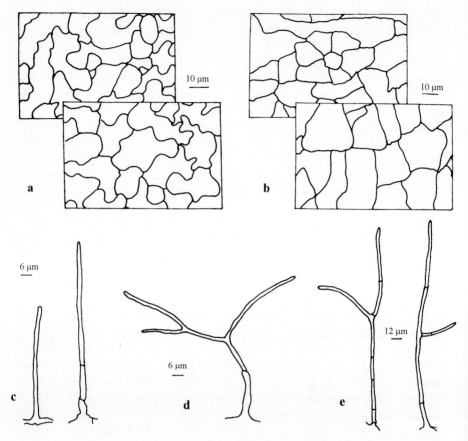

Fig. 6.2. Some diagnostic features of the most common *Tuber* species. Pseudoparenchymatous mantle with epidermoid cells such as *T. melanosporum* (**a**), and with angular cells such as *T. aestivum* (**b**). Simple cistidia *T. magnatum*-type (**c**), ramified cystidia, *T. macrosporum*-type (**d**) and *T. melanosporum*-type (**e**). Hair-type cistidia, *T. aestivum*-like, are not represented here

(Lanfranco et al. 1993, 1995; Paolocci et al. 1995a,b; Gandeboeuf et al. 1997). Screening of the genome of several *Tuber* species by microsatellite-primed PCR using the primers $(GTG)_5$ and $(TGTC)_4$, has recently been successfully used to discriminate between *T. magnatum* and *T. borchii* ECM irrespective of the host plants (Longato and Bonfante 1997). These methods further indicated that samples identified as *T. maculatum* Vittad. and *T. borchii* on the basis of their morphological features, obtained from the same geographical area, could be ascribed to a single species (*T. borchii*) with a certain level of intraspecific variability (Longato and Bonfante 1997). From the discussion above, it is clear that no single approach has so far proven completely satisfactory for the identification of *Tuber* ECM. It is thus probably wise to use integrated data obtained using several techniques for identification, as suggested by Agerer (1996).

6.5
Host Plant Growth Responses and Fungus-Derived-Benefits

6.5.1
Growth Responses

Growth improvement of plant ECM with *Tuber* species was noted qualitatively in several studies (Fassi and Fontana 1967; Fontana and Palenzona 1969; Fontana and Bonfante-Fasolo 1971; Giovannetti and Fontana 1981), with differential growth responses reported according to the host–*Tuber* combination (Chevalier and Grente 1974). It was not until 1983, however, that Zambonelli and Govi quantified such growth increases in several host–*Tuber* combinations. Later experiments using *T. aestivum* var. *uncinatum* (Chatin) Hollos with several host trees recorded significant increases in stem diameter and height with only some host species (Zambonelli and Govi 1990). Similar results were obtained by Pirazzi and Di Gregorio (1987) for four pine species (*P. brutia, P. pinea, P. nigra, P. halepensis*) ECM with five *Tuber* species; three black truffles [*T. melanosporum, T. brumale* f. *moschatum* (Ferry) Ceruti *T. aestivum*] and two white truffles (*T. borchii, T. maculatum*). All size and weight parameters increased significantly after 1 year, in some cases being two to three times greater for ECM plants than for non-ECM controls. In the second year the growth improvement remained; however, a reduction in root dry weight was detected in the ECM plants compared to controls. Stem diameter, height and aerial volume were significantly increased, while the increase in stem dry weight varied according to the host–*Tuber* combination. Overall, growth enhancement appeared to be related to the extent of ECM formation, with those taxa producing more dichotomous ECM causing better growth enhancement.

While these studies were conducted in the nursery, similar results have been obtained from field assessment of the growth of ECM seedlings following out-planting. Bencivenga and Venanzi (1990), for example, observed a significant increase in height and stem diameter of ECM plants after 3 years. Pirazzi (1992) obtained similar results using clones of *Populus* species. In both the nursery and field, *T. magnatum* ECM enhanced the basal area and volume of the stem, along with increasing branch production in the lower stem. The highest growth enhancement was obtained with *Populus alba*, a natural host of *T. magnatum*, while the least significant effect was observed in exotic poplar species *Populus deltoides*. Although *Tuber* ECM can thus enhance seedling growth, the ability of *Tuber* species to improve host plant nutrition has received little attention. What is clear, however, is that, besides the extramatrical mycelium, the cystidia that typically occur on the surface of *Tuber* ECM can contribute to the absorption and transfer of ions to the mantle (Le Disquet and Pargney 1994).

6.5.2
Nutritional Benefits to the Host

While there is no direct evidence of N absorption and transfer to the host in *Tuber* ECM, increased NH_4^+ and protein content has been reported in ECM compared with uninfected roots (Krupa et al. 1973; Guttenberger 1995). Increased N accumulation in shoots of *P. nigra* and *P. halapensis* infected with several *Tuber* species has also been reported (Pirazzi and Di Gregorio 1987). This effect appears to be connected to a strong increase in N anabolism in the ECM. Transmission electron microscopy has revealed abundant ribosomes and rough endoplasmic reticulum within the Hartig net hyphae, suggesting protein synthesis activity (Scannerini 1968). Altered patterns of N metabolism, characterised by accumulation of intermediates of the ornithine cycle, and presumed to be associated with mineral (largely K^+) deficiency, have been reported in axenic non-ECM roots (Krupa et al. 1973). This metabolic arrest is eliminated in *Tuber* ECM, resulting in a positive effect on protein synthesis and a specific increase in glutamine and glutamic acid concentrations. An increase in both free NH_4^+ and amides was also observed in the ECM roots, and this may indicate conversion of arginine, produced by urease activity, that efficiently transforms urea into aminic compounds, and which can be directly utilised in N metabolism. The correlated decrease in α-ketoglutarate, a precursor from the citric acid cycle pathway in the amino acid biosynthesis, further supports this explanation (Krupa et al. 1973). In axenic culture, *T. melanosporum* mycelium grows better with NH_4^+ than with NO_3^- as a sole N source; however, increased growth is observed when both N sources are mixed (Mamoun and Olivier 1991). The existence of the glutamine synthetase–glutamate synthase (GS-GOGAT) pathway for NH_4^+ assimilation has been indirectly confirmed in *Tuber* ECM by Dupré et al. (1993). Glutamate dehydrogenase (GDH) activity has also been detected, suggesting a situation similar to that demonstrated in the ECM ascomycete *Cenococcum geophilum* Fr., wherein both major NH_4^+ assimilation pathways operate to different extents, according to the physiological status of the mycelium (Martin et al. 1988).

Tuber species produce acid phosphatase (*p*Pase) activities which may play a role in the host P nutrition. *T. borchii* has been shown to produce *p*Pase activity during symbiosis with *P. pinea* (Pasquini et al. 1991). This *p*Pase activity in *P. pinea*–*T. borchii* ECM had two activity peaks at pH 3.5 and 5.0, while the activity in non-ECM roots showed a single peak at pH 5.0. The *T. borchii p*Pase was inhibited by tartrate, fluoride and molybdate ions, and decreased proportionally with external PO_4^- activity. *T. borchii* ECM were further shown to hydrolyse a variety of natural and synthetic phosphate esters, including phytate with two different K_m values of 0.22 mM at low substrate concentrations and 2.7 mM at higher concentrations (Pasquini et al. 1991). Alkaline *p*Pase has also been shown in *T. melanosporum* mycelium by isoelectric focusing (Cameleyre and Olivier 1993).

6.5.3
Non-nutritional Benefits to the Host

Tuber species which produce the hydnospheric effect can significantly reduce weed populations in the brûlé, leading to decreased competition for nutrients between host plant and weed roots. Changes to soil micro-organism communities within a brûlé (see above) may also have indirect benefits on host plant growth (Pacioni 1991). The domination of fluorescent pseudomonads in calcareous (Fe-deficient) black truffle habitats might also be advantageous to the host since they have the potential to increase Fe availability (Reid et al. 1986).

The hydnosphere is characterised by reduced soil fungal activity (Luppi-Mosca et al. 1970; Luppi-Mosca 1972; Luppi-Mosca and Fontana 1977). In vitro tests of the effect of *T. melanosporum* mycelium and water extracts from sporocarps on the growth of several soil-borne plant pathogens suggest a strong fungistatic effect towards a few pathogenic fungi, such as *Sclerotinia sclerotiorum* (Libert) de Bary, *Verticillium albo-atrum* Reinke & Berthold, *Fomes pinicola* (Sw.) Cooke and *Irpex pachyodon* (Pers.) Quél. (Fasolo-Bonfante et al. 1972). Other pathogens such as *Fomes annosus* (Fr.) Cooke [= *Heterobasidion annosum* (Fr.) Bref.] and *Polyporus squamosus* Fr. were only weakly inhibited by *T. melanosporum*. Similar water extracts have also been shown to reduce growth of four saprotrophic fungi, along with the ECM species *Boletus felleus* [= *Tylopilus felleus* (Bull.: Fr.) P. Karst] (Obert 1973). Such effects may be related to the production of courmarins by *Tuber* species. These secondary phenolic metabolites, including scopoletin, angelicin and bergaptene, have known antimicrobial effects, and are produced by both ECM and sporocarps of *Tuber* species (Tirillini and Granetti 1995; Tirillini and Stoppini 1996). While these data suggest that *Tuber* ECM may be capable of effecting a degree of biological control against some root pathogens, it must be stressed that this remains to be tested *in planta*.

The potential for *Tuber* ECM to protect host plants against toxic metals was investigated by Pirazzi (1993) using three *Pinus* species (*P. halepensis*, *P. nigra* and *P. pinea*) inoculated with *T. borchii*. His results indicated that ECM infection resulted in reduced contents of some metals in the shoots of host plants relative to uninfected controls. While *T. borchii* did not influence the absorption of Mn and Pb by the plant, Cu, Fe, Sr and Zn were accumulated by *T. borchii* ECM, reducing their translocation to the shoot. The accumulation of Fe in ECM roots seemed to occur to the detriment of the leaves, which showed signs of Fe deficiency, being one third of the usual concentration.

A further aspect of ECM benefit to the host might be a detoxicant action in favor of host roots. In the soil, primary metabolic activities occur in a lower percentage of O_2 and high levels, almost 2%, of CO_2 (Sequi 1989). These environmental conditions might result, as a consequence of activated glycolysis, in the production of methylglyoxal (MG), a substance with cytotoxic and genotoxic effects to cells (Thornalley 1990). Glyoxalase 1 (GLO1, EC 4.4.1.5) and glyoxalase 2 (GLO2, EC 3.2.16) are the enzymes involved in the MG scavenging, by using reduced glutathione (GSH) and releasing D-lactate and recycled GSH (Mannervik 1980; Thornalley 1990; Vander Jagt 1993). The necessity of this detoxicant activity was in fact recently detected in the uninfected roots of corn (*Zea mays*) where it seems correlated with enhanced resistance to cold stress (Amicarelli et al. 1997). The increase in GLO1 activity in response to glycolysis is probably connected to an enhancement of energy requirements. Truffle species are microaerobic organisms and normally they express a complex system of detoxicant enzymatic activities based on antioxidant and glutathione dependent enzymes, such as superoxide dismutase (SOD), catalase (CAT), glutathione perossidase Se-dependent (GSH-PX Se[+]), glutathione reductase (GSSG-RX) glyoxalase 1 and glyoxalase 2 (Amicarelli et al. 1999). Due to this enzymatic endowment, their mycelium might efficaciously assist mycorrhizal roots in MG scavenging. These results open a new field of research in ECM functioning.

6.6
Conclusions

Tuber has been studied extensively with regard to its taxonomy, morphology, ecology and truffle farming. The physiological and biological conditions for enhanced truffle production (in terms of soil structure and chemistry, edaphic conditions, farm practices and nursery methods) are thus well characterised. It is clear from the literature reviewed in this chapter, however, that information on functional aspects of host–*Tuber* symbioses is lacking. This probably reflects the fact that interest has so far been focused on the product (truffles), rather than on the producers, the plants in ECM symbiosis with *Tuber*. It is noteworthy that, with few exceptions, the literature cited in this chapter describes work conducted in France and Italy. The lack of mycelial cultures of *Tuber* species has served to confine scientific interest in truffle biology largely to countries historically and naturally involved in truffle production, where living material is readily available and industrial funds are available to support applied research. It is only during the past few years that a reliable *Tuber* strain has been available in axenic culture (*T. borchii*, ATCC 96540, Zambonelli et al. 1989). This provides standard material for study worldwide, and should allow for the inclusion of *Tuber* in comparative studies of ECM functioning that will lead to a greater understanding of interactions between *Tuber* and its host plants.

Acknowledgements. This review was supported by the grant CNR "*Tuber*: biotechnology of mycorrhization".

References

Agerer R (1996) Characterization of ectomycorrhizae: a historical overview. Descr Ectomyc 1:1–22

Amicarelli F, Poma A, Spanò L (1997) Glyoxalase-I activity in two lines of corn (*Zea mays*) with different sensitivity to cold stress. 3rd Int Conf "Oxygen, free radicals and environmental stress in plants" Abstr, p 67

Amicarelli F, Bonfigli A, Colafarina S, Cimini AM, Pruiti B, Cerù MP, Di Ilio C, Pacioni G, Miranda M (1999) Glutathione dependent enzymes and antioxidant defences in truffles: microaerobic environment living organism. Mycol Res (in press)

Awameh MS, Alsheikh A (1980a) Ascospore germination of black kamé (*Terfezia boudieri*). Mycologia 72:50–54

Awameh MS, Alsheikh A (1980b) Features and analysis of spore germination in the brown kamé (*Terfezia claveryi*). Mycologia 72:494–499

Balestrini R, Hahn MG, Bonfante P (1996) Location of cell-wall components in ectomycorrhizae of *Corylus avellana* and *Tuber magnatum*. Protoplasma 191:55–69

Barry D, Staunton S, Callot G (1994) Mode of the absorption of water and nutrients by ascocarps of *Tuber melanosporum* and *Tuber aestivum*: a radioactive tracer technique. Can J Bot 72:317–322

Barry D, Jaillard B, Staunton S, Callot G (1995) Translocation and metabolism of phosphate following absorption by ascocarps of *Tuber melanosporum* and *T. aestivum*. Mycol Res 99:167–172

Bartnicki-Garcia S (1968) Cell wall chemistry, morphogenesis and taxonomy of fungi. Annu Rev Microbiol 22:87–108

Bencivenga M, Granetti B (1989) Indagine preliminare sul contenuto in macro e microelementi del terreno e dei carpofori di *Tuber melanosporum* Vitt. Micol Ital 18:25–30

Bencivenga M, Granetti B (1990a) Analisi comparativa delle caratteristiche ecologiche di *Tuber magnatum* Pico e *Tuber melanosporum* Vitt. nell'Italia centrale. In: Bencivenga M, Granetti B (eds) Atti del II Congr Int sul Tartufo. Comunità Montana dei Monti Martani e del Serano, Spoleto, pp 433–434

Bencivenga M, Granetti B (1990b) Risultati produttivi di tartufaie coltivate di *Tuber melanosporum* Vitt. in Umbria. In: Bencivenga M, Granetti B (eds) Atti del II Congr Int sul Tartufo. Comunità Montana dei Monti Martani e del Serano, Spoleto, pp 313–322

Bencivenga M, Granetti B (1990c) Valutazione biometrica delle micorrize su *Ostrya carpinifolia* Scop prodotte da tartufi di varie specie. In: Bencivenga M, Granetti B (eds) Atti del II Congr Int sul Tartufo. Comunità Montana dei Monti Martani e del Serano, Spoleto, pp 265–270

Bencivenga M, Venanzi G (1990) Alcune osservazioni sull'accrescimento delle piante tartufigene in pieno campo. In: Bencivenga M, Granetti B (eds) Atti del II Congr Int sul Tartufo. Comunità Montana dei Monti Martani e del Serano, Spoleto, pp 241–248

Bencivenga M, Di Massimo G, Donnini D (1991) Produzioni e problemi di una tartufaia coltivata di *Tuber melanosporum* Vitt. Micol Ital 20:129–139

Bencivenga M, Di Massimo G, Donnini D (1992) Rapporto tra l'umidità, la temperatura del terreno e la produzione di sporocarpi in alcune tartufaie di *Tuber melanosporum*. Micol Veg Med 7:195–206

Beuchat LR, Brenneman TB, Dove CR (1993) Composition of the pecan truffle (*Tuber texense*). Food Chem 46:189–192

Blaschke H (1987) Vorkommen und Charakterisierung der Ektomykorrhizaassoziation *Tuber puberulum* mit *Picea abies*. Z Mykol 53:283–288

Botton B, Fourcy A, Bossy JP (1980) Localisation cellulaire du calcium chez ascomycète *Sphaerostilbe repens* par analyse directe à la microsonde électronique. Can J Bot 58:2395–2401

Bulawa CE (1993) Genetics and molecular biology of chitin synthesis in fungi. Annu Rev Microbiol 47:505–534

Bullini L, Biocca E, Chevalier G, Dupré C, Ferrara AM, Palenzona M, Sallicandro P, Urbanelli S (1994) Struttura clonale di alcune specie del genere *Tuber*. G Bot Ital 128:51

Callot G, Jaillard B (1996) Effect of structural characteristics of subsoil on the fruiting of *Tuber melanosporum* and other mycorrhizal fungi. Agronomie 16:405–419

Cameleyre I, Olivier JM (1993) Evidence for intraspecific isozymes variations among French isolates of *Tuber melanosporum* Vitt. FEMS Microbiol Lett 110:159–162

Ceruti A (1965) La tartuficoltura in Italia. Ann Accad Agric Torino 107:131–142

Ceruti A (1968) Biologia e possibilità di coltivazione dei tartufi. Proc I Congr Int sul Tartufo. Spoleto, Italy, pp 29–42

Chalvignac MA, Tysset C, Pochon J (1959) Action de la truffe (*Tuber melanosporum*) sur la microflore tellurique. Soc Fr Microbiol 96:355–3580

Chevalier G (1972) Obtention de cultures de mycélium de truffe à partir du carpophore et des mycorrhizes. C R Hebd Séances Acad Agric Fr 58:981–989

Chevalier G, Desmas C (1977) Synthéses des mycorhizes de *Tuber melanosporum* avec *Corylus avellana* sur agara partir de spores. Ann Phytopathol 9:531

Chevalier G, Dupré C (1990) Recherce et experimentation sur la truffe et la trufficulture en France. In: Bencivenga M, Granetti B (eds) Atti del II Congr Int sul Tartufo. Comunità Montana dei Monti Martani e del Serano, Spoleto, pp 157–166

Chevalier G, Frochot H (1990) Ecology and possibility of culture in Europe of the Burgundy truffle (*Tuber uncinatum* Chatin). Agric Ecosyst Environ 28:71–73

Chevalier G, Grente J (1974) Effectes des mycorhizes de *Tuber* sp. sur le développement de différents pins. Ann Phytopathol 6:218–219

Chevalier G, Poitou N (1990) Study of important factors affecting the mycorrhizal development of the truffle fungus in the field using plants inoculated in nurseries. Agric Ecosyst Environ 28:75–77

Chevalier G, Grente J, Pollacssek A (1973) Obtention de mycorhizes de différents *Tuber* par synthèse à partir de spores en conditions gnotoxéniques et à partir de cultures pures de mycélium en conditions axéniques et gnotoxéniques. Ann Phytopathol 5:107–108

Chevalier G, Mousain D, Couteaudier Y (1975) Ectomycorrhizal associations between *Tuber-aceae* and *Cistaceae*. Ann Phytopathol 7:355–356

Chevalier G, Giraud M, Bardet MC (1982) Interactions entre les mycorhizes de *Tuber melanosporum* et celles d'autres champignons ectomycorhiziens en sol favorables à la truffe. In: Gianinazzi S, Gianinazzi-Pearson V, Trouvelot A (eds) Mycorrhizae, an integral part of plants: biology and perspectives for their use. INRA Publ Colloq INRA 13:313–321

Chu-Chou M, Grace LJ (1983) Characterization and identification of mycorrhizas of Douglas fir in New Zealand. Eur J For Pathol 13:251

Citterio B, Cardoni P, Potenza L, Amicucci A, Stocchi V, Gola G, Nuti M (1995) Isolation of bacteria from sporocarps of *Tuber magnatum* Pico, *Tuber borchii* Vitt. and *Tuber maculatum* Vitt. Identification and biochemical characterization. In: Stocchi V, Bonfante P, Nuti M (eds) Biotechnology of ectomycorrhizae. Plenum Press, New York, pp 241–248

Comandini O, Pacioni G (1997) Mycorrhizae of Asian black truffles, *Tuber himalayense* and *T. indicum*. Mycotaxon 63:77–86

Condamy A (1876) Étude sur l'histoire naturelle de la truffe. Angoulême, France

Corocher N, Polimeni C, Girausi G, Papa G (1992) Sviluppo di un metodo immunoenzimatico (ELISA) per la caratterizzazione di ectomicorrize di *Tuber magnatum* e *Tuber albidum*. Micol Veg Med 7:151–158

De Angelis F, Arcadi A, Marinelli F, Paci M, Botti D, Pacioni G, Miranda M (1996) Partial structures of truffle melanins. Phytochemistry 43:1103–1106

Deffieux G, Vercauteren J, Baute MA, Neveu A, Baute R (1994) Bacterial conversion of the fungal ascopyrone T to 2(s)-dihydroascopyrone T (haliclonol). Phytochemistry 36:849–852

Donnini D, Bencivenga M (1995) Micorrize inquinanti frequenti nelle piante tartufigene. Nota 2 – Inquinanti in campo. Micol Ital 24:185–207

Dupré C, Chevalier G (1991) Analyse électrophorétique des protéines fungiques de différents *Tuber* en association ou non avec *Corylus avellana*. Cryptogam Mycol 12:243–250

Dupré C, Chevalier G, Palenzona M, Ferrara AM, Nascetti G, Mattiucci S, D'Amelio S, La Rosa G, Biocca E (1992) Differenziazione genetica di ascocarpi, miceli e micorrize di differenti specie di *Tuber*. Micol Ital 21:139–144

Dupré C, Chevalier G, Palenzona M, Biocca E (1993) Caractérisation des mycorhizes de différents *Tuber* par l'étude du polymorphisme enzymatique. Cryptogam Mycol 14:163–170

Fasolo-Bonfante P, Brunel A (1972) Caryological feature in a mycorrhizal fungus: *Tuber melanosporum* Vitt. Allionia 18:5–11

Fasolo-Bonfante P, Fontana A, Montacchini F (1971) Studi sull'ecologia del *Tuber melanosporum*. I. Dimostrazione di un effetto fitotossico. Allionia 17:7–54

Fasolo-Bonfante P, Ceruti-Scurti J, Obert F (1972) Interazione di *Tuber melanosporum* Vitt. con miceli di altri funghi. Allionia 18:53–59

Fassi B, Fontana A (1967) Sintesi micorrizica tra *Pinus strobus* e *Tuber maculatum*. I. Micorrize e sviluppo dei semenzali nel secondo anno. Allionia 13:177–186

Fogel R (1992) Evolutionary processes in truffles and false-truffles: evidence from distribution of hypogeous fungi in the Great Basin, USA. Micol Veg Med 7:13–30

Fogel R, Trappe JM (1978) Fungal consumption (mycophagy) by small animals. Northwest Sci 52:1–31

Fontana A (1971) Il micelio di *Tuber melanosporum* Vitt. in coltura pura. Allionia 17:19–23

Fontana A, Bonfante-Fasolo P (1971) Sintesi micorrizica di *Tuber brumale* Vitt. con *Pinus nigra* Arnold. Allionia 17:15–18

Fontana A, Giovannetti G (1979) Simbiosi micorrizica fra *Cistus incanus* ssp. *incanus* e *Tuber melanosporum* Vitt. Allionia 23:5–11

Fontana A, Palenzona M (1969) Sintesi micorrizica di *Tuber albidum* in coltura pura con *Pinus strobus* e pioppo americano. Allionia 15:99–104

Fonvieille JL, Touze-Soulet JM, Kulifaj M, Montant C, Dargent R (1990) Composition des ascospores de *Tuber melanosporum* et de leur paroi isolée. C R Acad Sci Paris 310 (Sér III):557–563

Frank AB (1885) Über die auf Wurzelsymbiose beruhende Ernährung gewisser Bäume durch unterirdische Pilze. Ber Dtsch Bot Ges 3:128–145

Fusconi A (1983) The development of the fungal sheath on *Cistus incanus* short roots. Can J Bot 61:2546–2553

Gandeboeuf B, Henrion B, Dupré C, Drevet P, Nicolas P, Chevalier G, Martin F (1995) Molecular identification of *Tuber* species and isolates by PCR-based techniques. In: Stocchi V, Bonfante P, Nuti M (eds) Biotechnology of ectomycorrhizae. Molecular approaches. Plenum Press, New York, pp 151–160

Gandeboeuf D, Dupré C, Roeckel-Drevet P, Nicolas P, Chevalier G (1997) Grouping and identification of *Tuber* species using RAPD markers. Can J Bot 75:36–45

Gea L, Normand L, Vian B, Gay G (1994) Structural aspects of ectomycorrhiza of *Pinus pinaster* (Ait.) Sol. formed by an IAA overproducer mutant of *Hebeloma cylindrosporum* Romagnesi. New Phytol 128:659–670

Genet P, Pargney JC (1994) Calcium localization in walls of *Tuber melanosporum* associated with *Corylus avellana*: advantages and limits of different techniques. J Trace Microprobe Tech 12:331–342

Giovannetti G, Fontana A (1981) Simbiosi micorrizica di *Tuber macrosporum* Vitt. con alcune *Fagales*. Allionia 24:13–17

Granetti B (1987) Micorrizazione di alcune specie di Salice con *Tuber magnatum* Pico. Ann Fac Agrar Univ Perugia 41:875–888

Granetti B (1994) I tartufi: biologia e tecniche di coltivazione. Micol Ital 23:63–68

Granetti B (1995) Caratteristiche morfologiche, biometriche e strutturali delle micorrize di *Tuber* di interesse economico. Micol Ital 24:101–117

Granetti B, Angelini P, Rubini A (1995a) Morfologia e struttura delle micorrize di *Tuber magnatum* Pico e *Tuber borchii* Vitt. con *Tilia platyphyllos* Scop. Micol Ital 24:27–34

Granetti B, Rubini A, Angelini P (1995b) Analisi comparativa morfo-biometrica e strutturale delle micorrize di *Tuber aestivum* Vitt. con alcune piante forestali. Micol Ital 24:48–63

Granetti B, Rubini A, Angelini P (1995c) Un parametro biometrico idoneo alla caratterizzazione delle micorrize di *Tuber borchii* Vitt. e di *Tuber magnatum* Pico. Micol Ital 24:17–23

Gregori G, De Paoli G (1992) Sintesi micorrizica con *Tuber magnatum* per "innesto radicale" su piante micropropagate di Tiglio e Pioppo. Inf Agrar 50:49–54

Gregori G, Tocci A (1985) Possibilità di produzione di piantine di *Alnus cordata*, micorrizate con *Tuber melanosporum* e *T. aestivum*. Ital For Montana 40:262

Grente J, Chevalier G, Pollacsek A (1972) La germination de l'ascospore de *Tuber melanosporum* et la synthèse sporale des mycorrhizes. CR Acad Sci Paris 275:743–746

Grente J, Delmas J (1974) Perspectives pour une trufficulture moderne. Grente and Delmas, Clermond-Ferrand, France

Guillemaud T, Raymond M, Callot G, Cleyetmarel JC, Fernandez D (1996) Variability of nuclear and mitochondrial ribosomal DNA of a truffle species (*Tuber aestivum*). Mycol Res 100:547–550

Guiochon P (1959) Germinations monospermes de la truffe *Tuber melonosporum* Vitt. Mushroom Sci 4:294–297

Guttenberger M (1995) The protein complement of ectomycorrhizas. In: Varma A, Hock B (eds) Mycorrhiza. Structure, function, molecular biology and biotechnology. Springer, Berlin Heidelberg New York, pp 59–78

Henrion B, Chevalier G, Martin F (1994) Typing truffle species by PCR amplification of the ribosomal DNA spacers. Mycol Res 98:37–43

Krupa S, Fontana A, Palenzona M (1973) Studies on the nitrogen metabolism in ectomycorrhizae: I. Status of free and bound amino acids in mycorrhizal root systems of *Pinus nigra* and *Corylus avellana*. Physiol Plant 28:1–6

Lanfranco L, Wyss P, Marzachì C, Bonfante P (1993) DNA probes for identification of the ectomycorrhizal fungus *Tuber magnatum* Pico. FEMS Microbiol Lett 114:245–252

Lanfranco L, Garnero L, Delpero M, Bonfante P (1995) Chitin synthase homologs in three ectomycorrhizal truffles. FEMS Microbiol Lett 134:109–114

Le Disquet I, Pargney C (1994) Characterization of the spinules wall of the truffle ectomycorrhizae and ultrastructural cytochemistry. J Trace Microprobe Tech 12:209–221

Li CY, Castellano MA (1985) Nitrogen-fixin bacteria isolated from within sporocarps of three ectomycorrhizal fungi. Proc VI North American Conf on Mycorrhizae, Bend, Oregon, pp 264

Lindsay WL (1974) Role of chelation in micronutrient availability. In: Corson EW (ed) The plant root and its environment. University Press, Charlottesville, pp 507–527

Longato S, Bonfante P (1997) Molecular identification of mycorrhizal fungi by direct amplification of microsatellite regions. Mycol Res 101:425–432

Luppi-Mosca AM (1972) La microflora della rizosfera delle tartufaie. III. Analisi micologiche di terreni tertufiferi francesi. Allionia 18:33–40

Luppi-Mosca AM, Fontana A (1977) Studi sull'ecologia del *Tuber melanosporum*. IV. Analisi di terreni tartufiferi dell'Italia centrale. Allionia 22:105–113

Luppi-Mosca AM, Gribaldi L, Jaredi-Sodano G (1970) La micoflora della rizosfera nelle tartufaie. II. Analisi micologiche di terreni tartufiferi piemontesi. Allionia 16:115–132

Maia LC, Yano AM, Kimbrough JW (1996) Species of Ascomycota forming ectomycorrhizae. Mycotaxon 52:371–390

Mamoun M, Olivier JM (1989) Dynamique des populations fongiques et bactériennes de la rhizosphère des noisetiers truffiers. II. Chélation du fer et répartition taxonomique chez les *Pseudomonas* fluorescents. Agronomie 9:345–351

Mamoun M, Olivier JM (1990) Dynamique des populations fongiques et bactériennes de la rhizosphère de noisetiers truffiers. III. Effet du régime hydrique sur la mycorhization et la microflore associée. Agronomie 10:77–84

Mamoun M, Olivier JM (1991) Effect of carbon and nitrogen-sources on the in vitro growth of *Tuber melanosporum* Vitt. Application to mycelial biomass production. Agronomie 11:521–527

Mamoun M, Olivier JM (1992) Effect of soil Pseudomonads on colonization of hazel roots by the ectomycorrhizal species *Tuber melanosporum* and its competitors. Plant Soil 139:265–273

Mamoun M, Olivier JM (1996) Receptivity of cloned hazels to artificial infection by *Tuber melanosporum* and symbiotic competitors. Mycorrhiza 6:15–19

Mamoun M, Olivier JM (1997) Mycorrhizal inoculation of cloned hazels by *Tuber melanosporum*: effect of soil disinfestation and co-culture with *Festuca ovina*. Plant Soil 188:221–226

Mamoun M, Poitou N, Olivier JM (1985) Étude des interactions fongiques entre *Tuber melanosporum* Vitt. et son environnement biotique. In: Gianinazzi-Pearson V, Gianinazzi S (eds) Aspects physiologiques et génétiques des mycorrhizes. INRA, Paris, pp 761–765

Mannervik B (1980) Glyoxalate I. In: Jakoby WB (ed) Enzymatic basis of detoxification. Academic Press, New York, pp 263–273

Marchisio V (1964) Sulla cariologia degli aschi e delle spore di *Tuber maculatum* Vitt. Allionia 10:105–113

Martin F, Stewart GR, Genetet I, Mourot B (1988) The involvement of glutamate dehydrogenase and glutamine synthetase in ammonia assimilation by the rapidly growing ectomycorrhizal ascomycete *Cenococcum geophilum* Fr. New Phytol 110:541–550

Mehmann B, Brunner I, Braus GH (1994) Nucleotide sequence variation of chitin synthase genes among ectomycorrhizal fungi and its potential use in taxonomy. Appl Environ Microbiol 60:3105–3111

Mello A, Nosenzo C, Meotto F, Bonfante P (1996) Rapid typing of truffle mycorrhizal roots by PCR amplification of the ribosomal DNA spacers. Mycorrhiza 6:417–421

Miranda M, Bonfigli A, Zarivi O, Aimola P, Lanza B, Pacioni G, Botti D, Ragnelli AM (1992) Biochemical and morphological correlation between melanogenesis and development in truffles. Micol Veg Med 7:75–80

Montacchini F, Caramiello-Lomagno R (1977) Studi sull'ecologia del *Tuber melanosporum*. II. Azione inibitrice su specie erbacee della flora spontanea. Allionia 22:81–85

Napoliello A, Pintozzi P, Verdoliva A (1990) Il tartufo in Campania. In: Bencivenga M, Granetti B (eds) Atti II Congr Int sul Tartufo. Comunità Montana Monti Martani e del Serano, Spoleto, pp 539–544

Nylund JE (1980) Symplastic continuity during Hartig net formation in Norway spruce ectomycorrhizae. New Phytol 86:373–378

Nylund JE (1987) The ectomycorrhizal infection zone and its relation to acid polysaccharides of cortical cell walls. New Phytol 106:505–516

Nylund JE, Unestam T (1982) Structure and physiology of ectomycorrhizae. I. The process of mycorrhizal formation in Norway spruce in vitro. New Phytol 91:63–79

Obert F (1973) Interazione del carpoforo di *Tuber melanosporum* con miceli di altri funghi. Allionia 19:43–44

Olivier JM, Mamoun M (1988) Dynamique des populations fongiques et bactérienne de la rhizosphère des noisetiers truffiers. I. Relation avec le status hydrique du sol. Agronomie 8:711–717

Pacioni G (1985) La coltivazione moderna e redditizia del tartufo. De Vecchi, Milano

Pacioni G (1989) Biology and ecology of the truffles. Acta Med Rom 27:104–117

Pacioni G (1990) Scanning electron microscopy of *Tuber* sporocarps and associated bacteria. Mycol Res 94:1086–1089

Pacioni G (1991) Effects of *Tuber* metabolites on the rhizospheric environment. Mycol Res 95:1355–1358

Pacioni G (1992) Il ruolo dello zolfo nel metabolismo dei tartufi. Micol Ital 21:71–76

Pacioni G, Pomponi G (1989) Chemotaxonomy of some Italian species of *Tuber*. Micol Veg Med 4:63–72

Pacioni G, Pomponi G (1991) Genotypic patterns of some Italian populations of the *Tuber aestivum-T. mesentericum* complex. Mycotaxon 42:171–179

Pacioni G, Bologna MA, Laurenzi M (1991) Insect attraction by *Tuber*: a chemical explanation. Mycol Res 95:1359–1363

Pacioni G, Frizzi G, Miranda M, Visca C (1993) Genetics of a *Tuber aestivum* population. Mycotaxon 47:93–100

Pacioni G, Ragnelli AM, Miranda M (1995) Truffle development and interactions with the biotic environment. Molecular aspects. In: Stocchi V, Bonfante P, Nuti M (eds) Biotechnology of ectomycorrhizae. Molecular approaches. Plenum Press, New York, pp 213–227

Palenzona M (1969) Sintesi micorrizica tra *Tuber aestivum* Vitt., *Tuber brumale* Vitt., *T. melanosporum* Vitt. e semenzali di *Corylus avellana* L. Allionia 15:121–131

Palenzona M, Chevalier G, Fontana A (1972) Sintesi micorrizica tra i miceli in coltura di *Tuber brumale, T. melanosporum, T. rufum* e semenzali di conifere e latifoglie. Allionia 18:41–52

Palenzona M, Biocca E, Nascetti G, Ferrara AM, Mattiucci S, D'Amelio S, Balbo T (1990) Studi preliminari sulla tipizzazione genetica (sistema gene-enzima) di specie del genere *Tuber*. In: Bencivenga M, Granetti B (eds) Atti del II Congr Int sul Tartufo. Comunità Montana dei Monti Martani e del Serano, Spoleto, pp 53–58

Paolocci F, Cristofari E, Angelini P, Granetti B, Arcioni S (1995a) The polymorphism of the rDNA region in typing ascocarps and ectomycorrhizae of truffle species. In: Stocchi V, Bonfante P, Nuti M (eds) Biotechnology of ectomycorrhizae. Plenum Press, New York, pp 171–184

Paolocci F, Angelini P, Cristofari E, Granetti B, Arcioni S (1995b) Identification of *Tuber* spp. and corresponding ectomycorrhizae throught molecular markers. J Sci Food Agric 69:511–517

Papa G, Botolo R, Cassani G (1992a) Componenti di *Tuber melanosporum* con attività fitotossica. Micol Veg Med 7:109–110

Papa G, Polimeni C, Mischiati P, Cantini Cortellazzi G (1992b) Immunological aspects of the characterization of *Tuber magnatum* and *Tuber albidum*. In: Read DJ, Lewis DH, Fitter AH, Alexander IJ (eds) Mycorrhizas in ecosystem: 395. CAB International, Cambridge, 395

Pargney JC (1990) Essais de caractérisation cytochimique des structures de l'interface au niveau du réseau de Hartig dans l'association ectomycorhizienne entre la truffe (*Tuber melanosporum*) et le noisetier (*Corylus avellana*) Can J Bot 68:2722–2728

Pargney JC, Brimont A (1993) L'interface plante–champignon dans les ectomycorhizes de truffes: étude ultrastructurale et microanalyses par perte d'énergie des électrons et par rayons X. Acta Bot Gall 140:803–818

Pargney JC, Brimont A (1995) Production of concentrated polyphenols by the root cap cells of *Corylus avellana* associated with *Tuber*: ultrastructural study and element localization using electron energy loss spectrometry and imaging. Trees 9:149–157

Pargney JC, Kottke I (1994) Microlocalization of calcium in the walls of truffle mycorrhizae by analytical transmission electron microscopy. J Trace Microprobe Tech 12:305–321

Pargney JC, Leduc JP, Dexheimer J, Chevalier G (1988) Étude ultrastructurale et cytochimique de l'association mycorhizienne *Tuber melanosporum/Corylus avellana*. In: Bencivenga M, Granetti B (eds) Atti del II Congr Int sul Tartufo. Comunità Montana dei Monti Martani e del Serano, Spoleto, pp 129–133

Parguey-Leduc A, Montant C, Kulifaj M (1987) Morphologie et structure de l'ascospore adulte du *Tuber melanosporum* Vitt. Cryptogam Mycol 2:173–202

Pasquini S, Panara F, Antonielli M (1991) Acid phosphate activity in *Pinus pinea–Tuber albidum* ectomycorrhizal association. Can J Bot 70:1377–1383

Pirazzi R (1990a) Micorrizzazione artificiale con miceli isolati "in vitro" di *Tuber melanosporum* Vitt. e *T. magnatum* Pico. In: Bencivenga M, Granetti B (eds) Atti del II Congr Int sul Tartufo. Comunità Montana dei Monti Martani e del Serano, Spoleto, pp 173–184

Pirazzi R (1990b) Produzione naturale di *Tuber* spp. in rimboschimento di cedro e prove di sintesi. In: Bencivenga M, Granetti B (eds) Atti del II Congr Int sul Tartufo. Comunità Montana dei Monti Martani e del Serano, Spoleto, pp 303–311

Pirazzi R (1992) Influenza delle ectomicorrize sull'accrescimento in serra e in campo di piante di pioppo. Micol Ital 21:27–36

Pirazzi R (1993) Studio dell'adsorbimento, accumulo, traslocazione di differenti metalli in alcune simbiosi micorriziche. Studi Sassar Sez III 35:405–414

Pirazzi R (1994) Sintesi micorrizica di *Tuber melanosporum* Vitt. e *T. aestivum* Vitt. con *Carya illinoensis* (wangenh) K. Loch. G Bot Ital 128:399

Pirazzi R, Di Gregorio A (1987) Accrescimento di conifere micorrizate con specie diverse di *Tuber* spp. Micol Ital 16:49–62

Pirazzi R, Lubrano L (1994) Micorrizzazione artificiale di *Populus alba* L. micropropagato e da talea con *Tuber magnatum* Pico. Micol Ital 23:78–82

Plattner I, Hall IR (1995) Parasitism of non-host plants by the mycorrhizal fungus *Tuber melanosporum*. Mycol Res 99:1367–1370

Plattner I, Grabher T, Hall I, Haselwandter K, Stöffler G (1991) Identification of ectomycorrhizal fungi by use of immunological techniques. In: Abstr 3rd Eur Symp on Mycorrhizas, Sheffield, p 206

Ragnelli AM, Pacioni G, Aimola P, Lanza B, Miranda M (1992) Truffle melanogenesis: correlation with reproductive differentiation and ascocarp ripening. Pigment Cell Res 5:205–212

Rauscher T, Agerer R, Chevalier G (1995) Ektomykorrhizen von *Tuber melanosporum*, *Tuber mesentericum* und *Tuber rufum* (Tuberales) an *Corylus avellana*. Nova Hedwigia 61:281–322

Reid CPP, Szanislo PJ, Crowley DE (1986) Siderophore involvement in plant iron nutrition. In: Swinburne TR (ed) Iron, siderophores and plant diseases. NATO-Plenum Publishers, London, pp 29–42

Rouquerol T, Payre H (1974–75) Observations sur le comportement de *Tuber melanosporum* dans un site naturel. Rev Mycol 39:107–117

Sappa F (1940) Ricerche biologiche sul *Tuber magnatum* Pico. La germinazione delle spore e caratteri della micorriza. Nuovo G Bot Ital 47:155–198

Scannerini S (1968) Sull'ultrastruttura delle ectomicorrize. II. Ultrastructura di una micorriza di Ascomicete: *Tuber albidum* x *Pinus strobus*. Allionia 14:77–95

Scannerini S, Bonfante-Fasolo P (1983) Comparative analysis of mycorrhizal association. Can J Bot 61:917–943

Scannerini S, Palenzona M (1967) Ricerche sulle ectomicorrize di *Pinus strobus* in vivaio. 3. Micorrize di *Tuber albidum* Pico. Allionia 13:187–194

Sequi P (1989) La fase gassosa. In: Sequi P (ed) Chimica del suolo. Pàtron Editore, Bologna, pp 215–219

Shaw PJA, Lankey K, Jourdan A (1996) Factors affecting yield of *Tuber melanosporum* in a *Quercus ilex* plantation in southern France. Mycol Res 100:1186–1178

Stocchi V (1997) Un progetto per incrementare la produzione di tartufi. Ric Soc 6:32–39

Strullu DC (1976) Recherches de biologie et de microbiologie forestières. Étude des relations nutrition–développement et cytologie des mycorhizes chez le Douglas (*Pseudotsuga menziesii* Mirb.) et les Abiétacées. PhD Thesis, Rennes University, Rennes

Strullu DC, Gourret JP (1980) Données ultrastructurales sur l'intégration cellulaire de quelques parasites ou symbiotes de plantes. II. Champignons mycorhiziens. Bull Soc Bot Fr Actual Bot 127:97–106

Talou T, Gaset A, Delmas M, Kulifaj M, Montant C (1990) Dimethyl sulphide: the secret for black truffle hunting by animals? Mycol Res 94:277–278

Thornalley PJ (1990) The glyoxalase system: new developments towards functional characterization of a metabolic pathway fundamental to biological life. Biochem J 269:1–11

Tirillini B, Granetti B (1995) Composti cumarino-simili in micorrize di *Quercus pubescens* Willd. con *Tuber magnatum* Pico and *T. borchii* Vitt. Micol Ital 24:179–184

Tirillini B, Stoppini AM (1996) Coumarins distribution in four truffle species. Mycotaxon 57:227–232

Trappe JM (1962) Fungus associates of ectotrophic mycorrhizae. Bot Rev 28:538–606

Vander Jagt DL (1993) Glyoxalase II: molecular characteristics, kinetics and mechanism. Biochem Soc Trans 21:522–527

Yun W (1990) First report of a study on *Tuber* species from China. In: Bencivenga M, Granetti B (eds) Atti del II Congr Int sul Tartufo. Comunità Montana dei Monti Martani e del Serano, Spoleto, pp 45–50

Zambonelli A, Branzanti MB (1987) Competizione fra due funghi ectomicorrizici: *Tuber albidum* e *Laccaria laccata*. Micol Ital 16:159–164

Zambonelli A, Govi G (1983) Micorrizazione in semenzaio di *Quercus pubescens* Willd. con specie del genere *Tuber*. Micol Ital 12:17–22

Zambonelli A, Govi G (1990) Studi sulle ectomicorrize di *Tuber aestivum* var. *uncinatum* Chatin. In: Bencivenga M, Granetti B (eds) Atti del II Congr Int sul Tartufo. Comunità Montana dei Monti Martani e del Serano, Spoleto, pp 247–255

Zambonelli A, Govi G (1991) Competizione fra *Tuber albidum* ed altri funghi. 3° contributo. Micol Ital 20:5–12

Zambonelli A, Govi G, Previati A (1989) Micorrizazione in vitro di piantine micropropagate di *Populus alba* con micelio di *Tuber albidum* in coltura pura. Micol Ital 18:105–111

Zambonelli A, Giunchedi L, Poggi Pollini C (1993a) An enzyme-linked immunosorbent assay (ELISA) for the detection of *Tuber albidum* ectomycorrhiza. Symbiosis 5:71–76

Zambonelli A, Poggi Pollini C (1991) Possibilità di applicazione del metodo E.L.I.S.A. per il riconoscimento delle micorrize di *Tuber albidum*. Micol Veg Med 7:172

Zambonelli A, Salomoni S, Pisi A (1993b) Caratterizzazione anatomorfologica delle micorrize di *Tuber* spp. su *Quercus pubescens* Willd. Micol Ital 22:73–90

Zambonelli A, Salomoni S, Pisi A (1995) Caratterizzazione anato-morfologica delle micorrize di *Tuber borchii*, *Tuber aestivum*, *Tuber mesentericum*, *Tuber brumale*, *Tuber melanosporum*, su *Pinus pinea*. Micol Ital 24:119–137

Chapter 7

Scleroderma

P. Jeffries

7.1
Introduction

Scleroderma is a common and widespread gasteromycete genus which produces macroscopic sporocarps termed "earthballs" amongst leaf litter, grass or on bare soil in or adjacent to forested areas (Fig. 7.1). These brown, leathery, rounded structures become dry and crack at maturity, and the dry, powdery basidiospores are dispersed by the wind blowing through fissures in the upper surface of the basidiome. The ectomycorrhizal (ECM) status of many *Scleroderma* species is proven, but in some cases the relationship is inferred rather than established, and several also appear capable of a free-living saprotrophic existence. *Scleroderma citrinum* Pers. (syn. *S. aurantium* Viall.: Pers., *S. vulgare* Horn.), for example, occurs as a saprotroph on rotting wood but has also been demonstrated to form typical ECM. Richter and Bruhn (1989a) used data collected from enzyme tests to suggest the genus was wholly mycorrhizal rather than partly saprotrophic. The basidiomes are not edible and there are reports of mild toxic effects on mammals following ingestion. Basidiomes of *Scleroderma* have been reported as hosts for fungicolous fungi. For example, basidiomes of *Xerocomus parasiticus* (Fr.: Fr.) Gilb. are found occasionally growing out from basidiomes of *S. citrinum*, but this unique relationship has not been reported for other *Scleroderma* species. Moribund basidiomes of *S. citrinum* are frequently overgrown by *Sepedonium chrysospermum* (Bull.: Fr.) Link, a fungicolous species which also grows on other species as well as the related ECM genera *Paxillus* and *Boletus* (Jeffries and Young 1994).

7.2
Taxonomy and Ecology

Scleroderma contains at least 25 species (Hawksworth et al. 1995; Sims et al. 1995) and is classified within the Sclerodermataceae (a family which also includes *Pisolithus*; see Chap 1). Species can be identified on the morphology

Research School of Biosciences, University of Kent, Canterbury, Kent CT2 6NJ, UK

Fig. 7.1. Typical sporocarps of *Scleroderma citrinum*

of the sporocarp and by the morphology of the basidiospores. Guzmán (1970) noted that basidiospore ornamentation within the genus falls into three categories (spiny, sub-reticulate and reticulate), and these were also used by Sims et al. (1995) in the construction of a key based on spore characters. Mature sporocarps of *Pisolithus* resemble those of *Scleroderma*, but can be distinguished easily from those of *Scleroderma* by the presence of distinct peridioles. However, when sporocarps are young, this distinction is lacking as immature *Scleroderma* sporocarps possess minute fragile walled peridioles which break down at an early stage, leaving no traces of their former existence (Sims et al. 1995). This can lead to misidentification when field-collected sporocarps are used as inoculum for forest nurseries.

In general, *Scleroderma* sporocarps appear early in the fruiting succession (*sensu* Mason et al. 1983) of ECM fungi associated with particular trees and are also often reported from areas of freshly cleared land. This is consistent with the fact that *Scleroderma* spp., along with *Paxillus* and *Pisolithus*, are primary colonisers of coal spoil and other mining waste (Ingelby et al. 1985; Newton 1992), possibly as a consequence of their prolific rhizomorph-forming ability which enables them to spread rapidly and colonise young root systems. Raman (1988) noted the occurrence of large numbers of sporocarps of *S. citrinum* in *Pinus patula* plantations in India over the first 4 years after planting, but numbers rapidly dropped to zero by the seventh year. The frequent occurrence of sporocarps of *S. citrinum* was also recorded by Rao et al. (1997) in 2-year-old Indian plantations of *Pinus kesiya*, although in this case smaller numbers were observed in 4-, 11- and 17-year-old plantations. A *Scleroderma* sp. was one of the most common ECM types on 2- to 7-month-old seedlings of *Shorea leprosula* at two sites in Malaysia (Lee and Alexander 1996), and in a comparison of sporocarps occurring below *Quercus robor* planted along roadsides in the Netherlands, Keizer and Arnolds (1994) reported *S. areolatum* associated with trees of young or medium age, whereas *S. citrinum* was recorded more frequently with old trees. Thoen and Ba (1989) observed the prolific fruiting of *Scleroderma* spp. under regenerating individuals of *Afzelia africana* or under seedlings grown in the nursery, and the latter phenomenon has also been observed for *S. leptopodium* Har. & Pat. under dipterocarp seedlings in Indonesia (Jeffries, unpubl. data.). Malajczuk (1987) considers that early-colonising fungi such as *Scleroderma* and *Pisolithus* are often replaced at canopy closure by "late-stage" fungi. However, Newton (1991) reports that *S. citrinum* and *Paxillus involutus* (Fr.) Fr. may also be dominant in mature (>50-year-old) plantations established on sandy soil. Failure to establish mycorrhizas from spore inoculum of *S. citrinum* led Fox (1986) to conclude that this fungus was a "late-stage" fungus, but this conclusion may need re-evaluating in view of the information discussed above. It is likely that the genus is able to act as a pioneer fungus in disturbed habitats, yet persist in well-drained soils where water stress or high temperatures may occur.

Scleroderma has a worldwide range, although species distribution seems to be variable. Like *Pisolithus*, it appears to be able to withstand higher temperatures than many of the typical temperate ECM fungi. For example, *S. citrinum* grows in axenic culture at 30°C and the proportion of *Scleroderma*-type mycorrhizas developing on *Betula* roots was directly related to mean annual and spring substrate temperatures at a coal spoil site in Scotland (Ingleby et al. 1985). Some species have a restricted geography, such as *S. echinatum* (Petri) Guzmán, which is apparently confined to Southeast Asia (Rifai 1987; Watling 1994; Watling and Lee 1995). Other species generally prefer more arid conditions (e.g. *S. meridionale* Dem. & Mal. in the Mediterranean), but on the whole soils of a more mineral status are preferred, certainly for fruiting. *S. verrucosum* Pers. is widespread, although it probably increases southwards in Europe and is a frequent constituent of tropical floras (Sims et al. 1995). *S. columnare* Berk. & Br. is also probably more tropical than temperate, whereas *S. citrinum* is recorded commonly from mainly northern hemisphere temperate regions (Sims et al. 1995). *S. areolatum* Ehrenb. and *S. cepa* Pers. appear to be widely distributed. Care must be taken, however, in the interpretation of these data as they may reflect the journeys of mycologists interested in recording macrofungi, rather than detailing true global distributions. As Watling and Abraham (1992) point out, the fact that a fungus is not recorded from a particular region may be because it has just not been collected, rather than it being truly absent!

7.3
ECM Formation

Scleroderma spp. generally form white mycorrhizas from clamped hyphae, although *S. areolatum* may lack clamps (Godbout and Fortin 1985). In axenic culture, *S. areolatum*, *S. cepa*, *S. columnare*, and *S. verrucosum* do not form clamps (Richter 1992), suggesting that they may all form clampless ECM in nature, and this seems a general feature of *Scleroderma* section *Aculeatispora*. Well-developed mycorrhizas are coralloid, but in younger associations both dichotomous and simple branching mycorrhizas are recorded. The surface of the mantle is usually covered by abundant, interwoven extramatrical hyphae, and white rhizomorphs or thick mycelial strands of undifferentiated hyphae are typically found emanating into the substrate. Mohan et al. (1993) described the ECM of *S. citrinum* on *Pinus patula* and drew attention to the rhizomorphs as being abundant and having distinctive cystidia-like hyphae radiating from the surface. Small sclerotia may be formed by some species [e.g. *S. verrucosum* (Ba and Thoen 1990)]. The presence of these in soil may explain the rapid colonisation of bait seedlings of *Afzelia africana* by *S. verrucosum* in contrast to the slower colonisation by *S. dictyosporum* Pat. which predominated by 24 weeks after baiting (Ba et al. 1991). *S. sinnamariense* Mont. is distinctive in that the hyphae in the rhizomorphs running from the sporocarps to the ECM are bright yellow, whilst the ECM themselves are dark brown (Lee et al. 1997).

7.3.1
Host–Fungus Specificity

A range of host genera has been reported, either from laboratory syntheses or field-based observations, indicative that most species have a wide host range (Table 7.1). Both coniferous and non-coniferous hosts are included and there is no apparent host specificity. Attempts by Dell et al. (1994) to synthesise ECM between *Pisolithus* or *Scleroderma* spp. and *Eucalyptus grandis*, *Allocasuarina littoralis* and *Casuarina equisetifolia* were successful in the former two cases, but all *Casuarina* seedlings receiving the *Scleroderma* inoculum died.

7.3.2
Developmental Aspects of the Fungus–Host Interaction

Limited data are available regarding the development of the symbiosis. This may be because, in general, *Scleroderma* spp. grow slowly in axenic culture, with hyphal extension rates typically around 0.3–0.8 mm day^{-1} on nutrient agar media at optimum temperatures. *S. sinnamariense* tends to grow faster than other species of *Scleroderma* (Sims 1996), but even this species grows more slowly than isolates of *Pisolithus*. Axenic growth of *Scleroderma* spp. on agar is often accompanied by the production of dark brown diffusates in the

Table 7.1. Representative examples of references to host genera with which *Scleroderma* spp. have been confirmed as forming ECM by mycorrhiza synthesis experiments

Host genus	*Scleroderma* sp.	Reference
Afzelia	S. dictyosporum	Ba and Thoen (1990)
	S. verrucosum	Ba and Thoen (1990)
Betula	S. citrinum	Waller and Agerer (1993); Ek et al. (1996)
Carya	S. bovista	Marx and Bryan (1969)
Eucalyptus[a]	S. cepa	Pryor (1956); Aggangan et al. (1996)
	S. dictyosporum	Garbaye et al. (1988)
	S. laeve	Malajczuk et al. (1982)
Hopea	S. columnare	Hadi et al. (1991)
Larix	S. citrinum	Richter and Bruhn (1990)
Picea	S. citrinum	Ek et al. (1996)
	S. hypogaeum	Molina and Trappe (1982)
Pinus	S. bovista	Richter and Bruhn (1989a)
	S. citrinum	Kannan and Natarajan (1987); Mohan et al. (1993); Colpaert et al. (1996); Parladé et al. (1996)
	S. cepa	Richter and Bruhn (1989a)
	S. laeve	Malajczuk et al. (1982)
	S. meridionale	Richter and Bruhn (1989a)
	S. polyrhizum	Richter and Bruhn (1989a); Duñabeita et al. (1996)
	S. texense	Ivory and Munga (1983)
	S. verrucosum	Chu-Chou (1979)
Pseudotsuga	S. citrinum	Parladé et al. (1996)
Populus	S. areolatum	Godbout and Fortin (1985)
	S. citrinum	Godbout and Fortin (1985)
Quercus	S. citrinum	Beckjord and McIntosh (1983)
Shorea	S. columnare	Hadi et al. (1991)
	S. dictyosporum	Supriyanto et al. (1993)
	S. sinnamariense	Lee et al. (1997)
Tsuga	S. hypogaeum	Kropp and Trappe (1982)
Uapaca	S. dictyosporum	Thoen and Ba (1989)

[a] See also summary listing by Malajczuk et al. (1982).

medium. In addition, the growth in axenic culture often slows with time and frequent subculture is necessary to prevent cultures losing viability over time. In a long-term study of the cold storage of basidiomycete cultures in sterile water, Richter and Bruhn (1989b) noted that *Scleroderma* spp. did not survive well and usually died within the first 9–24 months, a result consistent with the observations for *S. bovista* Fr. by Marx and Daniel (1976). Storage at room temperature in liquid media can be more effective (Richter and Bruhn 1989b).

7.3.3
Symbiotic Functioning

In order to investigate mechanisms underlying the beneficial effect on host plants, Antibus et al. (1992) determined the ability of several ECM fungi to produce phosphatases in axenic culture and to use inositol phosphate as a P source. Five isolates of *S. citrinum* were included from several forest types in a limited geographic area. Considerable intraspecific variation in acid phosphatase production was noted (as much as two-fold at pH 5.0 in some cases). The acid phosphatases from all isolates showed maximum activity between pH 4.5 and 5.0 and this sharply decreased above pH 6.0, which presumably reflects the ecological conditions under which the mycorrhizas are functional. Few other studies have examined enzyme production in relation to the symbiosis, although Botton and Dell (1994) used isozyme analysis to show that glutamate dehydrogenase and aspartate aminotransferase (AAT) activity was expressed by *S. verrucosum* in culture. The formation of eucalypt mycorrhizas was accompanied by a significant increase in AAT activity, and an additional AAT isozyme was noted in mycorrhizal roots compared to those observed in non-mycorrhizal roots.

7.4
Host Plant Growth Responses and Fungus-Derived Benefits

Scleroderma species have been used to increase the early growth of a number of tree species, both in the glasshouse and in the field. Like *Pisolithus*, *Scleroderma* species have the advantage over many other ECM fungi as inoculants as their spores can be collected *en masse* from mature basidiomes. Thus, spore inoculum is relatively easy to prepare from field-collected material and therefore represents a cheap form of inoculum if labour costs are low. Early work was done by Pryor (1956) who added basidiospores of *S. cepa* to heat-sterilised soil in pots. Abundant ECM formed on roots and growth of *Eucalyptus* species was stimulated. From these results he concluded that the absence of ECM on these eucalypts accounted for regeneration failures in Iraq and other parts of the world. Now that the presence of ECM fungi is known to be important in tree seedling establishment, De La Cruz and colleagues (Bartolome et al. 1988) have pioneered the use of spore tablets of both *Pisolithus* and *Scleroderma* spp. for inoculation of *Eucalyptus* and *Pinus* spp. in the Philippines. Growth responses have been demonstrated in both nursery and field trials and many thousands of inoculated seedlings have been outplanted in reforestation programmes in the last two decades. Field-collected sporocarps are brought to the laboratory and spores extracted and mixed with finely ground soil or clay substrates before being compressed in a commercial pharmaceutical tableting machine. Growth responses have been recorded in the field which were previously only obtained from expensive use of phosphorus fertilisers. More recently, similar technology has been applied to the growth of dipterocarps and height increases of over 22% relative to uninoculated controls have been recorded for seedlings of *Parashorea malanonan* after 2 months in the field following nursery inoculation with mycelial cultures of *S. sinnamariense* (Dodd et al. 1996). In Indonesia, ECM were successfully synthesised on seedlings and cuttings of the dipterocarp *Shorea leprosula* using chopped basidiomes of *S. columnare* (Hadi et al. 1991;

Omon et al. 1995). In contrast, spores of *S. columnare* and *S. dictyosporum* were mixed with clay, alcosorb or gelatin carriers and used as inoculum for *Shorea mecistopteryx* (Supriyanto et al. 1993). Inoculum of a *Scleroderma* sp. has also been formulated as spore tablets, spore capsules and as a spore powder for successful inoculation of several dipterocarp species (Hadi et al. 1991).

Castellano (1996) has collated reports of the use of both spore and vegetative mycelium inoculum of *S. citrinum* and *S. texense* Berk. and vegetative inoculum of *S. bovista* and *S. dictyosporum* on a range of hosts worldwide, but beneficial growth responses were only reported for some of these combinations. This is typical of many of the results reported for inoculation trials involving *Scleroderma* species. For example, in a study of field-planted *Quercus rubra* seedlings inoculated with six different ECM fungi, the growth of seedlings inoculated with *S. citrinum* was significantly greater (by 31–51% in height and 11–34% in diameter) than all other seedlings by the third growing season (Beckjord and McIntosh 1983). This difference was not apparent at the end of the first growing season but had become significant by year two. Examination of root cores in year three indicated the persistence of *S. citrinum* on the roots whereas in some fungal treatments the original inoculant had been replaced by indigenous species. The authors suggested that the fungus provided some physiological benefit in negating the detrimental effects of transplant shock during the establishment phase following outplanting. In a more extensive experiment involving different forms of ECM fungal inoculum, beneficial effects of inoculation with *S. citrinum* could not be demonstrated after two seasons on a replanted strip-mine site, whereas on a clear-cut site treatment differences were still evident (Beckjord and McIntosh 1984). It was noted, however, that control (uninoculated) seedlings had also become infected with *S. citrinum* or *Pisolithus arhizus* (Pers.) Rauss after two seasons in the field, possibly by hyphae radiating from treated plants or from indigenous inoculum.

Two Australian isolates of *Scleroderma* were shown by Dell et al. (1994) to stimulate the early growth of *Eucalyptus* such that dry weights of shoots from 90-day-old seedlings were between 13 and 32 times larger than uninoculated controls. *Allocasuarina* seedlings also benefited from inoculation, but growth responses were much smaller. In Spain, inoculation of containerised *Pinus pinaster* seedlings with spore suspensions of *S. citrinum* resulted in <40% of the plants being infected if 10^3 spores per seedling were used, with an average of 20% of short roots becoming mycorrhizal after 5 months (Parladé et al. 1996). When the spore concentration was increased to 10^5 spores per seedling or above, all the seedlings developed mycorrhizas and over 80% of short roots were infected. There was, however, no significant difference in heights of seedlings receiving the smaller or larger inocula and all were of sufficient height to be outplanted in the field. From field observations, Malajczuk (1987) reported the significantly greater growth of 1-year-old *Eucalyptus wandoo* seedlings associated with sporocarps of *S. verrucosum* compared to seedlings with no associated sporocarps, and this is supported by inoculation trials on eucalypt seedlings in the Congo by Garbaye et al. (1988). These latter authors used mycelial cultures of five exotic ECM fungi including *S. citrinum, S. dictyosporum* and *S. texense*. Of these three, *S. citrinum* and *S. texense* formed mycorrhizas in the nursery, but this only resulted in growth stimulation of the eucalypts in the case of *S. texense*. However, significant root colonisation by an indigenous *Scleroderma*-like fungus was noted in all seedlings, including controls. In contrast, growth stimulation of outplanted seedlings was recorded for both fungi, but only after 27 (*S. texense*) and 50 months (*S. citrinum*) in the field in contrast to much earlier stimulation (20 months) recorded for inoculation with a *Pisolithus* species. The abundant occurrence of an indigenous *Scleroderma*-type mycorrhiza was again noted at 50 months, leading to the suggestion (Garbaye et al. 1988) that native *Scleroderma* species were much more competitive than the introduced fungi and ought to be examined as potential inoculum.

S. verrucosum was one of the most effective fungi in a screening study of 16 different ECM fungi for beneficial responses in the growth of *Eucalyptus* spp. (Burgess et al. 1993). Seedlings inoculated with this fungus grew to 75% of the maximum height response recorded after 100 days (i.e. those inoculated with *Pisolithus* sp.) and had an increased production of short roots compared to non-mycorrhizal controls as well as increased concentration of P in the plant biomass. These responses were recorded at P fertiliser additions of $4\,mg\,P\,kg^{-1}$ sand substrate, but the benefits were insignificant if the P addition was raised to $12\,mg\,P\,kg^{-1}$ sand. In view of these results it is perhaps suprising that an isolate of *S. cepa* was ineffective in stimulating growth of *Eucalyptus urophylla* at a range of pH levels, in contrast to beneficial effects of *Pisolithus* or *Laccaria* (Aggangan et al. 1996).

In a comparative North American trial, *S. citrinum* proved to be more effective by far than *P. arhizus, Thelephora terrestris* Fr. or *Rhizopogon roseolus* (Corda) T.M. Fries in conferring beneficial growth responses to loblolly pine (*Pinus taeda*) over a range of levels of soil P (Ford et al. 1985). Seedlings were significantly larger at the 11-month harvest date than those colonised by any of the other fungi, and had the largest root systems and greatest degree of mycorrhizal colonisation. This fungus is common in the hot, dry regions of the SE United States and was suggested as an appropriate fungus for inoculation on this pine species. However, as the authors pointed out, this was a pot study and attempts to improve the growth of *P. taeda* in plantations established on similar soils, by application of P fertilisers, had failed because N was in fact the limiting nutrient in the field. The growth responses to P achieved in pots were achieved because N limitation had been removed by N fertiliser application at the start of the experiment. *S. citrinum* has also been used to inoculate *Pinus kesiya* and this resulted in greater growth of the seedlings than inoculation with other fungi (Sharma and Mishra 1988). Beckjord et al. (1986) produced inoculum of *S. citrinum* or *P. arhizus* by combining field-collected basidiospores with compressed sand and peat into "basidiospore chips". These were cold-stored for up to 5 years at 3°C, a treatment which had a detrimental effect on the infectivity of the *Scleroderma* preparations but not on those of *Pisolithus*.

7.5
Nutritional Benefits to the Host and Influence on Host Carbon Economy

Scleroderma species have not found favour in many laboratory studies of host–fungus physiology, presumably again because of their relatively slow growth rate. However, C and N flow between seedlings of *Picea abies* and *Betula pendula* connected via a mycorrhizal mycelium of *S. citrinum* was studied by Ek et al. (1996). ^{15}N-labelled NH_4^+ was supplied exclusively to the fungus, while the birch or spruce plant was continuously fed with ^{13}C-labelled CO_2 for 72 h. Initially, during the first 2 months of the experiment, the spruce plant was the sole C source for the mycelium. However, 8 months later, the mycelium was almost entirely dependent on the birch plant for current photosynthates. On average, 93% of the total amount of labelled C recovered in the external mycelium originated from the birch. Further, the transfer of ^{15}N by the *Scleroderma* to the plants was largely directed towards the birch. The root systems of both plants were heavily mycorrhizal and there is no explanation for why the birch should be the main source of C for the fungus and the main sink for its soil-derived N (Ek et al. 1996). There was no conclusive evidence for net transfer of C between the plants via the mycorrhizal mycelial bridge. This experiment is one of the few which concentrates on *Scleroderma* and emphasises the need for more physiological studies on the widespread symbioses involving this genus.

Growth depressions as a result of mycorrhizal formation have been recorded over 12 weeks for seedlings of *Pinus sylvestris* infected with *S. citrinum* (Colpaert et al. 1996). Root growth was affected more than shoot growth, resulting in an increase in root/shoot ratio. Of three ECM fungi tested [*S. citrinum*, *T. terrestris*, *Suillus bovinus* (L.: Fr.) O. Kuntze], *S. citrinum* produced the most mycelial biomass within the substrate, and thus presumably effected a greater C drain on the host. However, retention of N in the external mycelium was responsible for a significant reduction in the N concentration of the host plants; therefore growth depressions may have resulted from either increased below-ground C allocation and/or high nutrient retention by the mycobiont (Colpaert et al. 1996). Plants colonised by *S. citrinum* showed no increase in net assimilation rate following infection, in contrast to the situation with *S. bovinus*.

Seedlings of *Quercus* spp. inoculated with *S. citrinum* and fertilised with NH_4Cl developed significantly more ECM than unfertilised plants or those fertilised with $NaNO_3$ (Beckjord et al. 1983) and the plants were larger than non-inoculated fertilised controls. In further experiments in which different amounts of N and P fertiliser were added, plant growth responses varied, and in some cases inoculation with *S. citrinum* did not increase seedling biomass.

It is perhaps significant that all of these experiments were carried out in containerised plants grown over a 100- to 110-day period, whereas significant growth responses in the field were not recorded until the second season after transplanting (Beckjord and McIntosh 1983; see above).

The natural occurrence of mycorrhizas on *Q. robor* and *B. pendula* seedlings growing in established deciduous forest has been monitored after 24 weeks into their first growth season (Newton 1991; Newton and Pigott 1991). A total of 41 ECM morphotypes were noted, but *S. citrinum* ECM dominated the *Q. robor* seedlings (61% of ECM root tips) whereas *P. involutus* dominated the *B. pendula*. Although the occurrence of most ECM morphotypes was reduced by fertiliser treatments, those of *S. citrinum* and *P. involutus* increased in some of the field sites. The importance of strand formation in the "success" of both these fungi was emphasised.

7.6
Non-nutritional Benefits to the Host

In common with several other ECM fungi, the non-nutritional benefits of ECM formation by *Scleroderma* to the host have been focused on the protection from root disease and the amelioration of the effects of toxic metals. The former situation was exemplified by the demonstration in culture of antagonism by geographically distinct isolates of *S. bovista* towards *Pythium* species associated with necrosis of feeder roots of pecan (*Carya illinoensis*). The dominance of mycorrhizas formed by this fungus in pecan orchards treated with nematicides and fungicides was considered as evidence that the fungus could act as a deterrent to root infection (Marx and Bryan 1969). There are few other reports of this phenomenon that can be attributed to *Scleroderma*.

In contrast, the ability to tolerate metal-contaminated substrates and to ameliorate the effect of metal toxicity when in symbiosis is the most frequently reported non-nutritional benefit reported for the genus. For example, a locally isolated *Scleroderma* sp. from Hong Kong was shown by Tam (1995) to be tolerant to high concentrations of Al, Fe, Cu and Zn. An isolate of *S. citrinum* was also isolated from Cu-contaminated soil by Howe et al. (1997), but, unlike the other ECM fungi obtained from this site, this isolate was not particularly Cu-tolerant and did not contain Cu-binding proteins.

Jones and Hutchinson (1986, 1988) carried out a series of investigations using a range of ECM fungi isolated from a Cu- and Ni-contaminated site and *Betula papyrifera*, the only tree able to naturally colonise this area. They showed that *S. flavidum* E. & E. (syn. *S. cepa*) was the most beneficial ECM fungi in terms of increasing seedling biomass in the presence of high concentrations of Ni, and in increasing Ni tolerance. Ni concentrations were higher in roots of mycorrhizal plants, but similar or lower in shoots of mycorrhizal plants, compared to non-mycorrhizal equivalents. Surprisingly, this fungus was shown to be the most sensitive of four ECM fungi to Ni and Cu tested in agar culture (Jones and Hutchinson 1986). It was suggested that the higher concentrations of P measured in ECM seedlings may act as a binding site for Ni, thus reducing its bioavailability. Isolates of ECM fungi from a metal-contaminated site were no better than isolates from an unpolluted site in terms of benefits conferred (Jones and Hutchinson 1988). When similar experiments were carried out in the presence of Cu (Jones and Hutchinson 1986), none of the *Scleroderma* isolates increased growth of seedlings compared to non-mycorrhizal controls. Colpaert and Van Assche (1992a,b) compared a range of concentrations of Cd and Zn on the axenic growth of a range of ECM fungi, including *S. citrinum*, as well as their effects on Zn accumulation in mycorrhizal *P. sylvestris*. The isolates of *Scleroderma* showed poor tolerance to both Cd and Zn, but pine seedlings forming ECM with this fungus had much higher concentrations of Zn in the roots than those ECM with some of the other ECM fungi. The addition of Cd reduced the survival of mycelium and sporocarp production of *S. citrinum* (Colpaert and Van Assche 1993), and it was suggested that in high Cd concentrations the fungus might "parasitise" the host tree.

In summary, there is currently not much evidence that *Scleroderma* species are particularly tolerant of high metal concentrations, but there are few data available, and the strains used have not always come from metal-contaminated habitats. Further screening of isolates from contaminated sites may prove rewarding in the search for inoculants that can ameliorate the effects of metals on tree seedling establishment. *Scleroderma* species are early-stage fungi of broad host range and possess many of the attributes of *Pisolithus* spp. They show a similar potential for development as inoculants, particularly in stressed environments. It should be relatively easy to adapt nursery practices for the large-scale inoculation of tree seedlings. As an example, Flemington et al. (1987) studied the effects of thiram (a fungicide used in nurseries) and Al (used as a lubricant in nurseries) on in vitro growth of *S. citrinum* and *S. macrorhizon* (Wall.) (syn. *S. meridionale*). Both fungi were inhibited by thiram, but only *S. macrorhizon* was inhibited by Al, suggesting that only modification of the fungicide regime would be needed for the incorporation of *S. citrinum* inoculum into nursery practice.

References

Aggangan N, Dell B, Malajczuk N (1996) Effects of soil pH on the ectomycorrhizal response of *Eucalyptus urophylla* S. T. Blake seedlings. New Phytol 134:539–546

Antibus RK, Sinsabaugh RL, Linkins AE (1992) Phosphatase activities and phosphorus uptake from inositol phosphate by ectomycorrhizal fungi. Can J Bot 70:794–801

Ba AM, Thoen D (1990) First syntheses of ectomycorrhizas between *Afzelia africana* Sm. (Caesalpinioideae) and native fungi from West Africa. New Phytol 114:99–103

Ba AM, Garbaye J, Dexheimer J (1991) Influence of fungal propagules during the early stage of the time sequence of ectomycorrhizal colonization on *Afzelia africana* seedlings. Can J Bot 69:2442–2447

Bartolome HT, De La Cruz RE, Aggangan NS (1988) Pilot testing of mycorrhizal tablets for pines and eucalypts in the Philippines. Proc UNESCO Regional Workshop on Development and Production of Mycorrhizal Inoculants. Biotech UPLB College, Laguna

Beckjord PR, McIntosh MS (1983) Growth and fungal retention by field-planted Quercus rubra seedlings inoculated with several ectomycorrhizal fungi. Bull Torrey Bot Club 110:353-359

Beckjord PR, McIntosh MS (1984) Growth and fungal persistence by Quercus rubra inoculated with ectomycorrhizal fungi and planted on a clear-cutting and strip mine. Can J Bot 62:1571-1574

Beckjord PR, Melhuish JM, McIntosh MS, Hacskaylo E (1983) Effects of nitrogen fertilization on growth and ectomycorrhizal formation of Quercus alba, Q. rubra, Q. falcata and Q. falcata var. pagodifolia. Can J Bot 61:2507-2514

Beckjord PR, Melhuish JM, Hacskaylo E (1986) Ectomycorrhiza formation on sawtooth oak by inoculation with basidiospore chips of Pisolithus tinctorius and Scleroderma citrinum. J Environ Hortic 4:127-129

Botton B, Dell B (1994) Expression of glutamate dehydrogenase and aspartate aminotransferase in eucalypt mycorrhizas. New Phytol 126:249-257

Burgess TI, Malajzuk N, Grove TS (1993) The ability of 16 ectomycorrhizal fungi to increase growth and phosphorus uptake of Eucalyptus globulus Labill. and E. diversicolor F. Muell. Plant Soil 153:155-164

Castellano MA (1996) Outplanting performance of mycorrhizal inoculated seedlings. In: Mukerji KG (ed) Concepts in mycorrhizal research. Kluwer, Dordrecht, pp 223-301

Chu-Chou M (1979) Mycorrhizal fungi of Pinus radiata in New Zealand. Soil Biol Biochem 11:557-562

Colpaert JV, Van Assche JA (1992a) Zinc toxicity in ectomycorrhizal Pinus sylvestris. Plant Soil 143:201-214

Colpaert JV, Van Assche JA (1992b) The effects of cadmium and the cadmium-zinc interaction on the axenic growth of ectomycorrhizal fungi. Plant Soil 145:237-243

Colpaert JV, Van Assche JA (1993) The effects of cadmium on ectomycorrhizal Pinus sylvestris L. New Phytol 123:325-333

Colpaert JV, Van Laere A, Van Assche JA (1996) Carbon and nitrogen allocation in ectomycorrhizal and non-mycorrhizal Pinus sylvestris L. seedlings. Tree Physiol 16:787-793

Dell B, Malajzuk N, Bougher NL, Thomson G (1994) Development and function of Pisolithus and Scleroderma ectomycorrhizas formed in vivo with Allocasuarina, Casuarina and Eucalyptus. Mycorrhiza 5:129-138

Dodd JC, Jeffries P, De La Cruz R (1996) The use of mycorrhizal fungi in reforestation programmes in the Philippines. Final Report, EU Contract CI1*-CT91-0904, STD3 Programme, DG XII, Brussels

Duñabeita MK, Hormilla S, Salcedo I, Peña JI (1996) Ectomycorrhizae synthesized between Pinus radiata and eight fungi associated with Pinus spp. Mycologia 88:897-908

Ek H, Andersson S, Söderström B (1996) Carbon and nitrogen flow in silver birch and Norway spruce connected by a common mycorrhizal mycelium. Mycorrhiza 6:465-467

Flemington PM, Bruhn JN, Richter DL (1987) Effects of thiram and aluminium on in vitro growth of Scleroderma citrinum and S. macrorhizon. Phytopathology 177:1735 (Abstr)

Ford VL, Torbert JL Jr, Burger JA, Miller OK (1985) Comparative effects of four ectomycorrhizal fungi on loblolly pine seedlings growing in a greenhouse in a Piedmont soil. Plant Soil 83:215-221

Fox FM (1986) Groupings of ectomycorrhizal fungi of birch and pine, based on establishment of mycorrhizas on seedlings from spores in unsterile soils. Trans Br Mycol Soc 87:371-380

Garbaye J, Delwalle JC, Diangana D (1988) Growth response of eucalypts in the Congo to ectomycorrhizal inoculation. For Ecol Manage 24:151-157

Godbout G, Fortin JA (1985) Synthesised mycorrhizae of aspen: fungal genus level of structural characterization. Can J Bot 63:252-262

Guzmán G (1970) Monographía de género Scleroderma Pers. emend Fr. (Fungi-Basidiomycetes). Darwiniana 16:233-407

Hadi S, Fakuara Y, Setiadi Y, Prematuri R, Nuhamara ST (1991) Status of mycorrhiza research on dipterocarps in Indonesia. In: Proceedings, BIO-REFOR Pre-workshop, IUFRO/SPDC, Bogor, pp 75–81

Hawksworth DL, Kirk PM, Sutton BC, Pegler DN (1995) Ainsworth and Bisby's dictionary of the fungi, 8th edn. CAB International, Wallingford

Howe R, Evans RL, Ketteridge SW (1997) Copper-binding proteins in ectomycorrhizal fungi. New Phytol 135:123–131

Ingleby K, Last FT, Mason PA (1985) Vertical distribution and temperature relations of sheathing mycorrhizas of *Betula* spp. growing on coal spoil. For Ecol Manage 12:279–285

Ivory MH, Munga FM (1983) Growth and survival of container-grown *Pinus caribea* infected with various ectomycorrhizal fungi. Plant Soil 71:339–344

Jeffries P, Young TWK (1994) Interfungal parasitic relationships. CAB International, Wallingford

Jones MD, Hutchinson TC (1986) The effect of mycorrhizal infection on the response of *Betula papyrifera* to nickel and copper. New Phytol 102:429–442

Jones MD, Hutchinson TC (1988) Nickel toxicity in mycorrhizal birch seedlings infected with *Lactarius rufus* or *Scleroderma flavidum*. II. Uptake of nickel, calcium, magnesium, phosphorus and iron. New Phytol 108:461–470

Kannan K, Natarajan K (1987) Pure culture synthesis of *Pinus patula* ectomycorrhizae with *Scleroderma citrinum*. Curr Sci 56:1066–1068

Keizer PJ, Arnolds E (1994) Succession of ectomycorrhizal fungi in roadside verges planted with common oak (*Quercus robor* L.) in Drenthe, The Netherlands. Mycorrhiza 4:147–159

Kropp BR, Trappe JM (1982) Ectomycorrhizal fungi of *Tsuga heterophylla*. Mycologia 74:479–488

Lee SS, Alexander IJ (1996) The dynamics of ectomycorrhizal infection of *Shorea leprosula* seedlings in Malaysian rain forests. New Phytol 132:297–305

Lee SS, Alexander IJ, Watling R (1997) Ectomycorrhizas and putative ectomycorrhizal fungi of *Shorea leprosula* Miq. (Dipterocarpaceae). Mycorrhiza 7:63–81

Malajczuk N (1987) Ecology and management of ectomycorrhizal fungi in regenerating forest ecosystems in Australia. In: Sylvia D, Hung LL, Graham JH (eds) Proceedings of the 7th North American conference on mycorrhizae. Institute of Food and Agricultural Sciences, Gainesville, pp 118–120

Malajczuk N, Molina R, Trappe JM (1982) Ectomycorrhiza formation in *Eucalyptus*. I. Pure culture synthesis, host specificity and mycorrhizal compatibility with *Pinus radiata*. New Phytol 91:467–482

Marx DH, Bryan WC (1969) *Scleroderma bovista*, an ectotrophic mycorrhizal fungus of pecan. Phytopathology 59:1128–1132

Marx DH, Daniel WJ (1976) Maintaining cultures of ectomycorrhizal and plant pathogenic fungi in sterile water cold storage. Can J Microbiol 22:338–341

Mason PA, Wilson J, Last FT (1983) The concept of succession in relation to the spread of sheathing mycorrhizal fungi on inoculated tree seedlings in unsterile soil. Plant Soil 71:247–256

Mohan V, Natarajan K, Ingleby K (1993) Anatomical studies on ectomycorrhizas. III. The ectomycorrhizas produced by *Rhizopogon luteolus* and *Scleroderma citrinum* on *Pinus patula*. Mycorrhiza 3:51–56

Molina R, Trappe JM (1982) Patterns of ectomycorrhizal host specificity and potential amongst Pacific Northwest conifers and fungi. For Sci 28:423–457

Newton AC (1991) Mineral nutrition and mycorrhizal infection of seedling oak and birch. III. Epidemiological aspects of ectomycorrhizal infection and the relationship to seedling growth. New Phytol 117:53–60

Newton AC (1992) Towards a functional classification of ectomycorrhizal fungi. Mycorrhiza 2:75–79

Newton AC, Pigott CD (1991) Mineral nutrition and mycorrhizal infection of seedling oak and birch. II. The effect of fertilizers on growth, nutrient uptake and ectomycorrhizal infection. New Phytol 117:45–52

Omon RM, Fakuara Y, Supriyanto, Suhendang E (1995) Effects of some ectomycorrhizal fungi and culture media on the growth of *Shorea leprosula* Miq. cuttings (Abstract). In: Supriyanto, Kartana JT (eds) Biology and biotechnology of mycorrhizae. BIOTROP Special Publication, no 56, Bogor, 181 pp

Parladé J, Pera J, Alvarez IF (1996) Inoculation of containerized *Pseudotsuga menziesii* and *Pinus pinaster* seedlings with spores of five species of ectomycorrhizal fungi. Mycorrhiza 6:237–245

Pryor LD (1956) Ectotrophic mycorrhiza in renantherous species of *Eucalyptus*. Nature 177:587–588

Raman N (1988) Succession of ectomycorrhizal fungi in the colonization of *Pinus patula* plantations at Kodaikanal. In: Mahadevan A, Raman N, Natarajan K (eds) Proceedings of the 1st Asian conference on mycorrhizae. Centre for Advanced Studies in Botany, Madras, pp 101–102

Rao CS, Sharma GD, Shukla AK (1997) Distribution of ectomycorrhizal fungi in pure stands of different age groups of *Pinus kesiya*. Can J Microbiol 43:85–91

Richter DL (1992) Six species of *Scleroderma* (Gasteromycetes, Sclerodermatales) described from pure cultures. Mycotaxon 45:461–471

Richter DL, Bruhn JN (1989a) *Pinus resinosa* ectomycorrhizae: seven host–fungus combinations synthesized in pure culture. Symbiosis 7:211–228

Richter DL, Bruhn JN (1989b) Revival of saprotrophic and mycorrhizal basidiomycete cultures from cold storage in sterile water. Can J Microbiol 35:1055–1060

Richter DL, Bruhn JN (1990) *Scleroderma citrinum* (Gasteromycetes, Sclerodermatales) and *Larix decidua* form ectomycorrhizae in pure culture. Nova Hedwigia 50:355–360

Rifai MA (1987) Malasian *Scleroderma* (Gasteromycetes). Trans Jpn Mycol Soc 28:97

Sharma GD, Mishra RR (1988) Production of mass inoculum and inoculation techniques of ectomycorrhizal fungi in sub-tropical pine (*Pinus kesiya*). In: Mahadevan A, Raman N, Natarajan K (eds) Proceedings of the 1st Asian conference on mycorrhizae. Centre for Advanced Studies in Botany, Madras, pp 319–321

Sims K (1996) Growth physiology and systematics of some SE Asian ectomycorrhizal fungi, with additional reference to isozyme interpretations. PhD Thesis, University of Kent, Kent

Sims K, Watling R, Jeffries P (1995) A revised key to the genus *Scleroderma*. Mycotaxon 56:403–420

Sims K, Watling R, De La Cruz, R, Jeffries P (1997) Ectomycorrhizal fungi of the Philippines: a preliminary survey and notes on the geographic biodiversity of the Sclerodermatales. Biodiv Conserv 6:45–58

Supriyanto, Setiawan I, Omon RM (1993) Effect of *Scleroderma* sp. on the growth of *Shorea mecistopteryx* Ridi. In: Proceedings International Workshop BIO-REFOR. IUFRO/SPDDC, Yogyakarta, pp 186–188

Tam PCF (1995) Heavy metal tolerance by ectomycorrhizal fungi and metal amelioration by *Pisolithus tinctorius*. Mycorrhiza 5:181–187

Thapar HS (1988) Nutritional studies on ectomycorrhizal fungi of chir pine in culture. In: Mahadevan A, Raman N, Natarajan K (eds) Proceedings of the 1st Asian conference on mycorrhizae. Sivakami Publishers, Madras, pp 179–183

Thoen D, Ba AM (1989) Ectomycorrhizas and putative ectomycorrhizal fungi of *Afzelia africana* Sm. and *Uapaca guineenis* Müll. Arg. in southern Senegal. New Phytol 113:549–559

Waller K, Agerer R (1993) *Scleroderma citrinum*. In: Agerer R (ed) Colour atlas of ectomycorrhizae. Einhorn-Verlag, Schwäbisch Gmünd, 180 pp

Watling R (1994) Taxonomic and floristic notes on some Malaysian larger fungi. I. Malay Nat J 48:67–78.

Watling R, Abraham SP (1992) Ectomycorrhizal fungi of Kashmir forests. Mycorrhiza 2:81–87

Watling R, Lee SS (1995) Ectomycorrhizal fungi associated with members of the Dipterocarpaceae in Peninsular Malaysia. I. J Trop For Sci 7:657–669

Chapter 8

Amanita

Z. L. Yang[1], M. Weiß[2], I. Kottke[2], F. Oberwinkler[2], U. Nehls[3], M. Guttenberger[3] and R. Hampp[3]

8.1
Introduction

The genus *Amanita* is one of the largest basidiomycetous genera with, to date, more than 400 species described worldwide. Its type species, the fly agaric *Amanita muscaria* (L.: Fr.) Pers. (Fig. 8.1A–I), the very embodiment of a fungus for many, made its way even into the arts, a subject of mycology as well as of mythology. The nearly cosmopolitan genus has always been of interest to mankind, as it contains delicious edible species such as Caesar's mushroom, *A. caesarea* (Scop.: Fr.) Pers., but also deadly poisonous species such as the death cap, *A. phalloides* (Fr.) Link (Fig. 8.1S–U). Many *Amanita* species are known to form ectomycorrhizas (ECM) (Table 8.2). Some members of the genus, however, may not be involved in ECM associations, their sporocarps being found in the open field (Bas 1969; Bas and de Meijer 1993). It is assumed that *Amanita* species are locally restricted in their habitats (Singer 1986).

8.2
Taxonomy

Since Persoon established the genus in 1797 and Fries (1821) removed from it species with pink spore powder, which are nowadays included in the genus *Volvariella*, many mycologists have contributed to *Amanita* systematics, either splitting it into smaller genera or suggesting diverse infrageneric classification concepts. Because often authors proposed new taxa while neglecting already existing names, more than 50 section names have been published (Bas 1969), causing considerable taxonomic confusion.

Today most mycologists agree that *Amanita* in its broader sense is a natural taxon of Agaricales. The combination of a bilateral lamellar trama (Fig. 8.2A), the trama of stipe and cap with characteristically inflated terminal cells

[1] Kunming Institute of Botany, Academia Sinica, Heilongtan, Kunming 650204, P.R. China
[2] Universität Tübingen, Spezielle Botanik und Mykologie, Auf der Morgenstelle 1, 72076 Tübingen, Germany
[3] Universität Tübingen, Physiologische Ökologie der Pflanzen, Auf der Morgenstelle 1, 72076 Tübingen, Germany

Table 8.1. Infrageneric classification of the genus *Amanita*. (Yang 1997)

Subgenus *Amanita*
  Section *Amanita*
  Section *Caesareae*
  Section *Vaginatae*
Subgenus *Lepidella*
  Section *Amidella*
  Section *Lepidella*
  Section *Phalloideae*
  Section *Validae*

("*Amanita* structure"), a volva which is not or only basally gelatinized, and the schizohymenial ontogeny of the lamellae (Yang and Oberwinkler 1999) separates the genus from the presumably closely related genera *Limacella*, *Pluteus* and *Termitomyces*.

An infrageneric classification concept accepted by many mycologists is the system of Corner and Bas (1962) and Bas (1969), based upon morphological characters such as presence or absence of a bulb or annulus, volva shape, form of lamellulae, striation of the cap margin, together with the chemical character of the spore reaction in Melzer's reagent (e.g. Moser 1983). In contrast to the variability of the morphological characters, the anatomic structures are considerably uniform in the genus (Yang 1997) and, therefore, cannot be used for an infrageneric classification. In the following key we use the system of Yang (1997) (Table 8.1), a modification of the system of Corner and Bas.

8.2.1
Key to Subgenera and Sections (after Yang 1997)

1   Basidiospores inamyloid; cap margin usually radially sulcate; lamellulae nearly always truncate (subgenus *Amanita*) 2

1* Basidiospores amyloid; cap margin mostly not sulcate; lamellulae often attenuate (subgenus *Lepidella*) 4

2   Base of the stipe with bulb; volva usually friable, sometimes limbate .... section *Amanita*

2* Base of the stipe without bulb; volva usually saccate, sometimes friable 3

3   Annulus and clamp connections present .... section *Caesareae*

3* Annulus and clamp connections lacking .... section *Vaginatae*

4   Cap margin not appendiculate; pileipellis often deeply coloured; annulus membranous, seldom friable 5

4* Cap margin appendiculate; pileipellis rarely deeply coloured; annulus membranous to friable 6

5 Volva pulverulent or breaking up into flocks, warts, patches, scales, belts, or crusts on the cap or at the stipe base, sometimes volva circumscissile, limbate or bulb of stipe distinctly marginate .... section *Validae*

5* Volva saccate, forming a membranous sac at base of stipe and only occasionally few volva remains on the cap .... section *Phalloideae*

6 Volva circumscissile, limbate or breaking up into flocks, scales, warts, patches, belts or crusts on cap and at base of stipe, mostly without visible layering; lamellae not darkening strongly after drying .... section *Lepidella*

6* Volva saccate, with friable inner layer and forming scales, patches or powder on cap; lamellae often darkening strongly after drying, especially at lower temperatures .... section *Amidella*

The two subgenera *Amanita* and *Lepidella* are separated by the basidiospore reaction in Melzer's reagent: spores of species of subgenus *Lepidella* are amyloid in contrast to those of species of subgenus *Amanita*, which are inamyloid. Well correlated to this chemical character are two morphological characters. Lamellulae of members of subgenus *Lepidella* usually attenuate towards the stipe, whereas they are more or less truncate in subgenus *Amanita*; the cap margin is radially sulcate in subgenus *Amanita* in constrast to the species of subgenus *Lepidella*, the cap of which is usually ungrooved. The combination of these characters justifies the separation of the two subgenera, although it has not yet been verified by DNA analyses (Weiß et al. 1998). Interestingly, two exceptions have been reported concerning the presumably conservative character "basidiospore amyloidity": Bas (1969) as well as Yang (1997) found inamyloid basidiospores in *Amanita* species which can clearly be morphologically assigned to section *Amidella* of subgenus *Lepidella*. Short profiles of the *Amanita* sections are given below.

8.2.2
Subgenus *Amanita*

8.2.2.1
Section *Amanita*

Section *Amanita* containing the type species *Amanita muscaria* of the genus (Fig. 8.1A–I) comprises species with inamyloid spores and a conspicuous bulb at the stipe base. Some of its members are known to be poisonous mushrooms, but, unlike the deadly poisonous species of section *Phalloideae* containing bicyclic peptides (Wieland 1986), species of section *Amanita* frequently form psychotropic alkaloids (Eugster 1968) like muscarine which is

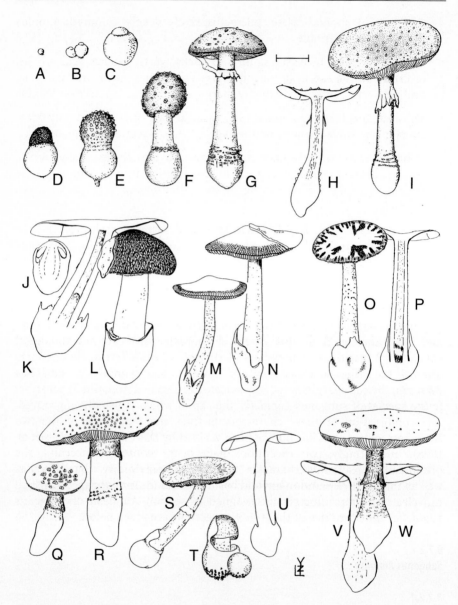

Fig. 8.1. Fruiting bodies of *Amanita* species at different ontogenetic stages. Bar 3 cm. A–I *Amanita muscaria*; J–L *A. yuaniana*; M and N *A. umbrinolutea*; O and P *A. clarisquamosa*; Q and R *A. solitaria*; S–U *A. phalloides*; V and W *A. excelsa*. (A–I, M, N, Q–W, Yang, original drawings; J–L, O, P, Yang 1997, with modifications)

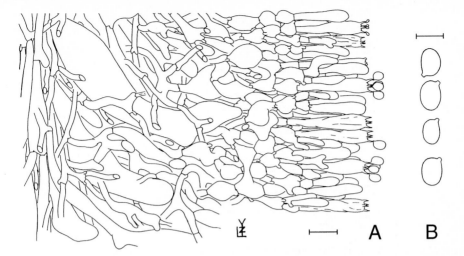

Fig. 8.2. *Amanita muscaria.* **A** Longitudinal section through lamella, showing hymenium, subhymenium, lateral stratum, and central stratum; bar 20 μm. **B** Basidiospores; bar 10 μm. (Yang and Oberwinkler 1999, with modifications)

the reason why these fungi have been used to kill flies (Singer 1986), for medicinal purposes, and as hallucinogenics (Eugster 1968).

8.2.2.2
Section Caesareae

Members of the section *Caesareae* are characterized by inamyloid spores, lacking a bulb at the stipe base and possessing an annulus as well as clamp connections. Some of its species are edible fungi such as Caesar's mushroom, *A. caesarea*, or *A. yuaniana* Z. L. Yang (Fig. 8.1J–L). The section was integrated into the section *Vaginatae* by Corner and Bas (1962). Recent molecular studies, however, support the separation of the *Caesareae* group from section *Vaginatae* (Fig. 8.3).

8.2.2.3
Section Vaginatae

The third inamyloid *Amanita* section *Vaginatae* was considered a genus of its own, *Amanitopsis* Roze (1876), indicating a distinct morphological appearance of its species. Slender and fragile fruiting bodies with sulcate cap, a prominent volva and the lack of bulb and annulus distinguish the *Vaginatae* from the fleshy *Caesareae* species. In contrast to the *Caesareae* species, the members of section *Vaginatae* lack clamp connections. A molecular marker for the section is a characteristic insertion of 13 nucleotides in the nuclear-coded gene for the large ribosomal subunit, which is

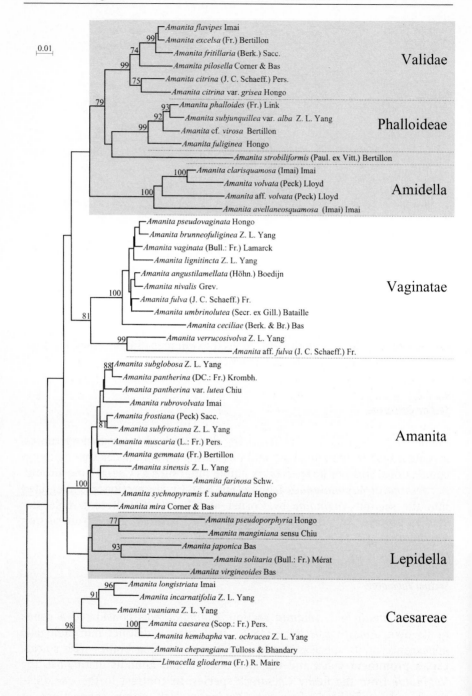

missing in all other sections of the genus (Weiß et al. 1998). As a representative of the section, *A. umbrinolutea* (Secr. ex Gill.) Bataille is shown in Fig. 8.1M,N.

8.2.3
Subgenus *Lepidella*

8.2.3.1
Section *Amidella*

Members of the section *Amidella*, as of all sections of subgenus *Lepidella*, have amyloid basidiospores. However, unlike species of the other sections of this subgenus, most *Amidella* species share a radially sulcate cap margin and more or less truncate lamellulae with the inamyloid *Amanita* species. Within the amyloid sections, section *Amidella* may be the most closely related to the sections of subgenus *Amanita*. As in section *Lepidella*, the margin of the cap is usually appendiculate, at least in young stages (see *A. clarisquamosa* (Imai) Imai; Fig. 8.1O,P).

8.2.3.2
Section *Lepidella*

This possibly heterogeneous section contains amyloid *Amanita* species the caps and stipes of which are often markedly ornamented with a variety of different forms of volva remains, from flocks to scales, warts, patches, belts or crusts (*A. solitaria* (Bull.: Fr.) Mérat; Fig. 8.1Q,R). As in species of section *Amidella*, the margin of the cap is appendiculate. In contrast to the other *Amanita* sections, the monophyly of section *Lepidella* could not be shown in molecular studies (Weiß et al. 1998).

8.2.3.3
Section *Phalloideae*

The most poisonous of all fungi belong to this *Amanita* section. *Amanita phalloides*, the death cap (Fig. 8.1S–U), and *A. subjunquillea* Imai contain lethal amatoxines and phallotoxines (Wieland 1986; Kawase et al. 1992).

◄───

Fig. 8.3. Phylogenetic relationships in the genus *Amanita*. Hypothesis derived by neighbor-joining analysis of an alignment over 587 base pairs of nuclear large subunit rDNA from 49 *Amanita* species using Kimura 2-parameter distances. Branch lengths are scaled in terms of expected numbers of nucleotide substitutions per site. Topology was rooted with *Limacella glioderma*. Numbers on branches are bootstrap values (1000 replicates, numbers rounded to next integers, values smaller than 70% not shown). Tree areas corresponding to subgenus *Lepidella* are shaded. For assignment of *Amanita strobiliformis*, *A. pseudoporphyria*, and *A. manginiana* sensu Chiu to sections, see text. (Weiß et al. 1998, with modifications)

Species of section *Phalloideae* possess a conspicuous bulb and a sac-like volva at the stipe base but only seldom sparse volva remains on the cap. Two species, *A. pseudoporphyria* Hongo and *A. manginiana* sensu Chiu, were separated from section *Phalloideae* in molecular phylogenetic analyses (Fig. 8.3). Unlike other *Phalloideae* species, they have inconspicuous bulbs and *A. manginiana* sensu Chiu, at least, is an edible fungus (Yang 1997). It seems that section *Phalloideae* should be restricted to morphologically typical, poisonous fungi to form a natural group.

8.2.3.4
Section *Validae*

Molecular phylogenetic analyses suggest that section *Validae* is the sister group of section *Phalloideae* (Fig. 8.3). Species of this group lack a saccate volva at the stipe base. Volva remains form at most irregular belts at the top of the bulb. Sometimes the volva is circumscissile or limbate, and the bulb is distinctly marginate, as visible in relatives of the false death cap, *A. citrina* (J. C. Schaeff.) Pers. Typically, there are conspicuous volva remains on the cap. In contrast to the species of section *Phalloideae*, no poisonous fungi have been detected in the *Validae* group – on the contrary, many *Validae* species are known to be delicious edible fungi, such as *A. excelsa* (Fr.) Bertillon (Fig. 8.1V,W).

8.2.4
Molecular Phylogenetic Studies

A group of 49 *Amanita* species have been studied by means of DNA sequence comparison in order to estimate phylogenetic relationships in the genus (Weiß et al. 1998). From these organisms, the 5' terminal domain of nuclear DNA coding for the ribosomal large subunit (LSU rDNA) was sequenced and analyzed, including DNA from *Limacella glioderma* (Fr.) R. Maire as an outgroup organism. The result obtained with neighbor-joining analysis is shown in Fig. 8.3. Well-supported groups are sections *Amanita*, *Caesareae*, *Vaginatae*, *Amidella*, *Validae*, and section *Phalloideae* s. str., excluding *A. pseudoporphyria* and *A. manginiana* sensu Chiu. The monophyly of section *Lepidella* was not confirmed. Three of the four *Lepidella* species studied, *A. japonica* Bas, *A. solitaria*, and *A. virgineoides* Bas, were closely linked, but without forming a group of their own; *A. strobiliformis* (Paul. ex Vitt.) Bertillon, the fourth species of this section in the study, was separated from the others and appeared to be linked to section *Phalloideae*. Sequence data for more species of section *Lepidella* have to be analyzed to obtain a more meaningful hypothesis about the phylogenetic relationships in this possibly heterogeneous group.

The molecular studies could not significantly verify or falsify the division of genus *Amanita* into the two subgenera *Lepidella* (areas shaded in Fig. 8.3) and *Amanita*. Although not supported by a good bootstrap value, all of the analyses performed by Weiß et al. (1998) yielded a group containing the amyloid sections *Validae, Phalloideae*, and *Amidella*. At this stage there is no reason to reject the two subgenera which reflect the distribution of morphological characters as well as the amyloidity of the basidiospores.

8.3
Host Specificity and Culture Characteristics

Criteria for the determination of the host specificity of ECM fungi were discussed by Trappe (1962). Based on these, morphological databases for the identification of ECM are available (Agerer 1987–1996; Ingleby et al. 1990; Agerer and Rambold 1996). In addition, molecular techniques allow for identification of ECM fungi from ribosomal DNA sequences (Bruns and Gardes 1993; Gardes and Bruns 1993; see also above). The most comprehensive list of host specificity of ECM fungi including 25 species of *Amanita* was provided by Trappe (1962). Employing the same criteria, this list is extended here to 30 species of *Amanita* in Table 8.2. Each species is assigned to a section according to Yang (1997).

Our knowledge of the host-range of ECM *Amanita* species is still very limited, thus only 30 out of some 400 species are included in this overview. Progress in this matter is also impeded by the lack of suitable culture conditions for many species of *Amanita*. Table 8.3 lists *Amanita* spp. which can be grown in sterile culture. With the exception of members of sections *Amidella* and *Vaginatae*, at least some members of each section can be grown in pure culture. Attempts to culture members of section *Vaginatae* (*A. crocea* (Quél.) Kühner & Romagn, *A. umbrinolutea*, or *A. vaginata* (Bull.: Fr.) Lamarck) on MMN-media have so far failed (M. Guttenberger, unpubl. data). As far as available, data on media composition, growth on different sources of C, N, and P, a suitable temperature range, sensitivity to fungitoxic substances, and conditions of long-term storage are also included. Information given in parentheses indicates limited effects on/responses of the isolate tested or variability between isolates. Contradictory data are probably due to differences between isolates, culture conditions, and evaluation criteria.

8.4
Anatomy and Ultrastructure of ECM

ECM formed by *A. muscaria* on different hosts in vitro show a white, hairy surface (Fig. 8.4a,e) owing to short, septate and sometimes branched emanating hyphae of about 2.5 µm in diameter (Godbot and Fortin 1985; Kottke et al. 1987; Ingleby et al. 1990; Mohan et al. 1993; Cripps and Miller 1995). The prominent hyphal sheath of up to 100 µm in depth is formed by fine interwoven hyphae. In young but well developed ECM, a loose net can be observed on the surface of the hyphal sheath (Fig. 8.4b). Hyphae become gradually more tightly interwoven close to the root cortex (Fig. 8.4c,d). As no specific differentiation occurs, the structure of the hyphal sheath may be considered as plectenchymatic (prosenchymatic) throughout. Clamp connections may be observed occasionally in the surface hyphae (Kottke 1986; Cripps and Miller 1995). Interestingly, a very similar structure was found in the hyphal sheath of ECM formed by *A. pantherina* (D.C.: Fr.) Krombh. (Cripps and Miller 1995) and *A. rubescens* (Pers.: Fr.) S. F. Gray (Kottke 1986), but clamp connections were missing in these types. When ageing, the hyphal sheath of *A. muscaria* ECM may be covered by a layer of moribund hyphae (Fig. 8.4f), resulting in a smooth surface. The hyphae are fixed together by a fibrillar matrix (Fig. 8.4g) which becomes more prominent between the moribund hyphae.

When grown with a high concentration of NH_4^+ in the substrate (e.g. MMN medium) *A. muscaria* ECM may display large amounts of vacuolar, metachromatic material in the hyphal sheath and in the Hartig net (Fig. 8.5a). N and P

Table 8.2. Host plants of *Amanita* spp.

A. aspera (Fr.) Quél. (*Validae*)
  Pseudotsuga menziesii (1), *Tsuga heterophylla* (1)
A. caesarea (Scop.: Fr.) Pers. (*Caesareae*)
  Pinus strobus (1), *Pinus virginiana* (1), *Castanea sativa* (1), *Castanopsis carlesii* (1), *C. hystrix* (1), *Fagus sylvatica* (1), *Quercus baronii* (1), *Q. faginea* (1), *Q. liaotungensis* (1), *Q. lusitanica* (1), *Q. petraea* (1), *Q. pubescens* (1), *Q. pyrenaica* (1), *Q. robur* (1), *Q. suber* (1)
A. ceciliae (Berk. & Br.) Bas (as *A. inaurata* Secr., *Vaginatae*)
  Pinus spp. (1), *Carpinus betulus* (1)
A. chlorinosma (Peck) Sacc. (*Lepidella*)
  Pseudotsuga menziesii (1)
A. citrina (J.C. Schaeff.) Pers. (= *A. mappa*, *Validae*)
  Abies alba (1), *Picea abies* (1, 1), *Pinus mugo* (1), *P. nigra* (1), *P. strobus* (1), *P. sylvestris* (1, 1), *P. virginiana* (1), *Fagus sylvatica* (1), *Quercus robur* (1)
A. crocea (Quél.) Kühner & Romagn. (*Vaginatae*)
  Betula sp. (1)
A. diemii Sing. (*Amanita*)
  Nothofagus dombeyi (1)
A. excelsa (Fr.) Bertillon (*Validae*)
  Picea abies (1)
A. flavorubescens Atk. (*Validae*)
  Pinus strobus (1), *P. virginiana* (1)
A. franchetii (Boud.) Fayod (*Validae*)
  Monotropaceae (2)
A. frostiana (Peck) Sacc. (*Amanita*)
  Pinus virginiana (1)
A. gemmata (Fr.) Bertillon (= *A. junquillea* Quél., *Amanita*)
  P. sylvestris (1), *Pseudotsuga menziesii* (1), *Tsuga heterophylla* (1), *Quercus* sp. (1)
A. gilberti Beaus. (*Amidella*)
  Quercus suber (1)
A. hyperborea Karst. (*Amanita*)
  Salix herbacea (3), *S. retusa* (3)
A. muscaria (L. ex: Fr.) Pers. (*Amanita*)
  Abies alba (1), *A. procera* (4, 5), *Larix decidua* (5, 1, 1), *L. occidentalis* (5), *Picea abies* (1, 1, 6), *P. sitchensis* (5, 5, 1, 6), *Pinus banksiana* (7), *P. cembra* (1), *P. contorta* (5, 5, 1, 6),

Table 8.2. *Continued*

P. echinata (**1**), *P. insignus* (**8**), *P. monticola* (**5**, 5), *P. mugo* (**1**), *P. nigra* (**1**), *P. ponderosa* (**5**, 1), *P. radiata* (**4**, 4, 1), *P. strobus* (**1**, 1), *P. sylvestris* (**1**, 1, 9, **6**), *P. taeda* (**1**), *P. virginiana* (**1**), *Pseudotsuga menziesii* (**5**, 5, 1), *Tsuga heterophylla* (**5**, 5, 1), *T. mertensiana* (**5**), *Arbutus menziesii* (**5**), *Arctostaphylos uva-ursi* (**5**), *Betula alba* (**8**), *B. nana* (**1**), *B. pendula* (**1**, 1, **10**, **11**, **12**), *B. pubescens* (**1**, **6**), *Eucalyptus calophylla* (**4**), *E. camaldulensis* (**4**), *E. dalrympleana* (**4**), *E. diversicolor* (**4**), *E. maculata* (**4**), *E. marginata* (**4**), *E. microcorys* (**4**), *E. obliqua* (**4**), *E. regnans* (**4**), *E. sieberi* (**4**), *E. st.-johnii* (**4**), *Fagus sylvatica* (**1**), *Populus tremula* x *tremuloides* (**13**), *P. trichocarpa* (**5**, 1), *Quercus* spp. (**1**), *Tilia europaea* (**8**), Monotropaceae (<u>2</u>)

A. nivalis Grev. (*Vaginatae*)
  Salix herbacea (**14**)
A. ovoidea (Bull.: Fr.) Link (*Amidella*)
  Pinus nigra (**1**), *Pinus* spp. (**1**), *Prunus spinosa* (**1**)
A. pantherina (D.C.: Fr.) Krombh. (*Amanita*)
  Abies alba (**1**), *Cedrus deodara* (**1**), *C. libanii* (**1**), *Picea abies* (**1**, 1), *P. engelmannii* (**1**), *P. sitchensis* (**1**), *Pinus banksiana* (**1**), *P. contorta* (**1**), *P. mugo* (**1**), *P. nigra* (**1**), *P. ponderosa* (**1**), *P. sylvestris* (**1**, 1), *Pseudotsuga menziesii* (**1**), *Carpinus betulus* (**1**), *Castanea dentata* (**1**), *Fagus sylvatica* (**1**), *Quercus faginea* (**1**), *Q. robur* (**1**), *Q. suber* (**1**), *Tilia cordata* (**1**)
A. phalloides (Fcr.) Link (*Phalloideae*)
  Picea abies (**1**), *Pinus nigra* (**1**), *P. radiata* (**4**), *P. sylvestris* (**1**), *Fagus sylvatica* (**1**), *Quercus faginea* (**1**), *Q. ilex* (**1**), *Q. robur* (**1**), *Q. suber* (**1**), Monotropaceae (<u>2</u>)
A. porphyria (Alb. & Schw.: Fr.) Fr. (*Validae*)
  Picea abies (**1**), *Pinus sylvestris* (**1**)
A. regalis (Fr.) Maire (*Amanita*)
  Pinus sylvestris (**15**), *Betula* sp. (**15**)
A. rubescens (Pers.: Fr.) S. F. Gray (= *A. rubens*, *Validae*)
  Larix laricina (**16**), *Pinus banksiana* (**1**), *P. divaricata* (**16**), *P. mugo* (**1**), *P. nigra* (**1**), *P. strobus* (**1**), *P. sylvestris* (**1**, **17**), *P. virginiana* (**18**, 1), *Castanea sativa* (**1**), *Corylus avellana* (**1**), *Fagus sylvatica* (**1**), *Quercus robur* (**1**), *Q.* sp. (**1**)
A. silvicola Kauff. (*Lepidella*)
  Pseudotsuga menziesii (**1**), *Allocasuarina littoralis* (**19**), *Betula* sp. (**20**), *Casuarina cunninghamiana* (**19**), *Polygonum viviparum* (**20**), *Salix* sp. (**20**)
A. spissa (Fr.) Kumm. (*Validae*)
  Pinus nigra (**1**), *Pinus*, sp (**1**), *Quercus* sp. (**1**)
A. strobiliformis (Paul ex Vitt.) Bertillon (= *A. solitaria*, *Lepidella*)
  Pinus virginiana (**1**), *Pseudotsuga menziesii* (**1**), *Carpinus betulus* (**1**), *Platanus acerifolia* (**1**), *Quercus robur* (**1**), *Quercus* sp. (**1**)
A. umbrinolutea (Secr. ex Gill.) Bataille (*Vaginatae*)
  Picea abies (M. Guttenberger, unpubl.)
A. vaginata (Bull. ex Fr.) Lamarck (*incl. A. malleata*, *Vaginatae*)
  Larix decidua (**1**), *Picea sitchensis* (**1**), *Pinus contorta* (**1**), *P. nigra* (**1**), *P. ponderosa* (**1**), *P. sylvestris* (**1**), *Pseudotsuga menziesii* (**1**), *Tsuga heterophylla* (**1**), *Betula pendula* (**1**), *Betula* sp. (**1**), *Carpinus betulus* (**1**), *Castanea sativa* (**1**), *Fagus sylvatica* (**1**), *Populus* sp. (**1**), *P. tremula* (**1**), *Quercus robur* (**1**), *Q.* sp. (**1**), *Salix herbacea* (**1**)
A. verna (Bull. ex Fr.) Lamarck (*Phalloideae*)
  Pinus strobus (**1**), *P. virginiana* (**1**)
A. virosa Bert. (*Phalloideae*)
  Pinus sylvestris (**1**)
A. vittadinii (Mor.) Vitt. (*Lepidella*)
  Quercus spp. (**1**), not ectomycorrhizal (**21**)

Table 8.3. Culture conditions for *Amanita* spp.

A. bisporigera Atk. (*Phalloideae*)
Media   MMN (22)
Storage   Cold storage in sterile water (22)
A. brunnescens Atk. (*Validae*)
Media   MMN (7, 22)
Temperature   Sensitive to 7°C, insensitive to 30°C (7)
Inhibitors   Tolerates benomyl, (cycloheximide), rose bengal, malachite green (23)
Storage   Cold storage in sterile water (22)
A. caesarea (Scop.: Fr.) Pers. (*Caesareae*)
                 No culture conditions specified (24)
A. citrina (J.C. Schaeff.) Pers. (= *A. mappa*, *Validae*)
Media   B IV (25), malt (26), MMN (7, 22), potato-dextrose (26), Ritter (27), N
                 sources: ammonium, acetamide, diethylamine, D-glucosamine,
                 glycine (27)
Temperature   Does not survive freezing at −10°C, some growth at 2–8°C (25),
                 sensitive to 7°C, insensitive to 30°C (7)
Inhibitors   Tolerates (benomyl), cycloheximide, rose bengal, sodium chloride (23)
Storage   Cold storage in sterile water (22)
A. flavoconia Atk. (*Validae*)
Media   Malt (26), MMN (7, 22), potato-dextrose (26)
Temperature   Sensitive to 7°C, insensitive to 30°C (7)
Inhibitors   Tolerates cycloheximide, rose bengal, malachite green, (sodium
                 chloride) (23)
Storage   Cold storage in sterile water (22)
A. frostiana (Peck) Sacc. (*Amanita*)
                 No culture conditions specified (24)
A. muscaria (L.: Fr.) Pers. (*Amanita*)
Media   B IV (24), Hagem (11, 28, 29, 30), malt (26), malt-extract (29), MMN (4, 5,
                 7, 22, 29, 31), MMNC (13), modified Norkrans (32), Norkrans' casein
                 hydrolysate (29), potato-dextrose (26, 29, 33), PT-5 (33), Ritter (27),
                 Wagner (9), growth on cellobiose > trehalose > fructose > glucose >
                 pectin > mannose > xylose > glycogen > mannitol (30), growth on
                 mannitol > trehalose > glucose = fructose (33), N sources: alanine,
                 peptides (di-alanine to hexa-alanine) (10), ammonium, (nitrate),
                 acetamide, diethylamine, D-glucosamine, glycine, (asparagine) (27)
Temperature   Sensitive to 7°C, (insensitive to 30°C) (7), does not survive freezing at
                 −10°C, some growth at 2–8°C (25)
Inhibitors   Tolerates (rose bengal, NaCl) (23)
Storage   Cold storage in sterile water (22), agar slants at 8°C (34)
A. polypyramis (Berk. & Curt.) Sacc. (*Lepidella*)
Media   Malt (26), potato-dextrose (26)
A. regalis (Fr.) Maire (*Amanita*)
Media   BW-medium, pH 4.6 (15)
A. rubescens (Pers.: Fr.) S. F. Gray (= *A. rubens*, *Validae*)
Media   B IV (25), Hagem (18, 30), Lindeberg (18), malt (26), MMN (7, 16, 17, 22),
                 Pachlewski (35), potato-dextrose (26), Ritter (27), growth on cellobiose
                 > trehalose > fructose > glucose > mannose > pectin > glycogen >
                 starch > mannitol (30), N sources: ammonium, nitrate, acetamide,
                 diethylamine, D-glucosamine, glycine, asparagine (27), growth on
                 inositol hexaphosphate as a source of phosphorus (17)
Temperature   Does not survive freezing at −10°C (25), sensitive to 7 and 30°C (7),
                 optimum at 29°C, maximum 32°C (18)

Table 8.3. *Continued*

Inhibitors	Tolerates (cycloheximide), rose bengal, malachite green, (sodium chloride) (23)
Storage	Cold storage in sterile water (22)
A. spissa (Fr.) Kumm. (*Validae*)	
Media	B IV (25)
Temperature	Does not survive freezing at −10°C (25)
A. velatipes Atk. (*Amanita*)	
Media	Malt (26), potato-dextrose (26)
A. virosa Bertillon (*Phalloideae*)	
Media	B IV (25)
Temperature	Suffers from freezing at −10°C (25)

References: boldface numbers indicate evidence from pure culture synthesis of ECM; under-lined numbers refer to molecular data. 1 Trappe (1962), 2 Cullings et al. (1996), 3 Kuhner (1972), 4 Malajczuk et al. (1982), 5 Molina and Trappe (1982), 6 Read et al. (1985), 7 Hutchison (1990b), 8 Watkinson (1964), 9 Rudawska et al. (1994), 10 Abuzinadah and Read (1989), 11 Mason (1975), 12 Mason et al. (1987), 13 Hampp et al. (1996), 14 Watling (1988), 15 Maijala et al. (1991), 16 Samson and Fortin (1986), 17 Antibus et al. (1992), 18 Hacskaylo et al. (1965), 19 Theodorou and Reddell (1991), 20 Gardes and Dahlberg (1996), 21 Bas (1969), 22 Richter and Bruhn (1989), 23 Hutchison (1990a), 24 Hacskaylo and Palmer (1955), 25 Moser (1958), 26 Campbell and Petersen (1975), 27 Lundeberg (1970), 28 Brown and Wilkins (1985), 29 Marx (1969), 30 Lamb (1974), 31 Molina (1981), 32 Ramstedt et al. (1987), 33 Taber and Taber (1987), 34 Guttenberger (1989), 35 El-Badaoui and Botton (1989).

were detected in the vacuolar material using electron-energy-loss spec-troscopy (EELS). N, which was first detected in the vacuoles of *Cenococcum geophilum–Pinus sylvestris* (Kottke et al. 1995), has been found recently to be stored in diverse ECM of *Picea abies* collected in forest soil (Kottke et al. 1998; Beckmann et al. 1998) and in mycelial cultures of *Xerocomus badius* (Fr.) Kühn. ex Gilb. and *Suillus bovinus* (L.: Fr.) O. Kuntze (Bücking et al. 1998). Although the vacuolar material may change in appearance during cryofixation (Orlovich and Ashford 1993; I. Kottke, unpubl. data) the gran-ules were present in the vacuoles of living hyphae, and P and N were detected independently of the fixation techniques (Bücking et al. 1998). We thus con-clude that *A. muscaria–Picea abies* ECM store nitrogen-containing com-pounds in large amounts in the presence of a surplus of NH_4^+.

Ultrastructural studies of *A. muscaria–P. abies* ECM revealed coenocytic and transfer cell-like structures in the Hartig net (Kottke and Oberwinkler 1987, 1989). Finger-like branching and rare septation of the hyphae, and intimate juxtaposition of the hyphal and cortical cell walls are also exhib-ited (Fig. 8.5a). The hyphal growth between the cortical cells is primarily endo-dermis oriented (Fig. 8.5a). Together with the rare septation of the hyphae, a channel-like system is initiated where solutes may be transported rapidly in both directions. Rough endoplasmic reticulum was found to be stretched in the supposed direction of the solute flow (Fig. 8.5a). The close contact of hyphal and cortical cell walls creates a shared apoplasm where solutes may be exchanged between the partners in both directions. This structure is

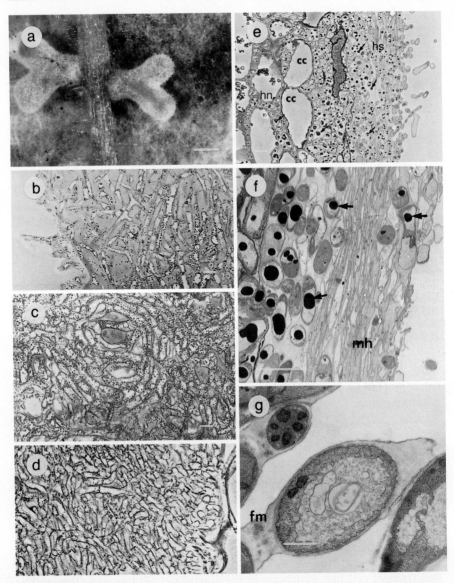

Fig. 8.4. Mycorrhizal structures I. **a** *Amanita muscaria–Pinus sylvestris* mycorrhizas obtained in axenic Petri dish cultures. Note the hairy surface. **b** Surface view of the net-like hyphal sheath. **c** Tangential section through the plectenchymatic hyphal sheath. **d** Tangential section through the plectenchymatic hyphal sheath close to the root cortex. **e** Longitudinal section through an *Amanita muscaria–Picea abies* mycorrhiza grown on MMN-agar plates. Note the short emanating hypha and metachromatic, vacuolar material (*arrows*) in the hyphal sheath (*hs*) and Hartig net (*hn*); *cc* cortical cell. **f** Cross section through the hyphal sheath of an *Amanita muscaria–Picea abies* mycorrhiza. The surface is covered by a layer of dying hyphae (*mh*). The living hyphae display vacuolar, metachromatic material (*arrows*). **g** Fibrillar matrix (*fm*) between hyphae of the sheath. Bars: **a** 100 μm; **b–f** 10 μm; **g** 1 μm

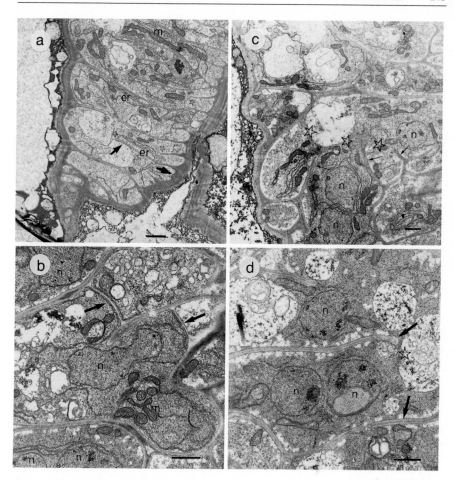

Fig. 8.5. Mycorrhizal structures II. **a** Ultrastructure of the Hartig net; multibranched hyphae growing toward the endodermis (from *left* to *right*), in intimate position to each other and to the walls of active cortical cells (*arrows*); septa are frequently absent (*asterisk*); mitochondria (*m*) and rough endoplasmic reticulum (*er*) are stretched in the assumed direction of solute flow. **b** True septum in the Hartig net (*arrow*) aligned by rough ER; *n* nuclei, *m* mitochondria. **c** and **d** Multinucleate, coenocytic compartments in the Hartig net; *n* nuclei, lack of septa (*asterisk*), intimate position of fungal cell walls (*arrows*). Bars: 1 μm

reminiscent of transfer cells found in plant tissues where fast, short-distance transport across membranes takes place (Gunning and Pate 1974). Septa are frequently absent between the dikaryotic hyphal cells. This results in a multinucleate, coenocytic Hartig net (Fig. 8.5c,d). Comparison of the ultrastructure of true septa (Fig. 8.5b) and of hyphae in the Hartig net (Fig. 8.5c) clearly shows that there are no "incomplete" septa (Kottke et al. 1987).

8.5
Physiology of *Amanita muscaria*

8.5.1
Quantification of the Fungus

In order to compare partner-specific properties in an ECM, it is necessary to have a measure of the relative abundance of both. Ergosterol is the most commonly used parameter to quantify the amount of fungal material in ECM. In *A. muscaria* its extraction and recovery is highly reproducible. Using HPLC, the lower limit of detection is below 0.1 µg. Reported values for the ergosterol content of mycelia of *A. muscaria* are between 1.3 and 17.6 mg g^{-1} DW. Based on the amount of ergosterol, fully developed *A. muscaria* ECM with *P. abies* share the dry weight between both partners at about equal amounts (Wallenda 1996). If partner-specific ribosomal RNA is taken as a measure, the ratio of fungus : root is also about 1 : 1. A comparison of ergosterol and chitin contents (another marker of fungal material) of a wider range of ECM was performed by Ekblad et al. (1998). These authors showed that the amount of ergosterol present is closely correlated with the viability of the fungal partner, yielding highest amounts in young and turgid ECM, while the amount of chitin was less dependent on ageing, and probably physiological activity. Thus, the ratio of chitin : ergosterol ranged from <20 to>100 : 1 in young and ageing ECM, respectively, and is suggested to be used as a measure for total and living fungal biomass.

8.5.2
Carbon Metabolism

8.5.2.1
Carbon Requirement for Growth

Carbohydrate dependency of ECM fungi has been tested by their ability to grow on media containing specific carbon sources. Major C sources are glucose and fructose (Jennings 1995). An exogenous supply of glucose has been shown to support ECM infection in a range of species (Theodorou and Reddell 1991). *Amanita muscaria* can obviously make no direct use of sucrose, while glucose and fructose are readily consumed (Salzer and Hager 1991). Apoplast invertase located in the host cell wall appears to be the only physiological means of making sucrose available to this fungus during symbiotic interaction. Although important for assimilate transfer, the activity of acid invertase in the *P. abies–A. muscaria* system is not related to the degree of fungal infection (Schaeffer et al. 1995). Rates of sucrose hydrolysis in non-ECM roots are, however, obviously high enough to support symbiotic requirements (Schaeffer et al. 1995). As a late-stage fungus, *A. muscaria* can also make use of proteins and peptides (Abuzinadah and Read 1986). Experiments with casein hydrolysate indicated, however, that *A. muscaria* can use this broth as a source of N (evidenced by increased protein content) but not of C (evidenced by reduced growth). Organic and amino acids such as citrate, pyruvate, glutamate or aspartate did not support fungal growth at all. Obviously this fungus lacks the ability to perform gluconeogenesis (Wingler 1995). With regard to organic acids, similar results were reported for *A. rubescens* (Palmer and Hackskaylo 1970).

8.5.2.2
Absorption of Carbohydrates

The main transport form for photoassimilates in higher plants is sucrose (Ziegler 1975; Giaquinta 1983). Sucrose unloading to root tissue has been shown for ECM-forming conifers (Komor 1983). The products of sucrose hydrolysis, glucose and fructose are obviously not absorbed at the same rates. A direct assay of sugar absorption employing [^{14}C]-labelled substrates and protoplasts from *A. muscaria* delivered K_M values for absorption of glucose and fructose of 1.25 and 11.3 mM, respectively. In addition, glucose uptake was only marginally affected by fructose, while glucose was highly competitive with regards to fructose (Chen and Hampp 1993).

For a further characterization of this monosaccharide uptake system a molecular approach was used (Nehls et al. 1998). Primers, designed against conserved regions of known fungal monosaccharide transporters, were used to amplify cDNA fragments from *P. abies–A. muscaria* ECM. These DNA fragments, in turn, were used to isolate full length cDNA clones from an ECM cDNA library. To date, one fungal gene coding for a monosaccharide uptake system has been identified. The full length cDNA clone (AmMST1) contains an open reading frame which codes for a protein of 520 amino acids with a molecular mass of 56455 daltons. The protein sequence revealed 12 putative transmembrane domains, arranged in two groups of six, each separated by a hydrophilic spacer of 65 amino acids, a pattern typical for the superfamily of transmembrane facilitators (Marger and Saier 1993). The best homology for AmMST1 was obtained with the *RCO3* monosaccharide transporter gene of *Neurospora crassa* (Madi et al. 1997) and to a lesser extent also to *SNF3* (Celenza et al. 1988) and *RGT2* (Özcan et al. 1996) from yeast.

Re-screening of the ECM cDNA library using the AmMST1 cDNA as a probe as well as using a different set of primers resulted in the identification of the same AmMST1 gene only. Furthermore, no additional monosaccharide transporters homologous to AmMst1 were detected by Southern blot analysis. We therefore conclude that AmMST1 represents the main, if not the only, monosaccharide import system for *A. muscaria* in ECM as well as for mycelia grown at elevated external glucose concentrations. This does not exclude the presence of further monosaccharide transporters expressed at a low level in ECM or at low external monosaccharide concentrations which could be of only limited identity to AmMst1. In comparison, in yeast, more than 20 monosaccharide transporters are encoded in the genome, and a number of these genes have been identified by using heterologous yeast probes (Reifenberger et al. 1995). Monosaccharide uptake systems are also present in other fungi (for a review see Jennings 1995), but in most cases the actual number of monosaccharide transporters is unknown. The function of the AmMST1 protein as an active monosaccharide transporter could be confirmed by heterologous expression of the full length cDNA in a yeast mutant, lacking a functional endogenous monosaccharide uptake system (Nehls et al. 1998; J. Wiese, unpubl. data). The uptake properties of the AmMST1 monosaccharide transporter resemble those obtained with *A. muscaria* protoplasts (J. Wiese, unpubl. data), again indicating that AmMST1 is the main monosaccharide uptake system.

A. muscaria hyphae grown at glucose concentrations of up to 5 mM expressed the AmMst1 gene at a basal level, while monosaccharide concentrations above 5 mM triggered a four-fold increase in the amount of the AmMst1 transcript that could not be further enhanced. Thus we assume that this change in AmMst1 expression is regulated by a threshold response mechanism depending on the extracellular concentration of monosaccharides. An increase in AmMST1 expression, comparable to that found in fungal mycelia cultivated at elevated monosaccharide concentrations, was also observed in symbiotic mycelia of ECM formed with both the gymnosperm *P. abies* and the angiosperm *Populus tremula* x *tremuloides* as plant partners. These data therefore suggest that the enhancement of AmMST1 expression in ECM is also sugar regulated in response to the in vivo concentration of monosaccharides at the fungus–plant interface (of at least 5 mM). In comparison, in the apoplast of barley leaves, glucose and fructose concentrations of approx. 5 mM each were obtained (Lohaus et al. 1995). While comparable data for roots are not available, the actual apoplastic concentration in root cell walls could be within the effective range. Glucose contents of *P. abies–A. muscaria* ECM were between 20 and 30 nmol mg^{-1} DW (Hampp et al. 1995), which can be calculated as an overall concentration of ca. 10–15 mM.

The increase in AmMST1 expression is a slow process. A transition from basal to maximal rate of expression of AmMST1 occurred within 18 h to 1 day of fungal culture. In yeast, in comparison, it takes only 90 min for the induction of the high affinity glucose transport system Hxt2 (Bisson and Fraenkel 1984). The rather slow response of *A. muscaria* hyphae is possibly due to the fact that soil-growing mycelia are exposed to low concentrations of carbohydrates (Wainwright 1993). Higher monosaccharide concentrations would only be encountered locally and for short periods. Under such conditions, monosaccharides could be exploited by the basal rate of AmMst1 expression and/or by a monosaccharide uptake system preferentially expressed at low external monosaccharide concentrations. In contrast to soil-growing hyphae, hyphae at the symbiotic interface are exposed to a continuous supply of plant-derived carbohydrates, delivering up to 30% of the photosynthetically fixed carbon (Söderström 1992). To maximize the carbohydrate flow into the fungus, monosaccharides must be quickly taken up and either metabolized or converted into compounds for intermediate or long-term storage (e.g. trehalose, mannitol, glycogen; see above). It could thus be assumed that both the extended lag phase for enhanced AmMST1 expression and its threshold response to elevated monosaccharide concentrations are adaptations of the ECM fungus to the conditions found at the symbiotic interface.

The uptake capacity of the *Amanita* monosaccharide transporter (K_M = 1.25 mM; Chen and Hampp 1993) is nearly at its maximum at ca. 5 mM glucose (threshold concentration for enhanced expression). This means that the fungus somehow recognizes when sugar transport is saturated. In due course the expression of AmMST1 is increased, resulting in additional transporter proteins in the plasma membrane and thus in a further increase of the monosaccharide import capacity of the hyphae. This type of regulation of AmMST1 expression results in a high capacity for monosaccharide uptake of the fungal hyphae in symbiosis and should thus significantly enhance the sink strength of the root system for carbohydrates in ECM compared to non-ECM plants.

8.5.2.3
Content of Sugars and Sugar Alcohols

Polyols such as glycerol, mannitol, and arabitol, in addition to trehalose and glycogen, have been detected in ECM (Bevege et al. 1975) and were labelled after feeding individual host plants with $^{14}CO_2$ (Söderström et al. 1988). Basidiomycetes such as *A. muscaria* preferably accumulate trehalose (a disaccharide) when grown in liquid culture. This is independent of the carbohydrate source (glucose or fructose; Wallenda 1992). Mannitol, a sugar alcohol accumulated in ascomycetes, is only present in traces. Highest amounts of glucose (up to 1300 nmol mg^{-1} DW) and trehalose (up to 660 nmol mg^{-1} DW) were detected in sporocarps of *A. muscaria* (Wallenda 1996).

8.5.2.4
Trehalose Metabolism

Trehalose is a dominant sugar in basidiomycetes such as *A. muscaria*. Despite its metabolic importance, there is little information about its metabolism. Trehalose-6-phosphate synthase is the key enzyme of trehalose synthesis. This enzyme catalyzes the formation of trehalose-6-phosphate from UDPglucose and glucose-6-phosphate. Its pH optimum in *A. muscaria* is 6.5 (Wallenda 1996), which indicates a cytosolic location. With regard to the hydrolysis of trehalose, there are mainly reports on the involvement of an acid trehalase. In *A. muscaria* this enzyme has a pH optimum of about 3.8 and an apparent K_m value of approx. 0.13 mM (Wallenda 1996; G. Wisser, unpubl.). Similar data are reported for non-ECM-forming fungi (see Wallenda 1996). The activity detected in *A. muscaria* from liquid culture is mainly apoplastic. A vacuolar

acid or cytosolic neutral trehalase has to be postulated, assuming that trehalose is not released from cells, hydrolyzed outside and reimported as
hexoses. We have, however, been unable to detect an intracellular trehalase of
sufficient activity (G. Wisser, unpubl.). The failure of attempts to detect neutral
trehalase in other fungi has been taken to suggest that the enzyme is labile
(D'Enfert 1997).

8.5.2.5
Regulation of Glycolysis

When mycelia of *A. muscaria* are cultured on glucose (2% initial concentration) for 3 weeks, a series of transients in metabolite pools can be observed
(Hoffmann et al. 1997). Absorption of glucose leads to an acidification of the
incubation medium (from pH >6.0 to <3.0). During a linear increase in fungal
dry weight between 5 and 12 days of culture, trehalose increases and reaches
a maximum when growth is halted due to a sharp decrease in external glucose.
From this point onwards trehalose is hydrolysed in order to deliver hexoses
for fungal metabolism. During the whole period, the level of glycogen is
relatively constant. This implies that glycogen constitutes a carbohydrate
reserve which will be only metabolized when other carbon sources are no
longer available. With regard to the regulation of fungal carbohydrate metabolism, the change in pool sizes of fructose-2,6-bisphosphate and of cyclic AMP
(cAMP) is of interest (Fig. 8.6). The highest content of cAMP occurred after
4 days, shortly before the transition into the phase of exponential growth.
With the onset of exponential growth, and with some lag in relation to cAMP,
the content of F26BP in the mycelium started to increase.

To some extent, this series of events resembles observations made during
the growth of yeast cells on glucose (François et al. 1984). Glucose added to a
suspension of *Saccharomyces cerevisiae* in stationary phase causes a transient
increase in the concentration of cAMP and a more persistent increase of
F26BP. In yeast, as in animal cells, F26BP is a positive effector of phospho-
fructokinase-1 and an inhibitor of fructose-1,6-bisphosphate phosphatase
(Hers and van Schaftingen 1982). Thus, an increase in the concentration of
F26BP will stimulate glycolysis and inhibit gluconeogenesis. In yeast, the effect
of glucose is mediated by the successive activation of adenylate cyclase
and of cAMP-dependent protein kinase, and by the phosphorylation of
PFK-2 by the latter enzyme. This causes an activation of PFK-2 and thus an
increase in the pool size of F26BP. From our data with *A. muscaria* we
suggest the same sequence for regulation of glycolysis in ECM-forming fungi
(Fig. 8.6). Upon transfer to glucose-containing medium and shortly before initiation of exponential growth we find a transient increase in cAMP, followed
by a more persistent increase in the amount of F26BP. As fungal PFK-1 is activated by small increases in the amount of F26BP (Schaeffer et al. 1996), glycolysis should be stimulated under these conditions. The parallel decrease in
the amount of glucose in the medium supports this assumption. However, the
rate of uptake of glucose by fungal cells, which has been shown to have a V_{max}

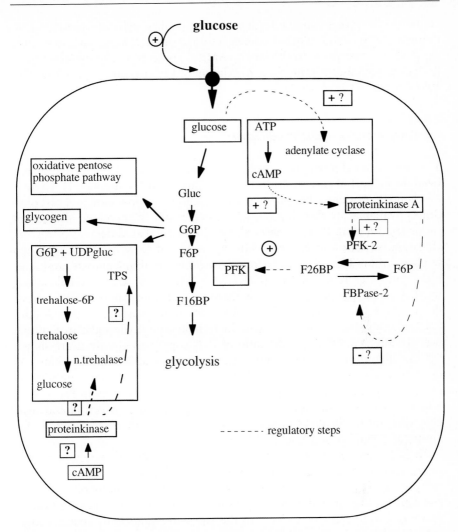

Fig. 8.6. Regulation of carbohydrate metabolism/glycolysis in *Amantia muscaria*. The scheme combines data obtained from *A. muscaria* cultures/mycorrhizas with regulatory steps identified in yeast, and thus constitutes primarily a working hypothesis. *cAMP* cyclic AMP; *FBP* fructose bisphosphate; *F6P/G6P* hexosemonophosphates; *gluc* glucose; *PFK* phosphofructokinase; (−)/(+), inhibiting/activating

of about 18 pmol $(10^6$ protoplasts$)^{-1}$ min^{-1} (Chen and Hampp 1993), obviously exceeds the substrate needs of glycolysis and allows for the storage of carbon in the form of trehalose and glycogen.

Thus, in contrast to the results for *S. cerevisiae* (François et al. 1987), an increase in trehalose and glycogen contents can already be seen during exponential growth of *A. muscaria*. In yeast, these fungal carbohydrates do not accumulate before the transition to the stationary phase. Therefore, with respect to the regulation of enzymes that metabolize trehalose and glycogen, there may also be differences between ectomycorrrhizal fungi such as *A. muscaria* and *S. cerevisiae*. In the latter, cAMP is involved in the metabolism of both carbohydrates via the phosphorylation of key enzymes (e.g. Hardy et al. 1994; Fernandez et al. 1996).

8.5.2.6
Carbohydrate Metabolism and Nutrient Supply

The interaction of N supply, carbohydrate availability, and ECM formation has been the subject of several studies (see Wallander and Nylund 1991). There is biochemical evidence that plants with an ample N supply have less carbohydrates available for distribution, which can result in increased shoot:root ratios (Radin et al. 1978; Stroo et al. 1988) and decreased degrees of mycorrhization. Variation of N supply affects the use of C by *A. muscaria*. When hyphae, cultured for 10 days in a liquid medium containing glucose, were supplied with N as NH_4^+ [330 mg $(NH_4)_2SO_4 l^{-1}$], the pool size of trehalose was most affected. The presence of N without an additional supply of a C source decreased trehalose by 50% within 24 h. Even in the presence of glucose (10 g l^{-1}), there was still a 25% decrease in the trehalose pool. Glycogen, in contrast, was not affected. Effects of the ratio of C:N supply on glycolysis are shown by the alteration of fungal pool sizes of fructose-2,6-bisphosphate. As described above, this compound activates fungal PFK but inhibits its counterpart, fructose-1,6-bisphosphatase (Schaeffer et al. 1996). Fructose-2, 6-bisphosphate increased significantly upon addition of N, which should favor glycolysis over gluconeogenesis, and hence the supply of C skeletons for amino acid synthesis (Wallenda 1996).

In ECM, the fungus is mainly affected by the reduced supply of host C under increased N supply. Under these conditions, N assimilation preferentially occurs in host tissues and only reduced amounts of photoassimilates are available for the fungal partner. This leads to a severe reduction in ergosterol and more so in trehalose, which in the end will significantly reduce fungal colonization of fine roots (Wallenda 1996). Under limited N supply, the partners of *Picea. abies–A. muscaria* ECM obviously share basic pathways necessary for amino acid synthesis. For N assimilation into glutamate, 2-oxoglutarate has to be provided. 2-Oxoglutarate is most probably synthesized from citrate by cytosolic aconitase and NADP-dependent isocitrate dehydrogenase (Chen and Gadal 1990). Consequently, citrate is withdrawn from the tricarboxylic acid cycle during N assimilation. This requires replenishment of cycle intermediates by means of oxaloacetate formation in anaplerotic (filling) reactions. Anaplerosis is possible via three enzymes, namely pyruvate carboxylase (PC), phosphoenolpyruvate carboxykinase (PEPCK) and phosphoenolpyruvate carboxylase (PEPC). PC and PEPCK are active in *A. muscaria*, while PEPC could only be detected in host tissue (Wingler et al. 1996). PC activity in *A. muscaria* was generally low (ca. 10 nmol mg^{-1} protein min^{-1}; pH optimum around 7.0; Wingler 1995). PEPCK was 25 times more active than this and had a similar pH optimum. The $K_{m\ (ADP)}$ of PEPCK was 92 µM, and that for phosphoenol pyruvate (PEP) was 1.2 mM, with a sigmoidal increase of activity with increasing PEP concentrations (Wingler 1995). This is probably due to the slow aggregation of subunits. ECM development on short roots decreased the activity of host PEPC by more than 75%, although "dilution" by the fungal biomass in ECM was only 35%. This reduction in activity was paralleled by a decreased content of PEPC protein. $^{14}CO_2$ labelling, on the other hand, revealed that in vivo CO_2 fixation was higher in ECM compared to non-ECM short roots (Wingler et al. 1996). We thus concluded that fungal carboxylases (most probably PEPCK) are important for anaplerotic carbon fixation during N assimilation in *A. muscaria* ECM of *P. abies*.

8.6
Influence on Host Plant Carbon Economy

8.6.1
Host Sucrose Metabolism

Sucrose production in source leaves depends on the rate of assimilate consumption in sink organs. There is experimental evidence that leaf photosynthesis of ECM seedlings can exceed that of non-ECM ones (Dosskey et al. 1991), and it has been shown that the formation of ECM can create a substantial increase in root sink strength (e.g. Hampp and Schaeffer 1995). In the *P. abies/A. muscaria* system this clearly affects the capacity for sucrose formation in source leaves by increasing the activation state of sucrose phosphate synthase and by decreasing the amount of fructose-2,6-bisphosphate, which releases the inhibition of fructose bisphosphatase (Hampp et al. 1995; Loewe et al. 1999).

8.6.2
Carbohydrate Pools in Mycorrhizal Roots

Mycorrhization of *P. abies* with *A. muscaria* alters carbohydrate pools of fine roots. The total amount of sucrose + glucose + fructose is about 30% higher in non-ECM fine roots, sucrose being the dominant sugar in both ECM and non-ECM fine roots (Rieger et al. 1992). A histochemical quantification of metabolite pools in different sections of mycorrhized root tips (*P. abies/ A. muscaria*) exhibited a distinct gradient. Compared with non-ECM short roots, the ECM samples always had significantly lower levels of sucrose in zones where the ergosterol content was highest. Trehalose, the fungus-specific disaccharide, behaved exactly opposite (Hampp et al. 1995). We take this as evidence that the conversion of host- into fungus-specific carbohydrates aids carbon allocation to the fungus by both sequestration and increasing the sucrose gradient between carbohydrate sources and the root sink in the host.

8.6.3
Effects of Elevated CO_2

The response of plants, and especially ECM-forming trees, toward elevated CO_2 depends on nutrient availability and the degree of utilization of photoassimilate. There are at least short-term responses of metabolism and carbon allocation (Eamus and Jarvis 1989; Mousseau and Saugier 1992), which apply mainly to plants with limited sink capacities. These plants do not take advantage of an increased supply of CO_2 because there is a feedback regulation (decreased gene expression) of carbon fixation by accumulating sugars (Stitt 1991; Koch 1996). ECM plants which are depleted of photoassimilates by their fungal partner should therefore perform better under conditions of increased CO_2. Indeed, seedlings of Pinus echinata exhibited a higher degree of root colonization by ECM fungi when grown under elevated CO_2 (Norby et al. 1987). Similarly, pine seedlings inoculated with the ECM-forming fungus Pisolithus tinctorius exhibited a much faster fungal growth under elevated CO_2 ($600\,\mu mol\,mol^{-1}$), while no effect on shoot biomass was found (Ineichen et al. 1995). ECM formed with A. muscaria showed only marginal CO_2-dependent responses. ECM formation stimulated net assimilation rate of Picea. abies and aspen plantlets and increased the capacity for sucrose formation via the enhanced activation state of sucrose phosphate synthase (key enzyme of sucrose formation). This was in parallel to decreased contents of starch and of fructose-2,6-bisphosphate, both indicative of a diversion of photoassimilates into sucrose formation (see above and Loewe et al. 1999). This response was largely independent of CO_2 supply. Increased CO_2 partial pressure, however, caused an increase in the amount of leaf starch. This implies that under these conditions the ECM seedlings could not use up all of the additional C supply, which caused some backing up of photoassimilates and their intermediate storage as starch. Such an effect might, however, depend on the "agressiveness" of the fungal strain.

8.6.4
Energy Metabolism

Trials to assess pool sizes of adenylates along fine roots at different degrees of ECM interaction were performed with ECM established between P. abies and A. muscaria (Namysl et al. 1991). Using freeze-dried material and dissecting techniques (see Rieger et al. 1992), four zones from tip to base of controls and infected fine roots were distinguished and compared. In this case, the ratio of ATP : ADP of zones involved in mutual interaction between both partners (Hartig net region) was higher compared to non-infected samples. This is in accordance with the assumption of an increased energy need in this interactive region and the higher rates of respiration found in ECM roots in general.

Acknowledgments. As far as our own results are presented, we gratefully acknowledge financial support from the Deutsche Forschungsgemeinschaft, the Bundesminister für Bildung, Wissenschaft, Forschung und Technologie (BMBF), the Projekt Europäisches Forschungszentrum für Maßnahmen zur Luftreinhaltung (PEF), and the Landesforschungsförderungsprogramm Baden-Württemberg.

References

Abuzinadah RA, Read DJ (1986) The role of proteins in the nitrogen nutrition of ectomycorrhizal plants. I. Utilisation of peptides and proteins by ectomycorrhizal fungi. New Phytol 103:481–493

Abuzinadah RA, Read DJ (1989) The role of proteins in the nitrogen nutrition of ectomycorrhizal plants. IV. The utilization of peptides by birch (*Betula pendula* L.) infected with different mycorrhizal fungi. New Phytol 112:55–60

Agerer R (ed) (1987–1996) Colour atlas of ectomycorrhizae. Einhorn Verlag, Schwäbisch Gmünd

Agerer R, Rambold G (1996) DEEMY v. 1.0 – a DELTA-based system for characterization and DEtermination of EctoMYcorrhizae. Inst Syst Bot, Sect Mycol, University of München, München

Antibus RK, Sinsabaugh RL (1993) The extraction and quantification of ergosterol from ectomycorrhizal fungi and roots. Mycorrhiza 3:137–144

Antibus RK, Sinsabaugh RL, Linkins AE (1992) Phosphatase activities and phosphorus uptake from inositol phosphate by ectomycorrhizal fungi. Can J Bot 70:794–801

Bas C (1969) Morphology and subdivision of *Amanita* and a monograph of its section *Lepidella*. Persoonia 5:285–579

Bas C, de Meijer AAR (1993) *Amanita grallipes*, a new species in *Amanita* subsection *Vittadiniae* from southern Brazil. Persoonia 15:345–350

Beckmann S, Haug I, Kottke I, Oberwinkler F (1998) N-Speicherung in den Mykorrhizen der Fichte. In: Raspe S, Feger KH, Zoettl HW (eds) Ökosystemforschung im Schwarzwald. Auswirkungen von atmogenen Einträgen und Restabilisierungsmaßnahmen auf den Stoffhaushalt von Fichtenwäldern. Verbundprojekt ARINUS. ecomed-Verlag, Landsberg, pp 325–335

Bevege DI, Bowen GD, Skinner MF (1975) Comparative carbohydrate physiology of ecto- and endomycorrhizas. In: Sanders FE, Mosse B, Tinker PB (eds) Endomycorrhizas. Academic Press, London, pp 149–174

Bisson LF, Fraenkel DG (1984) Expression of kinase-dependent glucose uptake in *Saccharomyces cerevisae*. J Bacteriol 159:1013–1017

Brown MT, Wilkins DA (1985) Zinc tolerance of *Amanita* and *Paxillus*. Trans Br Mycol Soc 84:367–369

Bruns TD, Gardes M (1993) Molecular tools for the identification of ectomycorrhizal fungi – taxon-specific oligonucleotide probes for suilloid fungi. Mol Ecol 2:233–242

Bücking H, Beckmann S, Heyser W, Kottke I (1998) Elemental contents in vacuolar granules of ectomycorrhizal fungi measured by EELS and EDXS. A comparison of different methods and preparation techniques. Micron 29:53–61

Campbell MP, Petersen RH (1975) Cultural characters of certain *Amanita* taxa. Mycotaxon 1:239–258

Celenza JL, Marshall-Carlson L, Carlson M (1988) The yeast SNF3 gene encodes a glucose transporter homologous to the mammalian protein. Proc Natl Acad Sci USA 85:2130–2134

Chen RD, Gadal P (1990) Structure, functions and regulation of NAD and NADP dependent isocitrate dehydrogenases in higher plants and in other organisms. Plant Physiol Biochem 28:411–427

Chen X-Y, Hampp R (1993) Sugar uptake by protoplasts of the ectomycorrhizal fungus *Amanita muscaria*. New Phytol 125:601–608

Corner EJH, Bas C (1962) The genus *Amanita* in Singapore and Malaya. Persoonia 2:241–304

Cripps CL, Miller OK (1995) Ectomycorrhizae formed in vitro by quaken aspen: including *Inocybe lacera* and *Amanita pantherina*. Mycorrhiza 5:357–370

Cullings KW, Szaro TM, Bruns TD (1996) Evolution of extreme specialisation within a lineage of ectomycorrhizal epiparasites. Nature 379:63–66

D'Enfert C (1997) Fungal spore germination: insights from the molecular genetics of *Aspergillus nidulans* and *Neurospora crassa*. Funct Gen Biol 21:163–172

Dosskey MG, Boersma L, Linderman RG (1991) Role for the photosynthate demand of ectomycorrhizas in the response of Douglas fir seedlings to drying soil. New Phytol 117:327–334

Eamus D, Jarvis PG (1989) The direct effects of increase in global atmospheric CO_2 concentration on natural and commercial temperate trees and forests. Adv Ecol Res 19:1–55

Ekblad A, Wallander H, Näsholm T (1998) Chitin and ergosterol combined to measure total and living fungal biomass in ectomycorrhizas. New Phytol 138:143–149

El-Badaoui K, Botton B (1989) Production and characterization of extracellular proteases in ectomycorrhizal fungi. Ann Sci For 46 (Suppl):728s–730s

Eugster CH (1968) Wirkstoffe aus dem Fliegenpilz. Naturwissenschaften 55:305–313

Fernandez J, Soto T, Vicente-Soler J, Cansado J, Gacto M (1996) Inhibition by polyols of the heat-shock-induced activation of trehalase in the yeast *Zygosaccharomyces rouxii*. Biochem Mol Biol Int 38:43–50

François J, van Schaftingen E, Hers H-G (1984) The mechanism by which glucose increases fructose 2,6-bisphosphate concentration in *Saccharomyces cerevisiae*. Eur J Biochem 145:187–193

François J, Eraso P, Gancedo C (1987) Changes in the concentration of cAMP, fructose 2,6-bisphosphate and related metabolites and enzymes in *Saccharomyces cerevisiae* during growth on glucose. Eur J Biochem 164:369–373

Fries E (1821) Systema mycologicum I. Gryphiswaldiae

Gardes M, Bruns TD (1993) ITS primers with enhanced specificity for basidiomycetes – application to the identification of mycorrhizae and rusts. Mol Ecol 2:113–118

Gardes M, Dahlberg A (1996) Mycorrhizal diversity in arctic and alpine tundra: an open question. New Phytol 133:147–157

Giaquinta RT (1983) Phloem loading of sucrose. Annu Rev Plant Physiol 34: 347–387

Godbout C, Fortin JA (1985) Synthesized ectomycorrhizae of aspen: fungal genus level of structural characterization. Can J Bot 63:252–262

Gunning BE, Pate J (1974) Transfer cells. In: Robards AW (ed) Dynamic aspects of plant ultrastructure. McGraw Hill, New York, pp 441–480

Guttenberger M (1989) Untersuchungen zur Biochemie der Pilz-Baumwurzel-Symbiose: Proteinanalytik im Mikromaßstab. PhD Thesis, University of Tübingen, Tübingen

Hacskaylo E, Palmer JG (1955) Hymenomycetous species forming mycorrhizae with *Pinus virginiana*. Mycologia 47:145–147

Hacskaylo E, Palmer JG, Vozzo JA (1965) Effect of temperature on growth and respiration of ectotrophic mycorrhizal fungi. Mycologia 57:748–756

Hampp R, Schaeffer C (1995) Mycorrhiza – carbohydrate and energy metabolism. In: Varma A, Hock B (eds) Mycorrhiza. Springer, Berlin Heidellary New York, pp 267–296

Hampp R, Schaeffer C, Wallenda T, Stülten C, Johann P, Einig W (1995) Changes in carbon partitioning or allocation due to ectomycorrhiza formation: biochemical evidence. Can J Bot 73 (Suppl 1):S548–S556

Hampp R, Ecke M, Schaeffer C, Wallenda T, Wingler A, Kottke I, Sundberg B (1996) Axenic mycorrhization of wild type and transgenic hybrid aspen expressing T-DNA indoleacetic acid-biosynthetic genes. Trees 11:59–64

Hardy TA, Huang D, Roach PJ (1994) Interactions between cAMP-dependent and SNF1 protein kinases in the control of glycogen accumulation in *Saccharomyces cerevisiae*. J Biol Chem 269:27907–27913

Hers H-G, van Schaftingen E (1982) Fructose 2,6-bisphosphate 2 years after its discovery. Biochem J 206:1–12

Hoffmann E, Wallenda T, Schaeffer C, Hampp R (1997) Cyclic AMP, a possible regulator of glycolysis in the ectomycorrhizal fungus *Amanita muscaria*. New Phytol 137:351–356

Hutchison LJ (1990a) Studies on the systematics of ectomycorrhizal fungi in axenic culture. IV. The effect of some selected fungitoxic compounds upon linear growth. Can J Bot 68:2172–2178

Hutchison LJ (1990b) Studies on the systematics of ectomycorrhizal fungi in axenic culture. V. Linear growth response to standard extreme temperatures used as a taxonomic character. Can J Bot 68:2179–2184

Ineichen K, Wiemken V, Wiemken A (1995) Shoots, roots and ectomycorrhiza formation of pine seedlings at elevated atmospheric carbon dioxide. Plant Cell Environ 18:703–707

Ingleby K, Mason PA, Last FT, Fleming LV (1990) Identification of ectomycorrhizas. ITE Research Publication No 5. HMSO, London

Jennings DH (1995) The physiology of fungal nutrition. Cambridge University Press, Cambridge

Kawase I, Shirakawa H, Watanabe M (1992) Deaths caused by *Amanita subjunquillea* poisoning and the distribution of this mushroom in Hokkaido. Trans Mycol Soc Jpn 33: 107–110

Koch KE (1996) Carbohydrate-modulated gene expression in plants. Annu Rev Plant Physiol Mol Biol 47:509–540

Komor E (1983) Phloem loading and unloading. In: Esser K, Kubitzki K, Runge M, Schnepf E, Ziegler H (eds) Progress in botany, vol 45. Springer, Berlin Heidelberg New York, pp 68–75

Kottke I (1986) Charakterisierung und Identifizierung von Mykorrhizen. I. Vergleich künstlich gezogener Mykorrhizen mit Formen vom Naturstandort. II. Zur Identität der "safrangelben" Mykorrhiza. In: Einsel G (ed) Das landschaftsökologische Forschungsprojekt Naturpark Schönbuch, DFG, VCH Verlagsgesellschaft, Weinheim, pp 463–485

Kottke I, Oberwinkler F (1987) Cellular structure and function of the Hartig net: coenocytic and transfer cell-like organization. Nord J Bot 7:85–95

Kottke I, Oberwinkler F (1989) Amplification of root–fungus interface in ectomycorrhizae by Hartig net architecture. Ann Sci For 46 (Suppl):737s–740s

Kottke I, Guttenberger M, Hampp R, Oberwinkler F (1987) An in vitro method for establishing mycorrhizae on coniferous tree seedlings. Trees 1:191–194

Kottke I, Holopainen T, Alanen E, Turnau K (1995) Deposition of nitrogen in vacuolar bodies of *Cenococcum geophilum* Fr. mycorrhizas as detected by electron energy loss spectroscopy. New Phytol 129:411–416

Kottke I, Qian XM, Pritsch K, Haug I, Oberwinkler F (1998) *Xerocomus badius–Picea abies*, an ectomycorrhiza of high activity and element storage capacity in acidic soil. Mycorrhiza 7:267–275

Kühner R (1972) Agaricales de la zone alpine. Amanitacées. Ann Sci L'Univ Besançon 12:31–38

Lamb RJ (1974) Effect of D-glucose on utilization of single carbon sources by ectomycorrhizal fungi. Trans Br Mycol Soc 63:295–306

Loewe A, Einig W, Shi L, Hampp R (1999) Mycorrhization and elevated CO_2 both increase the capacity for sucrose synthesis in source leaves of spruce and aspen. New Phytol. (in press)

Lohaus G, Winter H, Riens B, Heldt HW (1995) Further studies of the phloem loading process in leaves of barley and spinach. The comparison of metabolite concentrations in the apoplastic compartment with those in the cytosolic compartment and in the sieve tubes. Bot Acta 108:270–275

Lundeberg G (1970) Utilisation of various nitrogen sources, in particular bound soil nitrogen, by mycorrhizal fungi. Stud For Suec 79:1–87

Madi L, McBride SA, Bailey LA, Ebbole DJ (1997) Rco-3, a gene involved in glucose transport and conidiation in *Neurospora crassa*. Genetics 146:499–508

Maijala P, Fagerstedt KV, Raudaskoski M (1991) Detection of extracellular cellulolytic and proteolytic activity in ectomycorrhizal fungi and *Heterobasidion annosum* (Fr.) Bref. New Phytol 117:643–648

Malajczuk N, Molina R, Trappe JM (1982) Ectomycorrhiza formation in *Eucalyptus*. I. Pure culture synthesis, host specificity and mycorrhizal compatibility with *Pinus radiata*. New Phytol 91:467–482

Marger MD, Saier MH (1993) A major superfamily of transmembrane facilitators that catalyze uniport, symport and antiport. Trends Biol Sci 18:13–20

Marx DH (1969) The influence of ectotrophic mycorrhizal fungi on the resistance of pine roots to pathogenic infections. I. Antagonism of mycorrhizal fungi to root pathogenic fungi and soil bacteria. Phytopathology 59:153–163

Mason P (1975) The genetics of mycorrhizal associations between *Amanita muscaria* and *Betula verrucosa*. In: Torrey JG, Clarkson DT (eds) The development and function of roots. Academic Press, London, pp 567–574

Mason PA, Last FT, Wilson J, Deacon JW, Fleming LV, Fox FM (1987) Fruiting and successions of ectomycorrhizal fungi. In: Pegg GF, Ayres PG (eds) Fungal infections of plants. Cambridge University Press, New York, pp 253–268

Mohan V, Natarajan K, Ingleby K (1993) Anatomical studies on ectomycorrhizas. II. The ectomycorrhizas produced by *Amanita muscaria*, *Laccaria laccata* and *Suillus brevipes* on *Pinus patula*. Mycorrhiza 3:43–49

Molina R (1981) Ectomycorrhizal specificity in the genus *Alnus*. Can J Bot 59:325–334

Molina R, Trappe JM (1982) Patterns of ectomycorrhizal host specificity and potential among Pacific Northwest conifers and fungi. For Sci 28:423–458

Moser M (1958) Der Einfluß tiefer Temperaturen auf das Wachstum und die Lebenstätigkeit höherer Pilze mit spezieller Berücksichtigung von Mykorrhizapilzen. Sydowia 12:386–399

Moser M (1983) Die Röhrlinge und Blätterpilze (Polyporales, Boletales, Agaricales, Russulales). Kleine Kryptogamenflora. II b/2. Gustav Fischer, Stuttgart

Mousseau M, Saugier B (1992) The direct effect of increased CO_2 on gas exchange and growth of forest tree species. J Exp Bot 43:1121–1130

Namysl C, Rieger A, Hampp R, Dizengremel P (1991) Longitudinal distinction of adenine nucleotide pools in unmycorrhized and mycorrhized fine roots of spruce seedlings. Plant Physiol 96 (Suppl):1104

Nehls U, Wiese A, Guttenberger M, Hampp R (1998) Carbon allocation in ectomycorrhiza: identification and expression analysis of an *A. muscaria* monosaccharide transporter. Mol Plant Microb Int 11:167–176

Norby RJ, O'Neill EG, Hood WG, Luxmore RBJ (1987) Carbon allocation, root exudation and mycorrhizal colonization of *Pinus echinata* seedlings grown under CO_2 enrichment. Tree Physiol 3:203–210

Orlovich DA, Ashford AE (1993) Polyphosphate granules are an artefact of specimen preparation in the ectomycorrhizal fungus *Pisolithus tinctorius*. Protoplasma 173:91–102

Özcan S, Dover J, Rosenwald AG, Wölfl S, Johnston M (1996) Two glucose transporters in *Saccharomyces cerevisae* are glucose sensors that generate a signal for induction of gene expression. Proc Natl Acad Sci USA 93:12428–12432

Palmer JG, Hacskaylo E (1970) Ectomycorrhizal fungi in pure culture. I. Growth on single carbon sources. Physiol Plant 23:1187–1197

Persoon CH (1797) Tentamen dispositionis methodicae fungorum. PP Wolf, Lipsiae

Radin JW, Parker LL, Sell CR (1978) Partitioning of sugar between growth and nitrate reduction in cotton roots. Plant Physiol 62:550–553

Ramstedt M, Jirjis R, Söderhäll K (1987) Metabolism of mannitol in mycorrhizal and non-mycorrhizal fungi. New Phytol 105:281–287

Read DJ, Francis R, Finlay RD (1985) Mycorrhizal mycelia and nutrient cycling in plant communities. In: Fitter AH, Atkinson D, Read DJ, Usher MB (eds) Ecological interactions in soil. Blackwell, Oxford, pp 193–216

Reifenberger E, Freidel K, Ciriacy M (1995) Identification of novel HXT genes in *Saccharomyces cerevisae* reveals the impact of individual hexose transporters on glycolytic flux. Mol Microbiol 16:157–167

Richter DL, Bruhn JN (1989) Revival of saprotrophic and mycorrhizal basidiomycete cultures from cold storage in sterile water. Can J Microbiol 35:1055–1060

Rieger A, Guttenberger M, Hampp R (1992) Soluble carbohydrates in mycorrhized and non-mycorrhized fine roots of spruce seedlings. Z Naturforsch 47c:201–204

Roze E (1876) Essai d'une nouvelle classification des Agaricacées. Bull Soc Bot Fr 23:45–54

Rudawska M, Kieliszewska-Rokicka B, Debaud JC, Lewandowski A, Gay G (1994) Enzymes of ammonium metabolism in ectendomycorrhizal and ectomycorrhizal symbionts of pine. Physiol Plant 92:279–285

Salzer P, Hager A (1991) Sucrose utilization of the ectomycorrhizal fungi *Amanita muscaria* and *Hebeloma crustuliniforme* depends on the cell wall-bound invertase activity of their host *Picea abies*. Bot Acta 104:439–445

Samson J, Fortin JA (1986) Ectomycorrhizal fungi of *Larix laricina* and the interspecific and intraspecific variation in response to temperature. Can J Bot 64:3020–3028

Schaeffer C, Wallenda T, Guttenberger M, Hampp R (1995) Acid invertase in mycorrhizal and non-mycorrhizal roots of Norway spruce (*Picea abies* [L.] Karst.) seedlings. New Phytol 129:417–424

Schaeffer C, Johann P, Nehls U, Hampp R (1996) Evidence for an up-regulation of the host and a down-regulation of the fungal phosphofructokinase activity in ectomycorrhizas of Norway spruce. New Phytol 134:697–702

Schiebel G (1988) Lokalisierung und Charakterisierung primär energetisierter H⁺-Translokasen an Membranen von Ektomykorrhizapilzen (*Amanita muscaria* und *Hebeloma crustuliniforme*). PhD Thesis, University of Tübingen, Tübingen

Singer R (1986) The Agaricales in modern taxonomy. Koeltz Scientific Books, Koenigstein

Söderström B (1992) The ecological potential of the ectomycorrhizal mycelium. In: Read DJ, Lewis DH, Fitter AH, Alexander IJ (eds) Mycorrhizas in ecosystems. CAB International, Wallingford, pp 77–83

Söderström B, Finlay RD, Read DJ (1988) The structure and function of the vegetative mycelium of ectomycorrhizal plants. IV. Qualitative analysis of carbohydrate contents of mycelium interconnecting host plants. New Phytol 109:163–166

Stitt M (1991) Rising CO_2 levels and their potential significance for carbon flow in photosynthetic cells. Plant Cell Environ 14:741–762

Stroo HF, Reich PB, Schoettle AW, Amundson RG (1988) Effects of ozone and acid rain on white pine (*Pinus strobus*) seedlings grown in five soils. II. Mycorrhizal infection. Can J Bot 66:1510–1516

Sung S J-S, White LM, Marx DH, Otrosina WJ (1995) Seasonal ectomycorrhizal fungal biomass development in loblolly pine (*Pinus taeda* L.) seedlings. Mycorrhiza 5:439–447

Taber WA, Taber RA (1987) Carbon nutrition and respiration of *Pisolithus tinctorius*. Trans Br Mycol Soc 89:13–26

Theodorou C, Reddell P (1991) In vitro synthesis of ectomycorrhizas on Casuarinaceae with a range of mycorrhizal fungi. New Phytol 118:279–288

Trappe JM (1962) Fungus associates of ectotrophic mycorrhizae. Bot Rev 28:538–606

Wainwright M (1993) Oligotrophic growth of fungi – stress or natural state?. In: Jennings DJ (ed) Stress tolerance of fungi. Marcel Dekker, New York, pp 127–144

Wallander H, Nylund JE (1991) Effects of excess nitrogen on carbohydrate concentration and mycorrhizal development of *Pinus sylvestris* L. seedlings. New Phytol 119:405–411

Wallenda T (1992) Trehalose und andere Kohlenhydrate im Ektomykorrhizasystem *Amanita muscaria/Picea abies*. Diploma Thesis, University of Tübingen, Tübingen

Wallenda T (1996) Untersuchungen zur Physiologie der Pilzpartner von Ektomykorrhizen der Fichte (*Picea abies* [L.] Karst.) PhD Thesis, University of Tübingen, Tübingen

Watkinson JH (1964) A selenium-accumulating plant of the humid regions: *Amanita muscaria*. Nature 202:1239–1240

Watling R (1988) A mycological kaleidoscope. Trans Br Mycol Soc 90:1–28

Weiß M, Yang ZL, Oberwinkler F (1998) Molecular phylogenetic studies in the genus *Amanita*. Can J Bot 76:1070–1179

Wieland T (1986) Peptides of poisonous *Amanita* mushrooms. Springer, Berlin Heidelberg New York

Wingler A (1995) Bereitstellung von Kohlenstoffskeletten für die Stickstoffassimilation in der Fichte (*Picea abies* [L.] Karst.). PhD Thesis, University of Tübingen, Tübingen

Wingler A, Wallenda T, Hampp R (1996) Mycorrhiza formation on Norway spruce (*Picea abies*) roots affects the pathway of anaplerotic CO_2 fixation. Physiol Plant 96:699–705

Yang ZL (1997) Die *Amanita*-Arten von Südwestchina. Bibl Mycol 170:1–240

Yang ZL, Oberwinkler F (1999) Die Fruchtkörperentwicklung von *Amanita muscaria* (Basidiomycetes). Nova Hedwigia 68:411–468

Ziegler H (1975) Phloem transport. Nature of transported substances. In: Zimmermann MH, Milburn JA (eds) Enclyclopedia of plant physiology, New Series, vol 1. Transport in plants. Springer, Berlin Heidelberg New York, pp 59–100

Chapter 9

Paxillus

H. Wallander and B. Söderström

9.1
Introduction

Paxillus involutus is currently among the ectomycorrhizal (ECM) fungi used most frequently in experimental work in the laboratory. Between 1988 and 1997 more that 270 research articles containing information on this species were published. Most of these papers concern the mycorrhizal characteristics of this fungus. This situation is in striking contrast to that existing a few decades ago, as reflected in the quote by Laiho (1970): "*Paxillus involutus* has, indeed, been almost ignored by scientists working with mycorrhiza". Thus, almost all information about this fungus and its mycorrhizal status has been gathered during the last 25 years.

To be able to carry out controlled experimental work with ECM systems it must be possible to cultivate the fungal partner in pure culture. However, up until now, only a few taxa of ECM fungi could be cultivated in this way. One of the most easily cultivated ECM fungi is, in fact, *P. involutus*, which can explain why this species is among the most commonly used fungi in controlled laboratory experiments, both in pure culture and as ECM partners. In addition, the physiology of *P. involutus* has been studied more thoroughly compared with most other ECM fungi; nevertheless, our knowledge in this area is still rudimentary. The ability of the species to form functional ECM with a number of tree species (Trappe, 1962, and see below) has also contributed to its usefulness in experimental work.

The genus appears to be more common in temperate ecosystems than in tropical ones (Hahn 1996), and most *Paxillus* species seem to be saprophytes. Fischer (1995) suggested that there is a close affinity between the Coniophoraceae and the Paxillaceae. In the literature there is some confusion as to the authorship of *P. involutus*; for example, it has been referred to as *P. involutus* (Fr.) Fr., *P. involutus* (Batsch) Fr., and *P. involutus* (Batsch: Fr.) Fr. According to the prevailing rules of nomenclature, the proper designation of the fungus should be *Paxillus involutus* (Batsch: Fr.) Fr.

Department of Microbial Ecology, Ecology Building, University of Lund, 5-223 62 Lund, Sweden

9.2
ECM Formation

Malajczuk et al. (1990) developed a technique for aseptically synthesising ECM between *Eucalyptus* sp. and *Pisolithus tinctorious* (Pers.) Coker & Couch. This technique was used by Brun et al. (1995) to synthesise ECM between *P. involutus* and *Betula pendula*. Small seedlings are cultured in close contact with a fungal mycelium that grows on a cellophane-covered agar surface. This method allows synchronous production of large numbers of inoculated roots. ECM synthesis between *P. involutus* and *B. pendula* has also been achieved by supplementing soil with fresh basidiospores. *P. involutus* formed abundant ECM in coal spoils with this method, but ECM formation was poor in brown soil (Fox 1986a). Sclerotia formed by *P. involutus* can remain infective for up to 16 weeks after burial in the soil (Fox 1986b)

9.2.1
Host–Fungus Specificity

Harley and Smith (1983) suggested that *P. involutus* is a facultative ECM former able to form sporocarps without an established host relation, although all strains investigated by Laiho (1970) were ECM forming. Erland and Söderström (1991) reported that one isolate of *P. involutus* was able to grow as a saprophyte on non-sterile humus.

   P. involutus forms ECM with a number of different hosts (Duddridge 1987; Table 9.1) and has been classified as a broad-host-range fungus by Molina and Trappe (1982). There are many reports of *P. involutus* forming ECM associations with various *Alnus* species (Table 9.1), although Laiho (1970) was unable to confirm that *P. involutus* could establish a functional symbiosis with seedlings of this genus in laboratory experiments. In addition, he was unable to find *P. involutus–Alnus* ECM in the field. *Paxillus filamentosus* Fr., on the other hand, is common in *Alnus* forests and appears to be restricted to *Alnus* as a host plant (Bresinsky 1996). There is some evidence of host specificity at the isolate level in *P. involutus*, although this has not been investigated in great detail. Horan and Chilvers (1990) used three isolates of *P. involutus* in inoculation experiments with *Eucalyptus globulus*. Only the isolate collected from a eucalypt forest was attracted by compounds exuded by the host root and formed ECM. The other two *P. involutus* isolates, one collected in a pine forest and the other in a birch forest, failed to form ECM with the eucalypt seedlings. Fries (1985) identified three intersterility groups in *P. involutus* monokaryons collected from 25 sporocarps in a forest close to Uppsala, Sweden. One of these groups preferred conifers, while the other two preferred deciduous trees, especially birch.

Table 9.1. ECM associations formed between *P. involutus* and different hosts

Host	Reference
Allocasuarina littoralis	Theodorou and Reddell (1991)
Alnus glutinosa	Molina (1979, 1981); Arnebrant et al. (1993)
A. rubra	Molina (1979); Miller et al. (1991)
A. incana	Murphy and Miller (1994); Ekblad et al. (1995)
A. sinuate	Murphy and Miller (1994)
A. serrulata	Brunner et al. (1990)
Betula pendula	Grellier et al. (1984); Brun et al. (1995)
Castanea sativa	Strullu et al. (1986)
Dryas integrifolia	Melville et al. (1987)
Eucalyptus globulus	Horan and Chilvers (1990)
Fagus sylvatica	Finlay et al. (1989)
Picea abies	Marschner and Godbold (1995)
P. sitchensis	Sasa and Krogstrup (1991)
Pinus sylvestris	Laiho (1970); Turnau et al. (1994)
P. resinosa	Grenville et al. (1985)
P. contorta	Shaw et al. (1995)
P. strobus	Piche and Fortin (1982); Grenville et al. (1985)
P. pinaster	Pargney and Gourp (1991); Pera and Alvarez (1995)
Populus hybrid	Heslin and Douglas (1986)
Pseudotsuga menziesii	Jansen and de Vries (1990); Villeneuve et al. (1991)
Quercus robur	Lei and Dexheimer (1987)
Quercus suber	Branzanti and Zambonelli (1989)
Salix repens	Gale (1977)

9.2.2
Molecular Events

Certain molecular and biochemical events take place early in the formation of *P. involutus* ECM. Specific polypeptides (ectomycorrhizins) are formed in *Paxillus* ECM after only a few days of contact between the fungus and the ECM plant. In a study by Simoneau et al. (1993) some of these polypeptides accumulated in ECM roots during early stages of the infection, while others transiently accumulated during the first 4 days but were barely detectable in the ECM roots after 8 days. During ECM formation between eucalypts and *P. tinctorius*, Hilbert et al. (1991) found a strong downregulation of polypeptides that were present in the partners when they were not engaged in symbiosis. A similar downregulation of polypeptides was not found in ECM formed between *P. involutus* and *B. pendula* (Simoneau et al. 1993). There seems to be some similarity between these polypeptides and defence proteins investigated in other associations (Duchesne et al. 1989b; Simoneau et al. 1993). Timonen et al. (1996a) found abundant tubulin and actin after immunoblotting polypeptides from crude extracts of *P. involutus*–*P. contorta* ECM. The amount of fungal α-tubulin increased as the amount of fungal mycelium started to build up during the first 20 days of contact between the fungus and the host

(Timonen et al. 1996a). Amounts of plant and fungal actin were constantly elevated during ECM formation and remained high in the mature ECM. It was suggested that α-tubulin is a growth-related protein, while actin reflects the metabolic activity of the ECM (Timonen et al. 1996a).

Enzymatic activity is greatly influenced during the formation of ECM between *P. involutus* and its host roots. Blaudez et al. (1998) demonstrated that activities of several enzymes involved in the N metabolism, including glutamine synthetase (GS), aspartate aminotransferase (AAT) and malic dehydrogenase (MDH), increased and remained higher in *P. involutus*-colonised *B. pendula* roots, compared with non-colonised roots, during the first 21 days after contact between the two organisms. This suggests that processes connected with the N metabolic pathway are stimulated during the infection process.

Simoneau et al. (1993) found an increase in phenylalanine ammonia-lyase activity during an early stage of ECM formation between *P. involutus* and *B. pendula*. This enzyme is involved in the pathway of polyphenol synthesis. The activity of endoglucanase has also been reported to be increased in *P. involutus* which suggests that the fungus produces cell-wall degrading enzymes during formation of the Hartig net (Maijala et al. 1991). Amino acids synthesised in the ECM roots can also be used to produce tanniferous compounds which were formed in large amounts during ECM formation between *P. involutus* and *P. pinaster* (Pargney and Gourp 1991). Nutritional conditions can strongly influence ECM formation. Brun et al. (1995) found that the reduction of P combined with high N availability in the nutrient medium totally inhibited ECM formation between *P. involutus* and *B. pendula*. On the other hand, Ekblad et al. (1995) reported that low P availability in combination with high N availability greatly stimulated the production of mycelium by *P. involutus* growing in symbiosis with *Pinus* or *Alnus* seedlings.

9.2.3
Developmental Aspects of the Fungus–Host Interaction

A series of interactive events between the two symbionts is required for an ECM association to form. Initially, the fungus is probably attracted to host compounds (Horan and Chilvers 1990). Using semipermeable membrane filters (0.45 μm) between *P. involutus* inoculum and host roots Horan and Chilvers (1990) found that compatible strains of *P. involutus* were attracted by the apical region of eucalypt roots and grew through the membrane filters to form ECM with the roots. This never happened in cases where the fungus was exposed to a non-host root.

Slankis (1973) proposed that indolyl-3-acetic acid (IAA) produced by the fungus plays a central role in the formation of the ECM organ. Recently, Rudawska and Kieliszewska-Rokicka (1997) found that *P. involutus* strains good at producing IAA in pure culture were better at colonising pine seedlings than less efficient IAA producers. This suggests that IAA plays a role in ECM formation. In support of this view, Gea et al. (1994) found that IAA overproducing mutants of *Hebeloma cylindrosporum* Romagn. produced a larger mantle and a more extensive Hartig net compared to wild-type strains when inoculated on *Pinus pinaster* seedlings. In contrast, Wallander et al. (1994) found no indication that IAA was involved in ECM formation between *Pinus sylvestris* and *Laccaria bicolor* (Maire) Orton based on a comparison of amounts of IAA in ECM and non-ECM roots.

In *P. involutus–Betula pendula* ECM, the symbiotic organs start to form within a few days after inoculation. Brun et al. (1995) described structural changes in root tissue that occurred in connection with *P. involutus–B. pendula* ECM formation during the first 15 days after contact between the two organisms. After 4 days the roots were fully enveloped by hyphae, and a mantle had differentiated into two distinct layers. The first signs of Hartig net formation were also evident at this stage. After 8–10 days of contact a thick mantle had developed, and there was a recognisable paraepidermal Hartig net (Agerer 1991) characterised by a single layer of fungal hyphae between the root cells. After 15 days a periepidermal Hartig net (Agerer 1991) was evident. This structure is characterised by the presence of several hyphal rows between the epidermal cells. Simoneau et al. (1993) presented a similar description of structural changes during the early stages of ECM formation between *P. involutus* and *B. pendula*. Blaudez et al. (1998) and Simoneau et al. (1993) found that the ergosterol concentration in the roots increased linearly during the formation of the Hartig net. Furthermore, the ergosterol concentration in fully developed ECM corresponded to a fungal biomass accounting for 45% of the total root biomass.

9.2.4
Ageing in *P. involutus* ECM

ECM formed by *P. involutus* are usually rather short-lived compared with those formed by other ECM fungi. When *P. involutus* and *Piloderma croceum* Erikss. & Hjortst. were growing in competition when colonising *Pinus sylvestris* seedlings, the number of ECM formed by *P. involutus* reached a maximum after 2 months. After this time the numbers decreased. The number of ECM formed by *Piloderma croceum*, on the other hand, was still increasing after 6 months of growth with the two fungi present on the roots (Erland and Finlay 1992). In a study by Downes et al. (1992) ECM formed between *P. involutus* and *Picea sitchensis* were pale grey-buff and turgid with obvious extramatrical mycelium after 22–31 days. Many ECM showed repeated bursts of growth, leading to a beaded appearance. Most *P. involutus* ECM over 50 days of age had darkened, and the proximal cortex in some had started to collapse, leaving a lighter, turgid apical portion of variable length. In addition, their extramatrical mycelium had disappeared. After 132 days some ECM still had a light, turgid apex. The proportion of cortical cells containing nuclei decreased from 25 to 50% on day 22 to less than 15% on day 132. Cortical cell/Hartig net fluorescence declined markedly after 85 days. The rates at which the individual ECM passed through the morphological stages of ageing varied considerably from tip to tip, and morphological appearance was not a good indicator of age. Based on the actin measurements, Timonen et al. (1996a) concluded that 60-day-old *P. involutus–Pinus contorta* ECM were still metabolically active despite their withered appearance. Downes et al. (1992) hypothesised that ECM roots became non-functional (were no longer able to take up water and nutrients) once the final region of living cortical interface had disappeared. For most ECM formed by *P. involutus* and *Picea sitchensis* this stage was reached after 80 days, but some ECM became non-functional after only 30 days (Downes et al. 1992). Ekblad et al. (1998) measured chitin and ergosterol in ageing ECM formed between *P. involutus* and *Pinus sylvestris*. They found that the ergosterol concentration decreased markedly between 4 and 7 months of age when the ECM appeared brown and shrunken. The concentration of chitin, on the other hand, remained high in old ECM. It is likely that the increased vacuolation of fungal cells in old ECM results in a lower ergosterol concentration. Downes et al. (1992) reported that old *P. involutus–Pinus sitchensis* ECM had larger vacuoles, a lower cytoplasmic volume, cytoplasm with less fine granularity and cells with fewer organelles, compared with younger ECM. Ekblad et al. (1998) suggested that the combination of ergosterol and chitin concentration could be used to estimate both total and living fungal biomass.

9.3
Nutritional Aspects of Interactions Between *P. involutus* and its Hosts

9.3.1
Carbon Nutrition

Since *P. involutus* is easily cultured and rather fast growing, more physiological studies have been made on this fungus than on most other ECM fungi. Laiho (1970) made several basic studies on the physiology of the species, and, unlike many studies published later, he included a number of different strains (nine in most of his studies) which differed from each other in their physiological characteristics. He found that glucose was the best C source for most of his strains, and good growth was also observed on sucrose, maltose and dextrin as well as mannitol. Starch was also an acceptable C source, whereas no growth was observed on cellulose. Lactose, raffinose and citric acid were not suitable sources of C. Unlike Lahio (1970), Hughes and Mitchell (1995) did not find sucrose (or fructose) to be suitable for supporting the growth of their isolate, and Kieliszewska-Rokicka (1992) found that carboxymethyl cellulose was hydrolysed, indicating some cellulolytic activity. Amino acids, like alanine, can also be utilised as a C source by *P. involutus* (Chalot et al. 1994b). The conflicting results indicate that there are strong differences between isolates in this context. Very little is known, however, about C metabolism of *P. involutus* in intact ECM systems.

9.3.2
Nitrogen Nutrition

Nitrogen utilisation has been studied in some detail. Based on these studies it can be concluded that *P. involutus* is able to use NO_3^- or NH_4^+ as its sole N source (Laiho 1970; Lapeyrie et al. 1991; Finlay et al. 1992; Chalot et al. 1995; Keller 1996), although growth normally is somewhat better on NH_4^+. However, in cases where NO_3^- is the only N source, *P. involutus* seems to tolerate it better than many other ECM fungi (Keller 1996). Different isolates of *P. involutus* investigated by Laiho (1970) showed large variation in their capacity to utilise NO_3^-. Smith and Read (1997) suggested that natural selection would favour NO_3^- users in nitrifying environments. Results from pure culture experiments should, however, be evaluated with caution. Sarjala (1990, 1991) found no agreement between the nitrate reductase activity of *P. involutus* in pure culture and that of the fungus growing in association with a host. Glutamate has been shown to be the major sink for inorganic N (Ek et al. 1994a).

When *P. involutus* was grown in symbiosis with *B. pendula* or *Picea abies*, NH_4^+ was immediately assimilated to amino acids at fungal uptake sites (Ek et al. 1994a). Assimilation of inorganic N is an energy-consuming process, and Ek (1997) found that fungal respiration in *P. involutus–Pinus sylvestris* associations increased between 54 and 180% in response to N addition. Glutamine was the major form in which N was translocated to the host (Ek et al. 1994a). Nitrate, on the other hand, was translocated in unassimilated form to roots. More NH_4^+ than NO_3^- was taken up and transferred to the host (Ek et al. 1994a). Principal sinks for NH_4^+ in *P. involutus–Fagus sylvatica* ECM were alanine, aspartate/arginine and glutamine/glutamate, and 78% of the NH_4^+-N was incorporated into proteinaceous compounds. Detailed work on the uptake and assimilation of inorganic N by *P. involutus* growing in axenic cultures as well as in symbiosis has been performed by Chalot and co-workers (1994a,b, 1995). Using isotopic tracers, such as ^{15}N and ^{14}C, Chalot et al. (1994a,b) demonstrated that NH_4^+ was rapidly assimilated via the glutamine synthetase/GOGAT pathway in axenically grown *P. involutus* mycelium. Rudawska et al. (1994) also showed that the GS/GOGAT pathway dominated N assimilation in *P. involutus*. No glutamate dehydrogenase (NADP-GDH) activity could be detected in *P. involutus* (Chalot et al. 1994b; Rudawska et al. 1994). Nilsson et al. (1993) found that the uptake of inorganic N per unit root length was three-fold higher in pine roots colonised by *P. involutus* compared with non-ECM pine roots. The transfer of N from the fungus to the host depends strongly on the fungus–host combination. *P. involutus* has been shown to retain N in its external mycelium when growing in symbiosis with pine seedlings. In this way the fungus may suppress the growth of pine seedlings compared with non-ECM seedlings (Colpaert et al. 1992). Wallander et al. (1999) found that the proportion of N assimilated by the fungus that was allocated to shoots was smaller in seedlings colonised by *P. involutus* than in those colonised by *Suillus bovinus* (L.: Fr.) O. Kuntze.

A substantial portion (up to 70%) of the N in subsoils can be present as fixed NH_4^+ ions in clay complexes (Paul and Clark 1989). Paris et al. (1995) showed that *P. involutus* growing in axenic cultures had access to this N source in NH_4^+-saturated vermiculite. Ammonium ions were mobilised from the interlayer spaces by Mg^{2+} and Al^{3+} which displaced the N in the complexes. The presence of other organisms may also influence uptake of inorganic N by *P. involutus* ECM. Ek et al. (1994b) found that $^{15}NH_4^+$ was taken up better when collembola were present at moderate levels, which stimulated mycelial growth.

Of the eight amino acids that Laiho (1970) tested on his nine *P. involutus* isolates, he found that growth on six (leucine, tyrosine, proline, valine, histidine, lysine) was poor when each of them was used as the sole N source, whereas glutamate and arginine resulted in good growth with most of these isolates. Keller (1996) noted poor growth on asparagine and glycine. Uptake of glutamate and glutamine as well as alanine and aspartate in axenic cultures has subsequently been demonstrated by Chalot et al. (1995).

Chalot et al. (1994a,b, 1995) showed that the amino acids glutamate, glutamine and alanine are readily metabolised by glutamate synthetase, glutamine synthetase, and alanine aminotransferase, respectively. The uptake of these amino acids has a distinct pH optimum around 4.0. The total uptake of amino acids was unaffected by exogenously supplied NO_3^-, NH_4^+ or glucose (Chalot et al. 1995). Based on fractal analysis, Baar et al. (1997) concluded that *P. involutus* changed its foraging activity when NH_4^+ in axenic cultures was replaced by amino acids. The uptake of amino acids showed characteristics of active transport via a proton symport (Chalot et al. 1996). Interestingly, Chalot (1994b) showed that alanine is used as a respiratory substrate to form a number of intermediates of the tricarbolylic acid cycle, among which malate and pyruvate play a central role (Chalot et al. 1994a,b). This pathway was strongly repressed by aminooxyacetate, and they concluded that alanine aminotransferase plays an important role in the alanine metabolism of *P. involutus*. The amino acids glutamate, glutamine, aspartate and asparagine dominated the amino acid pool in birch roots colonised by *P. involutus* and in the associated mycelium (Finlay et al. 1989; Chalot et al. 1995). There may, however, be considerable intraspecific variation in the composition of the amino acid pool in this fungus. We (H. Wallander and B. Söderström, unpubl. data) found arginine to be the dominant amino acid in five strains of axenically grown *P. involutus* isolates collected in coniferous forests in northern and southern Sweden, while arginine was present in very low amounts in the isolate used by Finlay et al. (1989) and Chalot et al. (1995) which was isolated from a birch forest in Scotland. Free amino acids did not accumulate in external mycelium of *P. involutus* growing in symbiosis with birch seedlings. This may indicate that C supplied by the host retarded the senescence process (Blaudez et al. 1998). The synthesis of asparagine and glutamine would minimise C utilisation for N storage (Baxter et al. 1992).

P. involutus seems to be able to hydrolyse complex N compounds. Laiho (1970) showed that it grows on casein hydrolysate and peptone, and after the pioneering work of Abuzinadah et al. (1986) on protein usage by ECM fungi, Finlay et al. (1992) showed that one of their two tested isolates of *P. involutus* showed good growth on bovine serum albumen (BSA) and gliadin. Keller (1996), however, reported poor growth on BSA and gliadin. Differences between isolates can probably explain these discrepancies. Other macromolecules known to be hydrolysed by isolates of *P. involutus* include chitin (Hodge et al. 1995, 1996a,b) and DNA (Hutchison 1992).

Abuzinadah and Read (1989) demonstrated that *P. involutus* growing in symbiosis with *B. pendula* seedlings utilises amino acids and small peptides. *Amanita muscaria* (L.: Fr.) Pers. and *Hebeloma crustuliniforme* (Bull.: St.-Amans) Quél. were, however, more efficient than *P. involutus* in this respect. Recently, Näsholm et al. (1998) demonstrated that ECM plants were able to take up an intact amino acid (glycine) in the field. It has also been shown that more complex organic material can be used as an N source for ECM fungi growing in symbiosis in microcosms in laboratory experiments. Bending and Read (1995a,b) found that mycelia from *P. involutus* growing in association with *B. pendula* were able to colonise and utilise N from litter collected from the fermentation horizon of a birch forest. Activities of proteases and polyphenol oxidase increased after 28–50 days of colonisation and were still elevated after 90 days. ECM fungi were, however, less able to degrade lignolytic compounds compared with wood decomposers (Bending and Read 1995a,b), and Colpaert and van Tichelen (1996) found that ECM fungi (including *P. involutus*) growing in symbiosis with pine seedlings were unable to decompose fresh beech litter, whereas litter-decomposing fungi were able to gain N from the same litter (Colpaert and van Tichelen 1996). Uptake of both inorganic and organic N by *P. involutus* growing in symbiosis with *Pinus sylvestris* can be influenced by forestry practices such as liming. Thus, for example, the uptake of organic N was reduced when the pH of peat was raised from 4.1 to 5.9 through the addition of $CaCO_3$. This change was ascribed to the increased immobilisation (chemical or microbial) of N at the higher pH (Andersson et al. 1996).

In general, the input of large amounts of N to forest ecosystems hampers ECM fungi (Arnolds 1988). *P. involutus* seems to tolerate these negative influences of N better than other ECM species (Arnebrant 1994), although variation between isolates can be substantial (Wallander et al. 1999). Sporocarp formation has also been shown to increase in response to N fertilisation in coniferous forests (Ohenoja 1978). Assimilated N can be transferred between different hosts via an ECM mycelium. Arnebrant et al. (1993) showed that between 5 and 15% of the $^{15}N_2$ fixed by alder was translocated to pines via *P. involutus* mycelium (Arnebrant et al. 1993). Ekblad and Huss Danell (1995) studied the effect of nutrient availability on the translocation of N between *P. involutus* and *Alnus incana* via *P. involutus* mycelium and found that translocation was highest (9% of fixed N) when the pine seedlings were

N starved; under most nutritional conditions N transfer between alder and pine seedlings was insignificant.

9.3.3
Phosphorus Nutrition

P. involutus is able to use different forms of phosphorus, including KH_2PO_4, $Ca_3(PO_4)_2$ and iron phytate (Hilger and Krause 1989; Lapeyrie et al. 1991; McElhinney and Mitchell 1995). The most active acid phosphatases seem to be bound in the cell wall (McElhinney and Mitchell 1993), and when phosphate is in excess, it may accumulate as orthophosphate or polyphosphate (Grellier et al. 1989). Cairney and Smith (1993) found that *P. involutus* absorbed more ^{32}P-phosphate than five other ECM isolates tested. Dighton et al. (1993), on the other hand, found that *P. involutus* took up less ^{32}P than several (six isolates) other ECM fungi as well as saprotrophic fungi in axenic cultures (Dighton et al. 1993). In this study there was generally less variation in ^{32}P uptake within each fungal species than between species.

In many studies, *P. involutus* growing in symbiosis has been shown to improve the P status of the host (e.g. Ekblad et al. 1995; Andersson et al. 1996; Cumming 1996). This is most likely an effect of the increase in absorptive area, in the form of fungal hyphae. According to Read (1992) more than 1000 individual hyphae emerge from a single root tip colonised by *P. involutus*. The uptake of P by individual hyphae at the mycelial front of *P. involutus* was four times higher than that of dissected rhizomorphs of the same fungus when growing on peat (Timonen et al. 1996b). Ekblad et al. (1995) found that *P. sylvestris* seedlings colonised by *P. involutus* produced 660% more biomass compared with non-ECM seedlings when grown on P-deficient medium. Interestingly, *P. involutus* did not improve growth when N, K, S or Ca were in low supply, which indicates the importance of *P. involutus* in P nutrition. In the same experiment, *P. involutus* reduced the growth of *A. incana* in all nutrient regimes, including the P-limited one. Clearly, the effects of *P. involutus* (or any other ECM fungus) on the mineral nutrition of forest trees will vary greatly depending on the particular combination of host and mycobiont as well as on the nutritional conditions under which the trees are growing. The important influence that pH has on the uptake of inorganic P by *P. involutus* was demonstrated by Andersson et al. (1996). They found that the beneficial effects of *P. involutus* on P uptake by pine and birch seedlings were drastically reduced in peat limed to a pH of 6.1 compared with unlimed peat at a pH of 4. This was probably due to the reduced availability of P in the limed peat which could have been caused by the precipitation of P as Ca-phosphate or by microbial immobilisation of inorganic P at the higher pH.

ECM fungi are known to accumulate most of the P that they absorb in the fungal sheath (Harley 1989). Marschner and Godbold (1995) showed that in ECM of *P. involutus* the P content could be twice as high in young mycelia (<30 days old) compared with old ones (>50 days old). This is probably the result of reduced metabolic activity in older ECM and a lower influx of P due to a smaller amount of extramatrical mycelium.

It has also been suggested that ECM fungi, including *P. involutus*, improve P uptake by forest trees by producing acid phosphatases that degrade organic forms of P in the soil. *P. involutus* forms acid phosphatase in axenic cultures (Dighton 1983; Ho 1989). Nitrogen addition stimulated the activity of acid phosphatase in axenic cultures of *P. involutus* (Kieliszewska-Rokicka 1992), and McElhinney and Mitchell (1993) found that the most phosphatase activity in axenically grown *P. involutus* was membrane-bound and that only a small fraction was exuded into the medium. When growing in symbiosis, *P. involutus* does not always increase acid phosphatase activity. For example, in a hydroponic cultivation system, Cumming (1996) found that the activity of acid phosphatase in *P. involutus*-colonised pine roots dropped to less than 30% of that in non-ECM roots. On the other hand, Bending and Read (1995b), who studied the enzymatic activity of external mycelium of *P. involutus* colonising organic material collected from forest soil, found that acid phosphatase:phosphomonoesterase activity was enhanced during the first 30 days of colonisation. Thereafter, activity dropped to levels below that of non-colonised substrate. We still know little about the degree to which *P. involutus* and other ECM fungi contribute to the turnover of organic P in forest soil.

P. involutus is able to utilise sparingly soluble P sources, such as Ca_3PO_4 and Fe-phytate, in axenic cultures as well as in symbiosis (Lapeyrie et al. 1991; McElhinney and Mitchell 1995). The dissolution of Ca_3PO_4 is probably caused by oxalic acid produced by the fungus. The oxalate reacts with Ca_3PO_4, forming Ca-oxalate crystals while PO_4^- is released. Lapeyrie (1988) and Lapeyrie et al. (1987) demonstrated that *P. involutus* was able to synthesise oxalate from bicarbonate in pure culture, and NO_3^- as the N source enhanced the production of oxalate. This process was found to be important in calcareous soils in which P availability was low and NO_3^- was the N source (Lapeyrie et al. 1987; Lapeyrie 1988). Wallander et al. (1997) found that mycelium of *P. involutus* growing in symbiosis with pine roots improved the uptake of P from sparingly soluble P sources such as apatite (Wallander et al. 1997). This was, however, most likely an effect of the large external mycelium produced by *P. involutus*. In contrast to other tested ECM fungi [such as *Suillus variegatus* (Swartz: Fr.)] which reduced the pH and increased the amount of organic acids in the soil solution, *P. involutus* did not influence pH or the amount of dissolved organic acids, including oxalic acid (Wallander et al. 1997).

9.3.4
Other Minerals

Paris et al. (1995, 1996) found that axenically grown *P. involutus* had access to K trapped inside phlogopite (mica) interlayer spaces. This was established by increasing the interlayer openings and by interlayer K substitutions. Since the fungus was separated from the minerals by a cellophane sheath, soluble exudates must have been responsible for the mica vermiculitisation. Oxalic acid appeared to be involved in this process, and its exudation was stimulated by the simultaneous depletion of K and Mg from the nutrient media (Paris et al. 1995, 1996). Wallander and Wichman (1999) studied *P. involutus* growing in symbiosis with pine seedlings, but did not find any evidence that this fungus influences the uptake of K from biotite compared with non-ECM control seedlings. On the other hand, *S. variegatus* stimulated K uptake from biotite in the same experiment, and this fungus increased the amount of organic acids in the soil solution, in contrast to *P. involutus*.

9.4
Non-nutritional Aspects of Interactions Between *P. involutus* and its Hosts

9.4.1
Protection Against Pathogens

Evidence that ECM colonisation can decrease the susceptibility of their host to pathogen attacks has attracted a great deal of interest (Marx 1969; Smith and Read 1997). A number of studies indicate that *P. involutus* may act antagonistically against pathogenic fungi, *Fusarium–Pinus* being the most studied parasite–host relationship. In a series of papers, Duchesne et al. (1988a,b, 1989a) studied the interaction between *Pinus resinosa, P. involutus* and *Fusarium oxysporum* f. sp. *pini*. They found that the *P. involutus–Pinus* rhizosphere had general fungitoxic activity (Duchesne et al. 1988b), that exudates from *P. involutus* suppressed germination of *Fusarium* conidia (Duchesne et al. 1988a, 1989a,b) and that disease suppression was more pronounced in *P. resinosa* than in other pine species. They also suggested that the formation of oxalic acid by the ECM fungus was important in the suppression of *Fusarium* infection (Duchesne et al. 1989b). The morphological effects of *P. involutus* on *Fusarium* conidia were studied in more detail by Farquhar and Peterson (1990), and Chakravarty et al. (1990) studied the possibility of using a combination of fungicide applications and ECM in the integrated control of *Fusarium* damping-off in pine seedlings. These antagonistic effects were confirmed by Hwang et al. (1995) using *Pinus banksiana* as host and *Fusarium moniliforme* Sheldon as pathogen, whereas Chakravarty and Hwang (1991) could not find any evidence that *P. involutus* inhibited the growth of *F. oxysporum*. The possibility that effects may differ between isolates cannot be excluded, but, as noted by Duchesne et al. (1989b), the substrate might be a key factor in determining whether or not antagonism will occur. *Cylindrocarpon* (Buscot et al. 1992) and *Phytophthora* (Branzanti et al. 1994) have also been shown to be negatively affected by *P. involutus* in pure culture trials. *P. involutus* has also been demonstrated to suppress soil bacterial activity (Olsson et al. 1996; Olsson and Wallander 1999). Nurmiaho et al. (1997) found *P. involutus–P. sylvestris* ECM to be devoid of bacteria in contrast to non-ECM roots and roots colonised by *S. bovinus* which had abundant bacteria attached to their surfaces. On the other hand, the external mycelia produced by both *P. involutus* and *S. bovinus* were extensively colonized by bacteria.

9.4.2
Protection Against Toxic Heavy Metals

ECM associations seem to play a role in enhancing tree resistance to potentially toxic heavy metals in the soil. It seems clear that, at least in some circumstances, *P. involutus* can increase the tolerance to certain metals, but, in spite of substantial interest and research efforts, the mechanisms behind this protective effect are still unclear. In a study on Zn effects on birch, Wilkins and co-workers (Brown and Wilkins 1985a,b; Denny and Wilkins. 1987a,b) found that *P. involutus* increased Zn tolerance, and Denny and Wilkins (1987b) subsequently suggested that this effect was mainly due to the absorption of Zn to the hyphal surface, thereby lowering the Zn concentration in the soil solution. This hypothesis was supported by Colpaert and van Assche (1992) who found that the effectiveness of *P. involutus* in counteracting Zn toxicity was related to the amount of external mycelium. A similar hyphal-accumulating effect was found for Pb (Marschner et al. 1996), but there are also reports of *P. involutus* ECM failing to exclude Pb from root tissue (Jentschke et al. 1991, 1997). In soils, the most common heavy metal is Al. Hintikka (1988) reported that *P. involutus* showed some tolerance against Al in the culture medium; growth was observed at an Al^{3+} concentration of $15\,\mathrm{g}\,l^{-1}$ and was fairly good at $5\,\mathrm{g}\,l^{-1}$ which was equivalent to $100\,\mathrm{g}\;Al_2(SO_4)_3.16H_2O\,l^{-1}$. In a study of Norway spruce tolerance to Al, Wilkins and Hodson (1989) and Hodson and Wilkins (1991) showed that *P. involutus* in the rhizosphere (no proper infection) reduced the negative effects of Al on plant growth. Hentschel et al. (1993) showed that colonisation by *P. involutus* decreased the degree to which Al reduced shoot growth. However, in ECM plants that had been exposed to Al for 10 weeks, photosynthesis was, nevertheless, reduced. The mechanisms responsible for increased Al tolerance are still unclear. Turnau et al. (1994) found higher concentrations of cystine-rich proteins in mycelium sampled from heavy metal polluted soils, and Turnau et al. (1993) reported granules containing Al and P in the mycelium from *Pinus* ECM growing in polluted soil. The same authors also reported that Cd accumulated inside fungal vacuoles and also in another type of granule that contained high amounts of N.

Few data exist on the contents of heavy metals and their possible accumulation in *Paxillus* sporocarps. Hg was found not to accumulate in *P. involutus* sporocarps (Falandysz and Chwir 1997). The ^{137}Cs fallout after the Chernobyl accident in 1986 has resulted in a few studies of Cs concentrations in sporocarps. Both Tsvetnova and Shcheglov (1994) and Semerdzieva et al. (1992) found that many fungi accumulate Cs, and both studies found *P. involutus* to be one of the most efficient fungi in this respect. Tsvetnova and Shcheglov (1994) calculated a transfer factor (radioactivity in sporocarp/

radioactivity in soil) of 5057 for *P. involutus* and 1147 for *Xerocomus badius* (Fr.) Kühn. & Gilb., but only 5 for *A. muscaria* (3 km from Chernobyl). A general problem with comparing heavy metal studies is that the behaviour of a given metal differs depending on the ion species present. Thus, results can vary depending on the fungal isolate present as well as the chemical environment under which the experiments are performed.

9.4.3
Other Non-nutritional Benefits

ECM fungi can also affect the water relations and photosynthesis of their plant host. However, not many studies on such effects have been made with *Paxillus*. Experimental evidence that the fungus can improve water uptake was first presented by Read and Boyd (1986) who found that water can be transferred in hyphal strands over dry soil areas. The physiological significance of this observation was not fully elucidated. In a number of papers, Lehto (1990, 1992a,b,c) reported on how ECM colonisation with *P. sitchensis* was affected by soil moisture conditions. She showed that in drier soils *P. involutus* tended to show better ECM formation compared with other tested fungi (Lehto 1992b). Furthermore, when comparing ECM and non-ECM seedlings she found that stomatal conductance, net photosynthesis rate and water potential were all higher in ECM plants (Letho 1992a), but she interpreted this improved performance mainly as an effect of improved nutrient status. Boyle and Hellenbrand (1991) studied the effects of water potential on ECM and non-ECM seedlings of jack pine and black spruce by adding polyethylene glycol (PEG) as well as by growing axenic cultures of the fungi used. Of the five fungi used, *P. involutus* appeared to be less efficient in counteracting the negative effects of low water potential, as evaluated in terms of the apparent photosynthetic rate. However, *P. involutus* was not included in all their experiments. Generally, ECM plants performed better than non-ECM ones. In a field trial where inoculated oak (*Q. robur* and *Q. petraea*) seedlings were planted and the presence of the inoculated fungi was evaluated together with plant growth after 7 years, survival was found to be higher for *P. involutus* compared with *Laccaria laccata* (Scop.: Fr.) Cooke and *H. crustuliniforme* and the seedlings with *P. involutus* grew better (Garbaye and Churin 1997). The positive response was more pronounced in dry years.

In summary, there are indications that plants with *P. involutus* have a faster photosynthetic rate than uncolonised plants (Lehto 1992a; Garbaye and Churin 1997), which might be due to a fungal-induced enhancement of nutrient status. Plants colonised with *P. involutus* seem better able to withstand dry conditions, but *P. involutus* does not seem to be more efficient in this regard than many other common ECM fungi.

The possible importance of plant hormones in the ECM symbioses has been the subject of many publications and discussions, but few papers on this topic have dealt with *P. involutus*. Indolyl-3-acetic acid (IAA) occurs in sporocarps (Kieliszewska-Rokicka et al. 1995), and can be produced in pure cultures (Rudawska and Kieliszewska-Rokicka 1997), but its production seems to vary greatly between isolates. As mentioned above, *P. sylvestris* seedlings inoculated with isolates with a high potential to produce IAA developed more fine roots as well as more ECM roots compared with seedlings inoculated with isolates producing less IAA. As a consequence of their more developed root systems, the former seedlings also grew faster. Polyamine (putrescine and spermidine) levels in pine and spruce are also affected by ECM formation (Kytoviita and Sarjala 1997), although the consequences of this production are unclear.

9.5
Conclusions

P. involutus appears to have an unusually broad host range, both in nature and in laboratory experiments. It is an ECM fungus efficient in utilising complex organic nutrients, both in the form of proteins and amino acids, and in the uptake of ions of low mobility in the soil, such as phosphate. Furthermore, it appears to be active in reducing the damage caused by root-colonising pathogens. *P. involutus* seems to have a strong influence on other microorganisms in the soil, where antagonistic effects have been demonstrated against both pathogens and saprophytes. It also increases the tolerance of host plants to heavy metals and *P. involutus* is often seen fruiting in mine spoils. Compared with many other ECM fungi, *P. involutus* is relatively fast-growing. Not only does it produce a high mycelial biomass, but also it is fast in colonising root tips. ECM roots colonised by this fungus tend to be more short lived than root tips colonised by other ECM fungi. *P. involutus*, like most other ECM fungi, shows great variability in many respects, such as in its ability to utilise different nutrient sources. Much of the physiological work on this fungus has, however, been performed with very few isolates. Predictions concerning the performance of a specific fungus–host association are very difficult to make because of the great variability between different isolates of the same fungus. Future studies involving more isolates are necessary to allow better predictions.

References

Abuzinadah RA, Read DJ (1989) The role of proteins in the nitrogen nutrition of ectomycorrhizal plants. IV. The utilization of peptides by birch *Betula pendula* Roth. infected with different mycorrhizal fungi. New Phytol 112:55–60

Abuzinadah RA, Finlay RD, Read DJ (1986) The role of proteins in the nitrogen nutrition of ectomycorrhizal plants. II. Utilization of protein by mycorrhizal plants of *Pinus contorta*. New Phytol 103:495–506

Agerer R (1991) Characterisation of ectomycorrhizae. In: Norris JR, Read DJ, Varma AK (eds) Methods in microbiology, vol 23. Academic Press, London, pp 25–73

Andersson S, Jensen P, Söderström B (1996) Effects of mycorrhizal colonization by *Paxillus involutus* on uptake of Ca and P by *Picea abies* and *Betula pendula* grown in unlimed and limed peat. New Phytol 133:695–704

Arnebrant K (1994) Nitrogen amendments reduce the growth of extramatrical ectomycorrhizal mycelium. Mycorrhiza 5:7-15

Arnebrant K, Ek H, Finlay RD, Söderström B (1993) Nitrogen translocation between *Alnus glutinosa* L. Gaertn. seedlings inoculated with *Frankia* sp. and *Pinus contorta* Doug. ex Loud seedlings connected by a common ectomycorrhizal mycelium. New Phytol 124:231-242

Arnolds EJM (1988) The changing macromycete flora in the Netherlands. Trans Br Mycol Soc 90:391-406

Baar J, Comini B, Elferink MO, Kuyper TW (1997) Performance of four ectomycorrhizal fungi on organic and inorganic nitrogen sources. Mycol Res 101:523-529

Baxter R, Emes J, Lee LA (1992) Effects of an experimentally applied increase in ammonium on growth and amino-acid metabolism of *Sphagnum cuspidatum* Ehrh. ex Hoffm. from differently polluted areas. New Phytol 120:265-274

Bending GD, Read DJ (1995a) The structure and function of the vegetative mycelium of ectomycorrhizal plants. V. The foraging behaviour of ectomycorrhizal mycelium and the translocation of nutrients from exploited organic matter. New Phytol 130:401-409

Bending GD, Read DJ (1995b) The structure and function of the vegetative mycelium of ectomycorrhizal plants. VI. Activities of nutrient mobilizing enzymes in birch litter colonized by *Paxillus involutus* (Fr.) Fr. New Phytol 130:411-417

Blaudez D, Chalot M, Dizengremel P, Botton B (1998) Structure and function of the ectomycorrhizal association between *Paxillus involutus* and *Betula pendula*. New Phytol 138:543-552

Boyle CD, Hellenbrand KE (1991) Assessment of the effect of mycorrhizal fungi on drought tolerance of conifer seedlings. Can J Bot 69:1764-1771

Branzanti B, Zambonelli A (1989) Synthesis of mycorrhizas on *Quercus ruber* using *Hebeloma sinapizans* and *Paxillus involutus*. Agric Ecosyst Environ 28:35-40

Branzanti MB, Rocca E, Zambonelli A (1994) Effects of ectomycorrhizal fungi on *Phytophthora cambivora* and *Phytophthora cinnamomi*. Micoe Ital 23:47-52

Bresinsky A (1996) On *Leccinum subcinnamomeum, Rhizopogon pumilionus, Paxillus filamentosus, Paxillus rubicundulus*. Z Mykol 62:61-68

Brown MT, Wilkins DA (1985a) Zinc tolerance of *Amanita* and *Paxillus*. Trans Br Mycol Soc 84:367-369

Brown MT, Wilkins DA (1985b) Zinc tolerance of mycorrhizal *Betula*. New Phytol 99:101-106

Brun A, Chalot M, Finlay RD, Söderström B (1995) Structure and function of the ectomycorrhizal association between *Paxillus involutus* (Batsch) Fr. and *Betula pendula* Roth. I. Dynamics of mycorrhiza formation. New Phytol 129:487-493

Brunner IL, Brunner F, Miller OK (1990) Ectomycorrhizal synthesis with Alaskan *Alnus tenuifolia*. Can J Bot 68:761-767

Buscot F, Weber G, Oberwinkler F (1992) Interactions between *Cylindrocarpon destructans* and ectomycorrhizas of *Picea abies* with *Laccaria laccata* and *Paxillus involutus*. Trees 6:83-90

Cairney JWG, Smith SE (1993) Efflux of phosphate from the ectomycorrhizal basidiomycete *Pisolithus tinctorius*: general characteristics and the influence of intracellular phosphorus concentration. Mycol Res 97:1261-1266

Chakravarty P, Hwang SF (1991) Effect of an ectomycorrhizal fungus *Laccaria laccata* on *Fusarium* damping-off in *Pinus banksiana* seedlings. Eur J For Pathol 21:97-106

Chakravarty P, Peterson RL, Ellis BE (1990) Integrated control of *Fusarium* damping-off in red pine seedlings with the ectomycorrhizal fungus *Paxillus involutus* and fungicides. Can J For Res 20:1283-1288

Chalot M, Brun A, Finlay RD, Söderström B (1994a) Respiration of (^{14}C) alanine by the ectomycorrhizal fungus *Paxillus involutus*. FEMS Microbiol Lett 121:87-91

Chalot M, Brun A, Finlay RD, Söderström B (1994b) Metabolism of (^{14}C) glutamate and (^{14}C) glutamine by the ectomycorrhizal fungus *Paxillus involutus*. Microbiol 140:1641-1649

Chalot M, Kytoviita MM, Brun A, Finlay RD, Söderström B (1995) Factors affecting amino acid uptake by the ectomycorrhizal fungus *Paxillus involutus*. Mycol Res 99:1131-1138

Chalot M, Brun A, Botton B, Söderström B (1996) Kinetics, energetics and specificity of a general amino acid transporter from the ectomycorrhizal fungus *Paxillus involutus*. Microbiol 142:1749–1756

Colpaert JV, van Assche JA (1992) Zinc toxicity in ectomycorrhizal *Pinus sylvestris*. Plant Soil 143:201–211

Colpaert JV, van Tichelen KK (1996) Decomposition, nitrogen and phosphorus mineralization from beech leaf litter colonized by ectomycorrhizal or litter-decomposing basidiomycetes. New Phytol 134:123–132

Colpaert JV, van Assche JA, Luijtens K (1992) The growth of the extramatrical mycelium of ectomycorrhizal fungi and the growth response of *Pinus sylvestris* L. New Phytol 120:127–135

Cumming JR (1996) Phosphate-limitation physiology in ectomycorrhizal pitch pine (*Pinus rigida*) seedlings. Tree Physiol 16:977–983

Denny HJ, Wilkins DA (1987a) Zinc tolerance in *Betula* spp. III. Variation in response to zinc among ectomycorrhizal associates. New Phytol 106:535–544

Denny HJ, Wilkins DA (1987b) Zinc tolerance in *Betula* spp. IV. The mechanism of ectomycorrhizal amelioration of zinc toxicity. New Phytol 106:545–554

Dighton J (1983) Phosphatase production by mycorrhizal fungi. Plant Soil 71:455–462

Dighton J, Poskitt JM, Brown TK (1993) P influx into ectomycorrhizal and saprotrophic fungal hyphae in relation to P supply. Mycol Res 97:355–358

Downes GM, Alexander IJ, Cairney JWG (1992) A study of ageing of spruce *Picea sitchensis* Bong. Carr. ectomycorrhizas. II. Morphological and cellular changes in mycorrhizas formed by *Tylospora fibrillosa* Burt. Donk and *Paxillus involutus* (Batsch ex Fr.) Fr. New Phytol 122:141–152

Duchesne LC, Peterson RL, Ellis BE (1988a) Interaction between the ectomycorrhizal fungus *Paxillus involutus* and *Pinus resinosa* induces resistance to *Fusarium oxysporum*. Can J Bot 66:558–562

Duchesne LC, Peterson RL, Ellis BE (1988b) Pine root exudate stimulates the synthesis of antifungal compounds by the ectomycorrhizal fungus *Paxillus involutus*. New Phytol 108:471–476

Duchesne LC, Ellis BE, Peterson RL (1989a) Disease suppression by the ectomycorrhizal fungus *Paxillus involutus* contribution of oxalic acid. Can J Bot 67:2726–2730

Duchesne LC, Peterson RL, Ellis BE (1989b) The time-course of disease suppression and antibiosis by the ectomycorrhizal fungus *Paxillus involutus*. New Phytol 111:693–698

Duddridge JA (1987) Specificity and recognition in ectomycorrhizal associations. In: Pegg GF, Ayres PG (eds) Fungal infection of plants. Cambridge University Press, Cambridge, pp 25–44

Ek H (1997) The influence of nitrogen fertilization on the carbon economy of *Paxillus involutus* in ectomycorrhizal association with *Betula pendula*. New Phytol 135:133–142

Ek H, Andersson S, Arnebrant K, Söderström B (1994a) Growth and assimilation of NH_4 and NO_3 by *Paxillus involutus* in association with *Betula pendula* and *Picea abies* as affected by substrate pH. New Phytol 128:629–637

Ek H, Sjögren M, Arnebrant K, Söderström B (1994b) Extramatrical mycelial growth, biomass allocation and nitrogen uptake in ectomycorrhizal systems in response to collembolan grazing. Appl Soil Ecol 1:155–169

Ekblad A, Huss Danell K (1995) Nitrogen fixation by *Alnus incana* and nitrogen transfer from *A. incana* to *Pinus sylvestris* influenced by macronutrients and ectomycorrhiza. New Phytol 131:453–459

Ekblad A, Wallander H, Carlsson R, Huss Danell K (1995) Fungal biomass in roots and extramatrical mycelium in relation to macronutrients and plant biomass of ectomycorrhizal *Pinus sylvestris* and *Alnus incana*. New Phytol 131:443–451

Ekblad A, Wallander H, Näsholm T (1998) Chitin and ergosterol combined to measure total and living biomass in ectomycorrhizae. New Phytol 138:143–149

Erland S, Finlay RD (1992) Effects of temperature and incubation time on the ability of three ectomycorrhizal fungi to colonize *Pinus sylvestris* roots. Mycol Res 96:270–272

Erland S, Söderström B (1991) Effects of liming on ectomycorrhizal fungi infecting *Pinus sylvestris* L. III. Saprophytic growth and host plant infection at different pH values by some ectomycorrhizal fungi in unsterile humus. New Phytol 117:405–411

Falandysz J, Chwir A (1997) The concentrations and bioconcentration factors of mercury in mushrooms from the Mierzeja Wislana sand-bar, northern Poland. Sci Total Environ 203:221–228

Farquhar ML, Peterson RL (1990) Early effects of the ectomycorrhizal fungus *Paxillus involutus* on the root rot organism *Fusarium* associated with *Pinus resinosa*. Can J Bot 68:1589–1596

Finlay RD, Ek H, Odham G, Söderström B (1989) Uptake, translocation and assimilation of nitrogen from nitrogen-labelled ammonium and nitrate sources by intact ectomycorrhizal systems of *Fagus sylvatica* infected with *Paxillus involutus*. New Phytol 113:47–56

Finlay RD, Frostegård A, Sonnerfeldt AM (1992) Utilization of organic and inorganic nitrogen sources by ectomycorrhizal fungi in pure culture and in symbiosis with *Pinus contorta* Dougl. ex Loud. New Phytol 120:105–115

Fischer M (1995) On the order Boletales: isolation and characterization of DNA from fruiting bodies and mycelia. Z Mykol 61:245–260

Fox FM (1986a) Groupings of ectomycorrhizal fungi of birch and pine based on establishment of mycorrhizas on seedlings from spores in unsterile soils. Trans Br Mycol Soc 87:371–380

Fox FM (1986b) Ultrastructure and infectivity of sclerotia of the ectomycorrhizal fungus *Paxillus involutus* on birch (*Betula* spp.). Trans Br Mycol Soc 87:627–631

Fries N (1985) Intersterility groups in *Paxillus involutus*. Mycotaxon 24:403–410

Gale W (1977) Formation of mycorrhiza from cuttings application to *Salix repens* associated with *Paxillus involutus* and *Pisolithus arhizus*. Bull Jard Bot Nate Belg 47:91–98

Garbaye J, Churin JL (1997) Growth stimulation of young oak plantations inoculated with the ectomycorrhizal fungus *Paxillus involutus* with special reference to summer drought. For Ecol Manage 98:221–228

Gea L, Normand L, Vian B, Gay G (1994) Structural aspects of ectomycorrhiza of *Pinus pinaster* (Ait.) Sol. formed by an IAA-overproducer mutant *of Hebeloma cylindrosporum* Romagnesi. New Phytol 128:659–670

Grellier B, Strullu DG, Letouze R (1984) Micropropagation of birch (*Betula pendula*) and mycorrhizal formation in vitro. New Phytol 97:591–600

Grellier B, Strullu DG, Martin F, Renaudin S (1989) Synthesis in-vitro, microanalysis and phosphorus-31 NMR study of metachromatic granules in birch mycorrhizas. New Phytol 112:49–54

Grenville DJ, Peterson RL, Piché Y (1985) The development, structure and histochemistry of sclerotia of ectomycorrhizal fungi. II. *Paxillus involutus*. Can J Bot 63:1412–1417

Hahn C (1996) Studies in the genus *Paxillus*. I. *Paxillus gymnopus*. A new *Paxillus* from the Pacific rainforests of Columbia. Z Mykol 62:43–60

Harley JL (1989) The significance of mycorrhiza. Mycol Res 92:591–600

Harley JL, Smith SE (1983) Mycorrhizal symbiosis. Academic Press, London

Hentschel E, Godbold DL, Marschner P, Schlegel H, Jentschke G (1993) The effect of *Paxillus involutus* Fr. on aluminum sensitivity of Norway spruce seedlings. Tree Physiol 12:379–390

Heslin MC, Douglas GC (1986) Synthesis of poplar mycorrhizas. Trans Br Mycol Soc 86:117–122

Hilbert JL, Costa G, Martin F (1991) Ectomycorrhizin synthesis and polypeptide changes during the early stage of eucalypt mycorrhiza development. Plant Physiol 97:977–984

Hilger AB, Krause HH (1989) Growth characteristics of *Laccaria laccata* and *Paxillus involutus* in liquid culture media with inorganic and organic phosphorus sources. Can J Bot 67:1782–1789

Hintikka V (1988) High aluminum tolerance among ectomycorrhizal fungi. Karstenia 28:41–44

Ho I (1989) Acid phosphatase alkaline phosphatase and nitrate reductase activity of selected ectomycorrhizal fungi. Can J Bot 67:750–753

Hodge A, Alexander IJ, Gooday GW (1995) Chitinolytic enzymes of pathogenic and ectomycorrhizal fungi. Mycol Res 99:935–941

Hodge A, Alexander IJ, Gooday GW, Killham K (1996a) Carbon allocation patterns in fungi in the presence of chitin in the external medium. Mycol Res 100:1428–1430

Hodge A, Gooday GW, Alexander IJ (1996b) Inhibition of chitinolytic activities from tree species and associated fungi. Phytochemistry 41:77–84

Hodson MJ, Wilkins DA (1991) Localization of aluminum in the roots of Norway spruce *Picea abies* L. Karst. inoculated with *Paxillus involutus* Fr. New Phytol 118:273–278

Horan DP, Chilvers GA (1990) Chemotropism, the key to ectomycorrhizal formation. New Phytol 116:297–302

Hughes E, Mitchell DT (1995) Utilization of sucrose by *Hymenoscyphus ericae* (an ericoid endomycorrhizal fungus) and ectomycorrhizal fungi. Mycol Res 99:1233–1238

Hutchison LJ (1992) The taxonomic significance of extracellular deoxyribonuclease activity among species of ectomycorrhizal fungi. Mycotaxon 45:63–69

Hwang SH, Chakravarty P, Chang KF (1995) The effect of two ectomycorrhizal fungi, *Paxillus involutus* and *Suillus tomentosus*, and of *Bacillus subtilis* on *Fusarium* damping-off in jack pine seedlings. Phytoprotection 76:57–66

Jansen AE, de Vries FW (1990) Mycorrhizas on Douglas fir in the Netherlands. Agric Ecosyst Environ 28:197–200

Jentschke G, Fritz E, Godbold D (1991) Distribution of lead in mycorrhizal and non-mycorrhizal Norway spruce seedlings. Physiol Plant 81:417–422

Jentschke G, Fritz E, Marschner P, Rapp C, Wolters V, Godbold DL (1997) Mycorrhizal colonization and lead distribution in root tissues of Norway spruce seedlings. Z Pflanzen Boden 160:317–321

Keller G (1996) Utilization of inorganic and organic nitrogen sources by high-subalpine ectomycorrhizal fungi of *Pinus cembra* in pure culture. Mycol Res 100:989–998

Kieliszewska-Rokicka B (1992) Acid phosphatase activity in mycorrhizal and non-mycorrhizal Scots pine seedlings in relation to nitrogen and phosphorus nutrition. Acta Soc Bot Pol 61:253–264

Kieliszewska-Rokicka B, Rudawska M, Leski T (1995) Effects of acid rain and aluminium on ectomycorrhizal symbiosis: alterations of IAA-synthesizing activity in ectomycorrhizal fungi. Bulg J Plant Physiol 21:111–119

Kytoviita MM, Sarjala T (1997) Effects of defoliation and symbiosis on polyamine levels in pine and birch. Mycorrhiza 7:107–111

Laiho O (1970) *Paxillus involutus* as a mycorrhizal symbiont of forest trees. Acta For Fenn 106:5–72

Lapeyrie F (1988) Oxalate synthesis from soil bicarbonate by the mycorrhizal fungus *Paxillus involutus*. Plant Soil 110:3–8

Lapeyrie F, Chilvers GA, Bhem CA (1987) Oxalic acid synthesis by the mycorrhizal fungus *Paxillus involutus* (Batsch ex Fr.) Fr. New Phytol 106:139–146

Lapeyrie F, Ranger J, Vairelles D (1991) Phosphate-solubilizing activity of ectomycorrhizal fungi in-vitro. Can J Bot 69:342–346

Lehto T (1992a) Effect of drought on *Picea sitchensis* seedlings inoculated with mycorrhizal fungi. Scand J For Res 7:177–182

Lehto T (1992b) Mycorrhizas and drought resistance of *Picea sitchensis* Bong. Carr. I. In conditions of nutrient deficiency. New Phytol 122:661–668

Lehto T (1992c) Mycorrhizas and drought resistance of *Picea sitchensis* Bong. Carr. II. In conditions of adequate nutrition. New Phytol 122:669–673

Lehto TH (1990) Effects of mycorrhiza and drought on photosynthesis and water relations of Sitka spruce. Agric Ecosyst Environ 28:299–304

Lei J, Dexheimer J (1987) Preliminary results concerning the controlled mycorrhization of oak (*Quercus robur* L.) vitroplants. Ann Sci For 44:315–324

Maijala P, Fagerstedt KV, Raudaskoski M (1991) Detection of extracellular cellulolytic and proteolytic activity in ectomycorrhizal fungi and *Heterobasidion annosum* Fr. Bref. New Phytol 117:643–648

Malajczuk N, Lapeyrie F, Garbaye J (1990) Infectivity of pine and eucalypt isolates of *Pisolithus tinctorius* on roots of *Eucalyptus urophylla* in vitro. I. Mycorrhizal formation in model systems. New Phytol 114:627–631

Marschner P, Godbold DL (1995) Mycorrhizal infection and ageing affect element localization in short roots of Norway spruce (*Picea abies* (L.) Karst.). Mycorrhiza 5:417–422

Marschner P, Godbold DL, Jentschke G (1996) Dynamics of lead accumulation in mycorrhizal and non-mycorrhizal Norway spruce (*Picea abies* (L.) Karst). Plant Soil 178: 239–245

Marx DH (1969) The influence of ectotrophic ectomycorrhizal fungi on the resistance of pine roots to pathogenic infections. I. Antagonism of mycorrhizal fungi to pathogenic fungi and soil bacteria. Phytopathology 59:153–163

McElhinney C, Mitchell DT (1993) Phosphatase activity of four ectomycorrhizal fungi found in a Sitka spruce–Japanese larch plantation in Ireland. Mycol Res 97:725–732

McElhinney C, Mitchell DT (1995) Influence of ectomycorrhizal fungi on the response of Sitka spruce and Japanese larch to forms of phosphorus. Mycorrhiza 5:409–415

Melville LH, Massicotte HB, Peterson RL (1987) Morphological variations in developing ectomycorrhizae of *Dryas integrifolia* and five fungal species. Scanning Microsc 1:1455–1464

Miller SL, Koo CD, Molina R (1991) Characterization of red alder ectomycorrhizae, a preface to monitoring belowground ecological responses. Can J Bot 69:516–531

Molina R (1979) Pure culture synthesis and host specificity of red alder (*Alnus rubra*) mycorrhizae. Can J Bot 57:1223–1228

Molina R (1981) Ectomycorrhizal specificity in the genus *Alnus*. Can J Bot 59:325–334

Molina R, Trappe J (1982) Patterns of ectomycorrhizal host specificity and potential among Pacific Northwest conifers and fungi. For Sci 28:1223–1228

Murphy JF, Miller OK (1994) Mycorrhizal syntheses with *Alnus serrulata* (Ait.) Willd. Castanea 59:156–166

Näsholm T, Ekblad A, Nordin A, Giesler R, Högberg M, Bogberg P (1998) Boreal forest plants take up organic nitrogen. Nature 392:914–916

Nilsson MC, Högberg P, Zackrisson O, Wang F (1993) Allelopathic effects by *Empetrum hermaphroditum* on development and nitrogen uptake by roots and mycorrhizae of *Pinus sylvestris*. Can J Bot 71:620–628

Nurmiaho Lassila EL, Timonen S, Haahtela K, Sen R (1997) Bacterial colonization patterns of intact *Pinus sylvestris* mycorrhizospheres in dry pine forest soil: an electron microscopy study. Can J Microbiol 43:1017–1035

Ohenoja E (1978) Mushrooms and mushroom yield in fertilized forests. Ann Bot Fenn 15:38–46

Olsson PA, Wallander H (1999) Interactions between ectomycorrhizal fungi and the bacterial community in soils amended with various primary minerals. FEMS Microbiol Ecol 27:195–205

Olsson PA, Chalot M, Bååth E, Finlay RD, Söderström B (1996) Ectomycorrhizal mycelia reduce bacterial activity in sandy soil. FEMS Microbiol Ecol 21:77–86

Pargney JC, Gourp V (1991) Contribution to the study of mycorrhiza in *Pinus pinaster* and ultrastructure of the associations obtained with two basidiomycetes *Hebeloma cylindrosporum* Romagn. and *Paxillus involutus* Fr. Phytomorphology 41:1992

Paris F, Bonnaud P, Ranger J, Lapeyrie F (1995) In vitro weathering of phlogopite by ectomycorrhizal fungi. I. Effect of K^+ and Mg^{2+} deficiency on phyllosilicate evolution. Plant Soil 177:191–201

Paris F, Botton B, Lapeyrie F (1996) In vitro weathering of phlogopite by ectomycorrhizal fungi. Plant Soil 179:141–150

Paul EA, Clark FE (1989) Soil microbiology and biochemistry. Academic Press, San Diego

Pera J, Alvarez IF (1995) Ectomycorrhizal fungi of *Pinus pinaster*. Mycorrhiza 5:193–200

Piché Y, Fortin JA (1982) Development of mycorrhizae extramatrical mycelium and sclerotia on *Pinus strobus* seedlings. New Phytol 91:211–220

Read DJ (1992) The mycorrhizal mycelium. In: Allen MF (ed) Mycorrhizal functioning. Chapman and Hall, London, pp 102–133

Read DJ, Boyd R (1986) Water relations of mycorrhizal fungi and their host plants. In: Ayres PG, Boddy L (eds) Water, fungi and plants. Cambridge University Press, Cambridge, pp 287–304

Rudawska ML, Kieliszewska-Rokicka B (1997) Mycorrhizal formation by *Paxillus involutus* strains in relation to their IAA-synthesizing activity. New Phytol 137:509–517

Rudawska M, Kieliszewska-Rokicka B, Debaud JC, Lewndowski A, Gay G (1994) Enzymes of ammonium metabolism in ectoendomycorrhizal and ectomycorrhizal symbionts of pine. Physiol Plant 92:279–285

Sarjala T (1990) Effect of nitrate and ammonium concentration on nitrate reductase activity in five species of mycorrhizal fungi. Physiol Plant 79:65–70

Sarjala T (1991) Effect of mycorrhiza and nitrate nutrition on nitrate reductase activity in Scots pine seedlings. Physiol Plant 81:89–94

Sasa M, Krogstrup P (1991) Ectomycorrhizal formation in plantlets derived from somatic embryos of Sitka spruce. Scand J For Res 6:129–136

Semerdzieva M, Vobecky M, Tamchynova J, Tethal T (1992) Activity of caesium-137 and caesium-134 in several mushrooms of two different sites of Central Bohemia in 1986–1990. Ceska Mykol 46:67–74

Shaw TM, Dighton J, Sanders FE (1995) Interactions between ectomycorrhizal and saprotrophic fungi on agar and in association with seedlings of lodgepole pine (*Pinus contorta*). Mycol Res 99:159–165

Simoneau P, Viemont JD, Moreau JC, Strullu DG (1993) Symbiosis-related polypeptides associated with the early stages of ectomycorrhiza organogenesis in birch (*Betula pendula* Roth.). New Phytol 124:495–504

Slankis V (1973) Hormonal relationships in mycorrhizal development. In: Marks GC, Kozlowski TT (eds) Ectomycorhizae. Academic Press, New York, pp 231–298

Smith SE, Read DJ (1997) Mycorrhizal symbiosis. Academic Press, London

Strullu DG, Grellier B, Marciniak D, Letouze R (1986) Micropropagation of chestnut and conditions of mycorrhizal syntheses in vitro. New Phytol 102:95–102

Theodorou C, Reddell P (1991) In vitro synthesis of ectomycorrhizas on Casuarinaceae with a range of mycorrhizal fungi. New Phytol 118:279–288

Timonen S, Söderström B, Raudaskoski M (1996a) Dynamics of cytoskeletal proteins in developing pine ectomycorrhiza. Mycorrhiza 6:423–429

Timonen S, Finlay RD, Olsson S, Söderström B (1996b) Dynamics of phosphorus translocation in intact ectomycorrhizal systems: non-destructive monitoring using a beta-scanner. FEMS Microbiol Ecol 19:171–180

Trappe JM (1962) Fungus associates of ectotrophic mycorrhizae. Bot Rev 28:538–606

Tsvetnova OB, Shcheglov AI (1994) Cs-137 content in the mushrooms of radioactive contaminated zones of the European part of the CIS. Sci Total Environ 155:25–29

Turnau K, Kottke I, Oberwinkler F (1993) *Paxillus involutus–Pinus sylvestris* mycorrhizae from a heavily polluted forest. I. Element localization using electron energy loss spectroscopy and imaging. Bot Acta 106:213–219

Turnau K, Kottke I, Dexheimer J (1994) *Paxillus involutus–Pinus sylvestris* mycorrhizae from heavily polluted forest. II. Ultrastructural and cytochemical observations. Bot Acta 107:73–80

Villeneuve N, Le Tacon F, Bouchard D (1991) Survival of inoculated *Laccaria bicolor* in competition with native ectomycorrhizal fungi and effects on the growth of outplanted Douglas-fir seedlings. Plant Soil 135:95–108

Wallander H, Nylund JE, Sundberg B (1994) The influence of IAA, carbohydrate and mineral concentration in host tissue on ectomycorrhizal development on *Pinus sylvestris* L. in relation to nutrient supply. New Phytol 127:521–528

Wallander H, Wickman T, Jacks G (1997) Apatite as a P source in mycorrhizal and non-mycorrhizal *Pinus sylvestris* seedlings. Plant Soil 196:123–131

Wallander H, Dahlberg A, Arnebrant K (1999) Relationship between fungal uptake of ammonium and fungal growth in ectomycorrhizal *Pinus sylvestris* seedlings grown at high and low nitrogen availability. Mycorrhiza 8:215–223

Wallander H, Wichman T (1999) Biotite and microcline as potassium sources in ectomycorrhizal and non-mycorrhizal *Pinus Sylvestris* seedlings. Mycorrhiza 9:25–32

Wilkins DA, Hodson J (1989) The effects of aluminum and *Paxillus involutus* Fr. on the growth of Norway spruce *Picea abies* (L.) Karst. New Phytol 113:225–232

Chapter 10

Cantharellus

E. Danell

10.1
Introduction

Chanterelles (*Cantharellus* spp.) have been known since medieval times (Lobelius 1581) as some of the most appreciated edible mushrooms (Danell 1994a). The current world market is estimated at £1 billion (Watling 1997). The number of described *Cantharellus* species worldwide exceeds 70. The genus is known from every forested continent where it is associated with a wide range of host genera (Corner 1966, 1969; Donk 1969; Petersen 1979; Nuhamara 1987; Thoen and Ba 1989; Watling and Abraham 1992; Danell 1994a; Pegler et al. 1997). Its economical and gastronomical value, the recent development of cultivation techniques, its ability to form long-lived ecto-mycorrhizal (ECM) associations, its insect repellent traits, sporocarp phototropism, decline in central Europe, symbiosis with bacteria and phylogenetic distance from other basidiomycetes have caused an increased interest in this genus. These aspects are reviewed below, but the reader may also contact the scientists involved in current research via the internet: http://www.mykopat.slu.se/mycorrhiza/edible/home.phtml.

10.2
Taxonomy

Since Fries (1874), Patouillard (1900), Donk (1964) and Corner's (1966) classical monograph on cantharelloid fungi, Petersen (1971) and Romagnesi (1995) have revised the family Cantharellaceae using classical microscopic and chemical characters. However, recent sequence analysis of the nuclear large subunit rDNA by Feibelman et al. (1997) revealed only two valid genera within Cantharellaceae in North America and Europe (*Cantharellus* and *Craterellus*). In general *Cantharellus* contains species with fleshy sporocarps, while *Craterellus* species have thin and leathery sporocarps with dark pigments. These results are concordant with the results of Pine et al. (in prep.)

Department of Forest Mycology and Pathology, Swedish University of Agricultural Sciences, Box 7026, 75007 Uppsala, Sweden and Dahlman et al. (1998). *Craterellus sinuosus* (Fr.) Fr. [*Pseudocraterellus undulatus* (Pers.: Fr.)] and *Craterellus tubaeformis* (Fr.) Quél. [*Cantharellus tubaeformis* (Bull.: Fr.) Fr.] are examples of modern nomenclature.

Gomphus, Clavariadelphus and *Hydnum* are sometimes cited as close relatives of chantarelles (Corner 1957, 1966; Donk 1964; Petersen 1971; Persson and Mossberg 1997). Reijnders and Stalpers (1992), however, claimed that *Hydnum* has a different hymenophoral trama and lacks carotenoid pigments and is therefore not related to *Cantharellus*. According to Hibbett et al. (1997) who made a comparison based on nuclear and mitochondrial DNA sequences, *Cantharellus* is related to *Hydnum* but not to *Gomphus* or *Clavariadelphus*. Hibbett et al. (1997) also showed a vast phylogenetic distance between *Cantharellus* and most other basidiomycetes. *Cantharellus* is often placed in a separate order, Cantharellales (Hansen and Knudsen 1997; Pegler et al. 1997), which, according to the results of Hibbett et al. (1997) and Pine et al. (in prep.), needs further revision. Hibbett and Thorn (in prep.) divided the homobasidiomycetes into eight clades of which the Cantharelloid clade contains Cantharellaceae, Hydnaceae, Clavariaceae, Clavulinaceae and Corticiaceae. According to these authors, this clade contains approximately 170 described species of homobasidiomycetes. The phylogenetic distance between *Cantharellus* and model euagaric species such as *Schizophyllum* and *Laccaria* might also imply striking differences in genetics, physiology, and ECM formation. It is important to bear this in mind as we progress toward understanding the biology of the economically important *Cantharellus* genus.

10.2.1
The Genus *Cantharellus*

Based on the descriptions of Smith and Morse (1947), Corner (1966), Petersen (1971, 1973, 1985), Danell (1994a), Feibelman et al. (1997) and Pegler et al. (1997), *Cantharellus* can be described as a homobasidiomycete genus, with terricolous, fleshy and long-lived but not perennial gymnocarpic sporocarps. The pileus has a sterile top, which distinguishes it from the Clavariaceae. The hymenium is either smooth or folded, with ridges on the stem and pileus. The gill-like ridges differ from the true gills of the order Agaricales. The *Cantharellus* hymenium thickens as new basidia develop over the layer of older ones. By contrast, in the Agaricales the basidia form a monolayer. *Cantharellus* basidia are stichic (Juel 1916) and long, bearing long, curved sterigmata. Spores are smooth, white or yellow and of variable size. The number of spores per basidium varies between two and eight within the same sporocarp (nuclear migration studied by Maire 1902). The haploid chromosome number in *C. cibarius* Fr. is two (Juel 1916). No cystidia are present. Hyphae are monomitic and clamp connections are present. The species studied so far within *Cantharellus sensu* Feibelman et al. (1997) have large ITS sequences (Danell 1994b; Feibelman et al. 1997), 1400–1600 bp, and E. Danel

	End of small subunit	Beginning of ITS1	
C. formosus	GGAAGGATCA	ACCCCTGTG	ATAGNTTGTTTG
C. subalbidus	GGAAGGATCA	ACCCCTGTG	GGTATAGTGAGA
C. cibarius	GGAAGGATCA	ACCCCTGTG	GGTATAGTGAGA
C. pallens	GGAAGGATCA	ACCCCTGTG	GGTATAGTGAGA
Rhizopogon subcaerulescens	GGAAGGATCATTA	ACGAAT	ATAATTCGAGGG
Suillus spraguei	GGAAGGATCATTA	TCGAAT	TATAATCCGGCG
Fomitopsis rosea	GGAAGGATCATTA	TCGAGT	TTTAATTGGGTT

Fig. 10.1. rDNA ITS1 regions from three *Cantharellus* species compared with other basidiomycetes. GenBank accession numbers AF 044688, AF 044690, AF 044692, AF 044694

et al. (unpubl. data) have revealed a sequence at the start of the ITS1 region which is unique to the genus *Cantharellus sensu* Feibelman et al. (1997).

10.2.2
Pigments and Volatile Compounds

Bicyclic carotenoid pigments are characteristic of the subgenus *Cantharellus* (Arpin and Fiasson 1971; Gill and Steglich 1987). The density of pigments of in vitro cultures can, however, differ within a strain (Danell and Fries 1990). Albino varieties are sometimes found, but species like *C. pallens* Pilàt have pale sporocarps with a pigmented hymenium. Carotenoids are generally rare among agaric fungi (Gill and Steglich 1987). The purpose of most fungal pigments is unknown, but since carotenoids in some fungi have light-sensitive functions (Carlile 1970) it could be related to positive phototropism, as observed in glasshouse-produced sporocarps (E. Danell, unpubl. data). Mattila et al. (1994) reported "remarkably high" contents of vitamin D in *C. cibarius*. Since vitamin D2 is derived from ergosterol under the action of light, it may be that the observed phototropism is due to carotenoid photoreceptors important to vitamin D synthesis. Mattila et al. (1994) suggested that the funnel-like shape of chanterelle sporocarps favours light exposure. Pyysalo (1976) identified 13 volatile acids and 36 other volatile compounds from *C. cibarius*. Octenols (causing the characteristic smell of mushrooms in general), caproic acid and acetic acid were found in highest concentrations in this species. Breheret (1997) discovered high amounts of octa-1.3-dien emitted by *C. cibarius*. However, the key compounds for the characteristic apricot smell remain unknown.

10.3
Ecology

10.3.1
General Ecology

Most data concerning the ecology of species of the genus *Cantharellus sensu* Feibelman et al. (1997) are derived from *C. cibarius* (Danell 1994a). This species has a wide host range, including conifers and broad-leaved trees worldwide (Danell 1994a). Some specificity, however, may exist, since some physiological strains apear to be adapted to certain host groups (Danell 1994b). *C. cibarius* often occurs in older forests and plantations, and has never been reported from nurseries unless specifically cultivated (O'Dell et al. 1992; Danell 1994a; Danell and Camacho 1997). Preferred soils are well drained with a low N content and a pH range of 4.0–5.5 (Danell 1994a).

Many European countries include *Cantharellus* spp. and *Craterellus* spp. in their red lists of endangered species (Arnolds 1995; Larsson 1997). Between 1960 and 1980, the number of localities at which *C. cibarius* was recorded in the Netherlands decreased by 60% (Jansen and van Dobben 1987; Arnolds 1988, 1991, 1995). It must be emphasised that the observed decline in *C. cibarius* occurrence is based on observations of sporocarps, and the extent to which soil mycelia and ECM are also declining is unknown. The causes of this decline are not clear. In ECM species that spread mainly vegetatively, such as *C. cibarius* (Danell 1994a), clear-cutting may eliminate large clones by removing their carbohydrate supply, destroying the drought protective moss layers and liberating toxic levels of N. Recolonization by spores might take decades since observations in plantations show that sporocarps of *C. cibarius* are usually associated with older stands (Danell 1994a), while young seedlings can be easily colonized by hyphal suspensions in vitro or by root to root contact in a glasshouse (Danell 1994b; Danell and Camacho 1997) or theoretically in a stand with mixed age classes.

Removal of logs and other water-holding substrates (Norvell 1992; Molina et al. 1993) in forests might also decrease carpophore production. Since *C. cibarius* mycelium is confined to the upper 5–10 cm of soil (Danell 1994b), it is strongly exposed to any deposits. In vitro experiments on other ECM fungi have shown that excess N can decrease fungal biomass (Wallander and Nylund 1992). A decrease of sporocarps of *C. cibarius* due to the addition of fertilizers was noted by Nohrstedt (1994) and Menge and Grand (1978). Certain N (and S) compounds can lower soil pH, thereby altering the mobility of numerous toxic and essential elements (Arnolds 1988). The optimum pH for *C. cibarius* is 4.5–5.5 (field and in vitro data from Straatsma 1986; Jansen and van Dobben 1987), and it may be that the apparent decline in *C. cibarius* occurrence relates to soil acidification arising from N deposition. Arnolds (1991) and de Vries et al. (1995) showed that removal of the litter layer containing high amounts of N stimulated a partial recovery of *C. cibarius* sporocarp production.

Restricted picking of *C. cibarius* has been suggested (Jansen 1990; Ebert 1992), but according to Egli et al. (1990) Arnolds (1991), Danell (1994a) and Norvell (1995), picking has no negative impact on successive sporocarp formation of *Cantharellus* spp. For example, a *C. pallens* site was harvested each year between 1941 and 1980 without any decrease in sporocarp production (Jahn and Jahn 1986). While the impact of picking on long-distance dispersal is unknown, trampling in the area might have some negative influence (Egli et al. 1990).

The decline in Europe has resulted in a profitable import from Africa (Pegler et al. 1997) where chanterelles are highly appreciated (Buyck 1994; Härkönen et al. 1995). Chanterelles are also imported to Europe from North America (Schlosser and Blatner 1995), where their importance to the local economy has triggered an increased interest among scientists (Largent 1994; Pilz and Molina 1996). However, it is now clear that much of the *"C. cibarius"* data from the American Pacific Northwest were actually based on *C. formosus* Corner (Danell 1995; Redhead et al. 1997), leading to confusion among mushroom dealers, ecologists and physiologists. Indeed, European canning companies refused to use North American chanterelles since the texture was different from European sporocarps, though the scientific names were the same.

10.3.2
Predators and Parasites

In Finland, 120 species of dipterans were recorded by Hackman and Meinander (1979) as fungivorous on agaric sporocarps. Their study showed that, while <1% of *C. cibarius* sporocarps were infested with dipterans (mainly polyphagous limoniid larva), 40–80% of mature sporocarps of agaric taxa were infested by dipterans (mainly mycetophilid larvae). Polyphagous elaterid larvae (Coleoptera) are also occasionally found in *C. cibarius*. In view of the long life of *C. cibarius* sporocarps (31–84 days, according to Kälin and Ayer 1983; Norvell 1992; Danell 1994a) it is surprising that *C. cibarius* does not become heavily infested. Pang and Sterner (1991) and Pang et al. (1992) have described two metabolites formed in *C. cibarius* and *C. tubaeformis* as a response to injury. While the effects of these compounds on potential predators have yet to be tested, they may be involved in protecting against insect predators. Selection for insect repelling compounds may explain why sporocarps of *C. cibarius* do not need to grow rapidly (Danell 1994a). A further adaptation of long-lived sporocarps is the ability of *C. cibarius* sporocarps to produce new tissue at wound sites (Danell 1994a).

Mammals are known to eat chanterelles (Danell 1994a). North et al. (1997) suggested that chanterelles were among the most preferred epigeous fungi for animal consumption. Most epigeous and fleshy fungal species may lose 50% of their sporocarps due to predation of vertebrates and gastropods, and the remaining sporocarps are often infested by dipteran larvae (Lacy 1984).

However, chanterelles are not preferred by slugs and snails (Frömming 1954; Worthen 1988), which is another interesting difference to euagaric fungi.

A rare fungal parasite which forms sporocarps exclusively on the *Cantharellus* hymenium is the agaric *Entoloma pseudoparasiticum* Noordeloos. Helfer (1991) also reported *Hypomyces odoratus* G. Arnold (Pyrenomycetes) as a parasite on *C. cibarius*. Viral diseases may further affect chanterelle sporocarps (Blattny and Kralik 1968). Symptoms are enations on top of the caps, often affecting a whole cluster of sporocarps.

10.3.3
Interactions with Bacteria

In contrast to most agaric fungi, natural and cultivated sporocarps of *C. cibarius* contain large numbers of moulds and bacteria (Danell et al. 1993). Such contaminating organisms are seldom severely parasitic, rather they appear to be incidentally incorporated during sporocarp formation (Danell et al. 1993). Since the bacteria multiply between the hyphae in chanterelle sporocarps without causing deterioration, it has been suggested that the bacteria utilize fungal exudates for growth (Danell et al. 1993). Indeed, preliminary NMR analysis indicates exudation of amino acids, organic acids and sugars from sporocarp tissue (Rangel and Danell 1998). According to Danell (1994a) the bacteria do not act as mycorrhization helper bacteria (*sensu* Garbaye et al. 1990), but other possible mutualistic interactions remain to be investigated.

10.4
Mycelial Growth

Despite the potential economic value of cultivating chanterelles, few articles on chanterelle physiology exist in the literature. This may be partly explained by difficulties associated with establishing *C. cibarius* in axenic culture. Such problems arise from specific nutritional demands of the fungus, its inability to use complex carbohydrates like cellulose, its sensitivity to autoclaved media, its slow growth and the natural contamination of sporocarps with moulds and bacteria (Danell and Fries 1990; Danell et al. 1993). Fries (1979) was first to succeed in germinating spores of *C. cibarius* using a complex, but defined, medium. However, spore production, spore viability and incidence of germination in *C. cibarius* are very low (Fries 1979; Danell 1994a). Fries' medium was modified by Straatsma and van Griensven (1986), who obtained pure *C. cibarius* mycelium from sporocarps (Straatsma et al. 1985). The mycelium was later verified as *C. cibarius* by Straatsma et al. (1985) using dot-blot DNA hybridization. Moore et al. (1989) also used the technique of Straatsma in her ECM experiments. The species identity of mycelia isolated by Danell and Fries (1990) for routine ECM synthesis (Danell 1994a,b) was confirmed with PCR and RFLP analyses and later by sporocarp formation (Danell and Camacho 1997).

Since the degree of natural fungal and bacterial contamination in *C. cibarius* sporocarps is high (Danell et al. 1993), only strains whose species identity has been proven by PCR/RFLP should be used for physiological studies. Microscopical traits of importance for preliminary determinations during the purification process on Murashige and Skoog medium (Straatsma et al. 1986) are the slow growth rate (0.2–0.3 mm day^{-1}), the long lag period, frequent clamp connections, a hyphal thickness of 3 μm and the presence of yellow carotenoids (Danell 1994a). Growth rate on modified Fries' medium (Straatsma and van Griensven 1986) is 0.5 mm day^{-1} (Danell 1994a), which is the optimized rate according to Straatsma (1998).

It is unfortunate that some interesting publications about the physiology of *C. cibarius* are based on strains of uncertain species identity (Doak 1934; Sugihara and Humfeld 1954; Hattula and Gyllenberg 1969; Siehr et al. 1969; Torev 1969; Riffle 1971; Volz 1972; Ballero et al. 1991; Pachlewski et al. 1996; Strzelcyk et al. 1997). The strains used in some of these studies include M83, or NRRL 2370, which is the same as ATCC 13228 and CBS 155.69, which was later defined as not being *C. cibarius* (Stalpers, pers. comm.).

Straatsma et al. (1986) showed that the increased growth of *C. cibarius* mycelia in the presence of roots was also observed in the presence of 0.5% CO_2. Straatsma and Bruinsma (1986) explained that radioactively labeled CO_2, incorporated by the mycelium by anaplerotic CO_2 fixation, would spread into all metabolic pathways. Therefore they used products of heterotrophic carboxylation reactions to reveal the fate of fixed CO_2. The authors showed that malic acid, thiamin and Tween 80 replaced the CO_2 as a growth factor, indicating that CO_2 was assimilated by *C. cibarius* into the Krebs cycle and into the biosynthetic pathways of pyrimidines and fatty acids. Danell (1994a,b) also showed that elevated CO_2 levels are crucial for successful ECM formation in most *C. cibarius* strains. It is noteworthy that normal CO_2 levels at 20-cm depth in pine forest soils are 0.5–2% (Magnusson 1992). However, the O_2 concentration must not be too low, and according to Danell (1994a) *C. cibarius* does not survive waterlogging. According to Magnusson (1992) waterlogging in forest soils in spring might result in a 12% O_2 concentration and a 4% CO_2 concentration. Such a low O_2 concentration might have favoured the evolution of hydrophobic mycelia able to create air pockets, as described by Unestam (1991) and observed in chanterelles by Danell (1994a). Low O_2 concentrations might also be one reason why *C. cibarius* is rarely found in soils with poor drainage.

10.5
ECM Formation

Since chanterelles seldom form mycelial mats, it is very hard to find chanterelle ECM in the field. Our data on morphology and hosts are based on axenic systems and pot cultures (Danell 1994a,b; Danell and Camacho 1997). Doak (1934) and Moore et al. (1989) have reported successful ECM formation using classical techniques. However, for rapid large-scale production on a routine basis, Danell (1994a,b) developed a technique based on McLaughlin (1970) and Jentschke et al. (1991). Thus, adding a *C. cibarius* hyphal suspension to axenic *Pinus sylvestris* and *Picea abies* seedlings in culture units filled with acid-washed quartz sand resulted in ECM formation in 8 weeks. Important factors for growth were rapid gas exchange through filters, continuous drainage, automatic addition of a low concentration mineral solution with 0.2% glucose, and a CO_2 level between 0.2 and 0.4% (Danell 1994a). After 8–12 weeks, these seedlings can be transferred to open pots in the glasshouse (Danell 1994b). E. Danell (unpubl. data) showed that if the number of ECM root tips was more than 12 when transferred from the in vitro system to pots in the glasshouse, then the chanterelle mycelia survived the transfer. The number of chanterelle ECM (Fig. 10.2) increased during the 5-month study (E. Danell, unpubl. data). Alien ECM and sporocarps (e.g. *Sphaerosporella* and *Laccaria*) were detected in the pots, but these species occupied areas where the chanterelle mycelium had been absent from the beginning. The optimum temperature for mycelial growth was 20°C, and temperatures above 28°C hampered growth of the chanterelle strains tested. The colour of chantarelle ECM can be yellow, white or brown depending on age, water saturation and the presence of air pockets around ECM (Danell 1994b). ECM formed in vitro have thick yellow mantles and a deep Hartig net throughout the cortex (Danell 1994a,b). For chanterelle species analyses in

Fig. 10.2. Ectomycorrhiza of *Cantharellus cibarius* and *Pinus sylvestris* after 5 months in a glasshouse pot culture

glasshouse pots, it is important to run a PCR test on selected root tips as the colours vary and alien species may occupy portions of the root system.

10.6
Reproduction and Prospects for Commercial Cultivation

The first fruiting of *C. cibarius* in a glasshouse was achieved in 1996 (Danell and Camacho 1997). Several sporocarps (up to 3.5 cm tall) and numerous primordia of a Swedish *C. cibarius* isolate were obtained in pots with 16-month-old *P. sylvestris* seedlings (Fig. 10.3). The fruiting occurred randomly from April 1996 until January 1997, when the experiment was terminated (Danell 1997; Danell and Camacho 1997). The sporocarps were not attached directly to the roots, but emerged both on top of the pot and in drainage holes. They were directed towards the light sources, suggesting phototropic behaviour (see 10.2.2). Primordia which were disturbed by handling of the pot became malformed and later disappeared. This suggests that hyphae supporting young primordia are easily ruptured due to rough handling. Sometimes they were replaced by new sporocarps. All cultivated *C. cibarius* sporocarps contained fluorescent *Pseudomonas* bacteria as observed in the field, suggesting that this symbiosis is inevitable when the two organisms are present. Since the climatic conditions had been fixed over a period of 1 year, and fruiting occurred over a period of 10 months, it appeared

Fig. 10.3. Sporocarp of *Cantharellus cibarius* produced in a drainage hole of a pot in a glasshouse

that no external events triggered fruiting. This has also been observed in natural conditions where constant climate results in eternal production of *Suillus* sporocarps (Hedger 1986). Danell (1994a) hypothesized that *C. cibarius* forms sporocarps once nutrient storage has reached saturation. Hammond (1985) explained that for *Agaricus bisporus* (Lange) Imbach, morphogenesis begins once a certain intracellular level of carbohydrates is reached. These carbohydrates will be transformed to mannitol which should attract water and start cell elongation in the primordia.

Commercial harvesting of chanterelles in the glasshouse is hardly possible using the present technique. Instead, plantations of seedlings carrying selected mycelia might allow commercial harvests. Such orchards already exist for production of, for example, black truffles (*Tuber* spp.) (Giovanetti et al. 1994; and see Chap. 6; Chevalier and Frochot 1997). Large-scale inoculation of young seedlings with chanterelle mycelium can be achieved by root to root contact, using inoculated mother plants. With a view to developing chantarelle orchards, outplanting experiments and optimized large-scale inoculation are currently being studied by a Swedish company (Cantharellus AB), using a protected strain of *C. cibarius* (European Community Plant Variety Office File number 96/1089).

Production is of course dependent on, for example, soil type, climate, nutrient supply, competing mushroom species and water supply. These factors are not triggers for fruiting, but are essential for optimizing production. An irrigated 15×35-m field plot in southern Sweden produced a total of 17 kg (FW) of *C. cibarius* in 1992 (Danell 1994a), while Slee (1991) reported a production of 50 kg (FW) *C. cibarius* ha^{-1} year^{-1} in Scottish forests. The potential for production in areas with constant climate was observed by Hedger (1986). He reported production of *Suillus luteus* (L.: Fr.) S. F. Gary equivalent to 1000 kg (DW) ha^{-1} year^{-1} on the mountain Cotopaxi in Ecuador. The large production was partly explained by the year-round season, and that this exotic *Pinus radiata* plantation did not harbour many other mushroom species.

In order to establish a chanterelle orchard of 225 m^2, 100 seedlings should be planted in a square at 1.5 m distance from each other. As the growth rate of the mycelium is 15 cm year^{-1} in southern Sweden (Danell 1994a), mycelia from all trees can theoretically fuse in 5 years if the same strain has been used throughout. Based on results obtained in the glasshouse (Danell and Camacho 1997), sporocarp production is not expected until 2–3 years after outplanting. Longevity of individual mycelia in nature is currently being studied by Danell, using RAPD of mycelia obtained from sporocarps in order to estimate size and age. Population studies are also being performed by Dunham et al. (1998).

References

Arnolds E (1988) The changing macromycete flora in the Netherlands. Trans Br Mycol Soc 90:391–406
Arnolds E (1991) Decline of ectomycorrhizal fungi in Europe. Agric Ecosyst Environ 35: 209–244

Arnolds E (1995) Consecration and management of natural populations of edible fungi. Can J Bot 73(Suppl 1):S987–S998

Arpin N, Fiasson J-L (1971) The pigments of Basidiomycetes: their chemotaxonomic interest. In: Petersen RH (ed) Evolution in the higher Basidiomycetes. University of Tennessee Press, Knoxville, pp 63–99

Ballero M, Rescigno A, Sanjust di Teulada E (1991) Osservazioni sull'enzimologia di *Cantharellus cibarius* Fries. Micol Ital 2:9–12

Blattny C, Králík O (1968) A virus disease of *Laccaria laccata* (Scop. ex Fr.) Cooke and some other fungi. Ceská Mykol 22:161–166

Breheret S (1997) Étude des arômes produits par des carpophores de champignons supérieurs sauvages et par des cultures mycéliennes de Morhcella et de Pleurotus. PhD Thesis, Institut National Politechnique de Toulouse, Toulouse

Buyck B (1994) Ubwoba: les champignons comestibles de l'Ouest du Burundi. Publication Agricole No 34, Administration Generale de la Cooperation au Developpement, Brussels

Carlile MJ (1970) The photoresponses of fungi. In: Halldal P (ed) Photobiology of microorganisms. Wiley, London, pp 309–344

Chevalier G, Frochot H (1997) La truffe de Bourgogne. Pétrarque, Levallois-Perret

Corner EJH (1957) *Craterellus* Pers., *Cantharellus* Fr. and *Pseudocraterellus* gen. nov. Sydowia 1:266–276

Corner EJH (1966) A monograph of cantharelloid fungi. Oxford University Press, Oxford

Corner EJH (1969) Notes on cantharelloid fungi. Nova Hedwigia 18:783–818

Dahlman M, Danell E, Spatafora JW (1998) Molecular systematics of *Craterellus*: cladistic analysis of nuclear LSU rDNA sequence data. Abstr 2nd Int Conf on Mycorrhiza, Uppsala, 47 pp

Danell E (1994a) *Cantharellus cibarius*: mycorrhiza formation and ecology. Acta Universitatis Upsaliensis. Comprehensive Summaries of Uppsala Dissertations from the Faculty of Science and Technology 35 Uppsala, 75 pp

Danell E (1994b) Formation and growth of the ectomycorrhiza of *Cantharellus cibarius*. Mycorrhiza 5:89–97

Danell E (1995) Comparisons between Swedish *Cantharellus cibarius* and *Cantharellus* spp. in the Pacific Northwest, based on differences in RFLP patterns of the ITS region. Inoculum 46:10

Danell E (1997) Les progrès dans la maîtrise de la culture de la chanterelle, *Cantharellus cibarius*. Rev For Fr 49:215–221

Danell E, Camacho FJ (1997) Successful cultivation of the golden chanterelle. Nature 385:303

Danell E, Fries N (1990) Methods for isolation of *Cantharellus* species, and the synthesis of ectomycorrhizae with *Picea abies*. Mycotaxon 38:141–148

Danell E, Alström S, Ternström A (1993) *Pseudomonas fluorescens* in association with fruit bodies of the ectomycorrhizal mushroom *Cantharellus cibarius*. Mycol Res 97:1148–1152

Doak KD (1934) Fungi that produce ectotrophic mycorrhizae of conifers. Phytopathology 24:7

Donk MA (1964) A conspectus of the families of Aphyllophorales. Persoonia 3:199–324

Donk MA (1969) Notes on *Cantharellus* sect. *Leptocantharellus*. Persoonia 5:265–284

Dunham S, O'Dell T, Molina R, Pilz D (1998) Fine scale genetic structure of chanterelle (*Cantharellus formosus*) patches in forest stands with different disturbance intensities. Abstr 2nd Int Conf on Mycorrhiza, Uppsala, Sweden, 55 pp

Ebert H-J (1992) Zur Situation des Pilzschutzes und des Handels mit Pilzen in Vergangenheit und Gegenwart – Gedanken über den Werdegang bestehender Gesetze in Deutschland und deren Inhalt. Beilage Z Mykol 2:30–39

Egli S, Ayer F, Chatelain F (1990) Der Einfluss des Pilzsammelns auf die Pilzflora. Mycol Helv 3:417–428

Feibelman TP, Doudrick RL, Cibula WG, Bennett JW (1997) Phylogenetic relationships within the Cantharelaceae inferred from sequence analysis of the nuclear large subunit rDNA. Mycol Res 101:1423–1430

Fries E (1874) Hymenomycetes Europaei sive Epicriseos systematis mycologici 2nd edn. Stockholm

Fries N (1979) Germination of spores of *Cantharellus cibarius*. Mycologia 71:216–219

Frömming E (1954) Biologie der mitteleuropäischen Landgastropoden. Duncker and Humblot, Berlin

Garbaye J, Duponnois R, Wahl JL (1990) The bacteria associated with *Laccaria laccata* ectomycorrhizas or sporocarps: effect on symbiosis establishment on Douglas fir. Symbiosis 9:267–273

Gill M, Steglich W (1987) Pigments of fungi (Macromycetes). Fortschritte der Chemie organischer Naturstoffe 51. Springer, Vienna New York

Giovannetti G, Roth-Bejerano N, Zanini E, Kagan-Zur V (1994) Truffles and their cultivation. Hortic Rev 16:71–107

Hackman W, Meinander M (1979) Diptera feeding as larvae on macrofungi in Finland. Ann Zool Fenn 16:50–83

Hammond JBW (1985) The biochemistry of *Agaricus* fructification. In: Development biology of higher fungi. Cambridge University Press, Cambridge, pp 389–401

Hansen L, Knudsen H (1997) Nordic macromycetes vol 3. Nordsvamp, Copenhagen

Härkönen M, Saarimäki T, Mwasumbi L (1995) Edible mushrooms of Tanzania. Karstenia 35, Suppl

Hattula ML, Gyllenberg HG (1969) Adaptability to submerged culture and amino acid contents of certain fleshy fungi common in Finland. Karstenia 9:39–45

Hedger J (1986) *Suillus luteus* on the equator. Bull Br Mycol Soc 20:53–54

Helfer W (1991) Pilze auf Pilzfruchtkörpern. Libri Botanici 1. IHW-Verlag, Eching

Hibbett DS, Pine EM, Langer E, Langer G, Donoghue MJ (1997) Evolution of gilled mushrooms and puffballs inferred from ribosomal DNA sequences. Proc Natl Acad Sci USA 94: 12002–12006

Hibbett DS, Thorn RG Homobasidiomycetes. In: The mycota, vol 7. Springer, Berlin Heidelberg New York (in preparation)

Jahn H, Jahn M-A (1986) Konstanz und Fluktuationen der Pilzvegetation in Norra Warleda (Uppland). Beobachtungen auf einem schwedischen Bauernhof 1945–1980. Westfälische Pilzbriefe 10–11:352–378

Jansen AE (1990) Conservation of fungi in Europe. Mycologist 4:83–85

Jansen E, van Dobben HF (1987) Is decline of *Cantharellus cibarius* in the Netherlands due to air pollution? AMBIO 16:211–213

Jentschke G, Godbold DL, Hütterman A (1991) Culture of mycorrhizal tree seedlings under controlled conditions: effects of nitrogen and aluminium. Physiol Plant 81:408–416

Juel HO (1916) Cytologische Pilzstudien. I. Die Basidien der Gattungen *Cantharellus*, *Craterellus* und *Clavaria*. Nova Acta R Soc Sci Upps 4:3–34

Kälin I, Ayer F (1983) Sporenabwurf und Fruchtkoerperentwicklung des Goldstieligen Pfifferlings (*Cantharellus lutescens*) im Zusammenhang mit Klimafaktoren. Mycol Helv 1:67–88

Kocór M, Schmidt-Szalowska A (1972) Constituents of higher fungi. The sterols from *Cantharellus cibarius*. Bull Pol Acad Sci Chem 20:515–520

Lacy RC (1984) Ecological and genetic responses to mycophagy in Drosophilidae (Diptera). In: Wheeler Q, Blackwell M (eds) Fungus–insect relationships. Columbia University Press, New York, pp 286–301

Largent DL (1994) Influence of environmental factors on fruiting of edible, mycorrhizal mushrooms. Report prepared for California Department of forestry and fire protection. Humboldt State University, Arcata California

Larsson KH (1997) Swedish red data book of fungi 1997. Artdatabanken, SLU, Uppsala

Lobelius M (1581) Plantarum seu Stirpium Icones. Tomus II. Christophori Plantini, Antwerp, pp 273–276

Magnusson T (1992) Temporal and spatial variation of the soil atmosphere in forest soils of northern Sweden. PhD Thesis, Swedish University of Agricultural Sciences, Umeå

Maire R (1902) Recherches cytologiques et taxonomiques sur les basidiomycètes. Bull Soc Mycol Fr Suppl 18, Fasc 2:1–209

Mattila PH, Piironen VI, Uusi-Rauva EJ, Koivistoinen PE (1994) Vitamin D contents in edible mushrooms. J Agric Food Chem 42:2449–2453

McLaughlin DJ (1970) Environmental control of fruit body development in *Boletus rubinellus* in axenic culture. Mycologia 62:307–331

Menge JA, Grand LF (1978) Effect of fertilization on production of epigeous basidiocarps by mycorrhizal fungi in loblolly pine plantations. Can J Bot 56:2357–2362

Molina R, O'Dell T, Luoma D, Amaranthus M, Castellano M, Russell K (1993) Biology, ecology and social aspects of wild edible mushrooms in the forests of the Pacific Northwest: a preface to managing commercial harvest. USDA Forest Service General Technical Report PNW-GTR-309

Moore LM, Jansen AE, van Griensven LJLD (1989) Pure culture synthesis of ectomycorrhizas with *Cantharellus cibarius*. Acta Bot Neerl 38:273–278

Nohrstedt H-Ö (1994) Fruit-body production and ^{137}Cs-activity of *Cantharellus cibarius* after nitrogen- and potassium-fertilisation. For Res Inst Sweden, Rep No 2, Uppsala

North M, Trappe J, Franklin J (1997) Standing crop and animal consumption of fungal sporocarps in Pacific Northwest forests. Ecology 78:1543–1554

Norvell L (1992) Studying the effects of harvesting on chanterelle productivity in Oregon's Mt. Hood National Forest. In: de Gues N, Redhead S, Callan B (eds) Wild mushroom harvesting discussion session minutes. Pacific Forestry Centre, Victoria, pp 9–15

Norvell L (1995) Loving the chanterelle to death? The ten-year Oregon chanterelle project. McIlvainea 12:6–25

Nuhamara ST (1987) Mycorrhizae in agroforestry: a case study. Biotropia 1:53–57

O'Dell TE, Luoma DL, Molina RJ (1992) Ectomycorrhizal fungal communities in young, managed and old-growth Douglas-fir stands. Northwest Environ J 8:166–168

Pachlewski R, Strzelczyk E, Kermen J (1996) Studies of *Cantharellus cibarius* – a mycorrhizal fungus of pine and spruce. Acta Mycol 31:143–150

Pang Z (1993) Secondary fungal metabolites isolated from fruit bodies. PhD Thesis, University of Lund, Lund

Pang Z, Sterner O (1991) Cibaric acid, a new fatty acid derivative formed enzymatically in damaged fruit bodies of *Cantharellus cibarius* (Chanterelle). J Org Chem 56: 1233–1235

Pang Z, Sterner O, Anke H (1992) (8E)-10-Hydroxydec-8-enoic acid: its isolation from injured fruit bodies of *Cantharellus tubaeformis*, and synthetic preparation. Acta Chem Scand B46: 301–303

Patouillard N (1900) Essai taxonomique sur les familles et les genres des Hyménomycètes. PhD Thesis, University of Paris, Paris

Pegler DN, Roberts PJ, Spooner BM (1997) British chanterelles and tooth fungi. Royal Botanic Gardens, Kew

Persson O, Mossberg B (1997) The chanterelle book. Ten Speed Press, Berkeley

Petersen RH (1971) Interfamilial relationships in the clavarioid and cantharelloid fungi. In: Petersen RH (ed) Evolution in the higher Basidiomycetes. University of Tennessee, Knoxville, pp 345–374

Petersen RH (1973) Aphyllophorales II: the clavarioid and cantharelloid Basidiomycetes. In: Ainsworth GC, Sparrow FK, Sussman AS (eds) The fungi, vol IVB. Academic Press, New York, pp 351–368

Petersen RH (1979) Notes on cantharelloid fungi. IX. Illustrations of new or poorly understood taxa. Nova Hedwigia 31:1–23

Petersen RH (1985) Notes on clavarioid fungi. XIX. Colored illustrations of selected taxa, with comments on *Cantharellus*. Nova Hedwigia 42:151–169

Pilz D, Molina R (1996) Managing forest ecosystems to conserve fungus diversity and sustain wild mushroom harvests. USDA Forest Service General Technical Report PNW-GTR-371

Pine EM, Hibbett DS, Donoghue MJ (1999) Phylogenetic relationships of cantharelloid and clavarioid homobasidiomycetes inferred from mitochondrial and nuclear rDNA sequences. Mycologia (submitted)

Pyysalo H (1976) Identification of volatile compounds in seven edible fresh mushrooms. Acta Chem Scand B30:235–244

Rangel I, Danell E (1998) Exudates of *Cantharellus cibarius* as a possible source of nutrients for *Pseudomonas fluorescens*. Abstr 2nd Int Conf on Mycorrhiza, Uppsala, Sweden, 142 pp

Redhead S, Norvell L, Danell E (1997) *Cantharellus formosus* and the Pacific golden chanterelle harvest in western North America. Mycotaxon 65:285–322

Reijnders AFM, Stalpers JA (1992) The development of the hymenophoral trama in the Aphyllophorales and the Agaricales. Stud Mycol 34:1–109

Riffle JW (1971) Effects of nematodes on root-inhabiting fungi. In: Hacskaylo (ed) Mycorrhizae. US Government Printing Office, Washington, pp 97–113

Romagnesi H (1995) Prodrome a une flore analytique des hymenomycetes agaricoides III. Fam. Cantharellaceae Schroeter. Doc Mycol 25:417–424

Schlosser WE, Blatner KA (1995) The wild edible mushroom industry of Washington, Oregon and Idaho. J For 93:31–36

Siehr DJ, Chang C-K, Cheng HL (1969) The transformation of tryptamine and D-tryptophan by basidiomycetes in submerged culture. Phytochemistry 8:397–400

Slee RW (1991) The potential of small woodlands in Britain for edible mushroom production. Scott For 45:3–12

Smith AH, Morse EE (1947) The genus *Cantharellus* in the western United States. Mycologia 39:497–534

Stamets P (1993) Growing gourmet and medicinal mushrooms. Ten Speed Press, Berkely

Straatsma G (1986) Physiology of mycelial growth of the mycorrhizal mushroom *Cantharellus cibarius* Fr. PhD Thesis, Mushroom Exp Station, Horst

Straatsma G (1998) Slow growth of *Cantharellus cibarius* mycelium in culture inevitable? Abstr 2nd Int Conf on Mycorrhiza, Uppsala, Sweden, p 161

Straatsma G, Bruinsma J (1986) Carboxylated metabolic intermediates as nutritional factors in vegetative growth of the mycorrhizal mushroom *Cantharellus cibarius* Fr. J Plant Physiol 125:377–381

Straatsma G, van Griensven LJLD (1986) Growth requirements of mycelial cultures of the mycorrhizal mushroom *Cantharellus cibarius*. Trans Br Mycol Soc 87:135–141

Straatsma G, Konings RNH, van Griensven LJLD (1985) A strain collection of the mycorrhizal mushroom *Cantharellus cibarius*. Trans Br Mycol Soc 85:689–697

Straatsma G, van Griensven LJLD, Bruinsma J (1986) Root influence on in vitro growth of hyphae of the mycorrhizal mushroom *Cantharellus cibarius* replaced by carbon dioxide. Physiol Plant 67:521–528

Strzelczyk E, Dahm H, Pachlewski R, Rozycki H (1997) Production of indole compounds by the ectomycorrhizal fungus *Cantharellus cibarius* Fr. Pedobiology 41:402–411

Sugihara TF, Humfeld H (1954) Submerged culture of the mycelium of various species of mushroom. Appl Microbiol 2:170–172

Thoen D, Ba AM (1989) Ectomycorrhizas and putative ectomycorrhizal fungi of *Afzelia africana* Sm. and *Uapaca guineensis* Müll. Arg. in southern Senegal. New Phytol 113:549–559

Torev A (1969) Submerged culture of higher fungi mycelium on an industrial scale. Mushroom Sci 7:585–589

Unestam T (1991) Water repellency, mat formation, and leaf-stimulated growth of some ectomycorrhizal fungi. Mycorrhiza 1:13–20

Volz PA (1972) Nutritional studies on species and mutants of *Lepista*, *Cantharellus*, *Pleurotus* and *Volvariella*. Mycopathol Mycol Appl 48:175–185

Vries de BWL, Jansen E, van Dobben HF, Kuyper THW (1995) Partial restoration of fungal and plant species diversity by removal of litter and humus layers in stands of Scots pine in the Netherlands. Biodiv Conserv 4:156–164

Wallander H, Nylund J-E (1992) Effects of excess nitrogen and phosphorus starvation on the extramatrical mycelium of ectomycorrhizas of *Pinus sylvestris* L. New Phytol 120: 495–503

Watling R (1997) The business of fructification. Nature 385:299–300

Watling R, Abraham SP (1992) Ectomycorrhizal fungi of Kashmir forests. Mycorrhiza 2: 81–87

Worthen WB (1988) Slugs (*Arion* spp.) facilitate mycophagous drosophilids in laboratory and field experiments. Oikos 53:161–166

Chapter 11

Lactarius

L. J. Hutchison

11.1
Introduction

Species of *Lactarius* encompass one of the larger known genera of ectomyc-orrhiza (ECM)-forming basidiomycetes. Cosmopolitan in distribution, several hundred taxa are currently recognized which collectively play an important role as late-stage ECM colonizers of woody trees and shrubs found in arctic, montane, temperate and tropical ecosystems. Because of their high degree of host specificity, fastidious in vitro growth requirements, and their role as late-stage ECM colonizers, species of *Lactarius* have not been examined intensively by ECM researchers. Detailed information is lacking on the colonization and development of ECM on host root systems, and the subsequent physiology and ecology of the host–fungus symbiosis. This chapter attempts to summarize current knowledge in order to present a starting point for further studies on members of this important genus.

11.2
Taxonomy

Species of *Lactarius* are placed in the Russulales, an order whose members possess sporocarps that range from being hypogeous and gasteroid in morphology (e.g. *Zelleromyces, Macowanites*) (Pegler and Young 1979) to those that are epigeous with more recognizable agaricoid features (e.g. *Lactarius, Russula*) (Singer 1986). Regardless of differences in macromorphology, all members of the Russulales possess basidiospores that are ornamented with distinctive amyloid ridges and/or warts. Species of *Lactarius* produce fleshy sporocarps that have subdecurrent to decurrent lamellae, basidia which are four-spored (rarely two-spored), and a spore print which is white to pale yellowish to buff or cream. Pleurocystidia and cheilocystidia are commonly present and the sporocarp trama is intermixed with lactiferous hyphae which exude a latex from cut or broken tissue in fresh material. The latex is usually white, watery or brightly coloured, and, if white or watery, may change colour

Agriculture Canada Research Station, P.O. Box 3000, Lethbridge, Alberta T1J 4B1, Canada upon exposure to air depending on species. Conidial anamorphs are lacking (Hutchison 1989). Recently, Hutchison (1990a–d, 1991) and Hutchison and Summerbell (1990) examined axenic mycelial cultures of *Lactarius* and found various biochemical and physiological characters which were taxonomically useful.

Controversy surrounds the appropriate designation of a type species. Hesler and Smith (1979) recognize *Lactarius deliciosus* (L.: Fr.) S. F. Gray as the type, Pegler and Young (1979) consider *L. piperatus* (Scop.: Fr.) S. F. Gray as the type species, while Singer (1986) argues for *L. torminosus* (Schaeff.: Fr.) S. F. Gray as the type. Worldwide, the subgeneric classification of Singer (1986) is currently employed in which the genus has been divided into ten sections (*Panuoidei, Lactariopsidei, Polysphaerophori, Venolactaris, Dulces, Plinthogali, Albati, Russulares, Lactarius, Dapetes*). North Americans tend to follow Hesler and Smith (1979) who divided *Lactarius* into six subgenera (*Lactarius, Plithogalus, Lactifluus, Piperites, Tristes, Russularia*).

As no worldwide monograph exists, the total number of taxa is not known. However, Singer (1986) treated 122 species while Hesler and Smith (1979) recognized 200 species as occurring in North America alone. Undoubtedly the total number exceeds several hundred if tropical species are considered. Several regional taxonomic treatments are available: Burlingham (1908), Coker (1918), Hesler and Smith (1979) and Bills (1986) for temperate North America; Rendall (1980) for boreal eastern Canada; Knudsen and Borgen (1982, 1994), Laursen and Ammirati (1982), Ohenoja and Ohenoja (1993) for arctic North America; Guevara et al. (1987), Montoya et al. (1990, 1996) and Montoya and Bandala (1996) for Mexico; Pegler and Fiard (1978) for the Lesser Antilles; Lange (1928), Konrad (1935), Tuomikoski (1953), Neuhoff (1956), Heinemann (1960), Kühner (1975), Bon (1980), and Korhonen (1984) for Europe; McNabb (1971) for New Zealand; Heim (1955), Buyck and Verbeken (1995), and Verbeken (1995, 1996a,b, 1998a,b) for tropical Africa; Bills and Cotter (1989) and Saini and Atri (1993) for the Indian subcontinent.

11.3
Ecology

Melin (1924) was the first to successfully confirm experimentally the ECM status of *Lactarius* when he synthesized ECM in vitro between *L. deliciosus* and *Pinus montana* (= *P. mugo*). All members of the genus are now considered to be obligately ECM. Fruiting of *Lactarius* spp. is known to cease if colonized root systems are severed from the host tree by trenching (Romell 1938). In addition, in vitro investigations have found that *Lactarius* spp. are unable to degrade complex C compounds such as cellulose and lignin (Lamb 1974; Oort 1981; Hutchison 1990a) and as a result are unable to decompose leaf and needle litter (Lindeberg 1948; Mikola 1954), further suggesting their status as obligate ECM fungi. Although Lactarii produce polyphenol oxidases such as laccase and tyrosinase (Lindeberg 1948; Giltrap

1982; Hutchison 1990b), these enzymes may play a role in helping *Lactarius* species detoxify phenolic compounds which are released from the leaf and needle litter in northern forests (Kuiters 1990), rather than aiding in decomposition as is known for wood- and litter-decay fungi (Szklarz et al. 1989).

Sporocarps of most Lactarii are found on the forest floor, although a few species are known to fruit on rotting wood (Kropp 1982). Sporocarps of *Lactarius* spp. can be quite abundant under certain conditions. Ohenoja (1993) found that in forest sites in northern Finland, on average, 2021 sporocarps of several *Lactarius* spp. could be found per hectare over 133 sites, which represented 1.15 kg dry wt ha^{-1}. Dahlberg et al. (1997) found that *Lactarius* spp. constituted 10% of the ECM biomass associated with a 100-year-old *Picea abies* stand in southern Sweden. Further, Dahlberg et al. (1997) found that during their 6-year study, sporocarps of the four species of *Lactarius* examined [*L. camphoratus* (Bull.: Fr.) Fr., *L. necator* (Bull.: Fr.) Karsten, *L. rufus* (Scop.: Fr.) Fr., and *L. theiogalus* (Bull.: Fr.) S. F. Gray] were not present every year. Although climatic conditions influence productivity, man-made influences also have an effect. Kropp and Albee (1996) reported an increase in *L. deliciosus* sporocarps with an increase in disturbance in the forest ecosystem. Stimulation of *L. glyciosmus* (Fr.: Fr.) Fr. fruiting bodies could be achieved by removing leaf and needle litter from the forest soil (Romell 1938), although a reduction in *L. subdulcis* (Fr.) S. F. Gray sporocarps occurred when leaf litter was removed from beech forest soils (Tyler 1991). Wiklund et al. (1995) found that *L. theiogalus* sporocarps increased in *P. abies* stands as a result of irrigation while those of *L. rufus* and *L. necator* increased in number the first year of recovery after artificial drought.

Several species which possess a bright coloured latex [e.g. *L. deliciosus* and *L. sanguifluus* (Paulet: Fr.) Fr.] tend to be among the choice edibles of mushroom hunters. Those with a peppery taste [e.g. *L. vellereus* (Fr.: Fr.) Fr., *L. piperatus* and *L. chrysorheus* Fr.] can cause gastrointestinal upset due to sesquiterpenes which are released upon damage to fungal tissue (Benjamin 1995). It has been suggested that these compounds act as defence compounds against predatory insects (Daniewski et al. 1993, 1995) and animals (Camazine et al. 1983; Camazine and Lupo 1984). Although sporocarps of many fungal species are known to concentrate metals, there are no reported cases among species of *Lactarius* (Tyler 1980). However, it was found that radio-caesium accumulated in sporocarps of *L. rufus* collected in Europe following the Chernobyl nuclear accident (Dighton and Horrill 1988; Paulus and Reisinger 1990).

Observations on fruiting behaviour in natural forests and in plantations suggest that most, if not all, species of *Lactarius* are late-stage ECM colonizers as fruiting is generally observed in older forest stands (Last and Fleming 1985; Last et al. 1987; Malajczuk 1987; Termorshuizen 1991; Visser 1995). However, *L. deliciosus* was found fruiting in stands of *Pinus kesiya* of various age classes in India (Rao et al. 1997) and *L. rufus* was found fruiting in different aged stands of *Pinus sylvestris* in Finland, Estonia and the Netherlands

(Hintikka 1988; Kalamees and Silver 1988; Termorshuizen 1991). At present, there are no explanations to account for these anomalies. Further attributes indicative of a late-stage ECM colonizer status for *Lactarius* include extremely poor basidiospore germination in the presence of root exudates of hosts compared with early-stage ECM fungi (Ali and Jackson 1988), the inability or poor ability to colonize roots of host seedlings compared with early-stage ECM fungi (Fox 1983; Hutchison and Piché 1995), and a relatively high degree of host specificity.

Because of increased deposition of atmospheric pollutants, particularly N, and acidification of forest soils of Europe, many *Lactarius* spp. have been reported as becoming increasingly rare in recent years and some have even been reported as having become extinct (Arnolds 1988, 1989; Arnolds and Jansen 1992; Arnolds and de Vries 1993). Ohenoja (1988), Termorshuizen (1993), and Wiklund et al. (1995) artificially applied N fertilizers to forest soils and observed a significant reduction in the occurrence of *Lactarius* sporocarps in subsequent years.

11.4
ECM Formation

11.4.1
Host–Fungus Specificity

Field observations indicate that a relatively high degree of host specificity exists for most species of *Lactarius* (see Table 11.1), a condition commonly found among late-stage ECM colonizers. Because of difficulties in synthesizing ECM in vitro between various *Lactarius* spp. and tree seedlings, experimental confirmation of host specificity is scant. In vitro synthesis experiments by Molina and Trappe (1982a) found that an isolate of *L. deliciosus*, a conifer associate, could form ECM with *Larix*, *Picea*, *Pinus*, *Pseudotsuga* and *Tsuga*, but the same isolate could not form ECM with 11 species of *Eucalyptus* (Malajczuk et al. 1982). Hutchison and Piché (1995) found that an isolate of the host-specific *L. subpurpureus* Peck could form ECM only with its host *Tsuga canadensis* under in vitro synthesis conditions, but not with *Abies*, *Picea*, *Pinus*, *Larix*, *Betula* or *Alnus*. Molina and Trappe (1982b) pointed out, however, that under in vitro conditions, the ericaceous plants *Arbutus menziesii* and *Arctostaphylos uva-ursi* lacked mycorrhizal specificity and could form arbutoid mycorrhizae with a broad range of ECM fungi, including *L. deliciosus*.

11.4.2
Genetics of the Interaction

Nothing is currently known about the mating systems among members of the genus *Lactarius* as basidiospore germination in vitro is nil or extremely low

Table 11.1. Host associations of selected *Lactarius* species based on field observations

Taxa	Host range	Reference
Lactarius aspideus	*Salix*	Korhonen (1984), Arnolds (1989), Watling (1992)
Lactarius blennius	*Fagus*	Moser (1983), Korhonen (1984), Lange (1987)
Lactarius chelidonius	*Pinus*	Hesler and Smith (1979)
Lactarius cinereus	*Fagus*	Hesler and Smith (1979), Bills (1986)
Lactarius controversus	Salicaceae	Hesler and Smith (1979), Moser (1983), Bills (1986)
Lactarius deliciosus	Pinaceae	Hesler and Smith (1979)
Lactarius deterrimus	*Picea*	Moser (1983), Korhonen (1984), Lange (1987)
Lactarius eucalypti	*Eucalyptus*	Miller and Hinton (1986)
Lactarius lignyotus	Pinaceae	Hesler and Smith (1979), Bills (1986)
Lactarius necator	Wide host range	Moser (1983), Korhonen (1984), Lange (1987)
Lactarius obscuratus	*Alnus*	Hesler and Smith (1979), Moser (1983), Korhonen (1984)
Lactarius oculatus	Pinaceae	Bills (1986)
Lactarius pubescens	*Betula*	Hesler and Smith (1979), Moser (1983), Korhonen (1984)
Lactarius quietus	*Quercus*	Moser (1983), Korhonen (1984), Lange (1987)
Lactarius rufus	Pinaceae	Hesler and Smith (1979), Bills (1986)
Lactarius scrobiculatus	Pinaceae	Hesler and Smith (1979), Moser (1983), Korhonen (1984)
Lactarius subpurpureus	*Tsuga*	Hesler and Smith (1979), Homola and Czapowskyj (1981)
Lactarius torminosus	*Betula*	Hesler and Smith (1979), Moser (1983), Korhonen (1984)
Lactarius vinaceorufescens	*Pinus*	Hesler and Smith (1979), Homola and Czapowskyj (1981)
Lactarius volemus	Wide host range	Moser (1983), Korhonen (1984)

among them (Fries 1978; Ali and Jackson 1988), resulting in a lack of monokaryons for mating studies and genetic interaction studies. Žel et al. (1989), however, were successful in isolating and regenerating protoplasts from tissue of *L. piperatus*, a possible method for obtaining monokaryotic cultures for mating studies, and investigating genetic interactions with hosts, as suggested by Kropp and Fortin (1986) for *Laccaria bicolor* (R. Maire) Orton.

In one of the only studies examining the genetics of this fungus–host interaction, Last et al. (1984a) found that sporocarp production by *L. glyciosmus* in association with clones of *Betula* spp. was affected by host genotype during the first 6 years after transplanting. They found that in one soil type, sporocarps of *L. glyciosmus* were found only with *Betula pendula* (clone 12.5D) but not with clones of *B. pubescens* (9.3D and 9.3G), whereas in another soil type, sporocarps were found only with clone 9.3D of *B. pubescens* but not with clone 9.3G or with clone 12.5D of *B. pendula*.

11.4.3
Dispersal and Modes of Infection by the Fungal Symbiont

Asexual propagules are unknown among species of *Lactarius* (Hutchison 1989); therefore basidiospores act as sole propagules for dispersal, and function as one of two known methods of infection. However, the viability and persistence of *Lactarius* basidiospores are quickly lost after release from the sporocarps. Utilizing fluorescein diacetate staining, Torres and Honrubia (1994) showed that basidiospores of *L. deliciosus* and *L. sanguifluus* (Paulet: Fr.) Fr. quickly became non-viable. Miller et al. (1994) found that only about 10% of the basidiospores of *L. scrobiculatus* (Scop.: Fr.) Fr. persisted over-winter in the leaf litter and upper 3 cm of the soil fraction. As is typical for late-stage ECM colonizers, basidiospore germination among various *Lactarius* species is nil to extremely low (Fries 1978; Ali and Jackson 1988). However, Fries (1978) found that he could stimulate germination in vitro among basidiospores of *L. helvus* (Fr.: Fr.) Fr. in the presence of activated charcoal and a living *Rhodotorula glutinis* (Fresen.) Harrison colony, but only at a rate below 0.001%. Similarly, Ali and Jackson (1988) found a germination rate below 0.006% among basidiospores of *L. turpis* (Weinm.) Fr. [= *L. necator*] in the presence of root exudates of *B. pubescens*, but not with root exudates of six other ECM trees and five non-ECM plants. Fox (1983) found that an inoculum of basidiospores from *L. pubescens* (Schrad.) Fr. was unable to colonize roots of birch seedlings. Basidiospores of *Lactarius* species are probably stimulated to germinate immediately after release and only in the presence of root exudates from older host trees rather than from root exudates from seedling or older non-host trees.

Apart from the role of basidiospores, infection of host roots can occur through hyphal systems arising from an already established colony. Last et al. (1984b) found that *L. pubescens* colonies associated with *B. pendula* expanded at a rate of 24.8 cm year^{-1}. When seedlings of *B. pendula* were planted around an 11-year-old *B. pubescens* tree, most of the seedlings bore ECM of *L. pubescens* after 17 weeks, but seedlings which had been planted and inoculated with cores containing an inoculum of *L. pubescens* were found to lack such ECM (Fleming 1983). Fleming suggested that *L. pubescens* infects seedling roots by means of mycelial strands which must remain attached to the older parent tree in order to infect. Agerer (1986) reported the occurrence of hyphal strands in *L. deterrimus* Gröger. Field experiments have repeatedly shown the difficulty in successfully inoculating seedlings of *Betula* spp. with mycelial and basidiospore inocula of *L. pubescens* (Deacon et al. 1983; Fox 1983: Fleming 1985), and observations by Last and Fleming (1985) have shown an association of *L. pubescens* with outplanted *Betula* only after 7 years. This probably explains the difficulty with in vitro ECM synthesis experiments involving species of *Lactarius* and various tree seedlings. However, the addition of exogenous glucose stimulated the formation of ECM in vitro between *L. pubescens* and *B. pubescens* (Gibson and Deacon 1990) and between *L. subpurpureus* Peck and *Tsuga canadensis* (Hutchison and Piché 1995), suggesting the possibility that the amount of simple carbohydrates required by late-stage ECM colonizers can only be provided by older (and larger) trees rather than from seedlings, unlike the less fastidious requirements found among early-stage ECM colonizers.

11.4.4
Developmental Aspects of the Fungus–Host Interaction

Hyphal systems of *Lactarius* spp. are undoubtedly attracted to root exudates of appropriate hosts (Horan and Chilvers 1990) which help to stimulate vegetative growth and initiate subsequent infection. Sirrenberg et al. (1995) found that callus tissue of *P. abies* stimulated the doubling of vegetative growth of the spruce associate *L. deterrimus*. Lectins, which are found on the surface of fungal hyphae, have been shown to play an important role in host recognition and specificity among species of *Lactarius* as they bind to specific saccharides found on the cell walls of the host roots. Giollant et al. (1993) and Guillot et al. (1991, 1994) found that lectins from *L. deliciosus*, *L. deterrimus* and *L. salmonicolor* Heim & Leclair were different in molecular structure and bound only to specific oligosaccharides found on the root cells of their appropriate hosts (*Pinus*, *Picea* and *Abies* respectively) but not to a non-host. Münzenberger et al. (1990) reported that in the compatible interaction between *L. deterrimus* and *P. abies*, the resulting ECM had lower levels of phenolics than non-ECM roots. Giltrap (1982), however, observed the occurrence of cortex disruption in the roots of *Betula* when inoculated with *L. subdulcis* and *L. tabidus* Fr., suggestive of an incompatible interaction between these fungi and a possible non-host (Molina and Trappe 1982a).

Species of *Lactarius*, irrespective of host, typically form ECM that are smooth and shiny, white or dull to brightly coloured depending on the fungal species. They have a thick mantle (15–35 μm thick) that is multilayered and generally contains lactiferous hyphae, and possess a well-developed Hartig net. A list of morphological and anatomical descriptions of known *Lactarius* ECM is provided in Table 11.2, with most being based on field-collected material. Most reports, based on in vitro ECM synthesis, indicate a long time period required for formation of ECM among *Lactarius* species. Deacon et al. (1983) found that it took at least 20–26 weeks for *L. pubescens* to infect seedlings of *B. pubescens*, while Yamada and Katsuya (1995) found that two isolates of *L. chrysorheus* Fr. took at least 6–7 months to form ECM with seedlings of *Pinus densiflora*. Formation of ECM involving older trees, however, appears to take less time (Fleming 1983).

Table 11.2. Descriptions of selected *Lactarius* ECM

Fungus	Host	References
Lactarius acris	*Fagus sylvatica*	Agerer (1987), Brand (1991)
Lactarius alnicola	*Picea engelmannii*	Kernaghan et al. (1997)
Lactarius alpinus	*Alnus viridis*	Agerer (1987)
Lactarius badiosanguineus	*Picea abies*	Agerer (1987)
	Pinus cembra	Agerer (1987)
Lactarius blennius	*Fagus sylvatica*	Voiry (1981), Prevost and Pargney (1995)
Lactarius caespitosus	*Abies lasiocarpa*	Kernaghan et al. (1997)
Lactarius camphoratus	*Fagus sylvatica*	Brand (1991)
Lactarius chrysorheus	*Quercus petraea*	Luppi and Gautero (1967)
	Quercus robur	Palfner and Agerer (1996)
Lactarius decipiens	*Picea abies*	Gronbach (1988)
Lactarius deliciosus	*Abies lasiocarpa*	Kernaghan et al. (1997)
	Arbutus menziesii	Molina and Trappe (1982b)
	Arctostaphylos uva-ursi	Molina and Trappe (1982b)
	Pinus montana (= *P. mugo*)	Melin (1924)
	Pinus ponderosa	Riffle (1973)
Lactarius deterrimus	*Picea abies*	Agerer (1986), Münzenberger et al. (1986)
Lactarius flexuosus	*Quercus pubescens*	Luppi and Gautero (1967)
Lactarius glyciosmus	*Betula pendula*	Ingleby et al. (1990)
Lactarius helvus	*Picea abies*	Modess (1941)
	Pinus montana (= *P. mugo*)	Modess (1941)
	Pinus sylvestris	Modess (1941)
Lactarius lignyotus	*Picea abies*	Kraigher et al. (1995)
Lactarius lilacinus	*Alnus glutinosa*	Pritsch et al. (1997)
Lactarius mitissimus	*Picea abies*	Weiss (1991)
Lactarius obscuratus	*Alnus glutinosa*	Pritsch et al. (1997)
	Alnus rubra	Froidevaux (1973), Miller et al. (1991)
Lactarius omphaliformis	*Alnus glutinosa*	Pritsch et al. (1997)
Lactarius pallidus	*Fagus sylvatica*	Agerer (1987), Brand (1991)
Lactarius paradoxus	*Pinus banksiana*	Danielson (1984)
Lactarius picinus	*Picea abies*	Agerer (1986), Agerer (1987), Haug and Pritsch (1992)
Lactarius porninsis	*Larix decidua*	Agerer (1987)
Lactarius pubescens	*Betula pendula*	Ingleby et al. (1990)
Lactarius rubrocinctus	*Fagus sylvatica*	Agerer (1987), Brand (1991)
Lactarius rufus	*Picea abies*	Haug and Pritsch (1992)
	Picea sitchensis	Alexander (1981), Ingleby et al. (1990)
	Pinus montana (= *P. mugo*)	Modess (1941)
	Pinus sylvestris	Pachlewski (1968)
Lactarius sanguifluus	*Arctostaphylos uva-ursi*	Zak (1976)
Lactarius subdulcis	*Fagus sylvatica*	Brand and Agerer (1986), Agerer (1987)
		Prevost and Pargney (1995)
Lactarius scrobiculatus	*Picea abies*	Amiet and Egli (1991)
Lactarius serifluus	*Quercus robur*	Palfner and Agerer (1996)
Lactarius theiogalus	*Picea abies*	Gronbach (1988)
Lactarius vellereus	*Fagus sylvatica*	Brand and Agerer (1986), Agerer (1987)
	Quercus petraea	Luppi and Gautero (1967)

11.5
Host Plant Growth Responses and Fungus-Derived Benefits

11.5.1
Growth Responses

Alexander (1981) utilized in vitro ECM synthesis techniques to inoculate *Picea sitchensis* seedlings with *L. rufus* and found that after 14 weeks shoot height and total dry weight of the inoculated seedlings were 9.2 ± 0.9 cm and 174.2 ± 14.7 mg respectively, versus 5.8 ± 0.5 cm and 72.6 ± 8.3 mg respectively for the uninoculated controls. Jones and Hutchinson (1986), however, found that 18 weeks after inoculation, the total dry weight of seedlings of *Betula papyrifera* inoculated with *L. hibbardae* Peck and *L. rufus* was 212.95 ± 10.49 and 192.01 ± 5.99 mg respectively, versus 218.14 ± 15.04 mg for the uninoculated controls. One year after outplanting on amended oil-sands tailings, Danielson and Visser (1989) found that *Pinus banksiana* seedlings inoculated with *L. paradoxus* Beardslee & Burlingham had on average a total dry weight and height of 1457 mg and 5.5 cm respectively, versus 1282 mg and 6.1 cm respectively for the uninoculated controls. After three growing seasons, the shoot dry weight and height on average for inoculated seedlings was 11.1 g and 28.4 cm respectively, versus 6.3 g and 22.0 cm respectively for the uninoculated controls. Stenström and Ek (1990) reported that *P. sylvestris* seedlings inoculated with *L. rufus* and outplanted for 2.5 years had greater growth than the controls.

11.5.2
Nutritional Benefits to the Host

Currently, there are little data on the effect of N and P on growth enhancement of host plants mediated by *Lactarius* ECM. Danielson and Visser (1989) found that there were no significant differences in the percentage of N and P in the needles of *P. banksiana* inoculated with *L. paradoxus* as compared to controls after being outplanted for two growing seasons. Gibson and Deacon (1990) reported that low N and P levels in a nutrient mixture applied to *B. pubescens* inoculated with *L. pubescens*, compared to a full strength nutrient mixture, caused a reduction in the number of ECM observed and a reduction in the percentage of roots that were ECM.

In axenic culture, various researchers have found that species of *Lactarius* grow best with NH_4^+ as a source of N and poorly with NO_3^- (Jayko et al. 1962; Lundeberg 1970; Finlay et al. 1992; Keller 1996). Keller (1996) found variability among strains of *L. rufus* for growth on bovine serum albumin.

11.5.3
Non-Nutritional Benefits to the Host

Much has been written about the non-nutritional benefits of ECM fungi to tree seedlings, such as ameliorating the effect of toxic heavy metals (Wilkins 1991; Wilkinson and Dickinson 1995), reducing host water deficit under conditions of mild drought (Parke et al. 1983), and protecting seedlings from infection by soil-borne root pathogens (Marx 1972; Chakravarty and Unestam 1987). However, it is possible that late-stage ECM colonizers like *Lactarius* may be less beneficial to seedlings than early-stage colonizers such as *Pisolithus*, *Hebeloma* and *Laccaria* in this regard. A case in point is shown with the general inability of *Lactarius* spp. to reduce the uptake of toxic heavy metals into host seedlings. Jones and Hutchinson (1988a) found that *L. rufus* and *L. hibbardae* were both sensitive to Ni (25 mgl^{-1}) and although *L. hibbardae* showed tolerance to Cu (25 mgl^{-1}), *L. rufus* did not. Further, inoculation studies by Jones and Hutchinson (1986) showed that neither *L. hibbardae* nor *L. rufus* reduced Ni uptake into tissue of *B. papyrifera* compared to the uninoculated control. However, Jones and Hutchinson (1988b) did suggest that *L. rufus* appeared to enhance tolerance to Ni in *B. papyrifera* seedlings in the early stages of their experiment, but once the fungus was exposed to Ni, fungal growth ceased and it was ineffective in increasing Ni tolerance over long periods.

Species of *Lactarius* do not appear to be very tolerant of water stress. Coleman et al. (1989) found that *L. controversus* (Pers.: Fr.) Fr. and *L. rufus* had little or no growth below −0.4 and −1.0 MPa respectively. Wilson and Griffin (1979) found that *L. deliciosus* was more tolerant as it could withstand solute potentials below −10 MPa; however, no experiments were conducted with ECM seedlings. Agerer (1986) reported the occurrence of rhizomorphs in *P. abies–L. deterrimus* ECM which could possibly play a role in water transport (Duddridge et al. 1980).

Antagonism observed in vitro in dual agar culture between species of *Lactarius* and various filamentous fungi and bacteria suggested initially that species of *Lactarius* may play an important role in protecting host tree roots from a variety of soil-borne root pathogens. Marx (1969) found that an isolate of *L. deliciosus* inhibited a number of fungi pathogenic to pine such as *Phytophthora cinnamomi* Rands, *Rhizoctonia repens* Bernard and *Inonotus circinatus* (Fr.) Gilbertson. Pratt (1971) found that an Australian isolate of *L. deliciosus* also could inhibit *P. cinnamomi*. Krywolop (1971) obtained acetone and chloroform extracts from fresh mycelia of *L. deliciosus* and found that they inhibited *Bacillus cereus* Frankland & Frankland and *Escherichia coli* (Migula) Castellani & Chalmers. Park (1970) isolated a culture of an unknown *Lactarius* sp. from roots of *Tilia americana* seedlings which exhibited antagonism against a wide range of filamentous fungi. Hyppel (1968), however, found that one isolate of *L. deliciosus* and two isolates of *L. torminosus* did not inhibit *Heterobasidion annosum* (Fr.: Fr.) Bref. in dual culture. Further,

Bücking (1979) found that *L. semisanguifluus* Heim & Leclair ECM did not protect *P. abies* from infection from *H. annosum.*

Apart from the production of possible antagonistic metabolites, *Lactarius* species may protect the root systems of their hosts by providing a physical barrier in the form of the fungal mantle against soil-borne root pathogens. Further, Zhao and Guo (1989) observed that *L. deliciosus* could behave as a hyperparasite by penetrating and lysing the hyphae of the soil-borne root pathogen *Rhizoctonia solani* Kühn. However, this observation has not been confirmed.

11.6
Conclusions

Due to the fact that species of *Lactarius* are late-stage ECM colonizers, relatively little information is available regarding the interactions between various Lactarii and their plant hosts in comparison to early-stage ECM colonizers found in the genera *Hebeloma, Laccaria,* and *Pisolithus.* With an increasing awareness of their role in the ecosystem, however, more attention is being paid to *Lactarius* species and their interactions with mature trees. It is hoped that this contribution will stimulate further investigations of this important genus of ECM fungi.

Acknowledgement. Dr. Bradley R. Kropp (Utah State University) is thanked for reviewing an early draft of this manuscript.

References

Agerer R (1986) Studies on ectomycorrhizae. III. Mycorrhizae formed by four fungi in the genera *Lactarius* and *Russula* on spruce. Mycotaxon 27:1–59

Agerer R (1987) Colour atlas of ectomycorrhizae. Einhorn Verlag, Schwäbish Gmünd

Alexander IJ (1981) The *Picea sitchensis* + *Lactarius rufus* mycorrhizal association and its effects on seedling growth and development. Trans Br Mycol Soc 76:417–423

Ali NA, Jackson RM (1988) Effects of plant roots and their exudates on germination of spores of ectomycorrhizal fungi. Trans Br Mycol Soc 91:253–260

Amiet R, Egli S (1991) Die Ektomykorrhiza des Grubigen Milchlings (*Lactarius scrobiculatus* (Scop.: Fr.) Fr.) an Fichte (*Picea abies* Karst.). Schweiz Z Forstwes 142:53–60

Arnolds E (1988) The changing macromycete flora in the Netherlands. Trans Br Mycol Soc 90:391–406

Arnolds E (1989) A preliminary red data list of macrofungi in the Netherlands. Persoonia 14:77–125

Arnolds E, de Vries B (1993) Conservation of fungi in Europe. In: Pegler DN, Boddy L, Ing B, Kirk PM (eds) Fungi of Europe: investigation, recording and conservation. Royal Botanic Gardens, Kew, pp 211–230

Arnolds E, Jansen E (1992) New evidence for changes in the macromycete flora of the Netherlands. Nova Hedwigia 55:325–351

Benjamin DR (1995) Mushrooms: poisons and panaceas: a handbook for naturalists, mycologists and physicians. WH Freeman, New York

Bills GF (1986) Notes on *Lactarius* in the high-elevation forests of the southern Appalachians. Mycologia 78:70–79

Bills GF, Cotter HVT (1989) Taxonomy and ethnomycology of *Lactarius* sect. *Dapetes* (Russulaceae) in Nepal. Mem NY Bot Gard 49:192–197

Bon M (1980) Clé monographique du genre *Lactarius* (Pers. ex Fr.) S. F. Gray. Doc Mycol 10:1–85

Brand F (1991) Ektomykorrhizen an *Fagus sylvatica* – Charakterisierung und Identifizierung, ökologische Kennzeichnung und unsterile Kultivierung. Libri Bot 2:1–229

Brand F, Agerer R (1986) Studien an Ektomykorrhizen. VIII. Die Mykorrhizen von *Lactarius subdulcis*, *Lactarius vellereus* und *Laccaria amethystina* an Buche. Z Mykol 52:287–320

Bücking E (1979) Fichten-Mykorrhizen auf Standorten der Schwäbischen Alb und ihre Beziehung zum Befall durch *Fomes annosus*. Eur J For Pathol 9:19–35

Burlingham GS (1908) A study of the *Lactariae* of the United States. Mem Torrey Bot Club 14:1–109

Buyck B, Verbeken A (1995) Studies in tropical African *Lactarius* species. 2. *Lactarius chromospermus* Pegler. Mycotaxon 56:427–442

Camazine S, Lupo AT (1984) Labile toxic compounds of the Lactarii: the role of the lactiferous hyphae as a storage depot for precursors of pungent dialdehydes. Mycologia 76:355–358

Camazine SM, Resch JF, Eisner T, Meinwald J (1983) Mushroom chemical defense: pungent sesquiterpenoid dialdehyde antifeedant to opossum. J Chem Ecol 9:1439–1447

Chakravarty P, Unestam T (1987) Differential influence of ectomycorrhizae on plant growth and disease resistance in *Pinus sylvestris* seedlings. J Phytopathol 120:104–120

Coker WC (1918) The Lactarias of North Carolina. J Elisha Mitchell Sci Soc 34:1–61

Coleman MD, Bledsoe CS, Lopushinky W (1989) Pure culture response of ectomycorrhizal fungi to imposed water stress. Can J Bot 67:29–39

Dahlberg A, Jonsson L, Nylund J-E (1997) Species diversity and distribution of biomass above and below ground among ectomycorrhizal fungi in an old-growth Norway spruce forest in south Sweden. Can J Bot 75:1323–1335

Danielson RM (1984) Ectomycorrhizal associations in jack pine stands in northeastern Alberta. Can J Bot 62:932–939

Danielson RM, Visser S (1989) Host response to inoculation and behaviour of introduced and indigenous ectomycorrhizal fungi of jack pine grown on oil-sands tailings. Can J For Res 19:1412–1421

Daniewski WM, Gumulka M, Ptaszynska K, Skibicki P, Bloszyk E, Drozdz B, Stromberg S, Norin T, Holub M (1993) Antifeedant activity of some sesquiterpenoids of the genus *Lactarius* (Agaricales: Russulaceae). Eur J Entomol 90:65–70

Daniewski WM, Gumulka M, Przemycka D, Ptaszynska K, Bloszyk E, Drozdz B (1995) Sesquiterpenes of *Lactarius* origin, antifeedant structure–activity relationships. Phytochemistry 38:1161–1168

Deacon JW, Donaldson SJ, Last FT (1983) Sequences and interactions of mycorrhizal fungi on birch. Plant Soil 71:257–262

Dighton J, Horrill AD (1988) Radiocaesium accumulation in the mycorrhizal fungi *Lactarius rufus* and *Inocybe longicystis*, in upland Britain, following the Chernobyl accident. Trans Br Mycol Soc 91:335–357

Duddridge JA, Malibari A, Read DJ (1980) Structure and function of mycorrhizal rhizomorphs with special reference to their role in water transport. Nature 287:834–836

Finlay RD, Frostegård Å, Sonnerfeldt A-M (1992) Utilization of organic and inorganic nitrogen sources by ectomycorrhizal fungi in pure culture and in symbiosis with *Pinus contorta* Dougl. ex Loud. New Phytol 120:105–115

Fleming LV (1983) Succession of mycorrhizal fungi on birch: infection of seedlings planted around mature trees. Plant Soil 71:263–268

Fleming LV (1985) Experimental study of sequences of ectomycorrhizal fungi on birch (*Betula* sp.) seedling root systems. Soil Biol Biochem 17:591–600

Fox FM (1983) Role of basidiospores as inocula of mycorrhizal fungi of birch. Plant Soil 71:269–273

Fries N (1978) Basidiospore germination in some mycorrhiza-forming Hymenomycetes. Trans Br Mycol Soc 70:319–324

Froidevaux L (1973) The ectomycorrhizal association, *Alnus rubra* + *Lactarius obscuratus.* Can J For Res 3:601–603

Gibson F, Deacon JW (1990) Establishment of ectomycorrhizas in aseptic culture: effects of glucose, nitrogen and phosphorus in relation to successions. Mycol Res 94:166–172

Giltrap NJ (1982) Production of polyphenol oxidases by ectomycorrhizal fungi with special reference to *Lactarius* spp. Trans Br Mycol Soc 78:75–81

Giollant M, Guillot J, Damez M, Dusser M, Didier P, Didier E (1993) Characterization of a lectin from *Lactarius deterrimus.* Plant Physiol 101:513–522

Gronbach E (1988) Charakterisierung und Identifizierung von Ektomykorrhizen in einem Fichtenbestand mit Untersuchungen zur Merkmalsvariabilität in sauer beregneten Flächen. Bibl Mycol 125:1–216

Guevara G, Garcia J, Castillo J, Miller OK (1987) New records of *Lactarius* in Mexico. Mycotaxon 30:157–176

Guillot J, Giollant M, Damez M, Dusser M (1991) Isolation and characterization of a lectin from the mushroom, *Lactarius deliciosus.* J Biochem 109:840–845

Guillot J, Giollant M, Damez M, Dusser M (1994) Intervention des lectines fongiques dans les événements précoces de reconnaissance arbre/champignon au cours de la formation des ectomycorhizes. Acta Bot Gall 141:443–447

Haug I, Pritsch K (1992) Ectomycorrhizal types of spruce (*Picea abies* (L.) Karst.) in the black forest – a microscopical atlas. Kernforschungszentrum, Karlsruhe

Heim R (1955) Les Lactaires d'Afrique intertropicale. Bull Jard Bot Etat Brux 25:1–91

Heinemann P (1960) Les Lactaires. Bull Nat Belg 41:133–156

Hesler LR, Smith AH (1979) North American species of *Lactarius.* University of Michigan Press, Ann Arbor

Hintikka V (1988) On the macromycete flora in oligotrophic pine forests of different ages in south Finland. Acta Bot Fenn 136:89–94

Homola RL, Czapowskyj MM (1981) Ectomycorrhizae of Maine. 2. A listing of *Lactarius* with the associated hosts (with additional information on edibility). Maine Agric Exp Sta Bull No 779

Horan DP, Chilvers GA (1990) Chemotropism – the key to ectomycorrhizal formation? New Phytol 116:297–301

Hutchison LJ (1989) The absence of conidia as a morphological character in ectomycorrhizal fungi. Mycologia 81:587–594

Hutchison LJ (1990a) Studies on the systematics of ectomycorrhizal fungi in axenic culture. II. The enzymatic degradation of selected carbon and nitrogen compounds. Can J Bot 68:1522–1530

Hutchison LJ (1990b) Studies on the systematics of ectomycorrhizal fungi in axenic culture. III. Patterns of polyphenol oxidase activity. Mycologia 82:424–435

Hutchison LJ (1990c) Studies on the systematics of ectomycorrhizal fungi in axenic culture. IV. The effect of some selected fungitoxic compounds upon linear growth. Can J Bot 68:2172–2178

Hutchison LJ (1990d) Studies on the systematics of ectomycorrhizal fungi in axenic culture. V. Linear growth response to standard extreme temperatures used as a taxonomic character. Can J Bot 68:2179–2184

Hutchison LJ (1991) Description and identification of cultures of ectomycorrhizal fungi found in North America. Mycotaxon 42:387–504

Hutchison LJ, Piché Y (1995) Effects of exogenous glucose on mycorrhizal colonization in vitro by early-stage and late-stage ectomycorrhizal fungi. Can J Bot 73:898–904

Hutchison LJ, Summerbell RC (1990) Studies on the systematics of ectomycorrhizal fungi in axenic culture. Reactions of mycelia to diazonium blue B staining. Mycologia 82:36–42

Hyppel A (1968) Antagonistic effects of some soil fungi on *Fomes annosus* in laboratory experiments. Stud For Suec 64:1–17

Ingleby K, Mason PA, Last FT, Fleming LV (1990) Identification of ectomycorrhizas. HMSO, London

Jayko LG, Baker TI, Stubblefield RD, Anderson RF (1962) Nutrition and metabolic products of *Lactarius* species. Can J Microbiol 8:361–371

Jones MD, Hutchinson TC (1986) The effect of mycorrhizal infection on the response of *Betula papyrifera* to nickel and copper. New Phytol 102:429–442

Jones MD, Hutchinson TC (1988a) The effects of nickel and copper on the axenic growth of ectomycorrhizal fungi. Can J Bot 66:119–124

Jones MD, Hutchinson TC (1988b) Nickel toxicity in mycorrhizal birch seedlings infected with *Lactarius rufus* or *Scleroderma flavidum*. I. Effects on growth, photosynthesis, respiration and transpiration. New Phytol 108:451–459

Kalamees K, Silver S (1988) Fungal productivity of pine heaths in north west Estonia. Acta Bot Fenn 136:95–98

Keller G (1996) Utilization of inorganic and organic nitrogen sources by high-subalpine ectomycorrhizal fungi of *Pinus cembra* in pure culture. Mycol Res 100:989–998

Kernaghan G, Currah RS, Bayer RJ (1997) Russulaceous ectomycorrhizae of *Abies lasiocarpa* and *Picea engelmannii*. Can J Bot 75:1843–1850

Knudsen H, Borgen T (1982) Russulaceae in Greenland. In: Laursen GA, Ammirati JF (eds) Arctic and alpine mycology. University of Washington Press, Seattle, pp 216–238

Knudsen H, Borgen T (1994) The *Lactarius torminosus*-group in Greenland. Mycol Helv 2:49–56

Konrad P (1935) Les Lactaires. Bull Soc Mycol Fr 51:160–191

Korhonen M (1984) Suomen rouskut. Otava, Helsinki

Kraigher H, Agerer R, Javornik B (1995) Ectomycorrhizae of *Lactarius lignyotus* on Norway spruce, characterized by anatomical and molecular tools. Mycorrhiza 5:175–180

Kropp BR (1982) Fungi from decayed wood as ectomycorrhizal symbionts of western hemlock. Can J For Res 12:36–39

Kropp BR, Albee S (1996) The effects of silvicultural treatments on occurrence of mycorrhizal sporocarps in a *Pinus contorta* forest: a preliminary study. Biol Conserv 78:313–318

Kropp BR, Fortin JA (1986) Formation and regeneration of protoplasts from the ectomycorrhizal basidiomycete *Laccaria bicolor*. Can J Bot 64:1224–1226

Krywolop GN (1971) Production of antibiotics by certain mycorrhizal fungi. In: Hacskaylo E (ed) Mycorrhizae. USDA Forest Service Misc Publ No 1189. US Government Printing Office, Washington, pp 219–221

Kühner R (1975) Agaricales de la zone alpine. Genre *Lactarius*. Bull Soc Mycol Fr 91:5–69

Kuiters AT (1990) Role of phenolic substances from decomposing forest litter in plant–soil interactions. Acta Bot Neerl 39:329–348

Lamb RJ (1974) Effect of D-glucose on utilization of single carbon sources by ectomycorrhizal fungi. Trans Br Mycol Soc 63:295–306

Lange JE (1928) Studies in the Agarics of Denmark. Part VII. *Volvaria, Flammula, Lactarius*. Dan Bot Ark 5:1–44

Lange M (1987) Mykorrhiza-svampenes afhaengighed af vaertstrae og jordbund. Svampe 16:57–59

Last FT, Fleming LV (1985) Factors affecting the occurrence of fruitbodies of fungi forming sheathing (ecto-) mycorrhizas with roots of trees. Proc Indian Acad Sci (Plant Sci) 94:111–127

Last FT, Mason PA, Pelham J, Ingleby K (1984a) Fruitbody production by sheathing mycorrhizal fungi: effects of "host" genotypes and propagating soils. For Ecol Manage 9:221–227

Last FT, Mason PA, Ingleby K, Fleming LV (1984b) Succession of fruitbodies of sheathing mycorrhizal fungi associated with *Betula pendula*. For Ecol Manage 9:229–234

Last FT, Dighton J, Mason PA (1987) Successions of sheathing mycorrhizal fungi. Trends Ecol Evol 2:157–161

Laursen GA, Ammirati JF (1982) Lactarii in Alaskan Arctic tundra. In: Laursen GA, Ammirati JF (eds) Arctic and alpine mycology. University of Washington Press, Seattle, pp 245–276

Lindeberg G (1948) On the occurrence of polyphenol oxidases in soil-inhabiting Basidiomycetes. Physiol Plant 1:196–205

Lundeberg G (1970) Utilisation of various nitrogen sources, in particular bound soil nitrogen, by mycorrhizal fungi. Stud For Suec 79:1-95

Luppi AM, Gautero C (1967) Richerche sulle micorrize di *Quercus robur, Q. petraea* e *Q. pubescens* in Piemonte. Allonia 13:129-148

Malajczuk N (1987) Ecology and management of ectomycorrhizal fungi in regenerating forest ecosystems in Australia. In: Sylvia DM, Hung LL, Graham JH (eds) Mycorrhizae in the next decade: practical applications and research priorities. University of Florida, Gainesville, pp 118-120

Malajczuk N, Molina R, Trappe JM (1982) Ectomycorrhiza formation in *Eucalyptus*. I. Pure culture synthesis, host specificity and mycorrhizal compatibility with *Pinus radiata*. New Phytol 91:467-482

Marx DH (1969) The influence of ectotrophic mycorrhizal fungi on the resistance of pine roots to pathogenic infections. I. Antagonism of mycorrhizal fungi to root pathogenic fungi and soil bacteria. Phytopathology 59:153-163

Marx DH (1972) Ectomycorrhizae as biological deterrents to pathogenic root infections. Annu Rev Phytopathol 10:429-454

McNabb RFR (1971) The Russulaceae of New Zealand. 1. *Lactarius* DC. ex S. F. Gray. NZ J Bot 9:46-66

Melin E (1924) Zur Kenntnis der Mykorrhizapilze von *Pinus montana* Mill. Bot Not 77:69-92

Mikola P (1954) Metsämaan kantasienien kyvystä hajoittaa neulas-ja lehtikarikkeita. Comm Inst For Fenn 42:1-17

Miller OK, Hilton RN (1986) New and interesting agarics from Western Australia. Sydowia 39:126-137

Miller SL, Koo CD, Molina R (1991) Characterization of red alder ectomycorrhizae: a preface to monitoring below ground ecological responses. Can J Bot 69:516-531

Miller SL, Torres P, McClean TM (1994) Persistence of basidiospores and sclerotia of ectomycorrhizal fungi and *Morchella* in soil. Mycologia 86:89-95

Modess O (1941) Zur Kenntnis der Mykorrhizabildner von Kiefer und Fichte. Symb Bot Ups 5:1-146

Molina R, Trappe JM (1982a) Patterns of ectomycorrhizal host specificity and potential among Pacific Northwest conifers and fungi. For Sci 28:423-458

Molina R, Trappe JM (1982b) Lack of mycorrhizal specificity by the ericaceous hosts *Arbutus menziesii* and *Arctostaphylos uva-ursi*. New Phytol 90:495-509

Montoya L, Bandala VM (1996) Additional new records on *Lactarius* from Mexico. Mycotaxon 57:425-450

Montoya L, Guzmán G, Bandala VM (1990) New records of *Lactarius* from Mexico and discussion of the known species. Mycotaxon 38:349-395

Montoya L, Bandala VM, Guzmán G (1996) New and interesting species of *Lactarius* from Mexico including scanning electron microscope observations. Mycotaxon 57:411-424

Moser M (1983) Keys to Agarics and Boleti (Polyporales, Boletales, Agaricales, Russulales). Roger Phillips, London

Münzenberger B, Metzler B, Kottke I, Oberwinkler F (1986) Morphologische und anatomische Charakterisierung der Mykorrhiza *Lactarius deterrimus–Picea abies* in vitro. Z Mykol 52:407-422

Münzenberger B, Heilemann J, Strack D, Kottke I, Oberwinkler F (1990) Phenolics of mycorrhizas and non-mycorrhizal roots of Norway spruce. Planta 182:142-148

Neuhoff W (1956) Die Milchlinge (Lactarii). In: Die Pilze Mitteleuropas, Bd Iib. Julius Klinkhardt, Bad Heilbrunn

Ohenoja E (1988) Behaviour of mycorrhizal fungi in fertilized forests. Karstenia 28:27-30

Ohenoja E (1993) Effect of weather conditions on the larger fungi at different forest sites in northern Finland in 1976-1988. Acta Univ Oulu A 243:1-69

Ohenoja E, Ohenoja M (1993) Lactarii of the Franklin and Keewatin Districts of the Northwest Territories, Arctic Canada. Bibl Mycol 150:179-192

Oort AJP (1981) Nutritional requirements of *Lactarius* species, and cultural characters in relation to taxonomy. North-Holland Publishing, Amsterdam

Pachlewski R (1968) Badania nad grzybami mikoryzowymi sosny-*Lactarius rufus* (Scop. ex Fr.) Fr. i *Rhizopogon luteolus* Fr. et Nordh.-W naturalnych warunkach I w czystych kulturach. Pr Inst Badaw Lesm 365:173–187

Palfner G, Agerer R (1996) Die Ektomykorrhizen von *Lactarius chrysorrheus* and *L. serifluus* an *Quercus robur*. Sendtnera 3:119–136

Park JY (1970) Antifungal effect of an ectotrophic mycorrhizal fungus, *Lactarius* sp., associated with basswood seedlings. Can J Microbiol 16:798–800

Parke JL, Linderman RG, Black CH (1983) The role of ectomycorrhizas in drought tolerance of Douglas-fir seedlings. New Phytol 95:83–95

Paulus W, Reisinger A (1990) Die Auswikungen des Reaktorunfalls von Tschernobyl auf den Gehalt an radioaktiven Cäsium in den Fruchtkörpern der Mykorrhizapilzarten *Lactarius rufus* und *Xerocomus badius* in Fichtelgebirge. Z Mykol 56:279–284

Pegler DN, Fiard JP (1978) Taxonomy and ecology of *Lactarius* (Agaricales) in the Lesser Antilles. Kew Bull 33:600–628

Pegler DN, Young TWK (1979) The gasteroid Russulales. Trans Br Mycol Soc 72:353–388

Pratt BH (1971) Isolation of Basidiomycetes from Australian eucalypt forest and assessment of their antagonism to *Phytophthora cinnamomii*. Trans Br Mycol Soc 56:243–250

Prevost A, Pargney JC, (1995) Comparaison des ectomycorhizes naturelles entre le hêtre (*Fagus sylvatica*) et 2 lactaires (*Lactarius blennius* var. *viridis* et *Lactarius subdulcis*). I. Caracteristiques morphologiques et cytologiques. Ann Sci For 52:131–146

Pritsch K, Munch JC, Buscot F (1997) Morphological and anatomical characterization of black alder *Alnus glutinosa* (L.) Gaertn. ectomycorrhizas. Mycorrhiza 7:201–216

Rao CS, Sharma GD, Shukla AK (1997) Distribution of ectomycorrhizal fungi in pure stands of different age groups of *Pinus kesiya*. Can J Microbiol 43:85–91

Rendall D-L (1980) The genus *Lactarius* occurring in the southern boreal forest region of Ontario and Québec. M Sc Thesis, University of Toronto, Toronto

Riffle JW (1973) Pure culture synthesis of ectomycorrhizae on *Pinus ponderosa* with species of *Amanita*, *Suillus* and *Lactarius*. For Sci 19:242–250

Romell L-G (1938) A trenching experiment in spruce forest and its bearing on problems of mycotrophy. Sven Bot Tidskr 32:89–99

Saini SS, Atri NS (1993) Studies on genus *Lactarius* from India. Indian Phytopathol 46:360–364

Singer R (1986) The Agaricales in modern taxonomy, 4th edn. Koeltz Scientific Books, Koenigstein

Sirrenberg A, Salzer P, Hager A (1995) Induction of mycorrhiza-like structures and defence reactions in dual cultures of spruce callus and ectomycorrhizal fungi. New Phytol 130:149–156

Stenström E, Ek M (1990) Field growth of *Pinus sylvestris* following nursery inoculation with mycorrhizal fungi. Can J For Res 20:914–918

Szklarz GD, Antibus RK, Sinsabaugh RL, Linkins AE (1989) Production of phenol oxidases and peroxidases by wood-rotting fungi. Mycologia 81:234–240

Termorshuizen AJ (1991) Succession of mycorrhizal fungi in stands of *Pinus sylvestris* in the Netherlands. J Veg Sci 2:555–564

Termorshuizen AJ (1993) The influence of nitrogen fertilisers on ectomycorrhizas and their fungal carpophores in young stands of *Pinus sylvestris*. For Ecol Manage 57:179–189

Torres P, Honrubia M (1994) Basidiospore viability in stored slurries. Mycol Res 98:527–530

Tyler G (1980) Metals in sporophores of Basidiomycetes. Trans Br Mycol Soc 74:41–49

Tyler G (1991) Effects of litter treatments on the sporophore production of beech forest macrofungi. Mycol Res 95:1137–1139

Tuomikoski R (1953) Die *Lactarius*-Arten Finnlands. Karstenia 2:9–25

Verbeken A (1995) Studies in tropical African *Lactarius* species. 1. *Lactarius gymnocarpus* R. Heim ex Singer and allied species. Mycotaxon 55:515–542

Verbeken A (1996a) Studies in tropical African *Lactarius* species. 3. *Lactarius melanogalus* R. Heim and related species. Persoonia 16:209–223

Verbeken A (1996b) New taxa of *Lactarius* (Russulaceae) in tropical Africa. Bull Jard Bot Natl Belg 65:197–213

Verbeken A (1998a) Studies in tropical African *Lactarius* species. 5. A synopsis of the subgenus *Lactifluus* (Burl.) Hesler & A. H. Sm. Emend. Mycotaxon 66:363–386

Verbeken A (1998b) Studies in tropical African *Lactarius* species. 6. A synopsis of the subgenus *Lactariopsis* (Henn.) R. Heim Emend. Mycotaxon 66:387–418

Visser S (1995) Ectomycorrhizal fungal succession in jack pine stands following wildfire. New Phytol 129:389–401

Voiry H (1981) Classification morphologique des ectomycorhizes du chêne et du hêtre dans le nord-est de la France. Eur J For Pathol 11:284–299

Watling R (1992) Macrofungi associated with British willows. Proc R Soc Edinb 98B:135–147

Weiss M (1991) Studies on ectomycorrhizae. XXXIII. Description of three mycorrhizae synthesized on *Picea abies*. Mycotaxon 40:53–77

Wiklund K, Nilsson L-O, Jacobsson S (1995) Effect of irrigation, fertilization, and artificial drought on basidioma production in a Norway spruce stand. Can J Bot 73:200–208

Wilkins DA (1991) The influence of sheathing (ecto-) mycorrhizas of trees on the uptake and toxicity of metals. Agric Ecosyst Environ 35:245–260

Wilkinson DM, Dickinson NM (1995) Metal resistance in trees: the role of mycorrhizae. Oikos 72:298–300

Wilson JM, Griffin DM (1979) The effect of water potential on the growth of some soil Basidiomycetes. Soil Biol Biochem 11:211–212

Yamada A, Katsuya K (1995) Mycorrhizal association of isolates from sporocarps and ectomycorrhizas with *Pinus deniflora* seedlings. Mycoscience 36:315–323

Zak B (1976) Pure culture synthesis of bearberry mycorrhizae. Can J Bot 54:1297–1305

Žel J, Dermastia M, Gogala N (1989) The isolation and regeneration of protoplasts from the mycorrhizal fungi *Lactarius piperatus* and *Suillus variegatus*. Biol Vestn 37:93–100

Zhao Z, Guo X (1989) Study on hyphal hyperparasitic relationships between *Rhizoctonia solani* and ectomycorrhizal fungi. Acta Microbiol Sin 29:170–173

Chapter 12

Cenococcum

K. F. LoBuglio

12.1
Introduction

Cenococcum geophilum Fr. is one of the most frequently encountered ECM fungi in nature. This genus is cosmopolitan and is well recognized for its extremely wide host and habitat range. *C. geophilum* was originally described from its black sclerotia by J. Sowerby in 1800 under the name *Lycoperdon graniforme* Sow. Elias Fries introduced the genus *Cenococcum* and the species *C. geophilum* in 1825, and considered Sowerby's *L. graniforme* to be a synonym of *C. geophilum* Fr. (Fries 1825). In their monographic study of *Cenococcum* Fr., Ferdinandsen and Winge (1925) made the new combination *C. graniforme* (Sow.) Ferd. and Winge. The isolation of a "jet-black mycelium" that formed ECM was first identified by Hatch (1934) and provisionally named *Mycelium radicus nigrostrigosum* Hatch. The connection of these black ECM to *C. geophilum* was made by Linhell (1942). This black ECM was subsequently reported by many researchers on a wide variety of tree species in different parts of the world (Mikola 1948). *C. geophilum* Fr. is the sanctioned name.

12.2
Taxonomy

12.2.1
Cenococcum Geophilum

Cenococcum geophilum (*Cg*) exists as a sterile dematiaceous mycelium and thus lacks sexual or asexual spores, which are important taxonomic criteria in fungal classification. Thus, positive identification of *Cg* primarily relies on culture morphology and ECM characteristics. Based on the descriptions of Chilvers (1968) and Trappe (1962a) *Cg* ECM are black, usually monopodial structures but sometimes dichotomous, with thick, straight, unbranched,

Department of Plant and Microbial Biology, University of California, Berkeley, Berkeley, California 94720, USA black hyphae or seta radiating from the ECM surface (Fig. 12.1). The degree of setae emanating from the mantle surface can vary from sparse (Fig. 12.1A,B) to profuse (Fig. 12.1C–E). The mantle is generally 20–30 µm thick, composed of large diameter, darkly pigmented hyphae (5–10 µm) near the surface, and smaller diameter, paler hyphae (2.5–4 µm) adjacent to the host's epidermis. Surface hyphae of the mantle characteristically consist of clusters of isodiametric cells surrounded by radiating bundles of elongate cells forming a stellate or cephalothecoid pattern (Trappe 1971; Hawkworth 1986;

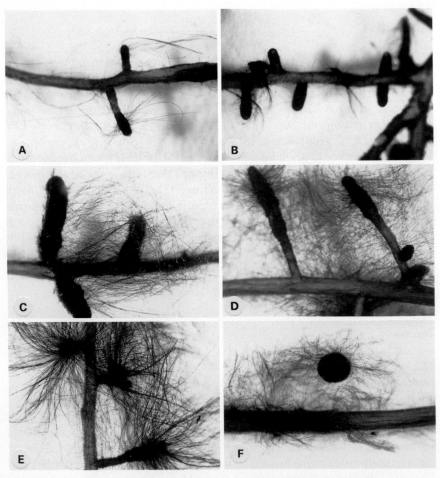

Fig. 12.1. ECM of *Cenococcum geophilum* in roots of 3-month-old seedlings of *Pinus resinosa*, synthesized in growth pouches. Presence of thick, dark, unbranched hyphae (seta) on root surface varies from limited: **A** isolate S8-1, ×13, and **B** isolate 349, ×13; to profuse: **C** isolate A181, ×13, **D** isolate CGGOLD, ×13, and **E** isolate WARR, ×13. **F** Sclerotium formed in growth pouch with mycorrhizae of isolate HUNT-6, ×13. (K. F. LoBuglio, unpubl. data; see LoBuglio et al. 1991 for origin of isolates)

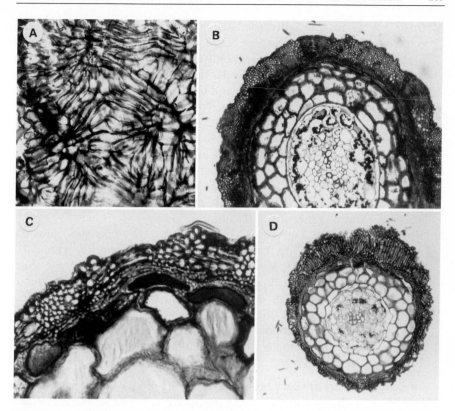

Fig. 12.2. Tangential and transectional sections of *Cenococcum geophilum* mycorrhizae showing mantle and Hartig net development. **A** Tangential section of mantle tissue from mycorrhiza of HUNT-1 depicting radiate growth pattern of mantle hyphae, ×2000. Transverse section of ECM showing characteristic mantle pattern, and Hartig net development: **B** isolate A181, ×130; **C** isolate ALB-1, ×2000. **D** Transverse section of ECM, isolate HUNT-6, demonstrating variable mantle development, ×130. (K. F. LoBuglio, unpubl. data; see LoBuglio et al. 1991 for origin of isolates)

Fig. 12.2a). This distinctive pattern is also reflected in cross section (Fig. 12.2b,c). It should be noted that the presence of a more disorganized mantle pattern in isolates that have the gross appearance of *Cg* ECM has been observed from ECM synthesized in vitro (Fig. 12.2d). Hyphae comprising the Hartig net typically consist of hyaline, thin-walled, smaller diameter hyphae (2 µm). It has been reported by Park (1970) that the aging process of *Cg* ECM can be associated with color change. According to Park (1970) the early stage of the *Cg* symbiosis is characterized by white and brown ECM while the black ECM characterize the most mature stage of ECM development. This has also been reported to occur in cultures of *Cg* (Lihnell 1942; Mikola 1948; Park 1970).

Trappe (1962a) described three morphologically distinct groups of *Cg* ECM based on the host species as follows:

1. Host species in the Salicaceae, Betulaceae (except *Corylus* spp.), and ectotrophic genera of Rosaceae. ECM are monopodial or occasionally branched. Often the mantle covers only the tip of the root and is ca. 15 μm thick, with the Hartig net never deeper than the third layer of cortical cells. Intracellular penetration is sparse and only in occasional cells. Hosts have both ectotrophic and endotrophic mycorrhizas, the latter often predominating. Short-root diameters are ca. 0.25 mm.
2. *Pinus* species. ECM are monopodial, dichotomous, or occasionally irregularly branched. Mantles usually cover all the short root and are 8–60 μm thick, ususally 20–30 μm. The Hartig net extends to the innermost layer of cortical cells and strong intracellular infection occurs throughout the cortex. Hosts rarely form endotrophic mycorrhizas. Short-root diameters are 0.35–1.0 mm (generally ca. 0.5 mm).
3. Host species in the Fagaceae and Pinaceae (except *Pinus* spp.) plus the genus *Corylus*. ECM are variously monopodial, racemose, irregularly branched, long or short. Mantles usually cover all or most of the short root and are 8–60 μm thick, generally 20–30 μm. The Hartig net extends to the innermost layer of cortical cells and strong intracellular infection occurs throughout the cortex. Some species form endotrophic mycorrhizas, but ectotrophic ones usually prevail. Depending on host species, short-root diameters range from 0.2 to 1.0 mm (generally 0.3–0.6 mm).

Cg mycelia in culture are dark brownish-black, the reverse being black to dark bluish-black, with mats at first fluffy becoming submerged, velvety, with a tightly compacted appearance. Growth is generally slow (2.2–2.4 cm in 1 month at 25°C) (Trappe 1962a; Miller and Miller 1983). Hyphae are yellowish when young, changing to brownish-black or bronze with age and are 1.5–8.2 μm in diameter (mostly 3.5–6.5 μm). They are generally stiff in appearance, but can be strongly sinuate, occasionally branching, anastomosing, and aggregating into short loosely arranged strands (but never becoming rhizomorphic). Septa are simple and cell walls are thick (ranging from 0.2 μm on young hyphae to 0.7 μm on older hyphal cells) and a hyaline gelatinous pellicle is present on hyphae which eventually breaks into regular or irregular dark brown to black papillae. Chlamydospores are usually intercalary when present (rarely terminal on setae) and are elliptical to globose to pyriform 8–12.8 × 6–8 μm, having 0.7-μm-thick walls at maturity. Sclerotia (0.05–7.0 mm diam.) can be produced in axenic culture or in association with ECM roots. These are spherical with a shiny-smooth surface and a black to dull roughened and brownish-black appearance. Surface hyphae form a cephalothecoid pattern, with thick black unbranched hyphae often radiating from the surface (Fig. 12.1f) (see Massicotte et al. 1992 for developmental details). Isolation of *Cg* mycelium is readily achieved from field-collected sclerotia (Trappe 1969).

Variation in cultural morphology, growth rate, and pH tolerance has been observed among isolates of *Cg* (Trappe 1962a; Cline et al. 1987; LoBuglio et al. 1991). Cline et al. (1987) observed significant variation in optimal temperature conditions for mycelial growth among 20 geographically distinct *Cg* isolates, with optima at 16, 21, and 27 °C. A correlation between optimal temperatures for mycelial growth and geographic origin of the isolates was not discernable. Trappe (1962a) noted a wide pH tolerance, between 3.4 and 7.3, among isolates of *Cg*. In vitro differences among sclerotium initiation and maturation have been observed also between different genotypes or ecotypes of *Cg* (Massicotte et al. 1992). In a study examining 70 isolates of *Cg* from both broad and regional geographic origins, variation in colony color, texture, and growth was reported (LoBuglio et al. 1991). As shown in Fig. 12.3 and described in Table 12.1, the isolates were grouped into six categories based on colony appearance after 10 weeks' growth on 2% malt agar. All isolates formed typical *Cg* ECM in growth pouches (LoBuglio et al. 1991).

12.2.2
Genetic Diversity

To see if the variation in *Cg* phenotype is mirrored by genotypic variation, restriction fragment length polymorphisms (RFLPs) were surveyed for the entire rDNA repeat unit of 70 *Cg* isolates chosen to sample geographic and host diversity and similarity (LoBuglio et al. 1991). Cluster analysis (UPGMA) of shared nuclear rDNA patterns indicated 32 unique genotypes and grouped *Cg* isolates into a broad range of clusters ranging from 44 to 100% similarity (LoBuglio et al. 1991). Limited correlation was observed among clusters from RFLP data and groupings based on culture morphology, geographic location, or host origin (LoBuglio et al. 1991). The amount of rDNA variation demonstrated in this study suggested that *Cg* is either a very heterogenous species or a fungal complex representative of a broader taxonomic rank (LoBuglio et al. 1991).

The *Cg* clusters defined by rDNA RFLPs (LoBuglio et al. 1991) were further examined by restriction mapping and nucleotide sequencing of PCR-amplified internal transcribed spacer (ITS) rDNA region (Shinohara 1994). The ITS was chosen because intraspecific variation in this region in other fungi has generally been low (Bruns et al. 1991; Baura et al. 1992; Chen et al. 1992; Lee and Taylor 1992; Yan et al. 1992; Gardes and Bruns 1993; LoBuglio et al. 1993; Seifert et al. 1995). The ITS from the same 70 *Cg* isolates analyzed by LoBuglio et al. (1991) was PCR amplified and digested with 11 restriction enzymes (Shinohara 1994). Digestion patterns of PCR amplified ITS sorted the isolates into 20 ITS groups. Much of the ITS variation was due to an approximately 500-bp addition found in 60 of the isolates. This 500-bp region has since been identified as a group-I intron (Rogers et al. 1993; Shinohara et al. 1996) located between the ITS5 (5′) and ITS1 (3′) primer sequences (White et al. 1990).

A representative from each of the 20 restriction groups (originating from different hosts and geographic origins) was sequenced by Shinohara (1994).

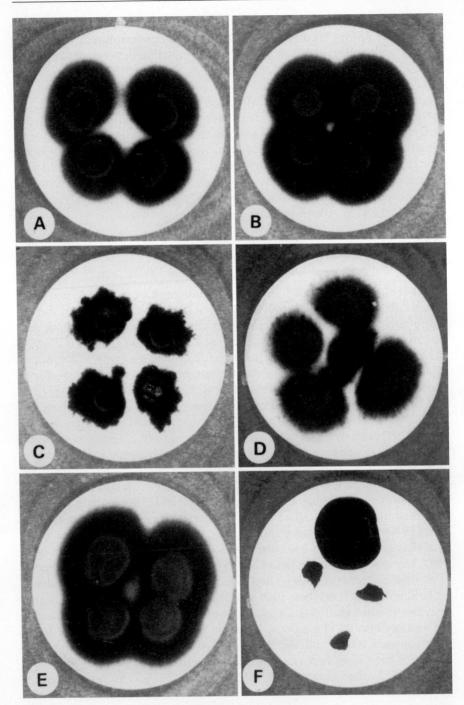

Fig. 12.3. Variation in colony morphology of 10-week-old cultures of *Cenococcum geophilum* on 2% Malt agar. **A** CGTAR; **B** 188; **C** R-26; **D** A181; **E** A145; **F** A166. (See LoBuglio et al. (1991) for origin of isolates)

Table 12.1. Variation in colony morphology among 10-week-old cultures of *Cenococcum geophilum* Fr. on 2% malt agar at 24°C

Group	Colony mycelium[a]						Plug mycelium[b]	
	Aerial mycelium	Texture	Color	Sclerotia	Margin	Reverse	Texture	Color
I	Sparse	Thin	Black	Scattered	Radiating, submerged	Black	Velvety to woolly	Brownish-black
II	Short	Thin	Black	–	Radiating, submerged	Black	Velvety	Black–dark brown
III	Short	Thin	Black–dark brown	–	Feather-like, submerged	Black–brown	Sparse, glaborous	Black–dark brown
IV	Dense	Woolly	Black–brown	–	Radiating, submerged	Black–brown	Woolly	Black–brown
V	Dense	Woolly	Black with grayish tint	–	Radiating submerged	Black	Woolly with colorless exudates	Black
VI	Dense	Velvety, elevated	Jet black	–	Even with compact black mycelium, submerged	Jet black	Velvety	Jet black

[a] Fungal mycelium growing beyond, and distinct from, the inoculum point (hyphal plug) and forming the fungal colony.
[b] Fungal mycelium growing directly from the surface of the inoculum plug (obtained from the margin of an estabished *Cenococcum* colony).

Table 12.2. Comparison of 17 variable sites found among 555 nucleotide base pairs of the ITS rDNA region for 5 isolates of *Cenococcum geophilum* Fr. Dots indicate identical nucleotides as compared to ITS sequence of isolate 52038. (K. F. LoBuglio, unpub. data)

ATCC No.[a]	Host	Origin																	
52038	Dwarf willow	Montana	T	T	C	C	C	T	A	C	T	C	G	A	C	C	A	C	C
58256	Slash pine	Georgia	C	G	T	.	.	A	C	A	T	.	.	G	T	T	G	.	.
	Kobresia	Colorado	.	.	.	T	.	.	.	.	.	.	A	G	.	T	G	.	.
46426	Douglas-fir	Oregon	.	.	.	.	.	.	.	.	.	.	A	G	.	T	G	.	.
38052	Loblolly pine	SE USA	.	.	.	.	.	.	G	.	C	T	A	G	.	T	G	T	T

[a] American type culture collection.

Phylogenetic analysis of ITS sequence data (excluding the intron region) indicated that over 96% sequence similarity exists among the *Cg* isolates chosen. As shown in Table 12.2, ITS sequence similarity is high among *Cg* isolates irrespective of host origin. The 555 nucleotide base pair ITS sequence from willow, pine, Douglas-fir, and sedge isolates had differences at only 17 nucleotide sites. Furthermore, the ITS rDNA sequence of an isolate originating from the sedge *Kobresia* (isolated by R. Mullen) was identical to that of Douglas-fir (K. F. LoBuglio, unpubl. data). The high sequence similarity found in the ITS (Shinohara 1994; K. F. LoBuglio, unpubl. data; Table 12.2) and in the nuclear small subunit rRNA gene (99.8–100%; LoBuglio et al. 1996) indicates that *Cg* is a monophyletic taxon and that rDNA RFLPs among different isolates represent variation in a single clade or lineage. The ability to isolate *Cg* DNA directly from ECM (LoBuglio et al. 1991; Gardes and Bruns 1993; Timonen et al. 1997) and the eventual development of molecular markers should contribute to our further understanding of the ECM diversity exhibited by *Cg*.

12.2.3
Phylogenetic Relationships with Other Ascomycetes

Since the mid-1800s, the genus *Elaphomyces* [a hypogeous ascomycete most recently classified in the Elaphomycetales by Trappe (1979)] has been considered the sexual state (teleomorph) of *Cg*, based on morphological and ecological traits [early work summarized by Ferdinandsen and Winge (1925)]. *Elaphomyces* and *Cg* are both ECM (Linhell 1942; Trappe 1971; Miller and Miller 1984), and *Cenococcum* ECM are often found embedded in peridia of *Elaphomyces* ascomata. In addition, the surface hyphae of the ECM mantle and sclerotia of *Cg* and the dark (dematiaceous) hyphae on the outer peridial layer of *Elaphomyces* species with blackish peridia both appear in tangential section as a "... mosaic of stellate hyphal clusters ..." (Trappe 1971) or cephalathecoid pattern (Hawksworth 1986). A definitive anamorph–teleomorph connection between these two morphs, however, is precluded by the absence of a sexual state in cultures of *Cg*, and the possibility of morphological homoplasy. For example, *Elaphomyces* spp. are not the only ascomycetes with "peridium walls composed of plates of radiating cells originating from a number of meristematic regions," known as cephalothecoid peridia (Hawksworth and Booth 1974); other genera have cephalothecoid peridia and could be closely related to *Cg* (LoBuglio et al. 1996).

Phylogenetic analysis of nucleotide sequence data has been useful in the integration of asexual (mitotic) fungal species with sexual (meiotic) species of fungi (Guadet et al. 1989; Berbee and Taylor 1992; LoBuglio and Taylor 1993; Rehner and Samuels 1994; Berbee 1996; Geiser et al. 1996). A recent study by LoBuglio et al. (1996) examined the phylogenetic relationship between *Elaphomyces* and *Cg*. Nucleotide sequence data were obtained from the 18S rRNA genic region of five *Cg* isolates (representing the range of RFLP variation observed in the entire rDNA repeat of 70 *Cg* isolates; LoBuglio et al. 1991) and three *Elaphomyces* species (with cephalothecoid peridia) and analyzed with sequences from 44 genera of ascomycetes in the classes Plectomycetes, Pyrenomycetes, Discomycetes, and Loculoascomycetes. Representatives of several Loculoascomycetes, including *Zopfia rhizophila* Rabenh., *Neotestudina rosatii* Segretain & Destombes, and *Phaeotrichum* sp., all of which have cephalothecoid peridia (Hawksworth 1986), were also included in this study. Molecular phylogenetic analysis (parsimony and distance methods) of the 18S rRNA gene phylogeny conflicts with the hypothesis that the genus *Elaphomyces* is the sexual state (teleomorph) of *Cg* or a close relative, and indicates two distinct evolutionary origins for these taxa; *Elaphomyces* is allied with the Plectomycetes while *Cg* is most closely associated with the Loculoascomycetes. *Cg* is positioned as a basal or intermediate lineage in between the Pleosporales and Dothidiales, with no close sexual relative among the taxa examined. If *Cg* is a Loculoascomycete, it is the first ECM Loculoascomycete.

Phialophora finlandia (Wang & Wilcox) and *Phialocephala fortinii* (Wang & Wilcox), also included in this study, were first isolated by H. E. Wilcox as sterile dark mycelia from *Pinus sylvestris* roots in Finland in 1975, and later described by Wang and Wilcox (1985) upon induction of conidiation. Since then these fungi have frequently been encountered in different environmental habitats (Wang and Wilcox 1985; Stoyke and Currah 1990; Stoyke et al. 1992; O'Dell et al. 1993; Harney et al. 1997). *P. finlandia* and *P. fortinii* are similar to *Cg* in that they are dematiaceous, asexual fungi and are commonly associated with plant roots, the former being ecto- or ectendomycorrhizal (Wilcox and Wang 1987a) and the latter reported as both mycorrhizal and pathogenic depending on the environmental conditions and host plant association (Wilcox and Wang 1987b; Stoyke et al. 1992). The phylogenetic position of these two taxa was not resolved, but it is clear that they are not closely related to *Cg* (LoBuglio et al. 1996). Parsimony and distance analysis positioned *P. fortinii* near the inoperculate discomycetes, the Helotiales, while *P.*

finlandia was positioned with the Helotiales by parsimony analysis, but distance analysis suggested a closer alliance with the operculate discomycetes, the Pezizales.

The absence of a close sexual relative to *Cg* could very well be a taxon sampling problem. Alternatively, *Cg* could represent an old and successful asexual lineage dating back to more than 150 million years ago when the Loculoascomycete lineage diverged from among the filamentous ascomycetes (Berbee and Taylor 1992). Other studies examining the evolution of asexual species have found them to be short, recently derived lineages from sexual species (LoBuglio et al. 1993; Geiser et al. 1996). K. F. LoBuglio and J. W. Taylor (unpubl. data) have investigated the hypothesis that *Cg* is truly an asexual fungus by analyzing DNA variation within two *Cg* populations (one in the New York Adirondack Mountains and a second in the Canadian Rocky Mountains) and conducting population genetic analyses testing for a clonal or recombining population biology structure. Results indicate the presence of recombination in both *Cg* populations. Whether this is due to a parasexual cycle in this fungus's life cycle or the presence of a cryptic sexual state has not been determined.

12.3
Ecology

12.3.1
Habitat Diversity

Cg is a cosmopolitan ECM fungus well known for its extremely wide habitat range. Intensive field sampling by Trappe (1964) identified the wide habitat range of *Cg*, and for the first time reported its occurrence in the southern hemisphere (Stellenbosch, South Africa). ECM of this species have been observed above the Arctic Circle in Alaska and Canadian High Arctic (75° 33'N, 84° 40'W) and as an important symbiont of trees at timber line in the Washington and Oregon Cascade mountain range (Haselwandter and Read 1982; Molina and Trappe 1984; Trappe 1988; Bledsoe et al. 1989). This contrasts with its occurrence in subtropical Florida and tropical Puerto Rico habiatats (Trappe 1962a, 1964). *Cg* is particularly noted for its drought resistance (Worley and Hacskaylo 1959; Mexal and Reid 1972; Pigott 1982; Coleman et al. 1989) and is often the predominant ECM type in soils of low moisture content (Trappe 1962a). However, this fungus is not limited to dry sites as it has been observed (albeit at a lower density) in poorly drained, wet soils (Trappe 1962a). Examination of the ECM status of spruce in urban settings indicated an infrequent occurrence of *Cg* ECM (Danielson and Pruden 1989). The drought resistance of *Cg* was demonstrated by Mexal and Reid (1972) using the osmoticum polyethylene glycol (PEG). It was determined that *Cg* was tolerant of low water potential, exhibiting maximum growth at 1.5 MPa. The extent of variation in water stress tolerance among *Cg* isolates was, however, not determined as only one isolate of *Cg* was tested in this study.

More recently, Coleman et al. (1989) examined the water-stress tolerance in pure culture of 55 isolates of 18 ECM fungal species, including nine isolates of *Cg*. Increasing water stress (-0.2 to -3.0 MPa) was accomplished by increasing solution PEG concentrations logarithmically. Three different growth responses to water stress were defined among the isolates tested: (1) type I response was typified by growth only occurring in the control treatment level (-0.2 MPa, no PEG); (2) the type II pattern was characterized by decreasing growth rates under increasing stress with the maximum growth rate always occurring in the control; and (3) type III patterns had maximum growth rates occurring at a greater stress level than the control. The *Cg* isolates were the only species group where all three growth patterns were observed. Based on the significant variation in growth rates of *Cg* isolates at different temperatures (Cline et al. 1987) and varying water potentials (Coleman et al. 1989), *Cg* seems to be a physiologically variable species.

12.3.2
Ecosystem Functioning

In addition to its broad habitat range, *Cg* has a pioneering capability and can be the dominant ECM fungus in forests of arctic, temperate and subtropical environments (Trappe 1962a,b, 1964, 1988; Read and Haselwandter 1981; Vogt et al. 1981a,b; Molina and Trappe 1982, 1984). Trappe (1962a) notes that *Cg* is among the first to invade newly formed soils on which ECM hosts become established, such as sand dunes, glacial moraines, volcanic ash, cinders, pumice, and clear-cut areas. For example, nearly all of the ECM of *Pinus contorta*, a pioneering species of sand flats and dunes along the Oregon coast (which are characteristically low in fertility and water-holding capacity) are that of *Cg* (Trappe 1962a). Similarly, Trappe (1988) observed that pioneering seedlings of *Pinus albicaulis* and *Larix lyalli* on recently exposed moraines in the Oregon Cascade mountains formed ECM predominately with *Cg*. The pioneering ability of *Cg* has been reported by many other researchers (Frydman 1957; Wright and Tarrant 1958; Dominik 1961); other fungi often invade in such situations and become more successful than *Cg* as environmental conditions gradually improve.

A glasshouse study by Borchers and Perry (1989) examined the hypothesis that hardwood species established in clear-cut forest environments benefit the future establishment of conifer seedlings. They found that Douglas-fir seedlings grown in soil beneath hardwood stands, under greenhouse conditions, had significantly greater height, weight, and ECM infection than seedlings grown in soil from open area soils. *Cg* and *Rhizopogon* dominated on seedlings grown in the hardwood soil, and an unidentified brown ECM dominated in the open area soils. These results suggest that *Cg* and *Rhizopogon* have an influence on the establishment of pioneering hardwoods in clear-cuts and on future conifer seedling establishment and growth. A study by Schoenberger and Perry (1982) of the effect of soil disturbance on growth and ECM of Douglas-fir and western hemlock seedlings suggested that Cg is especially sensitive to burning. They observed that the presence of Cg was significantly reduced in clear-cut sites recovering from burns 3–20 years prior as compared to unburned clear-cut sites.

In a recent study by Visser et al. (1995) changes of ECM and sporocarp production were examined in age sequence stands (6-, 41-, 65-, and 122-year-old stands) of Jack pine (*Pinus banksiana*) which had regenerated following wildfire. Cg was characterized as a "multi-stage fungus" as it was found in each age stand but was never dominant. Cg reached its greatest relative percent abundance in the 6-year-old stand when ECM diversity was the lowest, suggesting that Cg is more of a persistent ECM species rather than the most successful competitor.

Cg dominates in alpine ecosystems where environmental conditions can be extreme (Read and Haselwandter 1981; Vogt et al. 1981a,b; Trappe 1988). Trappe (1988) reported that in krummholz formed by coniferous species above 2700 m in the Cascades, over 90% of ECM were formed by Cg, while 100 m lower in a more sheltered area the proportion of *Cenococcum* ECM was less than half. Vogt et al. (1981a) examined the predominant ECM fungi in 23- and 180-year-old subalpine fir (*Abies amabilis*) stands which are characterized by a cool climate, reduced litter decomposition, and significant organic matter accumulation. The dominant ECM fungus infecting the root systems of these stands was Cg. The success of Cg in this habitat could be the result of the high levels of sclerotia biomass found in both stands, 3600–3700 kg ha^{-1}, and the autumnal burst of root growth coinciding with extensive germination of these sclerotia. In both *Abies* stands, the highest proportion of sclerotia was located in the A horizon. Vogt et al. (1981a) note that the competitive edge of Cg may also be due to the ability of this fungus to grow at low temperatures (1°C). The ECM biomass of Cg reached a maximum in autumn in the young stand, and in winter in the old stand. Over 80% of the ECM in both stands were found in the forest floor and A horizon throughout the study. However, there were seasonal fluctuations in the distribution of Cg ECM between the forest floor and A horizon throughout the year in the young stand while little fluctuation in ECM distribution was observed between the two soil horizons in the old stand.

The persistence of Cg sclerotia in soil for several years undoubtedly means there is a reservoir of fungal inoculum and can contribute to the success of seedling reestablishment. Massicotte et al. (1992) described the sclerotia as possessing a melanized rind of pseudoparenchymatous tissue surrounding a similar non-melanized internal tissue containing deposits of various reserve substances. This anatomy most likely contributes to the persistence of Cg sclerotia in the soil environment. The ability of sclerotia to survive for several years (even in low numbers) can provide sufficient inoculum to effectively colonize host species (Shaw and Sidle 1982). Kropp et al. (1985) found that containerized non-mycorrhizal western hemlock seedlings planted on a 5-year-old clear-cut site in Western Oregon formed *Cg* ECM within 6 months.

12.4
ECM Formation

This fungus is perhaps the least specialized of ECM fungi in respect to host species, forming ECM with many gymnosperms and angiosperms. *Cg* is ECM with all known ectotrophic hosts within its range, and in fact has been reported to form ECM with hosts such as *Juniperous communis, Acer pseudoplatatnus* and a number of herbaceous species thought otherwise to be endomycorrhizal (Lihnell 1939; Trappe 1962a, 1964). ECM of *Cg* have been reported to occur on over 200 tree species from 40 different genera of both angiosperms and gymnosperms, and include the Pteridophyte *Dryopteris filixmas* (Trappe 1962a, 1964; Chilvers 1968; Molina and Trappe 1982). *Cg* has been reported to be endomycorrhizal with *Stellaria holostea* (Trappe 1962a, 1964).

Possible evidence of host specificity among *Cg* isolates was observed from resynthesis experiments between three isolates of *Cg* and *Salix rotundifoia* seedlings from Barrow, Alaska (Antibus et al. 1981). Two of the *Cg* isolates originated from *Salix rotundifola* mycorrhiza at Cape Simpson, Alaska, and one isolate originated from *Pinus virginiana* roots in Maryland. Antibus et al. (1981) found that of the three isolates, only the isolate from Maryland repeatedly failed to form ECM with the *Salix* seedlings. This isolate may have lost its ability to form ECM through years of subculturing (although in the resynthesis experiments, prolific mycelial growth was noted on the surface of the *Salix* roots), or it may be incapable of penetrating the roots of *Salix rotundifola* (Antibus et al. 1981).

12.5
Developmental Aspects of the Fungus–Host Interaction

In searching for unique factors in the fungal cell wall which might have a role in the formation of ECM, Mangin et al. (1985) analyzed the chemical composition of hyphal walls from mycelia of *Cg*. Results from this study found that the cell wall of *Cg* was comprised of higher proportions of mannose compared to other non-mycorrhizal fungi. They suggested that mannose and galactose may be associated as galactomannan in the periphery of *Cg* cell walls; however, the role of this polymer has yet to be determined. A histochemical study of *Cg* cell wall structure by Paris et al. (1993) suggests that a "simplification" of *Cg* hyphal cell walls occurs in hyphae comprising the Hartig net. They observed that after staining with the Gomori-Swift test (which stains for cystine-rich proteins), hyphae in the outer fungal mantle appeared two-layered with a highly reactive external layer covering a poorly reactive internal layer. Hyphae of the Hartig net had only one poorly reactive layer, suggesting that the structure of the hyphal wall had become simpler. No difference in staining for polysaccharides was observed between mantle and hyphae of the Hartig net. The authors suggest that the simplified hyphal wall structure and decreased melanin in the Hartig net may be responsible for enhanced nutrient exchange between symbionts.

Trappe (1962a) followed the course of *Cg* ECM infection from synthesis experiments with two host species, *Pinus radiata* and *Pseudotsuga menziesii* var. *glauca*, and three different *Cg* isolates: he observed that *Cg* infected non-cutinized portions of both long and short roots independent of the presence of root hairs. Hyphal contact with host root systems led to a change in hyphal appearance, from straight, black hyphae to nearly colorless and ramifying, prior to penetration of the host epidermis. Hartig net development occurred in a radial pattern through to the endodermis. The hyphae comprising the Hartig net were interwoven, highly branched, usually hyaline, thin walled, and about 2 μm in diameter, and reaching up to 10 μm in length. Hartig nets occasionally developed hyphae as wide as 5 μm with thick dark walls which filled non-functioning cortical cells as large pseudoparenchymatic cells. Hartig nets were observed to grow >1 mm radially below the proximal end of the fungus mantle, and were limited to the innermost two layers of cortical cells. Intracelluar infection of cortical cells was observed, indicating that *Cg* forms ectendomycorrhizas as well as ECM. The Hartig net and intracellular infections were well developed prior to completion of mantles. The ramified growth of the first light-colored hyphae to penetrate the epidermis initiated mantle development. Dark, thick-walled hyphae overlaid this initial layer resulting in an unorganized mantle two or three hyphae thick. "Growth centers" were randomly initiated, which resulted in hyphae radiating out over the mantle surface to merge tip-to-tip with hyphae from another growth center. The development of these growth centers led to the characteristic "mosaic of radiate growths" (Trappe 1962a, 1971) or cephalothecoid (Hawksworth 1986) hyphal pattern observed in tangential and transverse section (Fig. 12.2).

12.6
Host Plant Growth Responses and Fungus-Derived Benefits

12.6.1
Growth Responses

A major benefit that ECM fungi confer to their host plants is increased access to nutrients accomplished through the development of the ECM mantle and external mycelium. The external hyphae emanating from the mantle surface of ECM roots form pathways for nutrient exchange between the soil and ECM roots, increasing the nutrient-absorbing surface for the host (Smith and Read

1997). A single hypha from a *Cenococcum–Pseudotsuga menziesii* mycorrhiza has been reported to extend 2 m into the soil and have at least 43 branch connections with other ECM, including those of *Tsuga heterophylla* (Trappe and Fogel 1977). Trappe (1962a) notes that from 200 to over 2000 hyphae emerge from the surface of most *Cg* ECM. The fungal mantle can increase the diameter of fine roots typically by 40–80 μm (Harley and Smith 1983) but is primarily thought of as a compartment for nutrient storage and exchange between the symbionts (Smith and Read 1997).

Rousseau et al. (1994) examined the contribution of *Pisolithus tinctorius* Pers. (Coker and Couch) and *Cg* ECM to the potential nutrient-absorbing surface of loblolly pine (*Pinus taeda*) as well as the effect on P uptake. They found that *P. tinctorius* stimulated some increase in root tip branching and surface area of fine roots, whereas *Cg* did not. It was suggested that the significant increase in nutrient-absorbing surface was provided by the external hyphae. Both *P. tinctorius* and *Cg* ECM led to increased P absorption compared to non-mycorrhizal roots (however, *P. tinctorius*-infected seedlings had significantly greater shoot dry weight, shoot P content and foliar P concentration than seedlings inoculated with *Cg*).

The fact that *Cg* ECM predominate in extreme environments such as sand dunes and alpine ecosystems is testimony to the beneficial nature of *Cg* as an ECM symbiont. However, conflicting results from inoculation studies have been reported on the effect of *Cg* on host growth. A comparison of ECM formation on slash pine by *Cg*, *Laccaria laccata* (Scop., Fr.) Berk. & Br. and *Suillus luteus* (L., Fr.) S. F. Gray in aseptic culture showed that *Cg* produced the least number of ECM, yet contributed as much to seedling growth as *L. laccata* which formed the most abundant ECM (Marx and Zak 1965). Kropp et al. (1985) reported that *Cg* significantly stimulated seedling top growth among outplanted *Tsuga heterophylla* seedlings. Yet Marx et al. (1978) reported the effect of *Cg* on host growth to be less effective when compared to the effect of other ECM species. As Graham and Linderman (1981) point out, in vitro isolate growth rate, which varies significantly among *Cg* isolates, may correlate with inoculation success of *Cg*.

12.6.2
Nutritional Benefits to the Host

Mejstrik and Krause (1973) investigated the efficiency of P uptake, from organic (P_o) and inorganic (P_i) sources, by *Pinus radiata* seedlings infected with *Cg* using ^{32}P. *Cg* also exhibited a stimulatory effect on incorporation of ^{32}P in organic form but exhibited reduced absorption of P from inorganic forms (H_3PO_4 in nutrient solution) as compared to non-mycorrhizal plants. Antibus et al. (1992) observed that *Cg* responded to growth on P_o with a much larger increase in uptake than the basidiomycetes examined. Growth on P_o, relative to P_i, resulted in a significant increase in mycelial phosphatase activity and ^{32}P uptake rates for *Cg*. An early study by Bowen and Theodorou (1967) demonstrated that *Cg* could solubilize P from rock phosphate. However, the contribution of P uptake from contaminating microorganisms on ECM surfaces was not ruled out.

As discussed by Smith and Read (1997), P uptake is most accurately measured in terms of inflow (uptake per unit length of root per unit time) or specific uptake rate (uptake per unit weight per unit time). A study by Heinrich and Patrick (1986) showed a definitive causal relationship between an increase of P inflow (from organic matter) in *Eucalyptus pilularis* when colonized by *Cg*. A significant correlation was observed between number of *Cg* ECM and both seedling dry weight and total seedling P content. In fact, the inflow of P for colonized roots was increased by 7.7×10^{-12} mol $P \, m^{-1} s^{-1}$ compared to non-mycorrhizal roots. A two-fold increase in plant growth and an increase in P inflow (3.2×10^{-12} mol $P \, m^{-1} s^{-1}$) was also seen by *Salix* seedlings colonized by *Cg* as compared to non-mycorrhizal seedlings.

The ability of *Cg* to utilize P from phosphomonoesters was demonstrated by Theodorou (1968). *Cg* was shown to use Ca and P phytates as sole sources of P. However, iron phytate (which is likely to be present in greater quantities in acid organic soil) was poorly utilized. Acid phosphatase (phosphomonoesterase) activity has been observed at the surface of *Cg* colonized roots (Antibus et al. 1981, 1986), and alkaline phosphatase has been isolated from cultures of *Cg* (Bae and Barton 1989). Production of acid phosphatase by *Salix* seedlings infected with *Entoloma sericeum* (Bull. ex Fr.) Kickx, *Hebeloma pusillum* J. Lange and *Cg* at a pH of 4.7 was compared by Antibus et al. (1981). In this study, *Cg* ECM of *Salix* were second to *Entoloma* ECM in terms of acid phosphatase activity. They noted that previous histochemical studies by Woolhouse (1969), Bartlett and Lewis (1973) and Williamson and Alexander (1975) found acid phosphatase activity of beech restricted to the mantle. Based on the findings of their study, Antibus et al. (1981) suggested that mantle thickness may influence acid phosphatase activity. *Entoloma*, which had the thickest mantle and greatest amount of attached mycelium, had the highest acid phosphatase activity. *Cg*, which had more of an intermediate mantle thickness, expressed less acid phosphatase activity than *Entoloma* but greater than *Hebeloma*, which had a relatively thin mantle. In 1986, Antibus et al. examined the effect of pH on acid phosphatase activity in *Cg*. They found that all but one of the *Cg* isolates demonstrated a distinct pH optimum at pH 4.5 or 5.0. A sharp decline above pH 5.0 was observed.

It has been suggested (but not confirmed) that uptake of P_i is followed by synthesis of a large amount of polyphosphate stored in the fungal vacuoles, which may be subsequently stored in the ECM mantle and ultimately translocated to the host (Smith and Read 1997). Martin et al. (1985), using ^{31}P NMR, observed the presence of relatively short-chain, soluble polyphosphate in the mycelium of *Cg*. Polyphosphate accumulation was low when growth was rapid in young mycelia and then linearly accumulated during the early and late stages of the stationary phase of growth, when P in the medium was relatively abundant compared with N. The mobilization of polyphosphate was observed when mycelia were transferred to low-P medium.

NH_4^+ is a well-utilized N source by *Cg* (Mikola 1948). Littke et al. (1984) examined biomass, hyphal density, and radial growth of five ECM fungi, including *Cg*, grown on solid high-N or low-N medium (where NH_4^+ was the N source). *Cg* was observed to grow more rapidly at low N concentrations, but at the expense of biomass production (Littke et al. 1984). Mikola (1948) further demonstrated the ability of *Cg* to utilize organic N. Casein hydrolysate and legumin were the best N sources (as measured by milligrams of mycelial dry weight), suggesting proteolytic ability by *Cg*. *Cg* was also shown to utilize nucleic acid, while urea, peptone, and aspartic acid proved poorer N sources (Mikola 1948). More recently, Abuzinadah and Read (1986) found that dry weight yield of *Cg* growing on bovine serum albumin (BSA) or gliadin as the sole N source was greater (between pH 3 and 6) than that obtained with NH_4^+. These authors note that the low pH optimum observed for proteolytic activity in their study suggests that an acid protease of broad spectrum is involved in the catalytic process, and that the N released is likely to be absorbed as amino-acid or small peptide units rather than NH_4^+. El-Badaoui and Botton (1989) isolated a protein-rich fraction from forest litter and observed that it induced greater proteolytic activity in *Cg*, *Amanita rubescens* Pers.: Fr. and *Hebeloma crustuliniforme* (Bull., St. Amans) Quel. than did BSA or gelatin. Mikola (1948) has demonstrated the ability of *Cg* to utilize most of the amino acids, including glutamic acid, as a sole N source.

It has been demonstrated that the pathway of primary NH_4^+ assimilation in rapidly growing cultures of *Cg* is through the concurrent activity of the NADP-glutamate dehydrogenase-glutamine synthase (GDH) and glutamine synthetase-glutamate synthase (GS) pathways, with a greater contribution from the GS pathway (Martin 1985; Martin et al. 1988). However, the assimilation of NH_4^+ by stationary phase (N-starved) *Cg* is primarily through the GDH pathway (Genetet et al. 1984). Arginine is a major free amino acid in *Cg* during its entire growth period and represents an important sink of assimilated nitrogen (Martin 1985; Martin et al. 1988). Martin et al. (1988) identified four pathways of N metabolism in rapidly growing cultures of *Cg* as follows: (1) glutamine synthesis, involving transfer of N to both amino and amido moieties; (2) glutamate formation; (3) transamination with pyruvate to yield alanine; and (4) transamination with oxaloacetate to yield aspartate.

Rodriguez et al. (1984) examined the mechanism of transmembrane Fe transport in *Cg* in an Fe-rich environment where excess metal chelate may occur. They observed (in liquid cultures of *Cg*) the release of reductants which reduced Fe(III). Furthermore, they observed that purified cell walls of *Cg* will bind reduced Fe(II), which may serve as a potential source of iron to adjacent cells. Hydroxamate siderophore iron chelators are known to be produced by *Cg* in low iron culture media (Szaniszlo et al. 1981).

12.6.3
Carbon Utilization and Flow

Mikola (1948) found that *Cg* utilized sucrose and maltose equally well as glucose, and that mycelial growth was promoted by dextrin, starch, mannitol, and lactose to a limited degree. Keller (1952) found that glucose, mannose, trehalose, cellobiose, and alpha-dextrin were equally effective C sources, while sorbitol, galactose, beta-dextrin, starch, insulin, and cellulose were not effective at all. In a study carried out by Palmer and Hacskaylo (1970), growth of six ECM species was examined on 39 soluble and 13 insoluble C sources. *Cg* was the only isolate to grow (albeit limited) on heterogeneous di- or trisaccharides, sucrose, melezitose, and raffinose. Giltrap and Lewis (1981) found that *Cg* utilized sucrose when grown in MES buffer, but not when grown in phosphate buffer. In their study, the sucrosyl-galactoside raffinose was poorly utilized, especially on phosphate buffer. As Lewis (1985) points out, however, the significance of in vitro carbohydrate utilization studies to the natural ECM condition is questionable.

Cg sclerotia can significantly contribute to the fungal biomass of forest situations, and thus represent an important source of assimilated C from host species. Vogt et al. (1982) estimated the biomass of ECM and decomposer fungi in a 180-year-old *Abies amabilis* stand to be 380 kg ha^{-1} year^{-1} of hypogeous sporocarps, 30 kg ha^{-1} year^{-1} of epigeous sporocarps, and 2700 kg ha^{-1} year^{-1} of *Cg* sclerotia. Similarly, significant *Cg* sclerotial biomass was reported by Fogel and Hunt (1979) (2785 kg ha^{-1} year^{-1}) in a 35- to 50-year-old Douglas-fir stand in western Oregon.

12.6.4
Production of Antibiotics

Krywolap and Casida (1964) identified an antibiotic produced by the mycelium of several *Cg* isolates. This antibiotic showed strong inhibition of Gram-positive bacteria and *Rhizobium*. Higher concentrations of the antibiotic inhibited Gram-negative bacteria as well as yeast and actinomycete species (Krywolap et al. 1964). An antibiotic resembling that which occurs in the mycelium of *Cg* in axenic culture was present in nature in the ECM, roots, and needles of white pine, red pine, and Norway spruce (Krywolap et al. 1964). This antibiotic was also detected in nursery seedlings of white and red pine lacking *Cg* ECM. However, it was noted that the soil in which these seedlings were growing contained large numbers of *Cg* sclerotia, which might be responsible for the production and absorption of this antibiotic by the seedlings. Although not conclusive, the production of antibiotics by *Cg*, and subsequent transport to the host, may offer some degree of protection to the host against bacterial invaders.

Acknowledgment. Sincere thanks to Hugh Wilcox for his helpful suggestions to this manuscript.

References

Abuzinadah RA, Read DJ (1986) The role of proteins in the nitrogen nutrition of ectomycorrhizal plants. I. Utilization of peptides and proteins by ectomycorrhizal fungi. New Phytol 103:481–493

Antibus RK, Croxdale JG, Miller OK, Linkins AE (1981) Ectomycorrhizal fungi of *Salix rotundifolia*. III. Resynthesized mycorrhizal complexes and their surface phosphatase activities. Can J Bot 59:2458–2465

Antibus RK, Kroehler CJ, Linkins AE (1986) The effects of external pH, temperature, and substrate concentration on acid phosphatase activity of ectomycorrhizal fungi. Can J Bot 64:2383–2387

Antibus RK, Sinsabaugh RL, Linkins AE (1992) Phosphatase activities and phosphorus uptake from inositol phosphate by ectomycorrhizal fungi. Can J Bot 70:794–801

Bae KS, Barton L (1989) Alkaline phosphatase and other hydrolases produced by *Cenococcum graniforme*, an ectomycorrhizal fungus. Appl Environ Microbiol 55:2511–2516

Bartlett EM, Lewis DH (1973) Surface phosphatase activity of mycorrhizal roots of beech. Soil Biol Biochem 5:249–257

Baura G, Szaro TM, Bruns TD (1992) *Gastrosuillus larcinus* is a recent derivative of *Suillus grevillei*: molecular evidence. Mycologia 84:592–597

Berbee ML (1996) Loculoascomycete origins and evolution of filamentous ascomycete morphology from 18s rRNA gene sequence data. Mol Biol Evol 13:462–470

Berbee ML, Taylor JW (1992) 18s ribosomal RNA gene sequence characters place the human pathogen *Sporothrix schenckii* in the genus *Ophiostoma*. Exp Mycol 16:87–91

Bledsoe C, Klein P, Bliss LC (1989) A survey of mycorrhizal plants on Truelove Lowland, Devon Island, N.W.T. Can J Bot 68:1848–1856

Borchers SL, Perry DA (1989) Growth and ectomycorrhiza formation of Douglas-fir seedlings grown in soils collected at different distances from pioneering hardwoods in southwest Oregon clear-cuts. Can J For Res 20:712–721

Bowen GD, Theodorou C (1967) Studies on phosphate uptake by mycorrhizas. Proc Int Union For Res Organ 5:116

Bruns TD, White TJ, Taylor JW (1991) Fungal molecular systematics. Annu Rev Ecol Syst 22:525–564

Chen W, Hoy JW, Schneider RW (1992) Species-specific polymorphisms in transcribed ribosomal DNA of five *Pythium* species. Exp Mycol 16:22–34

Chilvers GA (1968) Some distinctive types of eucalypt mycorrhiza. Aust J Bot 16:49–70

Cline ML, France RC, Reid CPP (1987) Intraspecific and interspecific variation of ectomycorrhizal fungi at different temperatures. Can J Bot 65:869–875

Coleman MD, Bledsoe CS, Lopushinsky W (1989) Pure culture response of ectomycorrhizal fungi to imposed water stress. Can J Bot 67:29–39

Danielson RM, Pruden M (1989) The ectomycorrhizal status of urban spruce. Mycologia 81:335–341

Dominik T (1961) Badanie mikotrofizmu zespolow roslinnych w Parku Narodwym w Pieninach i na skalce nad Lysa Polona w Tatrach ze szczegolnym uwzglednieniem mikotrofizmu sosny reliktowej. Inst Badawczy Lessn Prace 208:31–58

El-Badaoui K, Botton B (1989) Production and characterization of exocellular proteases in ectomycorrhizal fungi. Ann Sci For 46:728–730

Ferdinandsen C, Winge O (1925) *Cenococcum* Fr. A monographic study. Kgl Vet Landbohojsk Aarsskr 1925:332–382

Fogel R, Hunt G (1979) Fungal and arboreal biomass in a western Oregon Douglas-fir ecosystem: distribution patterns and turnover. Can J For Res 9:245–256

Fries E (1825) Systema Orbis Vegetablis I. Lundae. Typographia academica

Frydman I (1957) Mykotrofizm roslinnosci pokrywajacej gruzy i ruiny domow Wroclawia. Acta Soc Bot Polon 26:45–60

Gardes M, Bruns TD (1993) ITS primers with enhanced specificity for basidiomycetes – application to the identification of mycorrhizae and rusts. Mol Ecol 2:113–118

Geiser DM, Timberlake WE, Arnold ML (1996) Loss of meiosis in *Aspergillus*. Mol Biol Evol 13:809–817

Genetet I, Martin F, Stewart GR (1984) Nitrogen assimilation in mycorrhizas. Ammonium assimilation in the N-starved ectomycorrhizal fungus *Cenococcum graniforme*. Plant Physiol 76:395–399

Giltrap NJ, Lewis DH (1981) Phosphate inhibition of mycorrhizal fungi. New Phytol 87:669–675

Graham JH, Linderman RG (1981) Inoculation of containerized Douglas-fir with the ectomycorrhizal fungus *Cenococcum geophilum*. For Sci 27:27–31

Guadet J, Julien J, LaFay JF, Brygoo Y (1989) Phylogeny of some *Fusarium* species, as determined by large-subunit rRNA sequence comparison. Mol Biol Evol 6:227–242

Harley JL, McCready CC (1950) Uptake of phosphate by excised mycorrhizal roots of beech. New Phytol 49:388

Harley JL, Smith SE (1983) Mycorrhizal symbiosis. Academic Press, London

Harney SK, Rogers SO, Wang CJK (1997) Molecular characterization of dematiaceous root endophytes. Mycol Res 101:1397–1404

Haselwandter K, Read DJ (1982) The significance of a root–fungus association in two *Carex* species of high alpine plant communities. Oecologia 53:352–354

Hatch AB (1934) A jet-black mycelium forming ectotrophic mycorrhizae. Sven Bot Tidskr 28:369–383

Hawksworth DL (1986) The evolution and adaptation of sexual reproductive structures in the Ascomycotina. In: Rayner ADM, Brasier CM, Moore D (eds) Symposium of the British Mycological Society. Cambridge University Press, Cambridge, pp 179–189

Hawksworth DL, Booth C (1974) A revision of the genus *Zopfia* Rabenh. Mycol Pap 135:1–38

Heinrich PA, Patrick JW (1986) Phosphorus acquisition in the soil–root system of *Eucalyptus pilularis* Smith seedlings. II. The effect of ectomycorrhizas on seedling phosphorus and dry weight acquisition. Aust J Bot 34:445–54

Keller HG (1952) Untersuchungen über das Wachstum von *Cenococcum graniforme* (Sow.) Fer. et Winge auf verschiedenen Kohlenstoffquellen. Eidg Tech Hochsch Zurich Promotionsarb 2036:1–123

Kropp BR, Castellano MA, Trappe JM (1985) Performance of outplanted western hemlock (*Tsuga heterophylla* (Raf.) Sarg.) seedlings inoculated with *Cenococcum geophilum*. Tree Plant Not 36:13–16

Krywolap GN, Casida LE (1964) An antibiotic produced by the mycorrhizal fungus *Cenococcum graniforme*. Can J Microbiol 10:365–370

Krywolap GN, Grand LF, Casida LE (1964) The natural occurrence of an antibiotic in the mycorrhizal fungus *Cenococcum graniforme*. Can J Microbiol 10:323–328

Lee SB, Taylor JW (1992) Phylogeny of five fungus-like protoctistan *Phytophthora* species, inferred from the internal transcribed spacers of ribosomal DNA. Mol Biol Evol 9:636–653

Lewis DH (1985) Inter-relationships between carbon nutrition and morphogenesis in mycorrhizas. In: Gianinazzi-Pearson V, Gianinazzi S (eds) Physiological and genetical aspects of mycorrhizae. INRA, Paris, pp 85–100

Lihnell D (1939) Untersuchungen über die Mykorrhizen und Wurzelpilze von *Juniperus communis*. Symb Bot Ups 3:3

Linhell D (1942) *Cenococcum graniforme* als Mykorrizabildner von Waldbaumen. Symb Bot Ups 5:1–18

Littke WR, Bledsoe CS, Edmonds RL (1984) Nitrogen uptake and growth in vitro by *Hebeloma crustuliniforme* and other Pacific Northwest mycorrhizal fungi. Can J Bot 62:647–652

LoBuglio KF, Rogers SO, Wang CJK (1991) Variation in ribosomal DNA among isolates of the mycorrhizal fungus *Cenococcum geophilum* Fr. Can J Bot 69:2331–2343

LoBuglio KF, Pitt JI, Taylor JW (1993) Phylogenetic analysis of two ribosomal DNA regions indicates multiple independent losses of asexual *Talaromyces* state among asexual *Penicillium* species in subgenus *Biverticillium*. Mycologia 85:592–604

LoBuglio KF, Berbee ML, Taylor JW (1996) Phylogenetic origins of the asexual mycorrhizal symbiont *Cenococcum geophilum* Fr. and other mycorrhizal fungi among the ascomycetes. Mol Phylog Evol 6:287–294

Mangin F, Bonaly R, Botton, Martin F (1985) Chemical composition of hyphal walls of the ectomycorrhizal fungus *Cenococcum geophilum* In: Gianinazzi-Pearson V, Gianinazzi S (eds) Physiological and genetical aspects of mycorrhizae. INRA, Paris, pp 451–456

Martin F (1985) ^{15}N-NMR studies of nitrogen assimilation and amino acid biosynthesis in the ectomycorrhizal fungus *Cenococcum graniforme*. FEBS Lett 182:350–354

Martin F, Marchal JP, Timinska A, Canet D (1985) The metabolism and physical state of polyphosphates in ectomycorrhizal fungi. A ^{31}P nuclear magnetic resonance study. New Phytol 101:275–290

Martin F, Stewart GR, Genetet I, Mourot B (1988) The involvement of glutamate dehydrogenase and glutamine synthetase in ammonia assimilation by the rapidly growing ectomycorrhizal ascomycete, *Cenococcum geophilum*. New Phytol 110:541–550

Marx DH, Zak B (1965) Effect of pH on mycorrhizal formation of slash pine in aseptic culture. For Sci 11:66–75

Marx DH, Morris WG, Mexal JG (1978) Growth and ectomycorrhizal development of loblolly pine seedlings in fumigated and nonfumigated nursery soil infested with different fungal symbionts. For Sci 24:193–203

Massicotte HB, Trappe JM, Peterson RL, Melville LH (1992) Studies on *Cenococcum geophilum*. II. Sclerotium morphology, germination, and formation in pure culture and growth pouches. Can J Bot 70:125–132

Mejstrik VK, Krause HH (1973) Uptake of ^{32}P by *Pinus radiata* roots inoculated with *Suillus luteus* and *Cenococcum graniforme* from different sources of available phosphate. New Phytol 72:137–140

Mexal J, Reid CPP (1972) The growth of selected mycorrhizal fungi in response to induced water stress. Can J Bot 1579–1588

Mikola P (1948) On the physiology and ecology of *Cenococcum graniforme* especially as a mycorrhizal fungus of birch. Inst For Fenn Commun 36:1–104

Miller OK, Miller SL (1983) Description and identification of selected mycorrhizal fungi in pure culture. Mycotaxon 18:457–481

Miller SL, Miller OK (1984) Synthesis of *Elaphomyces muricatus* + *Pinus sylvestris* ectomycorrhizae. Can J Bot 62:2363–2369

Molina R, Trappe JM (1982) Patterns of ectomycorrhizal host specificity and potential among Pacific Northwest conifers and fungi. For Sci 28:423–458

Molina R, Trappe JM (1984) Mycorrhiza management in nurseries. In: Duryea ML, Landis TD (eds) Forest nursery manual: production of bareroot seedlings. Martinus Nijhoff/Dr W Junk, The Hague, pp 211–223

O'Dell TE, Massicotte HB, Trappe JM (1993) Root colonization of *Lupinus latifolius* Agardh. and *Pinus contorta* Dougl. by *Phialocephala fortinii* Wang & Wilcox. New Phytol 124:93–100

Palmer JG, Hacskaylo E (1970) Ectomycorrhizal fungi in pure culture. I. Growth on single carbon sources. Physiol Plant 23:1187–1197

Paris F, Dexheimer J, Lapeyrie F (1993) Cytochemical evidence of a fungal cell wall alteration during infection of *Eucalyptus* roots by the ectomycorrhizal fungus *Cenococcum geophilum*. Arch Microbiol 159:526–529

Park JY (1970) A change in color of aging mycorrhizal roots of *Tilia americana* formed by *Cenococcum graniforme*. Can J Bot 48:1339–1341

Pigott CD (1982) Fine structure of mycorrhiza formed by *Cenococcum geophilum* Fr. on *Tilia cordata* Mill. New Phytol 92:501–512

Read DJ, Haselwandter K (1981) Observations on the mycorrhizal status of some alpine plant communities. New Phytol 88:341–352

Rehner SA, Samuels GJ (1994) Taxonomy and phylogeny of *Gliocladium* analysed from nuclear large subunit ribosomal DNA sequences. Mycol Res 98:625–634

Rodriguez RK, Klemm DJ, Barton LL (1984) Iron metabolism by an ectomycorrhizal fungus, *Cenococcum graniforme*. J Plant Nutr 7:459–468

Rogers SO, Yan ZH, Shinohara M, LoBuglio KF, Wang CJK (1993) Messenger RNA intron in the nuclear 18s ribosomal RNA gene of deuteromycetes. Curr Genet 23:338–342

Rousseau JVD, Sylvia DM, Fox AJ (1994) Contribution of ectomycorrhiza to the potential nutrient-absorbing surface of pine. New Phytol 328–644

Schoenberger MM, Perry DA (1982) The effect of soil disturbance on growth and ectomycorrhizae of Douglas-fir and western hemlock seedlings: a greenhouse bioassay. Can J For Res 12:343–353

Seifert KA, Wingfield BD, Wingfield MJ (1995). A critique of DNA sequence analysis in the taxonomy of filamentous Ascomycetes and ascomycetous anamorphs. Can J Bot 73:S760–S767

Shaw CG, Sidle RC (1982) Evaluation of planting sites common to a southeast Alaska clear-cut. II. Available inoculum of the ectomycorrhizal fungus *Cenococcum geophilum*. Can J For Res 13:9–11

Shinohara ML (1994) Molecular evolutionary study of *Cenococcum geophilum*. PhD Dissertation, State University of New York, College of Environmental Science and Forestry, Syracuse, New York

Shinohara ML, LoBuglio KF, Rogers SO (1996) Group-I intron family in the nuclear ribosomal RNA small subunit genes of *Cenococcum geophilum* isolates. Curr Genet 29:377–387

Smith SE, Read DJ (1997) Mycorrhizal symbiosis. 2nd edn. Academic Press, New York

Stoyke G, Currah RS (1990) Endophytic fungi from the mycorrhizae of alpine ericoid plants. Can J Bot 69:347–352

Stoyke G, Egger KN, Currah RS (1992) Characterization of sterile endophytic fungi from the mycorrhizae of subalpine plants. Can J Bot 70:2009–2016

Szaniszlo PJ, Powell PE, Reid CPP, Cline GR (1981) Production of hydroxamate siderophore iron chelators by mycorrhizal fungi. Mycologia 73:1158–1174

Theodorou C (1968) Inositol phosphates in needles of *Pinus radiata* D. Don and the phytase activity of mycorrhizal fungi. Trans 9th Int Congr of Soil Science, Adelaide, vol III, pp 483–490

Timonen S, Tammi H, Sen R (1997) Characterization of the host genotype and fungal diversity in Scots pine ectomycorrhiza from natural humus microcosms using isozyme and PCR-RFLP analyses. New Phytol 135:313–323

Trappe JM (1962a) *Cenococcum graniforme* – its distribution, ecology, mycorrhiza and inherent variation. PhD Diss, University of Washington, Seattle, Washington

Trappe JM (1962b) Fungus associates of ectotrophic mycorrhizae. Bot Rev 28:538–606

Trappe JM (1964) Mycorrhizal host and distribution of *Cenococcum graniforme*. Lloydia 27:100–106

Trappe JM (1969) Studies on *Cenococcum graniforme*. I. An efficient method for isolation from sclerotia. Can J Bot 47:1389–1390

Trappe JM (1971) Mycorrhiza-forming Ascomycetes. In: Hacskaylo E (ed) Mycorrhizae: Proceedings of the First North American Conference on Mycorrhizae. USDA Forest Service, Misc Publ 1189

Trappe JM (1979) The orders, families, and genera of hypogeous ascomycotina (truffles and their relatives). Mycotaxon 9:297–340

Trappe JM (1988) Lessons from alpine fungi. Mycologia 80:1–10

Trappe JM, Fogel R (1977) Ecosystematic functions of mycorrhizae. Colo State Univ Range Sci Dept Sci Ser 26:205–214

Visser S (1995) Ectomycorrhizal fungal succession in jack pine stands following wildfire. New Phytol 129:389–401

Vogt KA, Edmonds RL, Grier CC (1981a) Dynamics of ectomycorrhizae in *Abies amabilis* stands: the role of *Cenococcum graniforme*. Holarct Ecol 4:167–173

Vogt KA, Edmonds RL, Grier CC (1981b) Biomass and nutrient concentrations of sporocarps produced by mycorrhizal and decomposer fungi in *Abies amabilis* stands. Oecologia 50:170–175

Vogt KA, Grier CC, Meier CE, Edmonds RL (1982) Mycorrhizal role in net primary production and nutrient cycling in *Abies amabilis* ecosystems in western Washington. Ecology 63:370–380

Wang CJK, Wilcox HE (1985) New species of ectendomycorrhizal and pseudomycorrhizal fungi: *Phialophora finlandia, Chloridium paucisporum,* and *Phialocephala fortinii.* Mycologia 77:951–958

White TJ, Bruns T, Lee S, Taylor J (1990) Amplification and direct sequencing of fungal ribosomal RNA genes for phylogenetics. In: Innis MA, Gelfand DH, Sninsky JJ, White TJ (eds) PCR protocols: a guide to methods and applications. Academic Press, New York, pp 315–322

Wilcox HE, Wang CJK (1987a) Ectomycorrhizal and ectendomycorrhizal associations of *Phialophora finlandia* with *Pinus resinosa, Picea rubens,* and *Betula alleghaniensis.* Can J For Res 17:976–990

Wilcox HE, Wang CJK (1987b) Mycorrhizal and pathological associations of dematiaceous fungi in roots of 7-month-old tree seedlings. Can J For Res 17:884–899

Williamson B, Alexander IJ (1975) Acid phosphatase localized in the sheath of beech mycorrhiza. Soil Biol Biochem 7:195–198

Woolhouse HW (1969) Differences in the properties of the acid phosphatases of plant roots and their significance in the evolution of edaphic ecotypes. In: Rorison IH (ed) Ecological aspects of the mineral nutrition of plants. Blackwell, Oxford, pp 357–380

Worley JF, Hacskaylo E (1959) The effect of available soil moisture on the mycorrhizal association of Virginia pine. For Sci 5:267–268

Wright E, Tarrant RF (1958) Occurrence of mycorrhizae after logging and slash burning in the Douglas-fir forest type. Pac Northwest For Range Exp Sta Res Note 160:1–7

Yan ZH, Rogers SO, Wang CJK (1994) Assessment of *Phialophora* species based on ribosomal DNA internal transcribed spacers and morphology. Mycologia 87:72–83

Chapter 13

Hysterangium

M. A. Castellano

13.1
Introduction

Hysterangium is a large genus among sequestrate genera and is commonly collected in large numbers in eastern and western North America, Australasia, and Europe. Several *Hysterangium* species are important members of ECM fungal communities, contributing significantly to sporocarp productivity and producing perennial mat-like structures in forest floor and mineral soil that occupy significant portions of the forested area. They occur in young and old forest stands alike across diverse habitats. This ecological importance and amplitude have only been recently recognized. Consequently, *Hysterangium* has received increased research attention with regard to its function in forested ecosystems. The difficulty of culturing from sporocarps and manipulation of pure cultures of *Hysterangium* has restricted physiological work.

13.2
Taxonomy and Phylogenetic Relationships

Hysterangium, a genus of sequestrate Basidiomycotina, is commonly placed in the Hysterangiaceae and, along with several lesser known genera (*Circulocolumella, Clathrogaster, Claustula, Clavarula, Gallacea, Gelopellis, Hoehneliogaster, Jaczewskia, Kobaysia, Maccagnia, Phallobata, Phallogaster, Phlebogaster, Protophallus, Protubera, Protuberella,* and *Rhopalogaster*), has been traditionally ascribed affinities to the Phallales. Because *Hysterangium* is commonly collected in eastern and western North America, Australasia, and Europe, it is often mistaken for other sequestrate Basidiomycota such as *Rhizopogon* or *Hymenogaster*, and occasionally for "eggs" of epigeous genera like *Phallus* or *Clathrus* species.

Carlo Vittadini (1831) established *Hysterangium* to accommodate three species; *H. clathroides, H. membranaceum,* and *H. fragile*. He described each as possessing smooth spores and resembling the egg stage of many

USDA Forest Service, Pacific Northwest Research Station, Forestry Sciences Laboratory, 3200 Jefferson Way, Corvallis, Oregon 97331, USA

Phallales. Unbeknown to Vittadini, and to many later workers, spores of all three species possess a fine ornamentation of minute to distinct verrucae (best seen under oil immersion) beneath the loosely to tightly adhering epispore (Fig. 13.1).

In 1843, the Tulasne brothers proposed an additional species (*H. stoloniferum*) which differed from their interpretation of Vittadini's previously published species. Unfortunately, the Tulasne brothers did not study Vittadini's material. *Hysterangium stoloniferum*, when compared with Vittadini's specimens, proves to be a synonym of *H. fragile* Vitt. (Castellano 1989). The Tulasne brothers (1851), again without studying material from Vittadini, proposed another new species, *H. pompholyx* and two new varieties of *H. clathroides*: *H. clathroides* var. *cistophilum* and *H. clathroides* var. *crassum*. This began a long history of misinterpreted species concepts that would plague future taxonomic study of this genus.

To add to the confusion, other European workers described new species and varieties of *Hysterangium* without studying material of previous workers (Berkeley 1844, 1848; Hesse 1884, 1891; Quélet 1886; Mattirolo 1900a,b; Bucholtz 1908; Patouillard 1914; Soehner 1949, 1952; Velenovsky 1939; Svrcek 1958). Still other researchers erroneously placed some new species into *Hysterangium* (Bresadola 1892; Malençon 1975), while others described supposed *Hysterangium* species that are not even sporocarpic in nature (De Toni 1888). Beginning in 1881, many new species of *Hysterangium* were described, especially from Australasia, North America, and South America. Unfortunately, often the material of previous authors again remained unstudied (Speggazini 1881; Massee 1898, 1899, 1901; Harkness 1899; Fischer 1908, 1938; Rodway 1918, 1920; Fitzpatrick 1913; Patouillard 1915; Lloyd 1921, 1922; Zeller and Dodge 1929; Cunningham 1934, 1938; Ito and Imai 1937; Zeller 1939, 1941; Corner and Hawker 1953; Cribb 1958; Horak 1964; Pacioni 1984; Beaton et al. 1985).

Zeller and Dodge (1929) were the only workers to study a majority of the available type specimens. Unfortunately, they had too little material of several important and widespread taxa (including *H. clathroides*, *H. coriaceum*, *H. setchellii*, and *H. crassum*) for accurate interpretation. This confusion caused many workers doing regional surveys in Europe to cite specimens of *H. coriaceum* as *H. clathroides* and workers in western North America to cite specimens of *H. coriaceum* as *H. separabile* and those of *H. setchellii* as *H. crassum*. Regional treatments of *Hysterangium* by Soehner (1952) and Gross et al. (1980) for Germany, by Svrcek (1958) for the former Czechoslovakia, and by Beaton et al. (1985) for Victoria, Australia, lack

---

Fig. 13.1. A Light micrograph of peridial hyphae of *Hysterangium setchellii* showing the oxalate crystals. **B** *Hysterangium coriaceum* sporocarps showing the distinct basal rhizomorph. **C** Light micrograph of basidiospores of *Hysterangium* sp. showing the appressed utricle. **D** Scanning electron micrograph of *Hysterangium separible* basidiospores showing the verrucose ornamentation present on some species

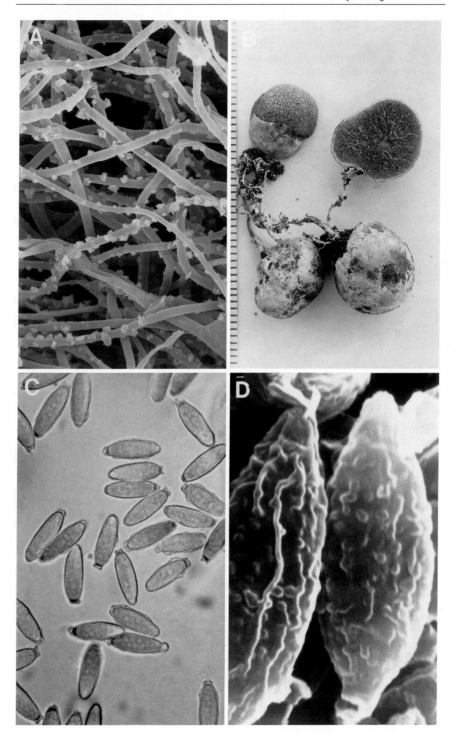

critical examination of authentic material and rely on the often misinterpreted species descriptions already present in the literature.

The world monograph of *Hysterangium* by Castellano (1989) marked the first time that all available type collections were studied and redescribed. The holistic approach of Castellano concluded that *Hysterangium sensu lato* was polyphyletic and in need of revision. He conserved 25 species names (out of 68 original names), raised two varieties to species rank, and proposed 11 new species. Thirteen species and seven varietal names were synonymized, 18 other species were transferred to nine other genera (Castellano and Beever 1994; Castellano and Muchovej 1995), and one new genus (*Trappea*) was erected to accommodate two species (Castellano 1990). Subsequent work has revealed many additional species and reflects the critical need to complete regional treatments of Mexico, the rest of North America, Europe, and Australasia.

Hysterangium has traditionally been considered as related to the phalloid fungi owing to its resemblance to the Phallales "egg" stage, possession of smooth spores, and green-tinted gleba (Rehsteiner 1892; Fischer 1900, 1933; Gäumann and Dodge 1928; Zeller 1939, 1948; Svrcek 1958; Dring 1980; Cunningham 1944; Hawker 1954; Gross et al. 1980; Beaton et al. 1985). This artificial grouping of characters, of which one character was commonly misinterpreted (smooth spores), led to a polyphyletic origin of the genus *Hysterangium sensu lato*. Morphologically, *Hysterangium sensu lato* resembles genera of Phallales in several ways. The white, separable peridium of many *Hysterangium* species resembles the volva in the Phallales. The cord-like basal rhizomorph of some *Hysterangium* spp. (Fig. 13.1) approaches in morphology that of some Phallales (Townsend 1954). A number of *Hysterangium* species develop the characteristic stinkhorn smell, and, in addition, the green-tinted gleba, which deliquesces at maturity in *Hysterangium*, is analogous to the slimy green-tinted spore mass of many Phallales.

The classification of *Hysterangium* and other sequestrate genera in the Phallales has been debated for some time. Rival hypotheses classify the Hysterangiaceae in the Boletales (Smith and Zeller 1966; Castellano 1989) or either closely related to (Dring 1980) or within the Phallales (Vittadini 1831; Miller 1983; Castellano 1989). The advent of DNA isolation and amplification techniques allowed testing of the phylogenies of all fungi. Recently, DNA techniques have confirmed the placement of *Hysterangium* within the Phallales (J. W. Spatafora, pers. comm.). Spatafora has also recognized the polyphylogeny of *Hysterangium sensu lato*. The inclusion of a presumably exclusively ECM genus within what has been considered a saprophytic family is intriguing. In fact, preliminary data from J. W. Spatafora (pers. comm.) suggest that *Hysterangium* may have given rise to some epigeous members of the Phallales. This work, however, needs critical follow-up to confirm these hypotheses.

13.3
Morphology and Anatomy of ECM

The ECM of four *Hysterangium* species have been described; all form bright white ECM, regardless of the host taxon. Molina and Trappe (1982a,b) described in some detail the ECM of *Hysterangium coriaceum* Hesse (as *H. separabile* Zeller) in pure culture synthesis with *Arbutus menziesii*, *Arctostaphylos uva-ursi*, *Pinus contorta*, *Pinus monticola*, *Larix occidentalis*, and *Tsuga heterophylla*. They described the ECM as white and densely pubescent with large wefts of emergent mycelia and stout, rope-like rhizomorphs. The adherence of abundant calcium oxalate crystals (Fig. 13.1) on extramatrical hyphae described subsequently by others was noted by neither Molina and Trappe (1982a,b) nor Acsai and Largent (1983).

Malajczuk et al. (1987) described the superficial ECM of *H. inflatum* with *Eucalyptus diversicolor* as having abundant calcium oxalate crystals on its hyphae although the fungal mantle was only one to five cells deep. These ECM root tips had the same thickness as uncolonized roots. They also noted that *H. inflatum* hyphae (mats) occupied up to 10% of the soil surface area in *E. diversicolor* forests in Western Australia. Griffiths et al. (1991a,b) presented a cursory description of *Hysterangium setchellii* Fischer collected from field soil under *Pseudotsuga menziesii*. They noted white, extremely tomentose ECM with abundant multibranched, stout, white rhizomorphs emanating from the mantle surface. Presence of calcium oxalate crystals on extramatrical hyphae was also lacking in their description.

Müller and Agerer (1996) provided the first comprehensive description of a *Hysterangium* ECM, that of *H. crassirhachis* Zeller & Dodge on *P. menziesii*. Their detailed description was extensively illustrated and noted the presence of clamp connections on the hyphae, a feature also common in the sporocarp (Castellano 1989). They described abundant "star-like crystals" that cover the surface of the outer hyphae of the rhizomorphs and those emanating from the mantle. These crystals represent calcium oxalate weddelite $CaC_2O_4 2H_2O$ as described by Graustein et al. (1977). Such oxalate production may be responsible for accelerated weathering of primary minerals and clays and may play an important role in the transport of Fe and Al (Graustein et al. 1977; Cromack et al. 1979; Griffiths et al. 1994).

13.4
Ecology

13.4.1
Geographical Distribution and Host Specificity

Although *Hysterangium* is distributed throughout the world except for Antarctica, most species display high levels of endemism and discrete host ranges (Castellano 1989). Nearly all *Hysterangium* species display either a northern or a southern hemisphere distribution, with most species restricted to a single continent (Castellano 1989; Alvarez et al. 1993; Castellano and Beever 1994; Castellano and Muchovej 1995). Of the more than 50 known *Hysterangium* species, only *H. coriaceum* is known to occur as part of the native flora in two continents, North America and Europe. Both *H. inflatum* and *H. gardneri* have been introduced with *Eucalyptus* spp. throughout the world (Chu-Chou and Grace 1982, 1983; Castellano and Muchovej 1995). Some *Hysterangium* species are a dominant component of the below-ground ECM network of *P. menziesii* forests in the Pacific Northwest (Cromack et al. 1979; Griffiths et al. 1996) and *Eucalyptus* forests in Australia (Malajczuk et al. 1987) and plantations in New Zealand (Castellano and Beever 1994).

The global biogeography of *Hysterangium* is consistent with it being either a relatively old taxon in geological terms or a more efficient disperser than currently appreciated. Owing to its dependency on mycophagy for dispersal, the former seems more likely. Given its global distribution pattern wherever ECM hosts occur, its ancestors probably existed when continental land masses were coalesced into larger continents. It appears, however, that *Hysterangium* has had three centers of evolution: North America (including Mexico), Western Europe, and Australia. Australia and North America have over 20 endemic species while Europe has 15 (Castellano 1989). This reflects the wide host ranging ability of the genus to associate with host genera in the Fagaceae, Myrtaceae, and Pinaceae.

There appear to be *Quercus*-specific associates on both the European and North American continents. In Europe, *H. clathroides, H. epiroticum*, and *H. membranaceum* associate only with Fagaceae; in North America, *H. separabile* is specific to *Quercus* spp. *Hysterangium coriaceum, H. crassirhachis*, and *H. setchellii* are dominant ECM fungal components in the vast expanse of *Pseudotsuga menziesii* forests of western North America. *H. setchellii* appears to be specific to *P. menziesii* in the field, but *H. coriaceum* and *H. crassirhachis* associate with most genera of Pinaceae. Molina and Trappe (1982a,b) synthesized ECM of *H. coriaceum* (as *H. separabile*) with *A. menziesii, A. uvaursi, L. occidentalis, P. contorta, P. monticola*, and *T. heterophylla. H. fallax* appears specific to the genus *Pinus*, particularly two-needled species, in western North America. In New Zealand, *H. neotunicatum* and *H. rugisporum* associate only with *Leptospermum* spp. and *Kunzea* spp. (Castellano and Beever 1994).

Problems with accurate identification of species within the genus prior to 1989 are exemplified by States (1984) who lists *H. separabile* as an associate of *Abies concolor, Cupressus* spp., *Pinus pinyonensis, Pinus ponderosa, Psuedotsuga menziesii*, and *Quercus gambelii*. Subsequent examination of his collections (M. A. Castellano, unpubl. data) reveal them to be a mix of *H. coriaceum* (probably with Pinaceae), *H. separabile* (probably with *Quercus*), a third, as yet undescribed, species associated only with *P. menzeisii*, and a fourth, as yet undescribed, species associated with *P. ponderosa*.

13.4.2
Habitats and Community Dynamics

The vast array of potential hosts for *Hysterangium* species is matched by the diverse array of known habitats. *Hysterangium* species occur in mineral soil, sand, pumice, and humus or in litter layers. A number of the *Hysterangium* species associated with Myrtaceae (e.g. *H. gardneri* and *H. inflatum*) fruit within the upper humus layers and occasionally in the litter layer. While many other sequestrate species occupy rotten wood habitats, there are no reports of such behaviour for any *Hysterangium* species. Various *Hysterangium* species fruit abundantly with many different ECM hosts throughout the world. Based on sporocarp presence, *H. setchellii*, *H. coriaceum*, and *H. crassirhachis* are common associates of *P. menziesii* in western North America. *Hysterangium separabile* is a common associate of *Quercus* species in western North America and Mexico. *H. affine*, *H. gardneri*, and *H. inflatum* are common associates of various *Eucalyptus* spp. throughout the world.

Fogel (1976, 1981) found that *H. setchellii* (as *H. crassum*) and *H. coriaceum* (as *H. separabile*) comprise a major portion of the sporocarpic biomass of sequestrate fungi in a western Oregon *P. menziesii* stand. *H. setchellii* and *H. coriaceum* produced up to 8617 and 3139 sporocarps ha^{-1}, and they yielded up to 1206 and 572 $g\,ha^{-1}$ (dry wt.) respectively. These two fungi accounted for over 70% of all sequestrate sporocarps and fungal dry weight produced. In a similar study of a single stand, Hunt and Trappe (1987) found that *H. setchellii* (as *H. crassum*) and *H. coriaceum* produced up to 3770 and 1833 sporocarps ha^{-1} and 842 and 204 $g\,ha^{-1}$ respectively. They accounted for over 80% of all sequestrate sporocarps found and over 50% of the dry weight produced. *H. crassirhachis* also was common but in much lower numbers and biomass. In a third study encompassing ten *P. menziesii* stands, Luoma et al. (1991) found that *H. coriaceum*, *H. crassirhachis*, and *H. setchellii* constitute 24–178, 0–84, and 5–60 $g\,ha^{-1}$ respectively, depending on age and habitat status of the stand. Luoma et al. (1991) also found that *H. coriaceum* has an affinity for wet old-growth and mesic young-growth stands, *H. crassirhachis* had an affinity for dry old-growth stands, and *H. setchellii* had an affinity for mesic old-growth stands. The *P. menziesii* stands studied by Fogel (1976) and Hunt and Trappe (1987) were located at less than 500-m elevation in the Coast Ranges of Oregon; Luoma et al. (1991) studied *P. menziesii* stands at 500 to 1500-m elevation in the western Cascades of Oregon. In both mature (ca. 100-year-old) and old-growth (>200-year-old) stands of *Abies* spp. in California, Waters et al. (1997) found that *H. coriaceum* and *H. crassirhachis* constitute a substantial portion of the standing crop of sequestrate fungi, particularly in the younger stands. North et al. (1997) found *H. coriaceum* to be a minor component of the sequestrate flora in three different age classes of *T. heterophylla* stands in Washington. An undescribed *Hysterangium* species [reported as *H. coriaceum* by States and Gaud (1997)] was the second most abundant sporocarp producing sequestrate species in *P. ponderosa* forests in Arizona.

In addition to the relative abundance of *H. setchellii* in the Oregon Coast Ranges, it appears to strongly influence site productivity. Perennial hyphal mats formed by *H. setchellii* in soils of a 40 to 65-year-old *P. menziesii* stand in Oregon can occupy up to 16.7% (mean of 9.6%) of the upper 10 cm of soil (Cromack et al. 1979). *H. setchellii* exudes large amounts of oxalic acid (Cromack et al. 1979), which may act in the chemical weathering of clay soils (Graustein et al. 1977; Malajczuk and Cromack 1982). Fogel and Hunt (1979) note the extremely high concentrations of Ca in ECM formed by *Hysterangium* spp., particularly *H. setchellii* (as *H. crassum*). Griffiths et al. (1987) found that *H. setchellii* hyphal mats are significantly higher in microbial biomass and respiration rate than adjacent non-mat soil. When respiration rate was compared by per unit of biomass, *H. setchellii* hyphal mats were over three times more active than adjacent non-mat soil. In addition, *H. setchellii* mat soils have significantly higher levels of phosphatase, protease, laminarinase, and peroxidase than non-mat soils. These enzymes are involved in the breakdown of organic material, including organic P and N, along with cell wall material. *H. setchellii* mat soil has also been reported to contain significantly higher levels of total N, P, K, Ca, Mg and various other micronutrients than non-mat soils (Entry et al. 1987, 1992). This is probably due to the high levels of oxalic acid produced by the fungus. Oxalic acid forms an insoluble precipitate with Ca, thereby increasing the concentration of other nutrients in solution (Graustein et al. 1977). *H. setchellii* hyphal mats also preferentially use organic N and are more efficient at obtaining and retaining atmospheric N than adjacent non-mat soil (R. Griffiths, pers. comm.). Cromack et al. (1989) found significantly greater microbial biomass and numbers of soil microarthropods (collembola, oribatid mites, nematodes and total mites), along with increased protozoan populations (flagellates, amoebae and ciliates), in *Hysterangium* mats compared to non-mat soils. Malajczuk et al. (1987) found *H. inflatum* to form hyphal mats similar in structure to those of *H. setchellii*, occupying up to 10% of the soil surface in *Eucalyptus diversicolor* stands in Australia.

13.4.3
Mycophagy of Sporocarps

Hysterangium is one of many sequestrate genera routinely consumed by forest-dwelling animals (Fogel and Trappe 1978). Since Parks (1919, 1922) reported the first mycophagy of *Hysterangium* by wood rats (*Neotoma* spp.), the phenomenon has been widely reported for other animal species. *Hysterangium* forms a significant portion of the diets of deer mice in western Oregon (Maser and Maser 1987a), yellow-pine chipmunk in north-eastern Oregon (Maser and Maser 1987b), western red-backed vole in south-west Oregon (Hayes et al. 1986), the northern flying squirrel in north-western

Oregon (Maser et al. 1985), Abert's squirrels in Arizona (Stephenson 1975) and the western gray squirrel in the western United States (Stienecker and Browning 1970). Rosentreter et al. (1997), however, found that *Hysterangium* species constituted only a small portion of the diet of northern flying squirrels in the interior conifer forests of central Idaho. Claridge and May (1994) also listed various *Hysterangium* spp. found in the diets of numerous Australian mammals including bettongs, potoroos and marsupial mice.

13.5
Physiological Aspects of Mycelia

Hysterangium species are difficult to isolate from sporocarp tissue owing to the extremely gelatinous nature of the gleba. Germination of *Hysterangium* spores has not been reported. The few *Hysterangium* species that have been isolated grow more slowly than most other ECM fungi in pure culture, regardless of the culture medium (Molina and Trappe 1982a). In fact, cultures of *Hysterangium* species die after only a few transfers over the course of a year or two. The slow growth restricts the amount of biomass produced so that physiological culture tests are difficult to accomplish. Slow growth in culture also limits tests for host compatibility in pure culture synthesis.

Many species of *Hysterangium* form unique perennial ECM mats capable of penetrating the mineral soil layer. As outlined above, *Hysterangium* mats function in soil mineralization (Cromack et al. 1979; Griffiths et al. 1994), N cycling (Griffiths et al. 1990; Aguilera et al. 1993) and cycling of other nutrients (Entry et al. 1992), and modification of polyphenolics (Bending and Read 1996). Denitrification rates and soil pH are lower in *Hysterangium* mat soils than in non-mat soils (Griffiths et al. 1990). The presence of the *Hysterangium* mats also strongly correlates with increased microarthropod biodiversity and populations (Cromack et al. 1989) and enhanced seedling establishment of *Pseudotsuga menziesii* under intact canopies (Griffiths et al. 1991a). Several

Table 13.1. Physiological traits of some *Hysterangium* species. (B. A. Caldwell, pers. comm.)

Species	Hydrolysis of fatty acid			Breakdown of tannic acid–protein complex[b,c]	Decoloration of polymeric dyes[c]	
	Ester[a]	RNA	Protein		RBB	R-478
Hysterangium sp. 1	+	+	+	+	+	+
H. coriaceum	+	+	+	+	+	+
H. crassirhachis	+	+	+	Nt	+	+
H. gardeneri	+	+	+	(+)	Nt	+
H. inflatum	+	+	+	Nt	Nt	Nt
H. setchellii	+	+	+	+	+	+

Nt not tested; RBB, Remazol Brilliant Blue.
[a] Tween 40 (Caldwell et al. 1991).
[b] (+), Bavendamm-like darkening, +, clearing of precipitate.
[c] Partially after Griffiths and Caldwell (1992).

Hysterangium species can hydrolyze fatty acid esters, protein, and nucleic acids, and appear to have some ability to degrade tannin–protein complexes and to decompose polymeric dyes (Table 13.1) that are putative indicator substrates for lignin breakdown (Caldwell et al. 1991; Entry et al. 1991; B. A. Caldwell, unpubl. data). Because a number of *Hysterangium* species form such mat-like structures in forest soils and these species show unique or heightened specific physiological capabilities, they are clearly an important guild of ECM fungi in ECM forests.

References

Acsai J, Largent DL (1983) Ectomycorrhizae of selected conifers growing in sites which support dense growth of bracken fern. Mycotaxon 16:509–518

Aguilera LM, Griffiths RP, Caldwell BA (1993) Nitrogen in ectomycorrhizal mat and non-mat soils of different-age Douglas-fir forests. Soil Biol Biochem 25:1015–1019

Alvarez IF, Parlade J, Trappe JM, Castellano MA (1993) Hypogeous mycorrhizal fungi of Spain. Mycotaxon 67:201–217

Beaton G, Pegler DN, Young TWK (1985) Gastroid Basidiomycota of Victoria State, Australia: 4. *Hysterangium*. Kew Bull 40:435–444

Bending GD, Read DJ (1996) Nitrogen mobilization from protein–polyphenol complex by ericoid and ectomycorrhizal fungi. Soil Biol Biochem 28:1603–1612

Berkeley MJ (1844) Notices of British fungi. Ann Mag Nat Hist 13:340–360

Berkeley MJ (1848) Notices of British fungi. Ann Mag Nat Hist 8:323–379

Bresadola J (1892) Fungi Tridentini 2:99

Bucholtz F (1908) Verbreitung der Hypogaeen in Russland. Bull Soc Imper Nat Moscou 4:467–470

Caldwell BA, Castellano MA, Griffiths RP (1991) Fatty acid esterase production by ectomycorrhizal fungi. Mycologia 83:233–236

Castellano MA (1989) The taxonomy of the genus *Hysterangium* (Basidiomycotina, Hysterangiaceae) with notes on its ecology. PhD Thesis, Oregon State University, Corvallis

Castellano MA (1990) The new genus *Trappea* (Basidiomycotina, Phallales): a segregate from *Hysterangium*. Mycotaxon 38:1–9

Castellano MA, Beever RE (1994) Truffle-like Basidiomycotina of New Zealand: *Gallacea, Hysterangium, Phallobata*, and *Protubera*. NZ J Bot 32:305–328

Castellano MA, Muchovej JJ (1995) Truffle-like fungi from South America: *Hysterangium sensu lato*. Mycotaxon 57:329–345

Chu-Chou M, Grace LJ (1982) Mycorrhizal fungi of *Eucalyptus* in the North Island of New Zealand. Soil Biol Biochem 14:133–137

Chu-Chou M, Grace LJ (1983) Hypogeous fungi associated with some forest trees in New Zealand. NZ J Bot 21:183–190

Claridge AW, May TW (1994) Mycophagy among Australian mammals. Aust J Ecol 19:251–275

Corner EJH, Hawker LE (1953) Hypogeous fungi from Malaya. Trans Br Mycol Soc 36:125–137

Cribb JW (1958) The Gasteromycetes of Queensland – IV. *Gautieria, Hysterangium* and *Gymnoglossum*. Pap Dep Bot Univ Queensl 3:153–159

Cromack K Jr, Sollins P, Graustein WC, Speidel K, Todd AW, Spycher G, Li CY, Todd RL (1979) Calcium oxalate accumulation and soil weathering in mats of the hypogeous fungus *Hysterangium crassum*. Soil Biol Biochem 11:463–468

Cromack K Jr, Fichter BL, Moldenke AM, Entry JA, Ingham ER (1989) Interactions between soil animals and ectomycorrhizal fungal mats. In: Edwards CA, Stinner BR, Stinner D, Rabatin S (eds) Biological interactions in soil. Elsevier, Amsterdam, pp 161–168

Cunningham GH (1934) The Gasteromycetes of Australasia. XVI. Hymenogastraceae, part I: the genera *Rhizopogon, Melanogaster* and *Hymenogaster*. Proc Linn Soc NSW 59:156–172

Cunningham GH (1938) The Gasteromycetes of Australasia, XVIII. Trans R Soc N Z 67:408–409

Cunningham GH (1944) The Gasteromycetes of Australia and New Zealand. McIndoe, Dunedin, New Zealand, 236 pp

De Toni GB (1888) In: Saccardo PA (ed) Sylloge Fungorum 7. Friedländer & Sohn, Berlin, pp 156–172

Dring DM (1980) Contributions towards a rational arrangement of the Clathraceae. Kew Bull 35:1–96

Entry JA, Rose CL, Cromack K Jr, Griffiths RP, Caldwell BA (1987) The influence of ecto-mycorrhizal mats on chemistry of a coniferous forest soil. In: Sylvia DM, Hung LL, Graham JH (eds) 7th North American Conference on Mycorrhizae. University of Florida, Gainsville, p 93

Entry JA, Donnelly PK, Cromack K Jr (1991) Influence of ectomycorrhizal mat soils on lignin and cellulose degradation. Biol Fert Soils 11:75–78

Entry JA, Rose CL, Cromack K Jr (1992) Microbial biomass and nutrient concentrations in hyphal mats of the ectomycorrhizal fungus *Hysterangium setchellii* in a coniferous forest soil. Soil Biol Biochem 24:447–453

Fischer E (1900) Phallineae, Hymenogastrineae, Lycoperdineae, Nidulariineae, Plectobasidiineae (Sclerodermineae). In: Engler A, Prantl K (eds) Natürliche Pflanzenfamilien, Teil 1. Wilhelm Engelmann, Leipzig, pp 276–346

Fischer E (1908) Zur Morphologie der Hypogaeen. Bot Ztg 66:141–168

Fischer E (1933) Unterklasse Eubasidii. In: Engler A, Prantl K (eds) Natürliche Pflanzenfamilien, Bd4. Wilhelm Engelmann, Leipzig, pp 1–122

Fischer E (1938) Hypogaeen–Studien. I. Zur Kenntnis der Gattung *Hysterangium*. Ber Schweiz Bot Ges 48:29–44

Fitzpatrick HM (1913) A comparative study of the development of the fruit-body in *Phallogaster*, *Hysterangium* and *Gautieria*. Ann Mycol 2:119–149

Fogel R (1976) Ecological studies of hypogeous fungi. II. Sporocarp phenology in a western Oregon Douglas-fir stand. Can J Bot 54:1152–1162

Fogel R (1981) Quantification of sporocarps produced by hypogeous fungi. In: Wicklow DT, Carroll GC (eds) The fungal community: its organization and role in the ecosystem. Marcel Dekker, New York, pp 553–568

Fogel R, Hunt G (1979) Fungal and arboreal biomass in a western Oregon Douglas-fir ecosystem: distribution patterns and turnover. Can J For Res 9:245–256

Fogel R, Trappe JM (1978) Fungus consumption (mycophagy) by small animals. Northwest Sci 52:1–31

Gäumann EA, Dodge CW (1928) Comparative morphology of fungi. McGraw-Hill, New York, 701 pp

Graustein WC, Cromack K Jr, Sollins P (1977) Calcium oxalate: occurrence in soils and effect on nutrient and geochemical cycles. Science 198:1252–1254

Griffiths RP, Caldwell BA (1992) Mycorrhizal mat communities in forest soils. In: Read DJ, Lewis DH, Fitter AH, Alexander IJ (eds) Mycorrhizas in ecosystems. CAB International, Wallingford, pp 98–105

Griffiths RP, Caldwell BA, Cromack K Jr, Castellano MA, Morita RY (1987) A study of chemical and microbial variables in forest soils colonized with *Hysterangium setchellii* rhizomorphs. In: Sylvia DM, Hung LL, Graham JH (eds) 7th North American Conference on Mycorrhizae. University of Florida, Gainsville, p 96

Griffiths RP, Caldwell BA, Cromack K Jr, Morita RY (1990) Douglas-fir forest soils colonized by ectomycorrhizal mats. I. Seasonal variation in nitrogen chemistry and nitrogen cycle transformation rates. Can J For Res 20:211–218

Griffiths RP, Castellano MA, Caldwell BA (1991a) Hyphal mats formed by two ectomycorrhizal fungi and their association with Douglas-fir seedlings: a case study. Plant Soil 134:255–259

Griffiths RP, Ingham ER, Caldwell BA, Castellano MA, Cromack K Jr (1991b) Microbial characteristics of ectomycorrhizal mat communities in Oregon and California. Biol Fert Soils 11:196–202

Griffiths RP, Baham JE, Caldwell BA (1994) Soil solution chemistry of ectomycorrhizal mats in forest soil. Soil Biol Biochem 26:331–337

Griffiths RP, Bradshaw GA, Marks B, Lienkaemper GW (1996) Spatial distribution of ectomycorrhizal mats in coniferous forests of the Pacific Northwest, USA. Plant Soil 180:147–158

Gross G, Runge A, Winterhoff W, Krieglsteimer GJ (1980) Bauchpilze (Gastromycetes S.L.) in der Bundesrepublik und Westberlin. Z Mykol Beih 2:1–220

Harkness HW (1899) Californian hypogeous fungi. Proc Calif Acad Sci 1:241–292

Hawker L (1954) British hypogeous fungi. Philos Trans R Soc Lond 237:429–546

Hayes JP, Cross SP, McIntire PW (1986) Seasonal variation in mycophagy by the western redbacked vole, *Clethrionomys californicus*, in southwestern Oregon. Northwest Sci 60:250–257

Hesse R (1884) *Hysterangium rubricatum*. Eine neue Hymenogastreenspecies. Prings Jahrb Wiss Bot 15:631–641

Hesse R (1891) Hypogaeen Deutschlands. 1. Die Hymenogastreen. Ludw., Hofsetter, Marburg, 133 pp

Horak E (1964) Fungi austroamericani V. Beitrag zur Kenntnis der Gattungen *Hysterangium* Vitt., *Hymenogaster* Vitt., *Hydnangium* Wallr. und *Melanogaster* Cda. in Südamerika (Argentinien, Uruguay). Sydowia 17:197–205

Hunt GA, Trappe JM (1987) Seasonal hypogeous sporocarp production in a western Oregon Douglas-fir stand. Can J Bot 65:438–445

Ito S, Imai S (1937) Fungi of the Bonin Islands. I. Trans Sapporo Mat Hist Soc 15:1–16

Lloyd CG (1921) Mycol Notes 65:1031

Lloyd CG (1922) Mycol Notes 66:1119–1120

Luoma DL, Frenkel RE, Trappe JM (1991) Fruiting of hypogeous fungi in Oregon Douglas-fir forests: seasonal and habitat variation. Mycologia 83:335–353

Malajczuk N, Cromack K Jr (1982) Accumulation of calcium oxalate in the mantle of ectomycorrhizal roots of *Pinus radiata* and *Eucalyptus marginata*. New Phytol 92:527–531

Malajczuk N, Dell B, Bougher NL (1987) Ectomycorrhiza formation in *Eucalyptus*. III. Superficial ectomycorrhizas initiated by *Hysterangium* and *Cortinarius* species. New Phytol 105:421–428

Malençon G (1975) Champignons hypogés du nord de l'Afrique II. Basidiomycetes. Rev Mycol 39:279–306

Maser C, Maser Z (1987a) Notes on mycophagy in four species of mice in the genus *Peromyscus*. Great Basin Nat 47:308–313

Maser Z, Maser C (1987b) Notes on mycophagy of the yellow-pine chipmunk (*Eutamius amoenus*) in northeastern Oregon. Murrelet 68:24–27

Maser Z, Maser C, Trappe JM (1985) Food habits of the northern flying squirrel (*Glaucomys sabrinus*) in Oregon. Can J Zool 63:1084–1088

Massee G (1898) Fungi exotici I. Kew Bull Misc Info, pp 113–136

Massee G (1899) Fungi exotici II. Kew Bull Misc Info, pp 164–185

Massee G (1901) Fungi exotici III. Kew Bull Misc Info, pp 150–169

Mattirolo O (1900a) Elenco dei fungi hypogaei raccolti nelle forests di Vallombrosa negli anni 1899–1900. Malpighia 14:247–270

Mattirolo O (1900b) Gli ipogei de Sardegna e di Sicilia. Malpighia 14:39–106

Miller OK (1983) Ectomycorrhizae in the Agaricales and Gastromycetes. Can J Bot 61:909–916

Molina R, Trappe JM (1982a) Patterns of ectomycorrhizal host specificity and potential among Pacific Northwest conifers and fungi. For Sci 28:423–458

Molina R, Trappe JM (1982b) Lack of mycorrhizal specificity by the ericaceous hosts *Arbutus menziesii* and *Arctostaphylos uva-ursi*. New Phytol 90:495–509

Müller WR, Agerer R (1996) *Hysterangium crassirhachis* Zeller & Dodge + *Pseudotsuga menziesii* (Mirb.) Franco. In: Agerer R, Danielson RM, Igli S, Ingleby K, Luoma D, Treu R (eds) Descriptions of ectomycorrhizae 1. Einhorn-Verlag, Swàbisch Gmünd, pp 29–34

North M, Trappe JM, Franklin J (1997) Standing crop and animal consumption of fungal sporocarps in Pacific Northwest forests. Ecology 78:1543–1554

Pacioni G (1984) *Hysterangium epiroticum* nov. sp. and other hypogeous macromycetes from S.S.R. (Albania). Nova Hedwigia 40:79–84

Parks HE (1919) Notes on California fungi. Mycologia 11:10–21

Parks HE (1922) The genus *Neotoma* in the Santa Cruz Mountains. J Mammal 3:241–253

Patouillard N (1914) Contribution à la flore mycologique hypogée du Jura. Bull Mycol Soc Fr 30:347–354

Patouillard N (1915) Champignons de la Nouvelle-Calédonie. (Suite) (1). Bull Soc Mycol Fr 31:31–35

Quélet L (1886) Enchridion fungorum in Europa nedia et praesertim in *Galbia vigentuim*. Octavio Doin, Lutetiae, 352 pp

Rehsteiner H (1892) Beiträge zur Entwicklungsgeschichte der Fruchtkörper einiger Gastromyceten. Bot Zeit 50:761–878

Rodway L (1918) Botanical notes. Proc Pap R Soc Tasmania 1917:105–110

Rodway L (1920) Notes and additions to the fungus flora of Tasmania. Proc Pap R Soc Tasmania 1919:110–116

Rosentreter R, Hayward GD, Wicklow-Howard M (1997) Northern flying squirrel seasonal food habits in the interior conifer forests of central Idaho, USA. Northwest Sci 71:97–102

Smith AH, Zeller SM (1966) A preliminary account of the North American species of *Rhizopogon*. Mem N Y Bot Gard 14:1–178

Soehner E (1949) *Hysterangium hessei* Soehner com. nov. und *Hysterangium coriaceum* var. *knappii* Soehner var. nov. Z Pilzkd 1949(3):29–32

Soehner E (1952) Bayerische *Hysterangium*-Arten. Sydowia 6:246–264

Spegazzini C (1881) Fungi *Argentini pugillus* IV. An Cient Argent 10:237

States JS (1984) New records of false truffles in pine forests in Arizona. Mycotaxon 19:351–367

States JS, Gaud WS (1997) Ecology of hypogeous fungi associated with ponderosa pine. I. Patterns of distribution and sporocarp production in some Arizona forests. Mycologia 89:712–721

Stephenson RL (1975) Reproductive biology and food habits of Abert's squirrels in central Arizona. MSc Thesis, Arizona State University, Tempe, Arizona

Stienecker W, Browning BM (1970) Food habits of the western gray squirrel. Calif Fish Game 56:36–48

Svrcek M (1958) Hysterangiales. In: Pilát A (ed) Gasteromycetes flora, CSR, Prague, pp 96–120

Townsend BB (1954) Morphology and development of fungal rhizomorphs. Trans Br Mycol Soc 222–231

Tulasne L-R, Tulasne C (1843) Champignons hypogés de la famille des Lycoperdaceas, observés dans les environs de Paris et les départements de la Vienne et d'Indre-et-Loire. Ann Sci Nat Ser 2, 19:373–381

Tulasne L-R, Tulasne C (1851) Fungi hypogaei. Histoire et monographie des champignons hypogés. Friedrich Klincksieck, Paris

Velenovsky J (1939) Novitates mycologicae. Societatis Botanicae Cechoslovacae Officium, Prague

Vittadini C (1831) Monographia Tuberaceum. Felicis Rusconi, Milan

Waters JR, McKelvey KS, Luoma DL, Zabel CJ (1997) Truffle production in old-growth and mature fir stands in northeastern California. For Ecol Manage 96:155–166

Zeller SM (1939) New and noteworthy Gasteromycetes. Mycologia 31:1–32

Zeller SM (1941) Further notes on fungi. Mycologia 33:196–224

Zeller SM (1948) Notes on certain Gasteromycetes, including two new orders. Mycologia 40:639–668

Zeller SM, Dodge CW (1929) *Hysterangium* of North America. Ann Mo Bot Gard 83–128

Chapter 14

Thelephora

J. V. Colpaert

14.1
Introduction

Thelephora terrestris is the best-known species of the basidiomycete genus *Thelephora*, and is a very common ectomycorrhizal symbiont in conifer tree nurseries over all continents. It has a somewhat dubious reputation, however, being considered to be a weak parasite in some instances. Researchers involved in ectomycorrhizal inoculation programmes undoubtedly have cursed this vigorously growing pioneer species because it easily outcompetes their inoculant ECM species. Nevertheless, this mycobiont is well adapted to the environmental conditions in modern nurseries and there is evidence that it confers some advantages to its young host plants in these nurseries. Despite its excellent growth on seedlings of many tree species, it is less common in experimental studies than many other species described in this book. However, its rather special ecological role makes it an interesting species for experimental work.

14.2
Taxonomy

The basidiomycete genus *Thelephora* has been monographed by Corner (1968, 1976). More than 50 species have been described. The genus belongs to the family of the Thelephoraceae ss. Stalpers, which, together with the Bankeraceae ss. Stalpers, belongs to the order of Thelephorales, a group of Aphyllophoraceous fungi (Stalpers 1993). The close relationship between genera of the two families was recently supported by determination of nuclear rDNA and mitochondrial rDNA sequences (Gardes and Bruns 1996; Hibbett et al. 1997; Taylor and Bruns 1997). The Bankeraceae is comprised of terrestrial stipitate genera, with fleshy pileate basidiocarps often with toothed or poroid hymenophores. All genera have ectomycorrhizal members, a lifestyle that is rather unusual among aphyllophoralean families. The biology of the related

Laboratory of Plant Ecology, Katholieke Universiteit Leuven, Kard. Mercierlaan 92, 3001 Leuven, Belgium

Thelephoraceae family is not well studied, but ectomycorrhizal (ECM) species have been found in the genera *Thelephora, Pseudotomentella, Tomentella* and *Tylospora* (Taylor and Alexander 1991; Agerer 1994, 1996; and see Chap. 15). However, the last genus is usually placed in the Corticiaceae. Species of the Thelephoraceae have clavaroid or small resupinate or effused-reflexed basidiocarps that are often found on solid substrata lying on the soil such as wood, twigs or stones. It is likely that these substrata are colonised by mycelium from the soil. Although most Thelephoraceae species have been presumed to be parasitic or saprotrophic, it is now accepted that the majority, if not all, *Thelephora* species from temperate regions are ECM formers (Stalpers 1993; Arnolds et al. 1995; Baici et al. 1995).

Stalpers (1993) remarked that the species concept in the genus *Thelephora* is traditionally very narrow, retaining only 45 species in his key to the Thelephorales. The nucleus has been the northern temperate alliance of *T. anthocephala* (Bull.: Fr.) Fr., *T. caryophyllea* (Schaeff.: Fr.) Fr., *T. palmata* Fr.: Fr., *T. terrestris* Ehrh.: Fr. and *T. vialis* Schw. It has led to the supposition that the genus is essentially northern temperate, although more than 30 species are confined to the tropics and the southern hemisphere (Corner 1968, 1976). *Thelephora* is, for example, well represented in the fagaceous forests in Southeast Asia, in tropical and subtropical Asia and Australasia, in the *Araucaria* forest in the Papuan Archipelago and in coastal *Casuarina* forests.

14.3
Ecology

Although the genus appears to contain at least two edible species, *T. vialis* and *T. aurantiotincta* Corner (Zhou and Wei 1986), *Thelephora* is best known among mycorrhiza researchers because of a single species, *T. terrestris* (*Tt*). It is a very common ECM symbiont in tree nurseries in many parts of the world (Weir 1921; Marx and Bryan 1970), especially in the northern hemisphere. Although it is now usually recognised as a mutualistic symbiont, the species was considered as a parasite early this century (Weir 1921). Indeed, sporocarps of *Tt* can completely envelope seedlings, having a suffocating effect. Examination of seedlings overrun by *Tt* has never shown any living above-ground part invaded by mycelium. Nevertheless, Weir (1921) suggested that seedlings in nurseries affected with *Tt* should be pulled up and burned.

Tt is a dominant ECM fungus in soils that have been subjected to sterilisation or fumigation in the field or in the glasshouse (Marx et al. 1970, 1984; Garbaye 1982). It is a pioneer mycobiont that produces large quantities of sporocarps. Sporulating sporocarps can be found from late spring until autumn and densities of 200 spores m^{-3} air have been recorded (Le Tacon et al. 1987). Other ECM *Thelephora* species are usually associated with trees in older mature forests, although *T. penicillata* Fr.: Fr. and *T. caryophyllea* have also been reported from nurseries (Weir 1921; Corner 1968). For most *Thele-*

phora species, production of sporocarps is not as obvious as for *T. terrestris*. In a natural stand of approximately 40-year-old *Pinus muricata*, DNA finger-prints from ECM showed the below-ground presence of a Thelephoroid species as well as of *Tomentella sublilacina* (Ellis & Holw.) Wakef. (Gardes and Bruns 1996). Although both species were quite dominant in the below-ground ECM community, they remained unrepresented in the above-ground fruiting record over a 3-year period. Furthermore, the observations made by Taylor and Bruns (1997) on ECM epiparasitism of members of the Thelephoraceae by the non-photosynthetic orchid *Cephalanthera austinae* suggest that the genus *Thelephora* might be much more represented in natural forest ecosystems than previously thought.

Tt occurs on a wide variety of soils, including non-fertilised and highly fertilised nursery soils, mineral or peaty soils, and in dry or wet conditions (Mason et al. 1983; Marx et al. 1984; Garbaye et al. 1986). Although often dominant on *Pinus* spp. in nurseries, *Tt* is not always present on container-grown conifer seedlings (Danielson and Visser 1990) or in tree nurseries on calcareous soils (Mousain et al. 1977). In New Zealand the species was apparently also absent from the majority of nurseries growing the exotic *Pinus radiata* (Chu-Chou 1979). It is likely that *Tt* and several other species from the northern temperate regions were introduced to Australasia (Corner 1968), as is probably also the case in central Africa (Ivory and Munga 1992). In more natural habitats, sporocarps of *Tt* are generally found on poor to mesotrophic sandy, rather acid soils with little humus. It is found on reforestated sites, in regenerating forests after severe disturbance, along roadsides and other disturbed areas (Keizer and Arnolds 1994; Visser 1995). Although *Tt* is considered to be poorly competitive with other ECM fungi, it can persist in older conifer or oak stands on sandy soils (Keizer and Arnolds 1994; Kårén and Nylund 1997). Sporocarp records, if reliable, suggest that most other *Thelephora* species are less common or have a narrower habitat preference than *Tt* (Corner 1968). The autecology of most species is poorly known and misdeterminations of fruiting bodies complicate the picture. Even the common resupinate forms of *Thelephora* spp. may be difficult to identify and are easily mistaken for *Tomentella* spp. (Arnolds et al. 1995).

Tt is regarded as an "early-stage" fungus: a fungus that is a dominant root coloniser of young trees growing on sites that have been treeless and thus symbiont-free for many years (Last et al. 1987). On such sites, *Tt* mycorrhizas that have been established in nurseries can remain dominant for several years in the young forest. The fungus colonises not only short roots in upper soil layers, but also roots in much deeper (below 0.5 m) mineral soil layers (Sylvia and Jarstfer 1997). In forests regenerating following wildfire disturbance, *Tt* is also present in the young regenerating stands, but it is not as dominant as in a first generation woodland (Visser 1995).

Early-stage fungi disseminate as spores and are able to colonise seedlings from spore inoculum when they are growing on mineral or peaty soils, even in unsterile conditions, but always in the absence of vegetative inoculum from late-stage fungi associated with mother trees (Mason et al. 1983; Fleming 1985). In general, the attributes of early-stage fungi are those of ruderal or "r" selected organisms (Last et al. 1987; Deacon and Fleming 1992; Smith and Read 1997). Ruderal species typically have a short vegetative phase followed by production of dispersal units, resulting in a widespread distribution; they also rapidly exploit an available resource. *Tt* certainly satisfies these criteria (Deacon and Fleming 1992). For example, non-mycorrhizal (NM) conifer seedlings colonising the debris and tephra deposited by the 1980 Mount St. Helens volcanic eruption were chlorotic and did not survive to the next growing season (Allen 1987). Only seedlings that became mycorrhizal with early-stage fungi were able to survive on the volcanic debris. In the blast and tephra fall areas, spores of *Tt* were captured at a rate of one per 24 h on wind traps at 1 m above ground level. Furthermore, the frequency of *Tt* sporocarps in roadside verges in the Netherlands planted with native *Quercus robur* has been reported to be similar for young and old trees (Keizer and Arnolds 1994). Such roadside verges with oak trees in the open landscapes in the Netherlands are characterised by the absence of a litter layer and regular disturbance. Sporocarp frequency, however, gives no reliable indication of ECM frequency below ground, so it is not clear to what extent this indicates similar levels of infection of old versus younger trees. In plantation stands of *Pseudotsuga menziesii*, *Tt* is present in young and old forests, but in this case a lack of compatible late-stage fungi might explain the persistent presence of *Tt* in the older stands (Jansen 1991). In mature stands of the exotic *Picea sitchensis* in the UK, *Tt* was also present (Thomas et al. 1983), although an isolate that originated from a forest seemed to be more adapted to the forest soil than an isolate from a nursery (Holden et al. 1983).

14.4
Life Cycle

Isolation of *Tt* in axenic culture is more difficult than for many other ECM fungi. Direct isolations from young basidiocarps are possible (J. V. Colpaert, unpubl. data), but the success rate is low. Attempts to germinate basidiospores of *Tt* and other Thelephoroid taxa on agar media are usually unsuccessful (Marx and Ross 1970; Stalpers 1993) unless conducted in the presence of host plant roots or activator yeast (Fries 1978; Birraux and Fries 1981). Isolations may also be made directly from ECM root tips (Marx et al. 1970). The fungus can probably not be preserved by freezing (Corbery and Le Tacon 1997). It is possible that *Tt* mycelia have a homothallic mating system since Fries (1987) did not find somatic incompatibility between different isolates in vitro, although this requires confirmation.

On agar plates, *Tt* mycelium grows very well in the agar due to its hydrophilic characteristics (Unestam and Sun 1995). In vitro growth on various culture media is, however, rather slow (0.5–1.0 mm day^{-1}) (Marx et al. 1970; Colpaert and Van Assche 1987). Sung et al. (1995) compared biomass production of nine ECM fungi, all from different genera, on liquid cultures of MMN and found lowest biomass production for their *Tt* isolate. The authors also found that the ergosterol concentration in *Tt* mycelium increased from 2.5 μg mg⁻¹ dry wt in young mycelium up to 5 μg mg⁻¹ in older mycelia. However, in association with conifers and under optimal growth conditions, the growth of the external mycelia is generally faster than in most other ECM genera (Coutts and Nicoll 1990a; Colpaert et al. 1992; Bending and Read 1995). Coutts and Nicoll (1990a) measured a mean extension rate of 3 mm day⁻¹ under natural conditions in summer for *Tt* in association with *P. sitchensis*. This rate decreased to 0.44 mm day⁻¹ in winter at a moment when roots were dormant, but it increased again as the temperature increased in spring. These observations show that the mycelium of *Tt* is a perennial structure (Read 1992). The fan of colonising hyphae is quite uniform and hyphal proliferation over nutrient-rich organic substrates to form patches or mats has not been observed in this species (Unestam 1991; Bending and Read 1995). The density of its external mycelium appears variable and probably depends on the isolate, the environmental conditions and host plant involved. Jones et al. (1990) recorded hyphal lengths of 319 m m⁻¹ of *Tt*-colonised root in *Salix viminalis* after growth of 12 weeks at a soil P concentration of 4 mg kg⁻¹. Read (1992) argues that this value is probably an underestimation of what might be found in a non-sterilised soil. With increasing P levels the hyphal length of *Tt* in the *S. viminalis* experiment decreased. At a P concentration of 21 mg kg⁻¹ it was only 103 m m⁻¹ of colonised root. When hyphal lengths were expressed per kilogram of soil, these hyphal densities were respectively 53 and 44 m kg⁻¹ (Jones et al. 1990). However, Colpaert et al. (1997) grew *Pinus sylvestris* seedlings in a semi-hydroponic system at two nutrient addition regimes with P as the growth limiting factor. They found highest ergosterol concentrations in the *Tt*-colonised perlite substrate at the highest nutrient addition regime, whereas the reverse was true for *Suillus luteus* (Fries) Gray.

Sporocarps of *Tt* are readily produced in the field and in experimental conditions on seedlings a few months old (Marx and Bryan 1970; Mason et al. 1982), often whenever moisture and temperature are suitable (Molina et al. 1992). In contrast to the water-repellent mycelia of many other ECM fungi, the external mycelium of *Tt* is hydrophilic and the cell walls of all hyphal structures are surrounded by a water film so that hyphae tend to wick water (Unestam 1991; Unestam and Sun 1995). The external hyphae of *Tt* can differentiate into linear organs (rhizomorphs), but the fungus does not appear to produce sclerotia (Ingleby et al. 1990). The rhizomorphs are only slightly differentiated and are the same color as the mycorrhizas, although thinner rhizomorphs have a lighter color than thick ones (150 μm). Rhizomorphs of *Tt* are more hydrophobic than the other structures of the external mycelium (Coutts and Nicoll 1990a). The central hyphae are somewhat thicker in diameter and have thinner walls than the surrounding outer ones (Agerer 1988). During ontogeny of the rhizomorphs the hyphae produce backward-growing side branches and anastomoses which are often combined with reversely oriented clamps. While rhizomorphs are frequently formed during ECM synthesis experiments, their importance under natural conditions is less clear (Agerer and Weiss 1989).

14.5
ECM Formation

14.5.1
Anatomy of ECM

Pine ECM with *Tt* ramify repeatedly in a dichotomous manner so that coralloid systems can develop. In associations with *Picea*, mycorrhiza can be fairly long, up to 2.5 cm, and sinuous with frequent, irregularly spaced, short side branches (Agerer and Weiss 1989; Ingleby et al. 1990). The color is usually white when young, turning fawn to darker shades of orange or grey–brown with age. Darker forms turning to reddish ochre/brown, purplish chesnut or cigar-brown also occur. The outer surface of the mantle is plectenchymatous with loosely woven hyphae, emanating hyphae and typically awl-shaped cystidia with abruptly thick walls with a distal clamp (Agerer and Weis 1989; Agerer 1991). The middle layers and inner surface of the mantle are densely plectenchymatous. Slightly different descriptions for the surface structure have been reported by Ingleby and Mason (1996) who characterise *Tt* mantles as having a smooth compact surface and by Mohan et al. (1993) who describe the outer surface as a loosely interwoven net prosenchyma. Chu-Chou and Grace (1983) have further described *Tt* ECM on *P. menziesii*, while Godbout and Fortin (1985) reported on *Populus–Tt* ECM and Miller et al. (1991) described the *Alnus–Tt* ECM.

The anatomy of the Thelephoroid orchid mycorrhizas of *Cephalanthera austinae* differs from the ECM of their host trees. Taylor and Bruns (1997) reported that *C. austinae* roots contained cortical cells with intracellular coils, or "pelotons", of distinctively brown, thick-walled, clamped hyphae, corresponding to typical orchid mycorrhizas (Smith and Read 1997). Neither mantle nor Hartig net was observed in the orchid roots, while both structures were well developed in the ECM roots of the associated trees. This illustrates the fundamental plant control over the anatomy of the symbiotic organ. The same conclusion can be drawn from the observations made by Zak (1976a,b) who described formation of typical arbutoid mycorrhizas between *Tt* and *Arbutus menziesii* or *Arctostaphylos uva-ursi*.

14.5.2
Fungus–Host Specificity

Trappe (1962) listed only one known host plant species (*P. sylvestris*) for *Tt*. Marx and Bryan (1969, 1970) extended the verified host range to 21 *Pinus* spp., *Picea abies* and *Pseudotsuga menziesii*. *Pinus* is certainly the most frequent host of *Tt*, but it can form ECM with a broad range of host plant genera, including some angiosperms (Molina et al. 1992; Table 14.1). It should be noted, however, that, while many genera have been shown to be hosts for *Tt* in synthesis experiment, not all such genera are frequently colonised with *Tt* in the field or nursery. All host plant genera that form ECM with *Tt* can associate with a much wider range of fungal associates, the highest specificity being found in the genus *Alnus* (Molina 1981; Miller et al. 1991). This should not disguise the fact that specificity in the interaction between host root and *Tt* can occur. Specificity between particular *Tt* isolates and host plants might explain, for example, why some workers have been unable to synthesise ECM

Table 14.1. Host genera forming mycorrhizas with *T. terrestris*. References reporting on in vitro synthesis experiments and on field (nursery) observations

Host genus	Type	In vitro synthesis	Field observation
Abies	ECM		Weir (1921)
Acacia	ECM		Theodorou and Reddell (1991)
Allocasuarina	ECM	Theodorou and Reddell (1991)	Theodorou and Reddell (1991)[a]
Alnus	ECM	Miller et al. (1991)	Miller et al. (1991)
Arbutus	ArM	Zak (1976b)	
Arctostaphylos	ArM	Zak (1976a)	Acsai and Largent (1983)
Betula	ECM	Fleming (1985)	Mason et al. (1982)
Castanea	ECM		Mousain et al. (1977)
Castanopsis	ECM	Tam and Griffiths (1994)	
Casuarina	ECM	Theodorou and Reddell (1991)	Theodorou and Reddell (1991)
Cephalanthera[a]	OM		Taylor and Bruns (1997)
Eucalyptus	ECM	Chan and Griffiths (1991)	Mousain et al. (1977)
Fagus	ECM		Mousain et al. (1977)
Hudsonia (Cistaceae)	ECM		Malloch and Thorn (1985)
Larix	ECM	Samson and Fortin (1986)	
Lithocarpus	ECM	Tam and Griffiths (1993)	
Picea	ECM	Thomas and Jackson (1979)	Thomas and Jackson (1979)
Pinus	ECM	Marx and Bryan (1969, 1970)	Trappe (1962)
Populus	ECM	Godbout and Fortin (1985)	Godbout and Fortin (1985)
Pseudotsuga	ECM	Marx and Bryan (1970)	Chu-Chou and Grace (1983)
Quercus	ECM	Mitchell et al. (1984)	Marx et al. (1984)
Salix	ECM	Jones et al. (1990)	Arnolds et al. (1995)
Tsuga	ECM	Kropp and Trappe (1982)	Kernaghan et al. (1995)

ECM, ectomycorrhizas; ArM, arbutoid mycorrhizas; OM, orchid mycorrhizas.
[a] Probably another *Thelephora* sp.

between *Tt* and certain plant genera (e.g. *Alnus, Eucalyptus*), while others have been successful (Malajczuk et al. 1982; Godbout and Fortin 1983; Chan and Griffiths 1991; Miller et al. 1991). Even isolates from different conifer species are not necessarily intercompatible. Two *Tt* cultures isolated from pine mycorrhizas from nurseries in the southern United States and Canada (Marx et al. 1970) were shown to be good symbionts for most *Pinus* species, but did not form or formed atypical mycorrhizas with some Asian pine species (Marx and Bryan 1970). Furthermore, the same isolates failed to form typical ECM with *Pseudotsuga menziesii* and *Picea abies*. This study thus revealed variation in both rate and extent of ECM formation by *Tt* isolates. The extent of ECM formation varied from a fully developed sheath and Hartig net in compatible associations, through mycorrhizas without a mantle but with a Hartig net, to interactions that formed no identifiable ECM structures. In another study, a *Tt* isolate from *Pseudotsuga menziesii* efficiently colonised *Pinus radiata* seedlings but was incompatible with 11 *Eucalyptus* species (Malajczuk et al. 1982). The authors concluded that several fungal associates of pine, such as *Tt*, were absent from *Eucalyptus* stands and vice versa, suggesting that both pine and eucalypt fungi exhibit substantial host specificity. However, fructification of *Tt* and typical *Tt* mycorrhizas have been observed on seedlings of *Eucalyptus dalrympleana* and *Eucalyptus macarthuri* in France (Mousain et al. 1977). Recently, *Tt* was also encountered in a glasshouse experiment in which *E. globulus* seedlings were inoculated with other compatible fungal isolates from South Australia and Tasmania (Ingleby and Mason 1996). It is possible that some *Tt* isolates infect exotic host genera. Generally, however, in laboratory experiments where *Tt* has been inoculated in vitro on *Eucalyptus*, infection percentages were rather low and growth stimulation and nutrient absorption were very poor (Dixon and Hiol-Hiol 1992a,b).

14.5.3
Developmental Aspects of the Fungus–Host Interaction

Birraux and Fries (1981) suggested that host plant roots seemed to release one or several compounds which induced spore germination in *Tt*. In their study, germination occurred when compatible roots and spores were separated by a dialysis membrane with a pore size of 2.4 nm, indicating that the molecular weight of these compounds was below 17000. The products could diffuse through agar, were relatively stable and not highly volatile; however, their chemical nature was not elucidated. *Tt* is known to produce substantial amounts of cytokinin-like substances both in liquid culture and in the presence of *P. abies* roots, although large variation exists between different *Tt* isolates (Wullschleger and Reid 1990; Kraigher et al. 1993). While root cultures of *P. abies* produced only traces of cytokinins, Kraigher et al. (1991) identified four predominant cytokinins produced by *Tt*. The production of three of these was increased in the presence of *P. abies* roots. Seedlings of *P. abies* inoculated with *Tt* also had considerably higher levels of cytokinins in their cotyledons than non-inoculated seedlings (Kraigher et al. 1993). The increase was probably partly due to an indirect effect of the mycobiont on the cytokinin synthesis in the needles. The ability of the fungus to affect host needle cytokinins was not dependent upon an ability to synthesise cytokinins in pure culture (Wullschleger and Reid 1990).

Gibberellin-like compounds have been detected in culture filtrates of *Tt* (Hanley and Greene 1987). Although their role in development and funct-ioning of the symbiosis is not clear, they may have an effect on the carbon allocation between the symbionts. Auxins are thought to influence the coloni-sation potential of ECM fungi as well as the morphogenesis of mycorrhizas (Ek et al. 1983). *Tt* produces IAA in in vitro cultures, although in soil it is pos-sible that other organisms in the rhizosphere enhance this process, produc-ing IAA and perhaps stimulating mycorrhiza formation (Garbaye 1994). However, the mechanisms involved in the interactions between bacteria and mycorrhizal development are complex and partly fungus-specific, since myc-orrhiza helper bacteria (MHB) from *Laccaria laccata* (Scop.: Fr.) Cooke ECM did not improve mycorrhiza formation and growth of *P. menziesii* colonised with *Tt* (Duponnois and Garbaye 1991; Garbaye and Duponnois 1992). Growth-promoting MHB typically associated with *Tt* have not been studied (Garbaye 1994), although Richter et al. (1989) isolated mycorrhizoplane-asso-ciated actinomycetes that stimulated growth of *Tt* in vitro. Conversely, Bowen and Theodorou (1979) found that bacteria did not stimulate the colonisation potential of *Tt*. Indeed, the presence of external mycelia of *Tt* has been shown to significantly reduce total bacterial activity in a sandy soil (Olsson et al. 1996).

Using an elicitor from *Tt*, Campbell and Ellis (1992a,b) showed enhanced phenolic metabolism in suspension-cultured cells of *Pinus banksiana*, as indicated by tissue lignification and accumulation of specific methanol-extractable compounds in the cells. Significant increases in activity of pheny-lalanine ammonia-lyase (PAL), the key enzyme of the phenylpropanoid metabolism, were also evident 6h after elicitation. Induction of this enzyme is a general defense mechanism in plant pathogenic relationships and it is questionable whether a strong activation of PAL also occurs when living mycelium colonises a pine root. Downregulation of plant defense mechanisms is more likely to occur during mycorrhiza formation.

14.5.4
Symbiotic Functioning

The mantles in *Tt* ECM are not water repellent (Unestam 1991; Unestam and Sun 1995), implying that apoplastic transport of water and soil solutes in the mantle and in the root cortex may be possible. If this is the case, then the fungus probably cannot fully control nutrient uptake in its host. Recent unpublished data from our laboratory indicate that intact pine root systems colonised with *Tt* have a lower affinity (high K_m) for uptake of phosphate than several other *Pinus*-ECM [*Paxillus involutus* (Batsch) Fr., *S. luteus, S. bovinus* (L.: Fr.) Roussel] associations. However, at higher external nutrient concentrations, the nutrient-absorbing capacity of *Tt*-colonised roots was comparable or even higher than those from the other plant–fungus associations studied. Only part of the nutrients assimilated by *Tt* are translocated to the host and considerable amounts of minerals are invested in fungal structures, in particular sporocarps (Colpaert et al. 1996). Jones et al. (1991) suggested that a temporal separation existed in the maximal fluxes of P and C between *Tt* and a host, *S. viminalis*. In the early stages of the infection relatively more P was supplied to the host than in the later stages, when carbon allocation to the fungus was larger (Durall et al. 1994b). The copious production of sporocarps that can occur over short periods in nurseries (Marx et al. 1984) can only be realised in relatively fertile soils with a high inorganic nutrient availability which support seedlings with a high carbon assimilation rate. Sporocarp production and thus nutrient retention in *Tt* mycelia colonising *P. sylvestris* seedlings was found to be larger at a relative growth rate (R_G) of 2.9% than at a lower R_G (Colpaert et al. 1996). Although the growth of most ECM fungi is stimulated when their hosts are nutrient limited (Wallander and Nylund 1992; Wallander 1995), a *Tt* isolate used in our laboratory has its highest hyphal substrate densities at a relatively high nutrient availability (J. V. Colpaert, unpubl. data). Furthermore, it has been reported elsewhere that in nurseries percentage infection by *Tt* can increase with increasing nutrient supply (Guehl and Garbaye 1990).

   Tt grows poorly on media containing sucrose. The addition of starter glucose can improve sucrose utilisation, but growth is still less than that on equivalent amounts of glucose (Hughes and Mitchell 1995). It therefore seems likely that *Tt* relies on apoplastic invertase activity for hydrolysis of sucrose effluxed from the plant.

14.6
Host Plant Growth Responses and Fungus-Derived Benefits

14.6.1
Growth Responses

Tt can rapidly colonise the whole root systems of compatible host plant species. A few weeks after inoculation all short roots can be infected, meaning that seedlings can react rapidly to colonisation. Growth responses to *Tt* infection can be positive or negative, depending on the carbon use efficiency of the mycorrhizal plants (Tinker et al. 1994). In glasshouse trials, there are many reports of *Tt*-mediated growth responses compared with uninoculated control plants (Thomas et al. 1982; Thomas and Jackson 1983). Percentage colonisation by *Tt* and host growth responses are, however, reduced as the soil P concentration increases (Ford et al. 1985; Jones et al. 1990). While intraspecific differences in growth stimulation have been observed in other fungal species (Burgess et al. 1994), there have been few comparative studies using multiple *Tt* isolates. Holden et al. (1983), however, compared two *Tt* isolates from a nursery and a forest. Although both isolates had a similar infection potential, the forest isolate had a significantly higher growth-promoting capacity. In contrast, inoculation in containers and field testing of *Picea sitchensis* and *Pseudotsuga menziesii* with five isolates of *Tt* indicated no intraspecific differences in growth responses (Jackson et al. 1995). Different isolates may also have differential effects on different host taxa; for example, Trappe (1977) found that an isolate of *Tt* markedly increased growth of *Tsuga heterophylla*, but that the same isolate decreased growth of *Pinus ponderosa* and *Pseudotsuga menziesii* under the same conditions. A *Tt* isolate from *Picea* has, however, been used to adapt *Populus* plants from in vitro tissue culture conditions to the glasshouse (Heslin and Douglas 1986). Although survival of the *Tt*-inoculated plants was lower than for uninoculated plants, high rates of colonisation were obtained as well as a doubling of shoot dry weight. It is noteworthy also that *Tt* often infects non-mycorrhizal control plants in growth studies, meaning that many workers have unwittingly compared host growth in the presence of *Tt* with other ECM taxa (Tyminska et al. 1986; Perry et al. 1989; Guehl and Garbaye 1990). In such experiments *Tt* generally enhanced host growth as well as other ECM fungi such as *Hebeloma*, *Laccaria* and *Pisolithus* (Bledsoe 1992; Le Tacon et al. 1992; Jackson et al. 1995).

Perry et al. (1989) found that *Tt* had a negative effect on competition between *Pinus ponderosa* and *Pseudotsuga menziesii*. When grown in equal mixture, there was mutual antagonism between the tree species when *Tt* was the mycobiont. In combination with *Rhizopogon*, *Laccaria* and *Hebeloma*, competition was reduced. In semi-hydroponic systems, *Tt* generally reduces host growth, but this is to be expected since nutrient uptake cannot be increased due to complete availability of the nutrients to all plants. The retention of nutrients by the fungus for its own metabolism or the increased belowground C allocation always result in small growth reductions in the host plants (Colpaert et al. 1996).

Tt is well adapted to nursery conditions and thus only vigorous pioneer (early-stage) fungi seem able to compete with *Tt* in the nursery or glasshouse (Fleming 1985; Tyminska et al. 1986; Henrion et al. 1994). For example, inoculation of pine with *Pisolithus tinctorius* (Pers.) Coker & Couch is more efficient when the nursery soil is fumigated, while *Tt* can be much more abundant in non-fumigated soil (Ruehle 1983). Nevertheless, in some nurseries seedlings inocluated with other ECM fungi can grow better than when naturally colonised with *Tt* (Marx et al. 1984; Le Tacon et al. 1985). It is also interesting to note that a dual, simultaneous co-inoculation of *Pinus patula* seedlings with *Tt* and *L. laccata* in a nursery resulted in better plant growth compared to individual inoculation (Sudhakara Reddy and Natarajan 1997). Fungicides may potentially affect ECM development in forest nurseries and in new plantations, and Marx et al. (1986) have shown that the systemic fungicide triadimefon can have negative effects on ECM development and basidiocarp formation by *Tt* in nurseries.

In the field, *Tt* generally becomes a poor competitor, particularly when outplanting sites contain indigenous populations of mycorrhizal fungi (Villeneuve et al. 1991; Smith and Read 1997). It is also possible that, in some instances, the growth of *Tt* is directly affected by phenolic compounds in soil or by allelopathic effects from other plants (Coté and Thibault 1988). There are, however, reports that Tt can enhance host growth in the field, for instance on poor organic soil or and on a site reforested with *P. menziesii* that was previously planted with *Larix* sp. (Le Tacon et al. 1992).

Tt cannot survive, even in the absence of competing species, on adverse sites subject to high acidity, high temperatures and heavy metals. In such situations responses to inoculation with better-adapted species such as *P. tinctorius* are generally better than with *Tt* (Smith and Read 1997). In vitro tests have confirmed the relatively low tolerance of most *Tt* isolates to high temperatures (Cline et al. 1987). In general, growth stimulation in *Tt*-inoculated plants is reported only in cool, temperate, and often wet climates. Nevertheless, *Tt* does occur in nurseries in warm climates (e.g. in South Africa, Kenya, Uganda and Puerto Rico) (Mikola 1970). Not all adverse environments for tree growth are, however, unsuitable habitats for *Tt* and some authors believe that growth-promoting *Tt* isolates should be included in inoculation programmes since they can provide benefits to seedlings planted on reforestation sites that lack symbiotic fungi (Kropp and Langlois 1990). In this regard, *Tt* is found in association with pine seedlings on black wastes from anthracite mines in Pennsylvania (Schramm 1966) and the species established and persisted with varying success on *P. banksiana* planted on amended oil sands tailings in Canada in conditions where growth of other fungi was poor (Danielson and Visser 1989; Danielson 1991). After two growing seasons, *Tt* seedlings in outplanting trials on the latter site were significantly heavier than uninoculated plants or seedlings inoculated with seven other ECM fungi. The degree of infection by *Tt*, however, decreased rapidly after three growing seasons, when indigenous fungi took over.

14.6.2
Nutritional Benefits to the Host

There are several reports that *Tt* can increase foliar P, N and K concentrations in seedlings compared with uninoculated controls or plants inoculated with some other ECM fungi (Thomas et al. 1982; Ford et al. 1985; Heslin and Douglas 1986; Coleman et al. 1990). Rapid P uptake in *Tt*-colonised willow resulted in a significant increase in P concentration in shoots and roots during the first weeks of colonisation (Jones et al. 1991). The P inflow rate in the willows was almost three times higher for ECM plants than for uninfected plants. Improved uptake of micro-nutrients has also been reported for oak species inoculated with *Tt* (Mitchell et al. 1984).

Tt can grow on different organic and inorganic N sources in axenic culture, although considerable variation exists between different isolates (Read et al. 1989; Finlay et al. 1992). Thus, growth on NH_4^+ or single amino acids is generally excellent, whereas growth on nitrate or soluble protein can vary from poor to very good. When grown in association with a host plant, *Tt* can assimilate NH_4^+ and thus improve host growth when N is a growth limiting factor. When NH_4^+ is replaced by a soluble protein, growth and N uptake in the host become very low, at least when an isolate is used which exhibits no protease activity in vitro (Finlay et al. 1992). *Tt* also appears to have little access to N present in complex organic matter from a fermentation horizon, to keratin or to more recalcitrant leaf litter (Bending and Read 1995; Colpaert and Van Tichelen 1996; Seith et al. 1996). Despite the N limitation, protease activity in the latter subtrate was very low for all ECM fungi tested, which seems to confirm that soluble peptides must be present for good enzyme induction (Colpaert and Van Laere 1996). Although it is tempting to suggest that pioneer fungi such as *Tt* have little access to more complex organic substrates, this should be done with caution. Some *Tt* strains exhibit proteolytic activity (Finlay et al. 1992) and it remains to be demonstrated that these ECM enzymes contribute in situ to the mineral nutrition of the fungi and their hosts (Dighton 1991). The apparently poor ability of *Tt* to degrade complex organic substrates, however, will at least restrict its access to organic N and P compounds. Some carbohydrate hydrolysing enzymes (e.g. β-glucosidases) are expressed by *Tt* and may be involved in establishment of the symbiosis.

Extracellular acid phosphatases are produced by mycorrhizas, rhizomorphs and external mycelia of *Tt* (Dinkelaker and Marschner 1992; Colpaert et al. 1997). The production of these enzymes seems to be controlled by the P status of the symbionts but is not induced in the presence of organic P. Extracellular phosphatases have little effect on poorly water-soluble organic P compounds, but more labile water-soluble organic P molecules are probably good substrates for these enzymes (Thomas et al. 1982; Colpaert et al. 1997).

14.6.3
Influence on Host Plant Carbon Economy

The effect of *Tt* on the C economy of the host appears to depend on a number of factors, including host plant species and environmental stresses. *Tt* has been shown, for example, to have a negative effect on the growth of *Eucalyptus* (associated with relatively low CO_2 assimilation), perhaps related to poor compatibility in combination with drought stress (Dixon and Hiol-Hiol 1992a). Jones et al. (1991) reported a two-fold increase in growth (associated with improved host P nutrition) and that 2.5 times more C was allocated below ground in *S. viminalis* when infected by *Tt*. In *P. sylvestris* seedlings grown at a low R_G under N limitation, almost 30% of net assimilated C was respired by the *Tt*-colonised root system compared to only 15% of a non-mycorrhizal root system (Colpaert et al. 1996). The increased below-ground respiration results from the presence of the mycelium in mycorrhizas, external hyphae and sporocarps. Net photosynthesis in the needles of the *Tt*-colonised plants was not signifcantly increased, probably due to lower N in mycorrhizal plants. Reid and Woods (1969) showed that ^{14}C fed to pine shoots was readily translocated to *Tt* mycorrhizas and to external hyphae up to a distance of 12 cm from the ECM tips. They further noted that movement of carbon between different root systems could take place.

14.6.4
Non-Nutritional Benefits to the Host

Tt appears to have a high tolerance to waterlogging. Short periods of flooding (up to a few hours) seem not to harm the fungus (Stenström 1991). Waterlogging over long periods during winter (149 days) can destroy external mycelium, except the rhizomorphs which retain their structure and thus provide a skeletal framework from which new hyphae can regenerate after draining (Coutts and Nicoll 1990b). The flooding tolerance of *Tt* may partly explain why the fungus flourishes in nurseries and forests where plants are flooded periodically. In contrast, *Tt* is not very drought tolerant (Dixon and Hiol-Hiol 1992a), at least when associated with *Eucalyptus* sp. *Tt* was, however, more drought tolerant than several other ECM fungi associated with *Picea* or *Pseudotsuga* (Coleman et al. 1990; Guehl et al. 1992; Lehto 1992). George and Marschner (1996) found that the external mycelium of *Tt* could not contribute significantly to plant water uptake in *P. abies* plants subjected to drought stress.

   Axenic culture studies indicate different degrees of sensitivity of *Tt* isolates towards most heavy metals (McCreight and Schroeder 1982; Colpaert and Van Assche 1987; Jones and Muehlchen 1994; Tam 1995), but in association with pine even isolates sensitive in vitro can be more tolerant than most other ECM fungi. In the presence of high Zn levels, for example, *Tt* survived best when compared to five other ECM fungi (Colpaert and Van Assche 1992). However, the *Tt* mycelium had a very low specific Zn-retaining capacity and transport of Zn to the shoots was significantly higher than in uninoculated plants or in any other pine–ECM association. In a recent experiment (K. K. Van Tichelen and J. V. Colpaert, unpubl. data), *Tt* associated with pine appeared relatively insensitive to Cu. At an external Cu concentration of 47 μM, growth of external mycelium and mycorrhizas was little affected, whereas all uninoculated roots of control plants did not survive. As with Zn, however, Cu transport to shoots was again higher in the *Tt*-inoculated plants. In contrast, Tt does not grow well in the presence of Cd (Colpaert and Van Asssche 1993). The above illustrates that nursery seedlings colonised with *Tt* probably cannot protect their host plant against heavy metal toxicity when planted on a strongly polluted soil. *Tt* is seldom reported in sporocarp surveys from metal-polluted sites, even in pioneer forest where competition from other ECM fungi is initially low (Rühling et al. 1984; Denny and Wilkins 1987; Rühling and Söderström 1990). Since *Tt* is a rapid coloniser of seedlings in nurseries, it probably competes with many feeder root pathogens for colonisation of the uninfected roots. Marx and Davey (1969) showed that *Tt* associated with *Pinus* spp. provided effective control of *Phytophthora cinnamomi* Rands infection and Zak and Ho (1994) report that *Tt* reduced root rot by *Rhizina undulata* Fr. in *Pseudotsuga menziesii*. Such protection may be afforded, as suggested by Marx (1972), by the penetration barrier of the ECM mantle.

References

Acsai J, Largent DL (1983) Fungi associated with *Arbutus menziesii, Arctostaphylos manzanita,* and *Arctostaphylos uva-ursi* in central and northern California. Mycologia 75:544–547

Agerer R (1988) Studies on ectomycorrhizae. XVII. The ontogeny of the ectomycorrhizal rhizomorphs of *Paxillus involutus* and *Thelephora terrestris* (Basidiomycetes). Nova Hedwigia 47:311–334

Agerer R (1991) Characterisation of ectomycorrhiza. In: Norris JR, Read DJ, Varma AK (eds) Methods in microbiology, vol 23. Academic Press, London, pp 25–73

Agerer R (1994) *Pseudotomentella tristis* (Thelephoraceae): eine Analyse von Fruchtkörper und Ektomykorrhizen. Z Mykol 60:143–157

Agerer R (1996) Ectomycorrhizae of *Tomentella albomarginata* (Thelephoraceae) on Scots pine. Mycorrhiza 6:1–7

Agerer R, Weiss M (1989) Studies on ectomycorrhizae. XX. Mycorrhizae formed by *Thelephora terrestris* on Norway spruce. Mycologia 81:444–453

Allen MF (1987) Re-establishment of mycorrhizas on Mount St. Helens: migration vectors. Trans Br Mycol Soc 88:413–417

Arnolds E, Jansen E (1992) New evidence for changes in the macromycete flora of the Netherlands. Nova Hedwigia 55:325–351

Arnolds E, Kuyper ThW, Noordeloos ME (1995) Overzicht van de paddestoelen in Nederland. (A survey of the macrofungi in the Netherlands.) Nederlandse Mycologische Vereniging, Wijster 872 pp

Baici A, Ricci G, Zecchin G (1995) Notes on *Thelephora atrocitrina* and *T. cuticularis*. Mycol Helv 7:71–81

Bending GD, Read DJ (1995) The structure and function of the vegetative mycelium of ectomycorrhizal plants. V. Foraging behaviour and translocation of nutrients from exploited organic matter. New Phytol 130:401–409

Birraux D, Fries N (1981) Germination of *Thelephora terrestris* basidiospores. Can J Bot 59:2062–2064

Bledsoe CS (1992) Physiological ecology of ectomycorrhizae: implications for field application. In: Allen MJ (ed) Mycorrhizal functioning. An integrative plant–fungal process. Chapman & Hall, New York, pp 424–437

Bowen GD, Theodorou C (1979) Interactions between bacteria and ectomycorrhizal fungi. Soil Biol Biochem 11:119–126

Burgess T, Dell B, Malajczuk N (1994) Variation in mycorrhizal development and growth stimulation by 20 *Pisolithus* isolates inoculated on to *Eucalyptus grandis* W. Hill ex Maiden. New Phytol 127:731–739

Campbell MM, Ellis BE (1992a) Fungal elicitor-mediated responses in pine cell cultures. I. Induction of phenylpropanoid metabolism. Planta 186:409–417

Campbell MM, Ellis BE (1992b) Fungal elicitor-mediated responses in pine cell cultures. II. Cell wall-bound phenolics. Phytochemistry 31:737–742

Chan WK, Griffiths DA (1991) The induction of mycorrhiza in *Eucalyptus microcorys* and *Eucalyptus torelliana* grown in Hong Kong. For Ecol Manage 43:15–24

Chu-Chou M (1979) Mycorrhizal fungi of *Pinus radiata* in New Zealand. Soil Biol Biochem 11:557–562

Chu-Chou M, Grace LJ (1983) Characterisation and identification of mycorrhizas of Douglas fir in New Zealand. Eur J For Pathol 13:251–260

Cline ML, France RC, Reid CPP (1987) Intraspecific and interspecific growth variation of ectomycorrhizal fungi at different temperatures. Can J Bot 65:869–875

Coleman MD, Bledsoe CS, Smit BA (1990) Root hydraulic conductivity and xylem sap levels of zeatin riboside and abscisic acid in ectomycorrhizal Douglas fir seedlings. New Phytol 115:275–284

Colpaert JV, Van Assche JA (1987) Heavy metal tolerance in some ectomycorrhizal fungi. Funct Ecol 1:415–421

Colpaert JV, Van Assche JA (1992) Zinc toxicity in ectomycorrhizal *Pinus sylvestris*. Plant Soil 143:201–211

Colpaert JV, Van Assche JA (1993) The effects of cadmium on ectomycorrhizal *Pinus sylvestris*. New Phytol 123:325–333

Colpaert JV, Van Tichelen KK (1996) Decomposition, nitrogen and phosphorus mineralization from beech leaf litter colonized with ectomycorrhizal or litter decomposing basidiomycetes. New Phytol 134:123–132

Colpaert JV, Van Assche JA, Luijtens K (1992) Relationship between the growth of the extramatrical mycelium of ectomycorrhizal fungi and the growth response of *Pinus sylvestris* plants. New Phytol 120:127–135

Colpaert JV, Van Laere A, Van Assche JA (1996) Carbon and nitrogen allocation between symbionts in ectomycorrhizal and non-mycorrhizal *Pinus sylvestris* L. seedlings. Tree Physiol 16:787–793

Colpaert JV, Van Laere A, Van Tichelen KK, Van Assche JA (1997) The use of inositol hexaphosphate as a phosphorus source by mycorrhizal and nonmycorrhizal Scots pine (*Pinus sylvestris*). Funct Ecol 11:407–415

Corbery Y, Le Tacon F (1997) Storage of ectomycorrhizal fungi by freezing. Ann Sci For 54:211–217

Corner EJH (1968) A monograph of *Thelephora*. Beih Nova Hedwigia 27:1–110

Corner EJH (1976) Further notes on cantharelloid fungi and *Thelephora*. Nova Hedwigia 27:325–342

Coté J-F, Thibault J-R (1988) Allelopathic potential of raspberry foliar leachates on growth of ectomycorrhizal fungi associated with black spruce. Am J Bot 75:966–970

Coutts MP, Nicoll BC (1990a) Growth and survival of shoots, roots, and mycorrhizal mycelium in clonal Sitka spruce during the first growing season after planting. Can J For Res 20:861–868

Coutts MP, Nicoll BC (1990b) Waterlogging tolerance of roots of Sitka-spruce clones and of strands from *Thelephora terrestris* mycorrhizas. Can J For Res 20:1894–1899

Danielson RM (1991) Temporal changes and effects of amendments on the occurrence of sheathing (ecto-) mycorrhizas of conifers growing in oil sands tailings and coal spoil. Agric Ecosyst Environ 35:261–281

Danielson RM, Visser S (1989) Host response to inoculation and behaviour of introduced and indigenous ectomycorrhizal fungi of jack pine grown on oil-sands tailings. Can J For Res 19:1412–1421

Danielson RM, Visser S (1990) The mycorrhizal and nodulation status of container-grown trees and shrubs reared in commercial nurseries. Can J For Res 20:609–614

Deacon JW, Fleming LV (1992) Interactions of ectomycorrhizal fungi. In: Allen MJ (ed) Mycorrhizal functioning. An integrative plant–fungal process. Chapman & Hall, New York, pp 249–300

Denny HJ, Wilkins DA (1987) Zinc tolerance in *Betula* spp. III. Variation in response to zinc among ectomycorrhizal associates. New Phytol 106:535–544

Dighton J (1991) Acquisition of nutrients from organic resources by mycorrhizal autotrophic plants. Experientia 47:362–369

Dinkelaker B, Marschner H (1992) In vivo demonstration of acid phosphatase activity in the rhizosphere of soil-grown plants. Plant Soil 144:199–205

Dixon RK, Hiol-Hiol F (1992a) Gas exchange and photosynthesis of *Eucalyptus camadulensis* seedlings inoculated with different ectomycorrhizal symbionts. Plant Soil 147:143–149

Dixon RK, Hiol-Hiol F (1992b) Mineral nutrition of *Pinus caribea* and *Eucalyptus camaldulensis* seedlings inoculated with *Pisolithus tinctorius* and *Thelephora terrestris*. Commun Soil Sci Plant Anal 23:1387–1396

Duponnois R, Garbaye J (1991) Mycorrhization helper bacteria associated with the Douglas fir–*Laccaria laccata* symbiosis: effects in aseptic and in glasshouse conditions. Ann Sci For 48:239–252

Durall DM, Jones MD, Tinker PB (1994b) Allocation of ^{14}C-carbon in ectomycorrhizal willow. New Phytol 128:109–114

Ek M, Ljungquist PO, Stenström E (1983) Indole-3-acetic acid production by mycorrhizal fungi determined by gas chromatography–mass spectrometry. New Phytol 94:401–407

Finlay RD, Frostegard A, Sonnerfeldt A-M (1992) Utilization of organic and inorganic nitrogen sources by ectomycorrhizal fungi in pure culture and in symbiosis with *Pinus contorta* Dougl. ex Loud. New Phytol 120:105–115

Fleming LV (1985) Experimental study of sequences of ectomycorrhizal fungi on birch (*Betula* sp.) seedling root systems. Soil Biol Biochem 17:591–600

Ford VL, Torbert JL jr, Burger JA, Miller OK (1985) Comparative effects of four mycorrhizal fungi on loblolly pine seedlings growing in a greenhouse in a Piedmont soil. Plant Soil 83:215–221

Fries N (1978) Basidiospore germination in some mycorrhiza-forming Hymenomycetes. Trans Br Mycol Soc 70:319–324

Fries N (1987) Somatic incompatibility and field distribution of the ectomycorrhizal fungus *Suillus luteus* (Boletaceae). New Phytol 107:735–739

Garbaye J (1982) Quelques aspects de la compétitivité des souches ectomycorhiziennes. In: Gianinazzi S, Gianinazzi-Pearson V, Trouvelot A (eds) Les mycorhizes, partie intégrante de la plante: biologie et perspectives d'utilisation. INRA, Paris, pp 303–312

Garbaye J (1994) Helper bacteria: a new dimension to the mycorrhizal symbiosis. New Phytol 128:197–210

Garbaye J, Duponnois R (1992) Specificity and function of mycorrhization helper bacteria (MHB) associated with the *Pseudotsuga menziesii–Laccaria laccata* symbiosis. Symbiosis 14:335–344

Garbaye J, Menez J, Wilhelm ME (1986) Les mycorhizes des jeunes chênes dans les pépinières et les régénérations naturelles du nord-est de la France. Acta Oecol Oecol Plant 7:87–96

Gardes M, Bruns TD (1996) Community structure of ectomycorrhizal fungi in a *Pinus muricata* forest: above- and below-ground views. Can J Bot 74:1572–1583

George E, Marschner H (1996) Nutrient and water uptake by roots of forest trees. Z Pflanzenernähr Bodenkd 159:11–21

Godbout C, Fortin JA (1983) Morphological features of synthesised ectomycorrhizae of *Alnus crispa* and *A. rugosa*. New Phytol 94:249–262

Godbout C, Fortin JA (1985) Synthesized ectomycorrhizae of aspen: fungal genus level of structural characterization. Can J Bot 63:252–262

Gogala N (1991) Regulation of mycorrhizal infection by hormonal factors produced by hosts and fungi. Experientia 47:331–340

Grogan HM, O'Neill JJM, Mitchell DT (1994) Mycorrhizal associations of Sitka-spruce seedlings propagated in Irish tree nurseries. Eur J For Pathol 24:335–344

Guehl JM, Garbaye J (1990) The effects of ectomycorrhizal status on carbon dioxide assimilation capacity, water-use efficiency and response to transplanting in seedlings of *Pseudotsuga menziesii* (Mirb) Franco. Ann Sci For 21:551–563

Guehl JM, Garbaye J, Wartinger A (1992) The effects of ectomycorrhizal status on plant–water relations and sensitivity of leaf gas exchange to soil drought in Douglas fir (*Pseudotsuga menziesii*) seedlings. In: Read DJ, Lewis DH, Fitter AH, Alexander IJ (eds) Mycorrhizas in ecosystems. CAB International, Wallingford, pp 323–332

Hanley KM, Greene DW (1987) Gibberellin-like compounds from two ectomycorrhizal fungi and the GA$_3$ response on Scots pine seedlings. Hortic Sci 22:591–594

Henrion B, Di Battista C, Bouchard D, Vairelles D, Thompson BD, Le Tacon F, Martin F (1994) Monitoring the persistence of *Laccaria bicolor* as an ectomycorrhizal symbiont of nursery-grown Douglas fir by PCR of the rDNA intergenic spacer. Mol Ecol 3:571–580

Heslin MC, Douglas GC (1986) Effects of ectomycorrhizal fungi on growth and development of poplar plants derived from tissue culture. Sci Hortic 30:143–149

Hibbett DS, Pine EM, Langer E, Langer G, Donoghue MJ (1997) Evolution of gilled mushrooms and puffballs inferred from ribosomal DNA sequences. Proc Natl Acad Sci USA 94:12002–12006

Holden JM, Thomas GW, Jackson RM (1983) Effect of mycorrhizal inocula on the growth of Sitka spruce seedlings in different soils. Plant Soil 71:313–317

Hughes E, Mitchell DT (1995) Utilization of sucrose by *Hymenoscyphus ericae* (an ericoid endomycorrhizal fungus) and ectomycorrhizal fungi. Mycol Res 99:1233–1238

Ingleby K, Mason PA (1996) Ectomycorrhizas of *Thelephora terrestris* formed with *Eucalyptus globulus*. Mycologia 88:548–553

Ingleby K, Mason PA, Last FT, Fleming LV (1990) Identification of ectomycorrhizas. Institute of Terrestrial Ecology, Natural Environment Research Council, HMSO, London

Ivory MH, Munga FM (1992) Ectomycorrhizal fungi in Kenya. In: Read DJ, Lewis DH, Fitter AH, Alexander IJ (eds) Mycorrhizas in ecosystems. CAB International, Wallingford, 383 pp

Jackson RM, Walker C, Luff S, McEvoy (1995) Inoculation and field testing of Sitka spruce and Douglas fir with ectomycorrhizal fungi in the United Kingdom. Mycorrhiza 5:165–173

Jansen AE (1991) The mycorrhizal status of Douglas fir in the Netherlands: its relation with stand age, regional factors, atmospheric pollutants and tree vitality. Agric Ecosyst Environ 35:191–208

Jones D, Muehlchen A (1994) Effects of the potentially toxic metals, aluminium, zinc and copper on ectomycorrhizal fungi. J Environ Sci Health A29:949–966

Jones MD, Durall DM, Tinker PB (1990) Phosphorus relationships and production of extramatrical hyphae by two types of willow ectomycorrhizas at different soil phosphorus levels. New Phytol 115:259–267

Jones MD, Durall DM, Tinker PB (1991) Fluxes of carbon and phosphorus between symbionts in willow ectomycorrhizas and their changes with time. New Phytol 119:99–106

Kåren O, Nylund J-E (1997) Effects of ammonium sulphate on the community structure and biomass of ectomycorrhizal fungi in a Norway spruce stand in southwestern Sweden. Can J Bot 75:1628–1642

Keizer PJ, Arnolds E (1994) Succession of ectomycorrhizal fungi in roadside verges planted with common oak (*Quercus robur* L.) in Drenthe, the Netherlands. Mycorrhiza 4:147–159

Kernaghan G, Berch S, Carter R (1995) Effect of urea fertilization on ectomycorrhizae of 20-year-old *Tsuga heterophylla*. Can J For Res 25:891–901

Kraigher H, Grayling A, Wang TL, Hanke DE (1991) Cytokinin production by two ectomycorrhizal fungi in liquid culture. Phytochemistry 30:2249–2254

Kraigher H, Strnad M, Hanke DE, Batic F (1993) Cytokiningehalte von Fichtennadeln (*Picea abies* [L.] Karst.) nach Inokulation mit zwei Stämmen des Mykorrhizapilzes *Thelephora terrestris* (Ehrh.) Fr. Forstwiss Centralbl 112:107–111

Kropp BR, Langlois GC (1990) Ectomycorrhizae in reforestation. Can J For Res 20:438–451

Kropp BR, Trappe JM (1982) Ectomycorrhizal fungi of *Tsuga heterophylla*. Mycologia 74:479–488

Last FT, Dighton J, Mason PA (1987) Successions of sheating mycorrhizal fungi. Trends Ecol Evol 2:157–161

Lehto T (1992) Effect of drought on *Picea sitchensis* seedlings inoculated with mycorrhizal fungi. Scand J For Res 7:177–182

Le Tacon F, Jung G, Mugnier J, Michelot P, Mauperin C (1985) Efficiency in a forest nursery of an ectomycorrhizal fungus inoculum produced in a fermentor and entrapped in polymeric gels. Can J Bot 63:1664–1668

Le Tacon F, Garbaye J, Carr G (1987) The use of mycorrhizas in temperate and tropical forests. Symbiosis 3:179–206

Le Tacon F, Alvarez IF, Bouchard D, Henrion B, Jackson RM, Luff S, Parlade JI, Pera J, Stenström E, Villeneuve N, Walker C (1992) Variations in field response of forest trees to nursery ectomycorrhizal inoculation in Europe. In: Read DJ, Lewis DH, Fitter AH, Alexander IJ (eds) Mycorrhizas in ecosystems. CAB International, Wallingford, pp 119–134

Malajczuk N, Molina R, Trappe JM (1982) Ectomycorrhiza formation in *Eucalyptus*. I. Pure culture synthesis, host specificity and mycorrhizal compatibility with *Pinus radiata*. New Phytol 91:467–482

Malloch D, Thorn RG (1985) The occurrence of ectomycorrhizae in some species of cistaceae in North America. Can J Bot 63:872–875

Marx DH (1972) Ectomycorrhizae as biological deterrents to pathogenic root infections. Annu Rev Phytopathol 10:429–454

Marx DH, Bryan WC (1969) Studies on ectomycorrhizae of pine in an electronically air-filtered, air-conditioned plant growth room. Can J Bot 47:1903–1909

Marx DH, Bryan WC (1970) Pure culture synthesis of ectomycorrhizae by *Thelephora terrestris* and *Pisolithus tinctorius* on different conifer hosts. Can J Bot 48:639–643

Marx DH, Bryan WC (1971) Influence of ectomycorrhizae on survival and growth of aseptic seedlings of loblolly pine at high temperature. For Sci 17:37–41

Marx DH, Davey CB (1969) The influence of ectotrophic mycorrhizal fungi on the resistance of pine roots to pathogenic infections. IV. Resistance of naturally occurring mycorrhizae to infections by *Phytophthora cinnamomi*. Phytopathology 59:559–574

Marx DH, Ross EW (1970) Aseptic synthesis of ectomycorrhizae on *Pinus taeda* by basidiospores of *Thelephora terrestris*. Can J Bot 48:197–198

Marx DH, Bryan WC, Grand LF (1970) Colonization, isolation, and cultural descriptions of *Thelephora terrestris* and other ectomycorrhizal fungi of shortleaf pine seedlings grown in fumigated soil. Can J Bot 48:207–211

Marx DH, Cordell CE, Kenney DS, Mexal JG, Artman JD, Riffle JW, Molina RJ (1984) Commercial vegetative inoculum of *Pisolithus tinctorius* and inoculation techniques for development of ectomycorrhizae on bare-root tree seedlings. For Sci Monogr 25:101

Marx DH, Cordell CE, France RC (1986) Effects of triadimefon on growth and ectomycorrhizal development of loblolly and slash pines in nurseries. Phytopathology 76:824–831

Mason PA, Last FT, Pelham J, Ingleby K (1982) Ecology of some fungi associated with an ageing stand of birches (*Betula pendula* and *B. pubescens*). For Ecol Manage 4:19–39

Mason PA, Wilson J, Last FT, Walker C (1983) The concept of succession in relation to the spread of sheathing mycorrhizal fungi on inoculated tree seedlings growing in unsterile soils. Plant Soil 71:247–256

McCreight, JD, Schroeder, DB (1982) Inhibition of growth of nine ectomycorrhizal fungi by cadmium, lead and nickel in vitro. Environ Exp Bot 22:1–7

Mikola P (1970) Mycorrhizal inoculation in afforestation. Int Rev For Res 3:123–196

Miller SL, Koo CD, Molina R (1991) Characterization of red alder ectomycorrhizae: a preface to monitoring belowground ecological responses. Can J Bot 69:516–531

Mitchell RJ, Cox GS, Dixon RK, Garret HE, Sander IL (1984) Inoculation of three *Quercus* species with eleven isolates of ectomycorrhizal fungi. II. Foliar nutrient content and isolate effectiveness. For Sci 30:563–572

Mohan V, Natarajan K, Ingleby K (1993) Anatomical studies on ectomycorrhizas. I. The ectomycorrhizas produced by *Thelephora terrestris* on *Pinus patula*. Mycorrhiza 3:39–42

Molina R (1981) Ectomycorrhizal specificity in the genus *Alnus*. Can J Bot 59:325–334

Molina R, Massicotte H, Trappe JM (1992) Specificity phenomena in mycorrhizal symbioses: community-ecological consequences and practical implications. In: Allen MJ (ed) Mycorrhizal functioning. An integrative plant–fungal process. Chapman & Hall, New York, pp 357–423

Mousain D, Pierson J, Allemand P, Augé P, Ferrandes P (1977) Observations sur la mycorrhization de diverses espèces ligneuses (dont 2 *Eucalyptus*) par le *Thelephora terrestris*. Ann Phytopathol 9:93–94

Olsson PA, Chalot M, Bååth E, Finlay RD, Söderström B (1996) Ectomycorrhizal mycelia reduce bacterial activity in a sandy soil. FEMS Microbiol Ecol 21:77–86

Perry DA, Margolis H, Choquette C, Molina R, Trappe JM (1989) Ectomycorrhizal mediation of competition between coniferous tree species. New Phytol 112:501–511

Read DJ (1992) The mycorrhizal mycelium. In: Allen MJ (ed) Mycorrhizal functioning. An integrative plant-fungal process. Chapman & Hall, New York, pp 102–133

Read DJ, Leake JR, Langdale AR (1989) The nitrogen nutrition of mycorrhizal fungi and their host plants. In: Boddy L, Marchant R, Read DJ (eds) Nitrogen, phosphorus and sulphur utilization by fungi. Cambridge University Press, Cambridge, pp 181–204

Reid CPP, Woods FW (1969) Translocation of ^{14}C-labeled compounds in mycorrhizae and its implications in interplant nutrient cycling. Ecology 50:179–181

Richter DL, Zuellig TR, Bagley ST, Bruhn JN (1989) Effects of red pine (*Pinus resinosa* Ait.) mycorrhizoplane-associated actinomycetes on in vitro growth of ectomycorrhizal fungi. Plant Soil 115:109–116

Ruehle JL (1983) The relationship between lateral-root development and spread of *Pisolithus tinctorius* ectomycorrhizae after planting of container-grown loblolly pine seedlings. For Sci 29:519–526

Rühling A, Söderström B (1990) Changes in fruitbody production of mycorrhizal and litter decomposing macromycetes in heavy metal polluted coniferous forests in north Sweden. Water Air Soil Pollut 49:375–387

Rühling A, Bååth E, Nordgren A, Söderström B (1984) Fungi in metal-contaminated soil near the Gusum brass mill, Sweden. Ambio 13:34–36

Samson J, Fortin JA (1986) Ectomycorrhizal fungi of *Larix laricina* and the interspecific and intraspecific variation in response to temperature. Can J Bot 64:3020–3028

Schramm JR (1966) Plant colonization studies on black wastes from anthracite mining in Pennsylvania. Trans Am Philos Soc 56:1–194

Seith B, George E, Marschner H, Wallenda T, Schaeffer C, Einig W, Wingler A, Hampp R (1996) Effects of varied soil nitrogen supply on Norway spruce (*Picea abies* [L.] Karst.). I. Shoot and root growth and nutrient uptake. Plant Soil 184:291–298

Smith SE, Read DJ (1997) Mycorrhizal symbiosis. 2nd edn. Academic Press, San Diego, 605 pp

Stalpers JA (1993) The aphyllophoraceous fungi. I. Keys to the species of Thelephorales. Stud Mycol 35:1–168

Stenström E (1991) The effects of flooding on the formation of ectomycorrhizae in *Pinus sylvestris* seedlings. Plant Soil 131:247–250

Sudhakara Reddy M, Natarajan K (1997) Coinoculation efficacy of ectomycorrhizal fungi on *Pinus patula* seedlings in a nursery. Mycorrhiza 7:133–138

Sung S-J S, White LM, Marx DH, Otrosina WJ (1995) Seasonal ectomycorrhizal fungal biomass development on loblolly pine (*Pinus taeda* L.) seedlings. Mycorrhiza 5:439–447

Sylvia DM, Jarstfer AG (1997) Distribution of mycorrhiza on competing pines and weeds in a southern pine plantation. Soil Sci Soc Am J 61:139–144

Tam PCF (1995) Heavy metal tolerance by ectomycorrhizal fungi and metal amelioration by *Pisolithus tinctorius*. Mycorrhiza 5:181–187

Tam PCF, Griffiths DA (1993) Mycorrhizal associations in Hong Kong Fagaceae. V. The role of polyphenols. Mycorrhiza 3:165–170

Tam PCF, Griffiths DA (1994) Mycorrhizal associations in Hong Kong Fagaceae. VI. Growth and nutrient uptake by *Castanopsis fissa* seedlings inoculated with ectomycorrhizal fungi. Mycorrhiza 4:169–172

Taylor AFS, Alexander IJ (1991) Ectomycorrhizal synthesis with *Tylospora fibrillosa*, a member of the Corticiaceae. Mycol Res 95:381–384

Taylor DL, Bruns TD (1997) Independent, specialized invasions of ectomycorrhizal mutualism by two nonphotosynthetic orchids. Proc Natl Acad Sci USA 94:4510–4515

Theodorou C, Reddell P (1991) In vitro synthesis of ectomycorrhizas on Casuarinaceae with a range of mycorrhizal fungi. New Phytol 118:279–288

Thomas GW, Jackson RM (1979) Sheathing mycorrhizas of nursery grown *Picea sitchensis*. Trans Br Mycol Soc 73:117–125

Thomas GW, Jackson RM (1983) Growth responses of Sitka spruce seedlings to mycorrhizal inoculation. New Phytol 95:223–229

Thomas GW, Clarke CA, Mosse B, Jackson RM (1982) Source of phosphate taken up from two soils by mycorrhizal (*Thelephora terrestris*) and non-mycorrhizal *Picea sitchensis* seedlings. Soil Biol Biochem 14:73–75

Thomas GW, Rogers D, Jackson RM (1983) Changes in the mycorrhizal status of Sitka spruce following outplanting. Plant Soil 71:319–323

Tinker PB, Durall DM, Jones MD (1994) Carbon use efficiency in mycorrhizas: theory and sample calculations. New Phytol 128:115–122

Trappe JM (1962) Fungus associates of ectotrophic mycorrhizae. Bot Rev 28:538–606

Trappe JM (1977) Selection of fungi for ectomycorrhizal inoculation in nurseries. Annu Rev Phytopathol 15:203–222

Tyminska A, Le Tacon F, Chadoeuf J (1986) Effect of three ectomycorrhizal fungi on growth and phosphorus uptake in *Pinus sylvestris* seedlings at increasing phosphorus levels. Can J Bot 64:2753–2757

Unestam T (1991) Water repellency, mat formation, and leaf-stimulated growth of some ecto-mycorrhizal fungi. Mycorrhiza 1:13–20

Unestam T, Sun Y-P (1995) Extramatrical structures of hydrophobic and hydrophilic ectomyc-orrhizal fungi. Mycorrhiza 5:301–311

Villeneuve N, Le Tacon F, Bouchard D (1991) Survival of inoculated *Laccaria bicolor* in compe-tition with native ectomycorrhizal fungi and effects on the growth of outplanted Douglas-fir seedlings. Plant Soil 135: 95–107

Visser S (1995) Ectomycorrhizal fungal succession in jack pine stands following wildfire. New Phytol 129:389–401

Wallander H (1995) A new hypothesis to explain allocation of dry matter between mycorrhizal fungi and pine seedlings in relation to nutrient supply. Plant Soil 168–169:243–248

Wallander H, Nylund J-E (1992) Effects of excess nitrogen and phosphorus starvation on the extramatrical mycelium of ectomycorrhizas of *Pinus sylvestris* L. New Phytol 120:495–503

Weir JR (1921) *Thelephora terrestris*, *T. fimbriata*, and *T. caryophyllea* on forest tree seedlings. Phytopathology 11:141–144

Wullschleger SD, Reid CPP (1990) Implication of ectomycorrhizal fungi in the cytokinin rela-tions of loblolly pine (*Pinus taeda* L.). New Phytol 116:681–688

Zak B (1976a) Pure culture synthesis of bearberry mycorrhizae. Can J Bot 54:1297–1305

Zak B (1976b) Pure culture synthesis of Pacific madrone ectendomycorrhizae. Mycologia 68:362–369

Zak B, Ho I (1994) Resistance of ectomycorrhizal fungi to rhizina root rot. Indian J Mycol Plant Pathol 24:192–195

Zhou T, Wei R (1986) Taxonomic study on two edible species of *Thelephora* in Yunnan, China. Acta Bot Yunnanica 8:295–297

Chapter 15

Resupinate Ectomycorrhizal Fungal Genera

S. Erland[1] and A. F. S. Taylor[2]

15.1
Introduction

Molecular methods have facilitated the identification of the ectomycorrhizal (ECM) mycobionts on single root tips and there are an increasing number of studies which compare ECM community structure both on the roots below and as sporocarps above ground (Mehmann et al. 1995; Gardes and Bruns 1996; Dahlberg et al. 1997; Jonsson et al. 1999). A feature common to most of these studies is the low correlation between species which are abundant as sporocarps and species which are abundant on the roots. In addition, many of the species which do occur abundantly on the roots have evaded identification despite access to reference libraries of RFLP-patterns (Kårén et al. 1997) and to identification guides for ECM (Agerer 1986–1997; Ingleby et al. 1990; Haug and Pritsch 1992) even where these have contained the majority of the ECM macromycetes fruiting in the studied areas. It is thought that some ECM fungi may not form sporocarps, while others may form inconspicuous sporocarps which are easily overlooked (Smith and Read 1997).

Many fungal species in the Corticiaceae and Thelephoraceae form thin (<1 mm) inconspicuous sporocarps, often resembling spiderwebs, directly on litter and soil debris (Fig. 15.1). These resupinate sporocarps are often not included in conventional sporocarp surveys or in RFLP databases. Two resupinate species which have been known as ECM fungi for a considerable time are *Piloderma fallax* (Libert) Stal. (syn. *P. croceum*; see Sect. 15.3.1) (Melin 1936; Björkman 1942; Erdtman 1948) and *Amphinema byssoides* (Pers.) Eriks. (Fassi and Vecchi 1962). More recently, other resupinate fungi have been shown to be important in a range of ecosystems. One example is *Tylospora fibrillosa* (Burt) Donk, which colonised 75% of the examined roots in a Sitka spruce forest in Scotland (Taylor and Alexander 1989).

At present, hard evidence on the mycorrhiza-forming abilities of most resupinate fungi is lacking, but several genera share the structural and

[1] Department of Microbial Ecology, Lund University, Ecology Building, 223 62 Lund, Sweden
[2] Department of Forest Mycology and Pathology, Swedish University of Agricultural Sciences, Box 7026, 750 07 Uppsala, Sweden

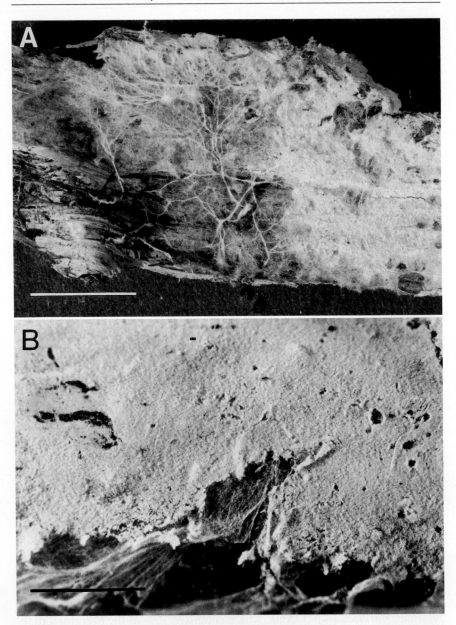

Fig. 15.1. Sporocarps of resupinate fungi. **A** Young sporocarp of *Piloderma fallax* with thin rhizomorphs, growing on wood fragment (*scale bar* 10 mm). (Photo by T. Hallingbäck from Eriksson et al. 1981, with permission). **B** Sporocarp of *Amphinema byssoides* showing granular hymenium and loose indeterminate edge with rhizomorphs (*scale bar* 3.5 mm). (Photo by S. Sunhede from Eriksson and Ryvardén 1973, with permission)

ecological characteristics which are common to most known ECM resupinate genera. These characteristics include soft, loosely attached basidiomata with well-developed rhizomorphs (with the exception of *Tylospora* spp.; see below) growing on a variety of substrates, producing basidiospores which are slow or fail to germinate on artificial media (K.-H. Larsson, pers. comm.). Current information on genera known or suspected to be, or which at least contain species, capable of forming ECM is summarised in Table 15.1. Around 100 species of resupinate fungi are considered to be ECM, with the majority belonging to the genus *Tomentella*.

Since little information is available on the mycorrhizal functioning of the majority of resupinate fungi, this chapter presents rather fragmentary data on selected species of resupinate fungi from the following five genera: *Amphinema*, *Piloderma* and *Tylospora* in the Corticiaceae and *Pseudotomentella* and *Tomentella* in the Thelephoraceae.

15.2
Amphinema Karst

15.2.1
Taxonomy and Ecology

Four species are currently recognised within the genus *Amphinema*: *A. arachispora* Burdsall & Nakasone; *A. byssoides* (Pers.) Eriks.; *A. diadema* Larsson & Hjortstam; and *A. tomentellum* (Bres.) M. P. Christiansen. Of these, only *A. byssoides* has been shown to be ECM (Ingleby et al. 1990; Haug and Pritsch 1992; Raidl 1997). It also appears to be the most widespread taxon, being recorded on *Pinus taiwanensis* from Taiwan (Haug et al. 1994), Korea (Jung 1991) and India (Thind and Dhingra 1985) and on *Pinus* sp. from Italy (Bernicchia 1971). Eriksson and Ryvarden (1973) state that *A. byssoides* is common in boreal forests of the *Cladina* or *Hylocomium* type, where it is one of the characteristic fungal species. In contrast to this apparent affinity for rather nutrient-poor habitats, *A. byssoides* is also commonly found associated with seedlings in nutrient-rich nursery soils (Fassi and Vecchi 1962; Thomas and Jackson 1979; Weiss and Agerer 1988; Danielson and Visser 1990; Grogan et al. 1994). Taylor and Brand (1992) reported that *A. byssoides* ECM increased in abundance in a *Picea abies* stand following the addition of lime. It was suggested that this increase was related to increased nutrient availability and an increase in soil pH.

A. byssoides readily colonises seedlings from forest soil inoculum (Danielson et al. 1984; Shishido et al. 1996a,b), suggesting that it could be regarded as an early-stage fungus (Dighton and Mason 1985). Further support for this assessment comes from reports that *A. byssoides* is a common associate of seedlings planted on reclamation sites (Danielson et al. 1984; Danielson 1991). However, its affinity for mature boreal forests would indicate that it is best regarded as a multi-stage ECM fungus (Danielson 1984).

Table 15.1. Genera of resupinate fungi which are known or suspected to form ectomycorrhizas

Family Genus (no. of sp.[a])	Mycorrhizal status	References for syntheses and/or descriptions
Corticiaceae		
Amphinema (4)	Confirmed	A. byssoides (Pers.: Fr.) J. Erikss. (Fassi and Vecchi 1962; Danielson 1984; Weiss 1991; Raidl 1997)
Byssocorticium (11)	Confirmed	B. atrovirens (Fr.) Bond & Sing. ex Sing. (Peyronel 1922; Boullard and Dominik 1960; Luppi and Gautero 1967; Brand 1991)
Byssoporia (1)	Confirmed	B. terrestris (D.C. ex Fries) Larsen & Zak (Killarmann 1927; Zak 1969; Froidevaux 1975; Zak and Larsen 1978)
Piloderma (10)	Confirmed	P. byssinum (P. Karst.) Jül. (Froidevaux and Amiet 1978) P. fallax (croceum) (Libert) Stal. (Nylund and Unestam 1982; Brand 1991; Raidl 1997)
Tylospora (2)	Confirmed	T. asterophora (Bonord.) Donk (Raidl 1997); T. fibrillosa (Burt) Donk (Taylor and Alexander 1991)
Athelia (76)	Some species suspected	K.-H. Larsson (pers. comm.)
Lindtneria (11)	Some species suspected	K.-H. Larsson (pers. comm.)
Trechispora (63)	Some species suspected	K.-H. Larsson (pers. comm.)
Thelephoraceae		
Pseudotomentella (20)	Confirmed	P. tristis (P. Karst.) M. J. Larsen (Agerer 1994)
Tomentella (60–70)	Confirmed	Tomentella, two spp. (Danielson and Visser 1989) T. crinalis (Fr.) M. J. Larsen (Kõljalg 1992) T. ferruginea (Pers.) Pat. (Raidl 1997)
Tomentellopsis (7)	Suspected	U. Kõljalg (pers. comm.)

[a] Estimates of species numbers within the Corticiaceae are taken from the Centraalbureau voor Schimmelcultures database, CBS, Baarn and Delft, the Netherlands. Estimates of species numbers within the Thelephoraceae are from U. Kõljalg (1996, pers. comm.).

15.2.2
ECM Formation

The host range of the genus *Amphinema* appears to be restricted to coniferous tree species. The sporocarps also demonstrate a preference for conifer wood as a substratum (Eriksson and Ryvarden 1973). Renvall (1995) examined the wood-rotting basidiomycetes associated with decomposing conifer trunks in northern Finland and found that *A. byssoides* sporocarps were associated exclusively with fallen trunks of *P. abies*. Haug et al. (1994) described an ectendomycorrhizal association on *Pinus taiwanensis*, stating that the fungal symbiont was *A. byssoides*. This is remarkable since all of the ectendomycorrhizal associations previously reported on *Pinus* sp. have been formed by fungi with ascomycetous affinities (see Danielson 1982).

15.3
Piloderma Jülich

15.3.1
Taxonomy and Ecology

Piloderma is a genus within the *Corticiaceae* which includes at present ca. ten species; of these only two have been shown to form ECM (Table 15.1). The taxonomy of *Piloderma* spp. has been subject to several nomenclatural changes and recently Larsen et al. (1997) submerged the well-known *P. croceum* Erikss. & Hjortst. and *P. bicolor* (Peck) *sensu* Jül. within *P. fallax* (Libert) Stal. They retained *P. byssinum* (P. Karst.) Jül. as a distinct species. Little information exists concerning *Piloderma* species other than *P. fallax* and *P. byssinum*. Therefore, generalisations made in this section concerning the ecology, geographic distribution and host range of the genus refer primarily to these two species.

   Piloderma spp. are a common component of the litter fungal flora of predominantly acid, boreal coniferous forests in Europe (Eriksson et al. 1981) and North America (Larsen 1983). Since *Piloderma* is often associated with acid coniferous soils, it is not surprising that several studies have shown a pH preference at the lower end of the pH interval. A pH optimum of between pH 4 and 5 has been found for growth of *P. fallax* in axenic culture (Hung and Trappe 1983; Erland et al. 1990). The colonisation potential of indigenous *P. fallax* in sieved and limed (CaO) forest humus also reached an optimum at around pH 5 and decreased to zero at pH values over 6.2 (Erland and Söderström 1990). Under field conditions, the pH colonisation optimum for pine seedlings planted in a mature forest which had been treated with lime and ash was also near pH 5 (Erland and Söderström 1991). Similarly, Antibus and Linkins (1992) found an increase in the colonisation by *P. fallax* when humus pH was increased by liming from 4 to around 5. Lehto (1994) on the other hand found a decrease in the occurrence of *P. fallax* in a limed *P. abies* forest in Finland, but unfortunately no information on humus pH values at the time of sampling are included.

   Visser (1995) investigated ECM fungal succession in jack pine (*Pinus banksiana*) stands following wildfire. Using the classification of Dighton and Mason (1985), Visser designated *P. byssinum* as a late-stage fungus.

P. byssinum ECM were absent from 6-year-old stands, occurred on 0.2% of the roots in 41-year-old stands and increased to 10% of the ECM in 65- and 122-year-old stands. Dahlberg (1990) found that the presence of a humus layer was important for formation of *P. fallax* ECM on pine seedlings planted at clearcut sites. Although it has been shown that *P. fallax* can colonise seedlings planted in sieved humus in the laboratory (Erland and Söderström 1990), other greenhouse studies have failed to get colonisation by *P. fallax* despite its presence in the undisturbed stand from which the soil was taken (Danielson and Visser 1989). The mycelial extension rate of *P. fallax* has been shown to increase several fold when the fungus is connected with the host plant (Erland et al. 1990). The importance of the host plant connection for efficient colonisation of new roots by *P. fallax* was demonstrated by Dahlberg and Stenström (1991) who found significantly higher colonisation levels on seedlings planted in a mature coniferous forest compared to a clearcut.

15.3.2
ECM Formation

P. fallax forms ECM with a wide range of coniferous hosts and has also been found in association with *Fagus sylvatica* (Brand 1991) and *Betula* sp. (A. F. S. Taylor and A. Dahlberg, unpubl.). Attempts have been made to synthesise ECM between *P. fallax* and other broad-leaved hosts. Herrmann et al. (1998) synthesised mycorrhiza with *Quercus robur*, while Martins et al. (1996) tested *P. fallax* with *Castanea sativa* and Cripps and Miller (1995) tested *P. fallax* with *Populus tremuloides*. However, no ECM developed in the former test, and in the latter a mantle but no Hartig net was formed.

Mycorrhiza formation in vitro between *P. fallax* and *Picea abies* has been described in detail by Nylund and Unestam (1982). They divided the process into the following sequential developmental phases: stimulation of fungal growth by root metabolites; formation of a hyphal envelope on the root; intercellular penetration by single hyphae; change in fungal morphology into labyrinthic tissue leading to Hartig net formation; extension of labyrinthic tissue to form a mantle. A more detailed study, specifically of the development of the Hartig net in *P. fallax* ECM, was also published by Nylund (1981). Nylund et al. (1982) observed that *P. fallax* regularly produced intracellular penetrations in senescent and dead cortical cells of the host roots.

Ramstedt et al. (1986) addressed the question of the role of carbohydrates in the ECM symbiosis and specifically studied mannitol metabolism in *P. fallax*. Enzyme activities of mannitol-1-phosphate dehydrogenase, mannitol dehydrogenase, glucose-6-phosphate dehydrogenase, hexokinase and mannitol-1-phosphatase indicated a mannitol inducible pathway in *P. fallax*. Mannitol did not inhibit glycolytic enzymes of the host. Since the host roots cannot metabolise mannitol, it was proposed that the function of the mannitol pathway in this symbiosis may be to transform host glucose and fructose to mannitol and then to store reducing power within the fungus.

15.3.3
Host Plant Growth Responses and Fungus-Derived Benefits

Piloderma ECM appear to be strongly associated with the organic soil layer and the ability to provide the host plant with nutrients of organic origin may be one of the most important symbiotic features of this fungus. Christy et al. (1982) examined the ECM status of *Tsuga heterophylla* seedlings which had established naturally on decaying logs. ECM fungi found on the seedlings included *P. fallax*, *Cenococcum geophilum* and four unidentified types. One-year-old ECM seedlings growing on logs had 60% longer shoots and 47% longer roots than first-year non-ECM seedlings.

The presence of an acid phosphatase on the surface of *P. fallax* ECM would indicate that organic P sources may be utilised by this fungus (Ho and Zak 1979; Antibus and Linkins 1992).

Inorganic NO_3^--N is usually present in very small amounts in the boreal forests (Adams and Attiwill 1982) where *Piloderma* is frequently found; however, it has been demonstrated that *P. fallax* can produce significant amounts of nitrate reductase and is able to utilise nitrate as a sole source of N (Sarjala 1990, 1991). Finlay et al. (1992) examined the ability of two isolates of *P. fallax* to use inorganic and organic N and found that both isolates grew as well on organic N sources as on NH_4^+, but growth on NO_3^- was reduced to about 50% of that on the other N sources.

To compare colonisation abilities of *P. fallax* and *Paxillus involutus*, Erland and Finlay (1992) planted a non-ECM bait plant between ECM pine seedlings in a microcosm. They observed a rapid colonisation of the bait plant by *P. involutus*, which subsequently declined as a more persistent colonisation by *P. fallax* developed. Recently, Herrmann et al. (1998) compared the actual events of mycorrhiza formation by *P. fallax* and *Paxillus involutus* with oak. They found that while *P. involutus* rapidly formed ECM and had no significant morphological effects on the host, *P. fallax* modified the entire plant development before a delayed mycorrhiza formation. Modification of the host included stimulation of the lateral root system and increased leaf surface of the host, but without corresponding weight increases. The authors attributed the modified plant growth to auxin production by the fungus. Furthermore, they concluded that *P. fallax* was a strong sink for assimilate as the larger leaf surface and enhanced potential for photoassimilation did not result in an increase in host biomass. In this respect *P. fallax* could be considered a late-stage fungus (Dighton and Mason 1985).

The bright yellow colour of *P. fallax* is due to an unusual pigment called corticrocin (Erdtman 1948). The pigment is degraded when the fungus is exposed to light, which is usually not the case in pigmented fungi (Ramstedt 1985). Gruhn and Miller (1991) studied pigmentation of ECM fungi in relation to their ability to produce tyrosinase during copper stress, suggesting that tyrosinase may act in detoxification by irreversibly binding to Cu-ions. They found that naturally brown pigmented fungi produced extracellular tyrosinase, while fungi with other pigments, such as *P. fallax*, failed to do so.

15.4
Tomentella Pers.: Pat. and *Pseudotomentella* Svrcek

15.4.1
Taxonomy and Ecology

The genera *Pseudotomentella* and *Tomentella*, along with other tomentelloid fungi, are located in the Thelephoraceae. Many members of this large family are well-known mycorrhizal formers (see Chap. 14), but the realisation that *Pseudotomentella* and *Tomentella* contain species capable of forming ECM is relatively recent (Danielson and Pruden 1989). *Tomentella* is by far the largest genus, but estimates of the number of species included in the genus have varied considerably. Larsen (1974) included 72 species, while Stalpers (1993) keyed out 69 species. In a recent revision of the genus, Kõljalg (1996) concluded that there were 43 valid species. It seems likely that new species will be found, particularly in the tropics, and the final number could be between 60 and 70 species (U. Kõljalg, pers. comm.).

Since the genus has often been included in inventories of wood-rotting fungi, considerable information is available concerning its geographic distribution (e.g. Renvall 1995). *Tomentella* spp. appear to be widespread in temperate North America (Larsen 1974) and Eurasia (Kõljalg 1996), but there are also records from more tropical locations: the Canary Islands (Larsen 1994), Korea (Jung 1994), Spain (Blanco et al. 1989), Iran (Saber 1987) and India (Thind and Rattan 1971).

15.4.2
ECM Formation

The first pure culture synthesis of *Tomentella* ECM was by Kõljalg (1992), using basidiospores of *T. crinalis* (Fr.) M. J. Larsen to inoculate *Pinus sylvestris* seedlings. The ECM were rather poorly formed and contained some intracellular hyphae, but it is likely that this was due to the high levels of exogenous carbon used in the growth medium. Raidl (1997) synthesised ECM on *P. sylvestris* non-sterile conditions using sporocarps and ECM of *T. ferruginea* (Pers.: Fr.) Pat. as inocula. The morphology of the ECM compared well with those formed under field conditions (Raidl and Müller 1996). *Pseudotomentella* is less well investigated with only the ECM from *P. tristis* described to date (Agerer 1994). The compilation of a database of DNA sequences from Tomentelloid fungi should rapidly increase the number of species known to form ECM (Kõljalg et al. 1998). The presence of *P. tristis* on the roots of

P. abies and *P. sylvestris* in northern Sweden has already been demonstrated using this database (Kõljalg et al. 1998).

Danielson et al. (1984) and Danielson and Pruden (1989) reported the occurrence of several unidentified species of *Tomentella* ECM occurring on *Pinus banksiana* and *Picea glauca* respectively. The ECM were characterised by pseudoparenchymatous mantles with epidermoid cells (type M: Agerer 1986–1997). The mantle of one group of these *Tomentella* ECM contained blue crystals which became green in KOH. This staining reaction is due to the presence of thelephoric acid which is an important taxonomic character in the Thelephoraceae (Bresinsky and Rennschmid 1971). The occurrence of thelephoric acid in the mantle of the unidentified "Piceirhiza nigra" ECM led Agerer et al. (1995) to conclude that the mycobiont involved was a member of the Thelephoraceae and was probably a species of *Tomentella*. "Piceirhiza nigra" ECM were found to comprise 28% of all ECM tips on limed *P. abies* plots in Germany (Brand et al. 1994).

The use of molecular techniques to investigate unknown ECM mycobionts has led to the discovery of other mycorrhizal *Tomentella* species. Gardes and Bruns (1996) used RFLP patterns to characterise the ECM community in a stand of *Pinus muricata* and found that *T. sublilacina* (Ellis & Holw.) Wakef. colonised ca. 15% of the roots examined. Taylor and Bruns (1997), using RFLPs and DNA sequence data, demonstrated that the fungal infection in the rhizomes of the achlorophyllous orchid *Cephalanthera austinae* was due to several members of the Thelephoraceae. Although no *Tomentella* species were positively identified, there was very strong evidence that at least some of the infections were from species in this genus. The fungi involved were also shown to simultaneously form ECM on tree roots in the vicinity of the orchid rhizomes. Taylor and Bruns (op. cit.) argued that the orchid acted as a cheater in the symbiotic association between the photosynthetic host and the ECM fungus, gaining photosynthetically derived carbon from the tree via the fungus. Chambers et al. (1998) sequenced the ITS region from an isolate of the fungus believed to be the sole mycobiont of *Pisonia grandis* (Nyctaginaceae). They demonstrated that this region had a high degree of homology with several members of the Thelephoraceae, in particular with an unknown *Tomentella* sp.

15.5
Tylospora Donk

15.5.1
Taxonomy and Ecology

The genus *Tylospora* includes two species, both of which have recently been shown to be ECM. Raidl (1997) synthesised ECM between *T. asterophora* (Bonord.) Donk and *P. abies* and Taylor and Alexander (1991) reported *T.*

fibrillosa (Burt.) Donk mycorrhizas on *Picea sitchensis*. Both of these studies used sporocarp fragments as inoculum. From repeated field observations on the occurrence of *T. fibrillosa* within woodlands, Watling (1981) first suggested that the species might be ECM.

Records of sporocarps are, however, available from earlier studies and Hjortstam et al. (1988) stated that both species are widely distributed in northern Europe, but are less common in the far north. This may reflect the distribution of the main host genus, *Picea* (see below). Sporocarps of *T. fibrillosa* have also been reported from Korea (Jung 1996) and the ECM have been found on *Picea* sp. in Taiwan (Haug et al. 1994) and Canada (R. M. Danielson, pers. comm.). This species may well be a symbiont of spruce in large parts of the world. The sporocarps of both species occur mostly on well-decayed wood but may develop on all kind of substrata on the forest floor (Hjortstam et al. 1988; Renvall 1995). Fruiting usually occurs during humid periods throughout most of the year.

15.5.2
ECM Formation

To date, *Tylospora* has been reported almost exclusively as a symbiont of spruce. Unidentified ECM described by Haug et al. (1986, type 7) and Gronbach (1988, "Piceirhiza guttata") on *P. abies* were later identified as being formed by *Tylospora* spp. *Tylospora* ECM have since been found in a number of studies in spruce forests across Europe: in Scotland (Taylor and Alexander 1989; Ryan and Alexander 1992), Sweden (Erland 1995; Dahlberg et al. 1997; Kårén 1997), Germany (Haug et al. 1996; Brand et al. 1994; Quian et al. 1998), Denmark and France (A. F. S. Taylor and D. J. Read, unpubl.). There is some evidence that *Tylospora* may form ECM with hosts other than spruce. Ryan and Alexander (1992) found *T. fibrillosa* ECM on both spruce and pine roots in a mixed stand in Scotland. *Tylospora*-type ECM were found on the roots of *Betula* sp. in an analysis of the mycorrhizal communities in a mixed spruce/birch/pine forest in northern Sweden (A. F. S. Taylor and A. Dahlberg, unpubl.).

The structure of the Hartig net, mantle and extramatrical hyphae of *T. fibrillosa/P. sitchensis* ECM have been described by Taylor and Alexander (1991) and the *T. fibrillosa/P. abies* ECM were described under "Piceirhiza guttata" by Gronbach (1989). A common feature of both *T. fibrillosa* (Taylor and Alexander 1991) and *T. asterophora* (Raidl 1997) ECM are that they form extensive extramatrical mycelia but lack rhizomorphs. Recently, Eberhardt et al. (1998) investigated the possibility of discriminating between *T. asterophora* and *T. fibrillosa* ECM using both morphological and molecular methods. Using DNA sequencing, Eberhardt et al. (1998) discovered a larger intraspecific variation in the ITS region of *T. fibrillosa* than had previously been detected by PCR/RFLP methods by Erland et al. (1994). The intraspecific variation in the ITS region of *T. asterophora* was, however, found to be more restricted than that of *T. fibrillosa* (Eberhardt et al. 1998).

Downes et al. (1992) studied ageing of *T. fibrillosa/P. abies* ECM in growth chambers and found that in the majority of ECM, the major decline in function (measured as a loss of cortical cell and Hartig net fluorescence with fluorescent diacetate [FDA]) took place 85 days after ECM formation. They concluded that this time period was in broad agreement with estimates of general ECM lifespan from biomass field studies.

15.5.3
Host Plant Growth Responses and Fungus-Derived Benefits

Carbon allocation and P uptake was studied in ageing *T. fibrillosa/P. sitchensis* ECM by Cairney and Alexander (1992a,b). They found that the allocation of current ^{14}C-labelled photosynthate was reduced in older compared to young ECM in the same root system. Levels of total soluble and insoluble carbohydrate were also lower in older compared to young ECM. Furthermore, they noted differences in levels of individual soluble sugars. Arabitol, glucose and mannitol were detected in significantly lower quantities in older ECM, whereas fructose and trehalose were present at similar levels in both older and young ECM. Absorption of P into excised ECM was reduced in older compared to young ECM (Cairney and Alexander 1992b). Individual young ECM labelled with ^{32}P also transferred more ^{32}P to the host than did older ECM in the same intact root system. The results suggest a progressive reduction in nutrient transfer to the host as the ECM tips age.

Sporocarps of *Tylospora* spp. commonly develop on well-rotted wood, perhaps suggesting wood decaying abilities which would provide access to organically bound nutrients for the host. Indeed, Cairney and Burke (1994) demonstrated that specific activities for Mn-peroxidase in *T. fibrillosa* were equivalent to, or greater than, those for known white rot basidiomycetes. Ryan and Alexander (1992) grew *T. fibrillosa* on bovine serum albumin as a sole source of N and recorded that the proteolytic ability of the fungus was intermediate between that of *Suillus variegatus* (Sw.: Fr.) O. Kuntze (a known protein fungus), which had the greatest ability, and *Lactarius rufus* (Scop.: Fr.) Fr., which had no ability to degrade the protein. Daehne et al. (1995) investigated the effect of liming upon aminopeptidase within several ECM species, including those formed by a *Tylospora* sp. They found that the levels of the enzyme were increased in ECM collected from Norway spruce plots which had been limed and attributed this to the increased levels of soluble organic N compounds in the soil following liming. It would appear that *Tylospora* sp., at least *T. fibrillosa*, has the potential to access some organically bound nutrients.

Jentschke et al. (1997) studied uptake of Pb, from soils with different Pb contents, into the roots of *P. abies* seedlings colonised by indigenous ECM fungi including *Tylospora* sp. No differences were found between ECM and non-ECM root tips with respect to Pb levels when plants were grown in soil with a high Pb content. It was concluded that the ECM, including those formed by indigenous *Tylospora* sp., could not exclude Pb from root tissues.

15.6
Conclusions

The known resupinate ectomycorrhiza formers appear to share a number of ecologically important features: they are predominantly associated with coniferous hosts and the organic soil layers; the extramatrical mycelia form rhizomorphs (with the exception of *Tylospora*); the fruit bodies form on decaying wood, but the ECM and mycelia can also be found growing inside pieces of rotting wood; in studies where ECM formed by resupinate fungi have been identified, they often represent a large proportion of the community on the roots; in studies with a successional perspective they are usually considered as late- or multi-stage fungi. *T. fibrillosa* may have some ligni-nolytic capacity and both *Tylospora* and *Piloderma* can utilise organic N. Other fungi mentioned in this review should be investigated for these abili-ties as it seems likely that they will possess similar qualities. In a study in northern Finland, Renvall (1995) found that resupinate ectomycorrhiza formers associated with well-decayed conifer logs and suggested that they contribute to the decomposition of the last remnants of the fallen logs, thereby linking the decomposition of wood with the germination of seed and the early growth of seedlings which commonly become established on these logs.

There are good reasons to believe that some of the abundant, unidentified mycorrhizal types found in recent community studies in the boreal conifer-ous forest may be formed by resupinate fungi. Efforts should be made to include such species in reference material. As suggested by Watling (1981), "we should look further afield within our taxonomic framework to assess the vast array of resupinate fungi as potential symbionts".

Acknowledgement. We thank Leif Ryvarden for permission to use the plates of the resupinate fruit bodies.

References

Adams MA, Attiwill PM (1982) Nitrogen mineralization and nitrate reduction in forests. Soil Biol Biochem 14:197–202
Agerer R (1986–1997) Colour atlas of ectomycorrhizae. Einhorn-Verlag, Schwäbisch Gmünd
Agerer R (1994) *Pseudotomentella tristis* (Thelephoraceae): eine Analyse von Fruchtkörper und Ektomykorrhizen. Z Mykol 60:143–158
Agerer R, Klostermeyer D, Steglich W (1995) Piceirhiza nigra, an ectomycorhiza on *Picea abies* formed by a species of Thelephoraceae. New Phytol 131:377–380
Antibus RK, Linkins AE (1992) Effects of liming a red pine forest floor on mycorrhizal numbers and mycorrhizal and soil acid phosphatase activities. Soil Biol Biochem 24:479–487
Bernicchia A (1971) Mycological notes of some Corticiaceae found on wood. Arch Bot Biogeogr Ital 47:69–81

Björkman E (1942) Über die Bedingungen der Mykorrhizabildung bei Kiefer und Fichte. Symb Bot Ups 6:1–191

Blanco MN, Hjortstam K, Manjon JL, Moreno G (1989) Mycological studies from Monfrague Natural Park Extremadura Spain II. Aphyllophoralles. Cryptogam Mycol 10:217–226

Boullard B, Dominik T (1960) Recherches comparatives entre le mycotrophisme du *Fagetum carpaticum* de Babia Góra et celui d'autres Fageta précédent étudies. Zeszyty Nauk. Wyzszej Szkoly Roln. Szezcinie 3:20 pp

Brand F (1991) Ektomykorrhizen an *Fagus sylvatica*. Charakterisierung und Identifizierung, ökologische Kennzeichnung und unsterile Kultivierung. Libri Bot 2:1–228

Brand F, Taylor AFS, Agerer R (1994) Quantitative Erfassung bekannter Ektomykorrhizen in Fichtenversuchsflächen nach Behandlung mit saurer Beregnung und Kalkung. BMFT-Bericht, Bonn

Bresinsky A, Rennschmid A (1971) Pigmentmerkmale, Organisationsstufen und systematische Gruppen bei Höheren Pilzen. Ber Dtsch Bot Ges 84:313–329

Cairney JWG, Alexander IJ (1992a) A study of ageing of spruce *Picea sitchensis* Bong. Carr. ectomycorrhizas. II. Carbohydrate allocation in ageing *Picea sitchensis/Tylospora fibrillosa* Burt. Donk ectomycorrhizas. New Phytol 122:153–158

Cairney JWG, Alexander IJ (1992b) A study of ageing of spruce *Picea sitchensis* Bong. Carr. ectomycorrhizas. III. Phosphate absorption and transfer in ageing *Picea sitchensis/Tylospora fibrillosa* Burt. Donk ectomycorrhizas. New Phytol 122:159–164

Cairney JWG, Burke RM (1994) Fungal enzymes degrading plant cell walls: their possible significance in the ectomycorrhizal symbiosis. Mycol Res 98:1345–1356

Chambers SM, Sharples JM, Cairney JWG (1998) Towards a molecular identification of the *Pisonia* mycobiont. Mycorrhiza 7:319–321

Christy EJ, Sollins P, Trappe MJ (1982) First year survival of *Tsuga heterophylla* without mycorrhizae and subsequent ectomycorrhizal development on decaying logs and mineral soil. Can J Bot 60:1601–1605

Cripps CL, Miller OK (1995) Ectomycorrhizae formed in vitro by quaking aspen: including *Inocybe lacera* and *Amanita pantherina*. Mycorrhiza 5:357–370

Daehne J, Klingelhoefer D, Ott M, Rothe GM (1995) Liming induced stimulation of the amino acid metabolism in mycorrhizal roots of Norway spruce (*Picea abies* (L.) Karst.). Plant Soil 173:67–77

Dahlberg A (1990) Effect of a humus cover on the establishment and development of mycorrhiza on containerized *Pinus sylverstris* L. and *Pinus contorta* ssp. *latifolia* Engelm. after outplanting. Scand J For Res 5:103–112

Dahlberg A, Stenström E (1991) Dynamic changes in nursery and indigenous mycorrhiza of *Pinus sylvestris* seedlings planted out in forest and clearcuts. Plant Soil 136:73–86

Dahlberg A, Jonsson L, Nylund J-E (1997) Species diversity and distribution of biomass above- and below-ground among ectomycorrhizal fungi in an old-growth Norway spruce forest in south Sweden. Can J Bot 75:1323–1335

Danielson RM (1982) Taxonomic affinities and criteria for identification of the common ectendomycorrhizal symbiont of pines. Can J Bot 60:7–18

Danielson RM (1984) Ectomycorrhizal association in Jack pine stands in north eastern Alberta. Can J Bot 42:932–939

Danielson RM (1991) Temporal changes and effects of amendments on the occurrence of sheathing mycorrhizas of conifers grown in oil sands tailings and coal spoil. Agric Ecosyst Environ 35:261–281

Danielson RM, Pruden M (1989) The ectomycorrhizal status of urban spruce. Mycologia 81:335–341

Danielson RM, Visser S (1989) Effects of forest soil acidification on ectomycorrhizal and vesicular-arbuscular mycorrhizal development. New Phytol 112:41–48

Danielson RM, Visser S (1990) The mycorrhizal and nodulation status of container-grown trees and shrubs reared in commercial nurseries. Can J For Res 20:609–614

Danielson RM, Zak JC, Parkinson D (1984) Mycorrhizal inoculum in a peat deposit formed under a white spruce *Picea glauca* stand in Alberta, Canada. Can J Bot 62:2557-2560

Dighton J, Mason PA (1985) Mycorrhizal dynamics during forest tree development. In: Moore D, Casselton LA, Wood DA, Frankland JC (eds) Developmental biology of higher fungi. Cambridge University Press, Cambridge, pp 117-139

Downes GM, Alexander IJ, Cairney JWG (1992) A study of ageing of spruce *Picea sitchensis* Bong. Carr. ectomycorrhizas. I. Morphological and cellular changes in mycorrhizas formed by *Tylospora fibrillosa* Burt. Donk and *Paxillus involutus* Batsch. ex Fr. Fr. New Phytol 122:141-152

Eberhardt U, Walter L, Kottke I (1998) Molecular and morphological discrimination between *Tylospora fibrillosa* and *Tylospora asterophora* mycorrhizae. Abstr ICOM II, Uppsala, Sweden, July 5-10, 1998

Erdtman H (1948) Corticrocin, a mycorrhiza pigment. Nature 160:331

Eriksson J, Ryvarden L (1973) The Corticiaceae of north Europe, vol 2. Fungiflora A/S, Oslo

Eriksson J, Hjortstam K, Ryvarden L (1981) The Corticiaceae of north Europe, vol 6. Fungiflora A/S, Oslo

Erland S (1995) Abundance of *Tylospora fibrillosa* ectomycorrhizas in a south Swedish spruce forest measured by RFLP analysis of the PCR-amplified rDNA ITS region. Mycol Res 99:1425-1428

Erland S, Finlay RD (1992) Effects of temperature and incubation time on the ability of three ectomycorrhizal fungi to colonise *Pinus sylvestris* roots. Mycol Res 96:270-272

Erland S, Söderström B (1990) Effects of liming on ectomycorrhizal fungi infecting *Pinus sylvestris* L. I. Mycorrhizal infection in limed humus in the laboratory and isolation of fungi from mycorrhizal roots. New Phytol 115:675-682

Erland S, Söderström B (1991) Effects of lime and ash treatments on ectomycorrhizal infection of *Pinus sylvestris* L. seedlings planted in a pine forest. Scand J For Res 6:519-526

Erland S, Söderström B, Andersson S (1990) Effects of liming on ectomycorrhizal fungi infecting *Pinus sylvestris* L. II. Growth rates in pure culture at different pH values compared to growth rates in symbiosis with the host plant. New Phytol 115:683-688

Erland S, Henrion B, Martin F, Glover LA, Alexander IJ (1994) Identification of the ectomycorrhizal basidiomycete *Tylospora fibrillosa* Donk by RFLP analysis of the PCR-amplified ITS and IGS regions of ribosomal DNA. New Phytol 126:525-532

Fassi B, Vecchi E (1962) Researches in ectotrophic mycorrhizae of *Pinus strobus* in nurseries. Allionia 8:133-152

Finlay RD, Frostegård Å, Sonnerfelt AM (1992) Utilization of organic and inorganic nitrogen sources by ectomycorrhizal fungi in pure culture and in symbiosis with *Pinus contorta* Dougl. ex Loud. New Phytol 120:105-115

Froidevaux L (1975) Identification of some Douglas fir mycorrhizae. Eur J For Pathol 5: 212-216

Froidevaux L, Amiet R (1978) Les Hymenomycetes resupines mycorrhiziques dans le bois pourri. Schweiz Z Pilzkd 56:9-14

Gardes M, Bruns TD (1996) Community structure of ectomycorrhizal fungi in a *Pinus muricata* forest: above- and below-ground views. Can J Bot 74:1572-1583

Grogan HM, O-Neill JJM, Mitchell DT (1994) Mycorrhizal associations of Sitka spruce seedlings propagated in Irish tree nurseries. Eur J For Pathol 24:335-344

Gronbach E (1988) Charakterisierung und Identifizierung von Ektomykorrhizen in einem Fichtenbestand mit Untersuchungen zur Merkmalsvariabilität in sauer beregneten Flächen. Bibl Mycol 125:1-216

Gronbach E (1989) Piceirhiza guttata. In: Agerer R (ed) Colour atlas of ectomycorrhizae. plate 32. Einhorn-Verlag, Schwäbisch Gmünd

Gruhn CM, Miller OK (1991) Effect of copper on tyrosinase activity and polyamine content of some ectomycorrhizal fungi. Mycol Res 95:268-272

Haug I, Pritsch K (1992) Ectomycorrhizal types of spruce (*Picea abies* (L.) Karst.) in the Black Forest. Kernforschungszentrum, Karlsruhe

Haug I, Kottke I, Oberwinkler F (1986) Licht- und electro-nemmikrokopische Untersuchungen von Mykorrhizen der Fichte (*Picea abies* (L.) Karst.) in Verikalprofilen. Z Mykol 52:373–392

Haug I, Weber R, Oberwinkler F, Tschen J (1994) The mycorrhizal status of Taiwanese trees and the description of some ectomycorrhizal types. Trees 8:237–253

Herrmann S, Munch JC, Buscot F (1998) A gnotobiotic culture system with oak microcuttings to study specific effects of mycobionts on plant morphology before, and in the early phase of, ectomycorrhiza formation by *Paxillus involutus* and *Piloderma croceum*. New Phytol 138:203–212

Hjortstam K, Larsson K-H, Ryvarden L (1988) The Corticiaceae of north Europe, vol 8. Fungiflora A/S, Oslo, pp 1584–1587

Ho I, Zak B (1979) Acid phosphatase activity of 6 ectomycorrhizal fungi. Can J Bot 57:1203–1205

Hung LL, Trappe JM (1983) Growth variation between and within species of ectomycorrhizal fungi in response to pH in vitro. Mycologia 75:234–241

Ingleby K, Mason PA, Last FT, Fleming LV (1990) Identification of ectomycorrhizas. Institute of Terrestrial Ecology research publication no 5. HMSO, London

Jentschke G, Fritz E, Marschner P, Wolters V, Godbold DL (1997) Mycorrhizal colonization and lead distribution in root tissues of Norway spruce seedlings. Z Pflanzenernaelur Bodenkd 160:317–321

Jonsson T, Kokalj S, Finlay RD, Erland S (1999) Ectomycorrhizal community structure in a limed spruce forest. Mycol Res 103:501–508

Jung HS (1991) Fungal flora of Kwanak mountain. Proc Coll Nat Sci (Seoul) 16:35–71

Jung HS (1994) Floral studies on Korean wood-rotting fungi: II. On the flora of the Aphyllophorales (Basidiomycotina). Kor J Mycol 22:62–99

Jung HS (1996) Taxonomic study on Korean Aphyllophorales (II). Some unrecorded species. Kor J Mycol 24:228–236

Kårén O (1997) Effects of air pollution and forest regeneration methods on the community structure of ectomycorrhizal fungi. Doct Thesis, Swedish University of Agricultural Sciences, Uppsala, Sweden

Kårén O, Högberg N, Dahlberg A, Jonsson L, Nylund J-E (1997) Inter- and intra-specific variation in the ITS region of rDNA of ectomycorrhizal fungi in Fennoscandia as detected by endonuclease analysis. New Phytol 136:313–325

Killarmann S (1927) Über zwei seltene Polyporaceae in Bayern. Hedwigia 67:125–130

Kõljalg U (1992) Mycorrhiza formed by basidiospores of *Tomentella crinalis* on *Pinus sylvestris*. Mycol Res 96:215–220

Kõljalg U (1996) *Tomentella* (Basidiomycota) and related genera in temperate Eurasia. Fungiflora A/S, Oslo

Kõljalg U, Dahlberg A, Taylor AFS, Larsson E, Hallenberg N, Larsson K-N, Stenlid J (1998) Molecular identification of tomentelloid and closely related mycorrhizal fungi. Abstr ICOM II, July 5–10, 1998, Uppsala, Sweden

Larsen MJ (1974) A contribution to the taxonomy of the genus *Tomentella*. Mycol Mem 4:1–14

Larsen MJ (1983) On *Piloderma bicolor* in North America and its relationship to *Piloderma byssinum*. Mycologia 75:1092–1093

Larsen MJ (1994) *Tomentella oligofibula* sp. nov. (Aphyllophorales. Thelephoraceae s. str.) from the Canary Islands. Mycotaxon 52:109–112

Larsen MJ, Smith JE, McKay D (1997) On *Piloderma bicolor* and the closely related *P. bysinnum*, *P. croceum* and *P. fallax*. Mycotaxon 63:1–8

Lehto T (1994) Effects of liming and boron fertilization on mycorrhizas of *Picea abies*. Plant Soil 163:65–68

Luppi AM, Gautero C (1967) Ricerche sulle micorrize di *Quercus robur*, *Q. petraea* & *Q. pubescens* in Piedmonte. Allionia 13:129–148

Martins A, Barroso J, Pais MS (1996) Effect of ectomycorrhizal fungi on survival and growth of micropropagated plants and seedlings of *Castanea sativa* Mill. Mycorrhiza 6:265–270

Mason PA, Last FT, Pelham J, Ingelby K (1982) Ecology of some fungi associated with an ageing stand of birches (*Betula pendula* and *Betula pubescens*). For Ecol Manage 4:19–39

Mehmann B, Egli S, Braus GH, Brunner I (1995) Coincidence between molecularly and morphologically classified ectomycorrhizal morphotypes and fruitbodies in a spruce forest. In: Stocchi V, Bonfante P, Nuti M (eds) Biotechnology of ectomycorrhizae. Plenum Press, New York, pp 229–239

Melin E (1936) Methoden der experimentellen Untersuchungen mykotropher Pflanzen. In: Alberhalden E (ed) Handbuch der biologischen Arbeitsmethoden, sect II. Urban & Schwartzenberg, Berlin, pp 1015–1108

Nylund JE (1981) Symplastic continuity during Hartig net formation in Norway spruce *Picea abies* ectomycorrhizae. New Phytol 86:373–378

Nylund JE, Unestam T (1982) Structure and physiology of ectomycorrhizae. 1. The process of mycorrhizal formation in Norway spruce in vitro. New Phytol 91:63–79

Nylund JE, Kasimir A, Arveby AS (1982) Cell wall penetration and papilla formation in scenescent cortical cells during ectomycorrhiza synthesis in vitro. Physiol Plant Pathol 21:71–74

Peyronel B (1922) Nuovi casi di rapporti micorize tra basidiomiceti e fanerogame arboree. Soc Bot Ital Bull 1:7–14

Quian XM, Kottke I, Oberwinkler F (1998) Influence of liming and acidification on the activity of the mycorrhizal communities in a *Picea abies* (L.) Karst. stand. Plant Soil 199: 99–109

Raidl S (1997) Studien zur Ontogenie an Rhizomorphen von Ektomykorrhizen. Bibl Mycol 169:1–184

Raidl S, Müller WR (1996) *Tomentella ferruginea* (Pers.) Pat. + *Fagus sylvatica* L. Descr Ectomycol 1:161–166

Ramstedt M (1985) Physiology of the ectomycorrhizal fungus *Piloderma croceum*, with special emphasis on mannitol metabolism. Doct Thesis, Uppsala University, Uppsala, Sweden

Ramstedt M, Niehaus WG JR, Söderhäll K (1986) Mannitol metabolism in the mycorrhizal fungus *Piloderma croceum*. Exp Mycol 10:9–18

Renvall P (1995) Community structure and dynamics of wood-rotting Basidiomycetes on decomposing conifer trunks in northern Finland. Karstenia 35:1–51

Ryan EA, Alexander IJ (1992) Mycorrhizal aspects of improved growth of spruce when grown in mixed stands on heathlands. In: Read DJ, Lewis DH, Fitter AH, Alexander IJ (eds) Mycorrhizas in ecosystems. CAB International, Wallingford

Saber M (1987) Contribution to the knowledge of Aphyllophoralles collected in Iran. Iran J Plant Pathol 23:21–36

Sarjala T (1990) Effect of nitrate and ammonium concentration on nitrate reductase activity in five species of mycorrhizal fungi. Physiol Plant 79:65–70

Sarjala T (1991) Effect of mycorrhiza and nitrate nutrition on nitrate reductase activity in Scots pine seedlings. Physiol Plant 81:89–94

Shishido M, Petersen DJ, Massicotte HB, Chanway CP (1996a) Pine and spruce seedling growth and mycorrhizal infection after inoculation with plant growth promoting *Pseudomonas* strains. FEMS Microbiol Ecol 21:109–119

Shishido M, Massicotte HB, Chanway CP (1996b) Effect of plant growth promoting *Bacillus* strains on pine and spruce seedling growth and mycorrhizal infection. Ann Bot (Lond) 77:433–441

Smith SE, Read DJ (1997) Mycorrhizal symbiosis, 2nd edn. Academic Press, New York

Stalpers JA (1993) The Aphyllophoralles fungi I. Keys to the species of the Thelephorales. Stud Mycol 35:1–168

Taylor AFS, Alexander IJ (1989) Demography and population dynamics of ectomycorrhizas of Sitka spruce fertilised with nitrogen. Agric Ecosyst Environ 28:493–496

Taylor AFS, Alexander IJ (1991) Ectomycorrhizal synthesis with *Tylospora fibrillosa*, a member of the Corticiaceae. Mycol Res 95:381–384

Taylor AFS, Brand F (1992) Reaction of the natural Norway spruce mycorrhizal flora to liming and acid irrigation. In: Read DJ, Lewis DH, Fitter AH, Alexander IJ (eds) Mycorrhizas in ecosystems. CAB International, Wallingfond, pp 237–245

Taylor DL, Bruns DT (1997) Independent, specialized invasions of ectomycorrhizal mutualism by two nonphotosynthetic orchids. Proc Natl Acad Sci USA 94:4510–4515

Thind KS, Dhingra GS (1985) Thelephoroid fungi of the eastern Himalayas, India I. Res Bull Panjab Univ Sci 36:165–174

Thind KS, Rattan SS (1971) The Thelephoraceae of India. Part 4. The genus *Tomentella*. Indian Phytopathol 24:32–42

Thomas GW, Jackson RM (1979) Sheathing mycorrhizas of nursery grown *Picea sitchensis*. Trans Br Mycol Soc 73:117–125

Visser (1995) Ectomycorrhizal fungal succession in jack pine stands following wild fire. New Phytol 129:389–401

Watling R (1981) Relationships between macromycetes and the development of higher plant communities. In: Wicklow DT, Carroll GC (eds) The fungal community, its organisation and role in the ecosystem. Marcel Dekker, New York, pp 427–458

Weiss M (1991) Studies on ectomycorrhizae. XXXIII. Description of three mycorrhizae synthesised on *Picea abies*. Mycotaxon 60:53–77

Weiss M, Agerer R (1988) Studien an Ektomykorrhizen. XII. Drei nicht identifizierte Mykorrhizen an *Picea abies* (L.) Karst. aus einer Baumschule. Eur J For Pathol 18:26–43

Zak B (1969) Characterization and classification of mycorrhizae of Douglas fir. I. *Pseudotsuga menziesii* + *Poria terrestris* (blue- and orange-staining strains). Can J Bot 47:1833–1840

Zak B, Larsen MJ (1978) Characterization and classification of mycorrhizae of Douglas-fir. III. *Pseudotsuga menziesii* + *Byssoporia* (Poria) *terrestris* vars. *lilacinorosea, parksii*, and *sublutea*. Can J Bot 56:1416–1424

Index